D0926613

Communications Technology
and Social Policy:
Understanding the New
"Cultural Revolution"

Communications Technology and Social Policy

UNDERSTANDING THE NEW "CULTURAL REVOLUTION"

Edited by

GEORGE GERBNER
LARRY P. GROSS
WILLIAM H. MELODY

A WILEY-INTERSCIENCE PUBLICATION

JOHN WILEY & SONS, New York • London • Sydney • Toronto

Library of Congress Cataloging in Publication Data:
Main entry under title:

Communications technology and social policy.

"A Wiley-Interscience publication."
Based on papers discussed at a symposium held at the Annenberg School of Communications, University of Pennsylvania, in March 1972.
Includes bibliographies.
1. Communication—Addresses, essays, lectures.
2. Communication—Social aspects—Addresses, essays, lectures. 3. Power (Social sciences)—Addresses, essays, lectures. 4. Telecommunication—Addresses, essays, lectures. I. Gerbner, George, ed. II. Gross, Larry P., ed. III. Melody, William H., ed. IV. Title.

HM258.C589   301.14   73-7563
ISBN 0-471-29670-8

Printed in the United States of America

10 9 8 7 6 5 4 3 2

CONTRIBUTORS

PIER L. BARGELLINI

Comsat Laboratories

JEROME S. BRUNER

Oxford University

JAMES W. CAREY

Institute of Communication Research, University of Illinois

FORREST P. CHISMAN

The John and Mary R. Markle Foundation

JONA COHN

Communications Division, Motorola, Inc.

PETER COWAN

Joint Unit for Planning Research, University College, London

PHILIP ELLIOTT

Centre for Mass Communication Research, University of Leicester, England

ARNOLD W. FRUTKIN

National Aeronautics and Space Administration

RICHARD GABEL

Office of Telecommunications, U.S. Department of Commerce

DENNIS GABOR

Nobel Laureate in Physics, Imperial College of Science and Technology, University of London

GEORGE GERBNER

The Annenberg School of Communications, University of Pennsylvania

OLIVER GRAY

The Urban Coalition

BERTRAM M. GROSS

Department of Urban Affairs, Hunter College

LARRY P. GROSS

The Annenberg School of Communications, University of Pennsylvania

MICHAEL GUREVITCH

Department of Sociology and Communication, Hebrew University, Jerusalem

ARTHUR D. HALL, III

Arthur D. Hall, Inc.

JAMES D. HALLORAN

Centre for Mass Communication Research, University of Leicester, England

NICHOLAS JOHNSON

Federal Communications Commission

MARK L. HINSHAW

Department of Urban Affairs, Hunter College

BRUCE R. JOYCE

Teachers College, Columbia University

ELIHU KATZ

The Communications Institute, The Hebrew University, Jerusalem

DIETER KIMBEL

Organization for Economic Cooperation and Development, Paris

JAMES T. MARTIN

IBM Systems Research Institute

ARMAND MATTELART

Center of Studies on National Reality, Catholic University, Santiago, Chile

RICHARD L. MEIER

Environmental Design, University of California, Berkeley

WILLIAM H. MELODY

The Annenberg School of Communications, University of Pennsylvania

STEFAN NEDZYNSKI

Postal Telegraph and Telephone International, Geneva, Switzerland

KAARLE NORDENSTRENG

Institute of Journalism and Mass Communications, University of Tampere, Finland

DAVID R. OLSON

The Ontario Institute for Studies in Education, Toronto, Canada

EDWARD L. PALMER

Children's Television Workshop

EDWIN B. PARKER

Institute for Communication Research, Stanford University

WILBUR L. PRITCHARD

Comsat Laboratories

JOHN J. QUIRK

Author, Chicago

HERBERT I. SCHILLER

Communications Program, Third College, University of California, San Diego

THEODORA SKLOVER

Open Channel, New York City

RALPH LEE SMITH

The Mitre Corporation

DALLAS W. SMYTHE

The Communications Program, University of Saskatchewan, Regina, Canada

TAPIO VARIS

Institute of Journalism and Mass Communications, University of Tampere, Finland

MELVIN M. WEBBER

Institute of Urban and Regional Development, University of California, Berkeley

JOHN W. WENTWORTH

Educational Development Engineering, RCA

GEORGE R. WHITE

Research and Engineering, Xerox Corporation

PREFACE

Leaders and pioneers in communications technology, economics, education, urban and international communications, and social communications research from many countries examine the implications of an anticipated new surge in industrial, political, and cultural power in this volume. This surge is the likely result of developments in communications technology.

Man is the symbol-using animal. Of all the environments that he creates (and pollutes), the climate of symbols and messages is the most crucial to his humanity. The means that shape our symbol systems are on the threshold of a new industrial revolution. The force that propels this technological takeoff fuses such elements as the "wired nation" of broadband communications; the encapsulated message modules of cassettes; the storage, retrieval, switching, analytical, and symbol-transforming capacities of computers; and the intimate immediacy of a "global village" of strangers created through communication satellites. The new fusion is imminent at a time when we are attempting to grasp the full consequences of the first communications revolution which, from printing to television, has transformed the quality of life and the life of the mind beyond all historical precedents of cultural change.

Rapid technological change brings an expanded awareness and renewed attention to any field of endeavor. In a sense, the study of society's evolution has traced the uncertain footsteps of changing technology down uncharted paths of private economic development. In communications, we are on the brink of an interacting series of technological changes of a magnitude that will, in comparison, virtually dwarf all past changes. We know that the consequences of the new communications technologies—economic, political, social, and cultural—will be enormous. But we do not know whether their net effects will be optimal or even beneficial in satisfying human needs, or whether society might prefer to pursue a different course if it were aware of and could select from alternative paths.

The new technologies will provide enormous increases in communications capacities and capabilities. But for whom? For what kinds of communication? For what purposes? What will be the social and cultural implications of such things as computerized data banks of personal information, satellite broadcasting from one nation to others, and interactive local broadband cable systems connected to versatile home terminals? Are we likely to drown in a sea of incomprehension from the information overload? If not, how do we devise a system of priorities among the many communication and information sets competing for attention? Must we rely almost exclusively on others to select, classify, summarize, and interpret for us the events, ideas, and opinions of the world's communities? If so, whom do we entrust with such unique positions of power? And by what standards of public accountability will such performance be judged?

Perhaps more than at any other time in our history, information and control over communications can be directly associated with economic and political power. The new technologies provide a potential not only for changing the structure of present power relationships but also for substantially expanding the aggregate power of those in control of communication and information systems. Yet we have not learned to discipline technology. We have accepted the beneficial and detrimental consequences of its apparently deterministic process of development with little more than an abiding faith in an ideology that defines technological change as technological advance and equates the latter with human betterment. As a result, "The masters of today, unlike the tyrants of past ages, are the technological determinants of human life which man has created and which now master him and his whole system."*

Most technological development has been undertaken in the pursuit of private economic gain. The public interest in achieving societal benefits has been entrusted to Adam Smith's "invisible hand" of market competition, which was supposed to bring an equality between the entrepreneur's private gain and society's economic welfare. In recent years, we have come to realize that the failure of competitive markets to recognize social costs requires the application of governmental regulatory standards. However, governmental policy making has been mostly concerned with operations within the static technological state and not with the rate and direction of technological change. Moreover, public policies have been based upon the narrow considerations of short-run private economic consequences to the exclusion of long-range social and cultural implications.

Public decisions have been made with little reliable information about alternatives and consequences. Concerns voiced about long-range cultural impact are often lost in utopian myths, in the ritual din of the cult of technology, or in nostalgia for an idyllic past that never was. We believe that it is premature to celebrate, too late to regret, and too dangerous to ignore what has been set in motion. We must learn how to navigate—and channel—the cultural currents of our own making, and to steer a course based on

* C. West Churchman, *Challenge to Reason*. New York: McGraw-Hill, 1968, p. 80.

reason and evidence toward goals of our own choosing. Those involved in communications service, scholarship, research, and technics have a special responsibility for calling a halt to the policy of drifting with the tide and to the process of deciding by default.

This volume discusses the issues and problems of directing the tidal wave of communications technology toward the achievement of public interest objectives. By providing an awareness of the deep economic, social, and cultural implications of the new technologies, we believe that this book will expand the analytical horizons of scholars, policy makers and students cf communications, and all readers interested in what these new developments portend. We also hope that by bringing a wide range of approaches to bear upon common issues of great public significance, we can help specialists in technology, information science, economics, politics, education, urban studies, international relations, and, of course, communications and cultural studies, to enrich their understanding with relevant knowledge from outside their own specialties.

The book has been organized into six parts. Each part is a self-contained unit addressed to a particular set of issues. But each part also relates to similar issues examined from different perspectives in other parts of the book by scholars from different disciplines.

Part I, "What Can the New Technology Do? Dimensions of Change," describes in layman's terms, the fundamental characteristics and capabilities of new and developing technologies. Potential new applications of the technologies for providing new kinds of communication services are outlined and major benefits and serious problems associated with applications of the new technologies are developed.

Part II examines "Institutional powers and controls; The Direction of Change." Technological change may reinforce or significantly alter the distribution of power among communications institutions. This part analyzes the effects of specific changes in communications technology on the policy foundations of government regulation and the economic structure of communications industries. Apparent limitations of existing institutional arrangements on the formulation of public policy and the direction of application of the new technologies are evaluated. Opportunities are assessed for changing the structure of communications institutions so that they will be more global in their perspective and more responsive to public interest requirements in directing the application of new technologies.

Part III is concerned with "Communication and Education for the Full Employment of Human Potential." Attention is directed both to the relevance of our inherited educational institutions for meeting the requirements of a humanistic education in the society of tomorrow and to the traditional conceptions of knowledge, intelligence, and competence conveyed through these institutions. The opportunities and limitations of communications technologies for enhancing the educational function in society are examined.

Part IV discusses "Urban Communications." Of all the new communications technologies, the one likely to have the greatest impact in industrialized nations is the broadband cable that can create not only the wired city but also the wired nation. Broadband cable provides enormous communications capacity into the home for cable antenna television (CATV) and a wide range of delivery and interactive services. Potentially it can change the role and functions of the city and substantially alter cultural patterns. Many analysts view it as the salvation for a whole spectrum of serious urban problems. The chapters in this part examine urbanization and advanced communication systems as well as pioneering developments in the utilization of initial cable systems.

Part V deals with the issue of "Global Communications; Cultural Explosion or Invasion?" A global communication systems can provide forces for cohesion or instruments of cultural invasion. This section examines the developmental process in developed and developing countries, focusing on problems with the existing flow of international communications. Particular attention is paid to the imbalance among countries in the generation and transmission of information and messages.

Part VI, "Tracking the Future," is directed to the need for, and the task of, developing measures for assessing the present environment and a foundation for predicting the consequences of new communications technologies. New forms of social accounting can permit us to track social and cultural trends and provide a basis for developing an improved understanding of social and cultural processes. Indicators of those aspects of the communications environment relevant to issues of public policy can lead to more informed policy decisions that reflect consideration of the social and cultural consequences of new technology.

It is important to raise the question of what is really new in the new "cultural revolution," because one of our rituals is instant obsolescence, or the cult of the new. Novelty is something we manufacture on the assembly line. The "new" avant garde is a standardized posture modeled in "underground" mass media (a contradiction in terms) and the Sears catalog. "Revolutionary" is likely to describe the latest version of an outmoded gadget or concept. The routine spell of the latest headline, the regular chill of the FLASH!, and the generally predictable details that daily nourish our well-cultivated assumptions about the world dull our ability to recognize what is really new. We cannot even ask what is really new without first asking what *is*, really? For what is new and real is a transformation in our organs shaping social reality.

We are now paying the piper for neglecting to consider the social and environmental consequences of air and water pollution in directing the course of a number of technologies introduced in the past. As a result, the concept of "technology assessment" has been recognized for those technologies for which we have found directly observable consequences. For such applications of technology as the SST, the Alaska pipeline, automobile exhaust systems, and nuclear power plants, environmental impact must be accounted for and considered in public policy decision. But the social and cultural implications of communications technology cannot be directly observed. We lack understanding of the processes and the measurement mechanisms to guide our analyses. We hope that this book will provide a foundation for taking steps down the long road toward overcoming these limitations.

We thank the Communications Workers of America for assisting with the funding of a symposium in March 1972 at the Annenberg School of Communications, University of Pennsylvania, at which these papers were discussed in draft form. Kiki Schiller and Craig Aronoff provided editorial assistance. We also thank our colleagues and students who reviewed many of the chapters and by doing so, became responsible for their shortcomings. We shall take credit for the rest, and thereby begin a new era in prefatory communication.

GEORGE GERBNER
LARRY P. GROSS
WILLIAM H. MELODY

Philadelphia, Pennsylvania
February 1973

CONTENTS

Communications Technology
and Social Policy:
Understanding the New
"Cultural Revolution"

What Can the
New Technology Do?
The Dimensions of Change

CONTENTS

INTRODUCTION

ARTHUR D. HALL, III

Inevitably, the future is a coalescence of what society (and subdivisions thereof) wants and what technology can provide, acting within economic, legal, and other constraints of the times. One can move to satisfy the needs of society from either direction: by defining what society wants, or presenting what technology can do. By starting with the latter, one runs the risk of presuming to say what society wants, but this is not intended, even though the technologists who have contributed to this section, being socially responsible, do have opinions about the uses to which technology should be put.

Part I provides summaries of current communications technology, general forecasts of the technology, and probable applications. The major dimensions of change in communications technology are outlined by authors who are directly involved with the course of technological change.

James Martin analyzes the interrelations between communications and the computer in his chapter, focusing his discussion on the role of computers in systems involving human communication. Arthur Hall directs his attention to the technology for providing switched-telecommunications services that permit connections upon demand between unique pairs of points in nationwide and worldwide communication networks. Special consideration is given to the Picturephone technology. Potential relationships to the "wired nation" concept, interactive computer terminals, and the graphic arts are developed.

In his chapter, John Wentworth examines broadcasting technologies. He reviews the current status and future prospects of message dissemination from the few to the many, and discusses recent developments that will shape the future of broadcasting services. Jona Cohn addresses the often neglected area of mobile communications in his chapter directed to communication with man on the move. Emphasis is placed on frequency spectrum issues, growth of mobile applications, and new functional capabilities such as automatic vehicle locators, mobile teleprinters, wireless telephones, and mobile computer access.

George White discusses the role and future of graphics systems as elements of information systems. He reviews the evolution of graphics systems and presents projections based on technological advances. W. L. Pritchard and P. L.

3

Bargellini examine communications satellites in their chapter. They discuss foreseeable advances in satellite technology as well as traffic growth patterns. In addition, they address the future role of broadcast satellites and their implications.

In the final chapter, Nobel Laureate Dennis Gabor addresses the issue of social control through communications. He discusses and illustrates the great potential of communications technology for alleviating a wide range of human problems, while cautioning against the potential abuse of communications technology.

Although many probable communications services and systems are described in the chapters that follow, the most important technologies appear to be those listed below.

> Broad-bandwidth (in the order of 300 MHz) bidirectional coaxial cable systems for urban distances.
> Satellite systems for intercity and intercontinental distances.
> Essentially "zero cost" digital logic and memory (including cassettes).
> User-oriented and human-dominated computer terminals.
> Greatly increased power of data processing computers and networks of computers.
> New mobile radio capabilities.

In a society essentially like that of today, assuming no major catastrophies, the major applications of the new technologies will likely be:

> A "cashless" society (already begun).
> Broadband interactive (two-way) communications.
> Urban coaxial cable head ends interconnected by satellites, eventually for worldwide broad-band communications.
> Information "utilities," serving to provide remote access to libraries, data banks, and so forth.
> Man-machine symbiosis on a wide scale.
> Continuous, instant, and universal contact with any person who wants it, for safety, efficiency, and so forth.

These and other applications will have many operational and economic impacts that are not too difficult to deduce. Many other impacts of a cultural, legal, and ethical nature are not readily foreseen. Social research has not yielded the tools and measures for deducing the consequences. Some possible operational and economic impacts include:

> Education that can be individualized to accomodate the variabilities of learners and teachers.
> Deurbanization will be possible at less cost.
> The tradeoff from less transportation of people to more communications among them will be accelerated.

The pace of business will be faster.

Society will be safer from crime and fraud.

Many medical services can be provided more cheaply and more widely by telemedicine techniques of all kinds.

Organizations and communities will be redefined and reorganized more frequently and more readily.

There will be greater interdependence with greater specialization. However, the potential benefits will bring dangers and pitfalls. For example:

The computer can make errors, "cybervalence," already on the increase, can cause loss of privacy.

There can be more depersonalization and breakdown of interpersonal relations through more universal use of more modes of communications.

Increasing turnover of job types, "drop outs," and expanded retraining requirements.

Information pollution.

Needs and expectations, especially in poorly developed regions, outpacing resources and capabilities.

Society will be more "vulnerable" to accidents, sabotage, and so on, through more and more sensitive points of pressure.

There will be greater extremes both of stress and of boredom.

There is a great need for new institutional mechanisms for gaining the benefits while avoiding the dangers of new technology. It is quite apparent, even at the community level, that there is very little dialogue for matching community needs to what technology can do. At the national and international levels, even making the strong assumption of compatible assemblies of technologists and social researchers and analysts, there is no continuing institution to address these important problems. Even the first and most fundamental step, that of formulating and testing goals, has yet to be taken.

Business, governments, unions, universities, and foundations all are playing appropriate roles, but none seems equipped to do what seems needed: to forecast the technology, to trace out the consequences for all people and groups, to formulate plans for making the best uses of technology, and to reach consensus among all affected groups. Far better means are needed for anticipating and planning for the future shock of technology-caused unemployment, loss of freedom through cybervalence, and other potential consequences. Communications research, operating at all required levels in society, is urgently needed if we are to make the wisest uses of technology.

CHAPTER 1

Communication and Computers*

JAMES T. MARTIN

During the years ahead, the growth of the computer industry and the acceptance of computer methods will depend to a large part on the success achieved in establishing effective man-machine communications. We can now foresee how men in all walks of life may use computers. If our foresight is correct, the computer revolution is only the beginning, and when it has run its course it will have profoundly changed the condition of man.

To achieve these dreams, however, two changes are necessary: we must learn to use the world's transmission media efficiently for computer data, we must learn how to design man-computer dialogues that are truly effective for a wide diversity of men.

In the first two decades of its history, the data processing industry has paid little attention to effective man-machine dialogue. The prime focus of technicians has been on the efficient use of the central processing unit. This is hardly surprising in view of the expense and remarkable capabilities of this unit. Where they were used at all, terminals and their languages have often been a peripheral adjacent to the computer and its languages. Systems have been designed from the inside out.

More and more, in the next decade, *man* must become the prime focus of system design. The computer is there to serve him, to obtain information for him, and to help him do his job. The ease with which he communicates with it will determine the extent to which he uses it. Whether or not he uses it powerfully will depend on the man-machine language available to him and how well he is able to understand it. To be effective, systems will have to be designed from the outside in. Instead of being a peripheral consideration, the terminal operator will become the tail that wags the whole dog.

* The material in this paper has been published in Prentice-Hall books listed in the references.

Computing power has been steadily dropping in cost. In 1955, about 100,000 program instructions could be executed for $1. In 1960, the same dollar bought 1 million, and by 1970, 100 million program instructions. In other words, it is going up by a factor of 10 every five years. Twenty years of computer history have brought a 10,000-fold increase in power. The promise of mass-produced large-scale integration (LSI) circuitry suggests that this increase will continue. If the increase continues to be exponential, as in the past, the dollar in 1980 will buy 100 times as many computer instructions as in 1970. The cost of *people* will increase instead of decrease, and the proposed uses for the available people will become increasingly complex. The computer industry will be forced to become increasingly concerned with the usage of people instead of with the computer's intestines.

As yet, little has been done to produce man-machine dialogues that are as natural as possible for the individual and efficient for him to use. The average manager is not going to spend a two-week training course learning how to communicate with his terminal, nor is he going to remember strings of mnemonics. Furthermore, he is going to become disgruntled when the machine spits back error messages at him or when its responses are unintelligible. It is essential that we provide him with information only when he needs it, and that we present it in the clearest and most digestible form.

At best, a man-computer dialogue must be so seductive that the man is drawn into it to explore, fascinated, what the machine has to offer. It must be clear to the man what he intends to do, and if he makes mistakes these must be circumvented by the machine so that the dialogue is immediately restored to the right track.

In rare instances today one can find an example of a man-computer dialogue so successful that, like a good novel, the man can hardly leave it alone. A hospital system that interviews patients, a corporate forecasting system on which an executive can challenge with a light pen the assumptions made, and a telephone voice answer-back system for diagnosing stomach ailments fall into this category. It is often the case that the man whose inspiration led to such a dialogue is not a programmer. He may be a behavioral psychologist or a man with a professional understanding of communication media.

Not enough people have as yet come to regard the computer terminal as a communication medium. A terminal for top management is not a machine to "help management make instant decisions." It would be better regarded as "a comprehension machine" designed to enable them to better understand the complex relationships between factors that they can nominate.

As the potentials of man-computer dialogue become better realized, the applications for it will become innumerable and will spread into all walks of life. If the years ahead bring a better understanding of man-computer communication, they will probably also bring the telecommunication links that make it possible on a massive scale.

The fastest growing area of the computer industry is teleprocessing, and the most rapidly growing use of the world's telecommunication links is for data transmission. The reason is the power and versatility that the interlinking of computers can bring and the potential benefits to the individual of having this power at his fingertips.

Data transmission may become as indispensible to city-dwelling man as his electricity supply. He will employ it in his home, in his office, in shops, and in his car. He will use it to pay for goods, to teach his children, and to obtain information, transportation, and items from the shops; he will use it from his home to obtain stock prices and football scores; he will use it to seek protection in crime-infested streets.

Data transmission will drop in cost. Long-distance costs will drop much more than short-distance costs. Intercontinental links will drop much more than national links. The microwave links and cables that provide today's telephone service have enormous data-handling capacity.

I was recently at a conference in England where it was decided to demonstrate some points by remotely using a computer in the United States. A teletype machine was employed as the terminal, and the communication link, a telephone line, was obtained with no difficulty. At the end of half an hour's use, I examined the printout and estimated that about 3000 characters had been transmitted in total to and from the computer.

The voice line was capable of transmitting 4800 bits per second (bps) with ease (more than this rate could be obtained with sophisticated modems). In half an hour then it could transmit $1800 \times 4800 = 8,640,000$ bits, and in fact it had sent only $3000 \times 7 = 21,000$ bits of data. One could say that the efficiency of the way we had used it was $21,000/8,640,000 = 0.0024$—a very poor way to use such an expensive facility.

Voice lines are now being constructed using PCM techniques in which one telephone channel becomes equivalent to 56,000 bps in each direction. This could transmit 1800×5600 bits in half an hour, in each direction. The efficiency with the use of time sharing could then be said to be

$$\frac{21,000}{2 \times 1800 \times 56,000} = 0.0001$$

In an industry with the logic capability of the computer industry, an efficiency of 0.0001 ought not to survive long. If we can push the efficiency up to 0.25 we have an improvement of 2500 times, and on the analog-voice line used at 4800 bps, we have a 100-fold improvement. This can be done if we can arrange for different users to share the transmission capacity simultaneously. The computer, after all, was being time shared. We must time share the lines also.

The need for sharing is even more pronounced with other transmission facilities. The microwave links, coaxial cables, and even wire pairs that form today's telecommunication links can be made to carry very much higher bit rates than those mentioned previously. The wire pair of the Bell System T1 Carrier, in widespread use today, transmits a useable 1,344,000 bps in each direction, and the T2 Carrier wire pair transmits 6 million bps.

The sharing of transmission lines can be carried out in a variety of different ways. There are basically two problems associated with it. The first is the technical problem of combining different transmissions on the same line. This is relatively easily solved. If several separate transmissions are sent over the same line at the same time, this is referred to as multiplexing. Many different multiplexers are available for data transmission and all achieve a high level of efficiency—0.8 to 0.9, for example.

The second problem is that of bringing together a sufficient number of users to fill the group of channels that has been derived from he line. Sometimes there are enough users within one organization. They can be brought together to share a lease line. Often this is not the case, however.

When it is not the case, the only way to achieve the efficiency that multi- plexing and concentration can bring is to have a public data network. Public data networks are now planned in several countries. However, it is question- able whether their rate of development will be fast enough for the rapidly growing computer industry. In some cases it is questionable whether the ver- satility of the network will be enough for the diverse types of future computer dialogue.

Some users of future data networks will want to operate at today's teletype speeds. However, for many man-machine dialogues such a speed is intolerably slow and speeds closer to human reading and scanning speeds are needed—say 4800 bps. Dialogues involving graphics or responses in the form of photo- graphic images will need much higher transmission rates. In many systems, instead of obtaining a slow channel for a lengthy time, a computer will need a very fast channel for a small fraction of a second. Today's microwave links and cables have the transmission capacity to combine these requirements provided that an appropriate organization of the network is planned.

Looking at the eventual future of a telecommunications plant, it seems cer- tain that it will become fully digital, with digital switching. Analogue signals will be carried in digital form and large-scale integration logic will be used throughout the network. The cost of transmission with such a network will become a fraction of today's cost, and the capability of the network will be much greater.

How to progress from today's analogue plant to tomorrow's digital plant presents a problem. To make the transition it may be necessary to make investments that will make the operation of today's telephone plant tempo- rarily more expensive, as shown in Figure 1. Taking course *B* rather than course *A* will result in lower costs in years ahead, at the expense of higher costs in the immediate future. However, if course *B* represents the building of

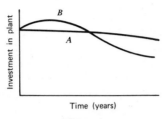

Figure 1.

a digital network with telephone, data, and other facilities integrated, it will provide much greater capability than course A in the years ahead.

To develop atomic energy, massive government funding was necessary. This has been true in other high-cost technologies such as aircraft and space. In a sense it has been true with automobiles if one counts the cost of highways, and a car without roads is like a Picturephone set without communication lines.

REFERENCES

Martin, James *Telecommunications and the Computer*. Englewood Cliffs, N.J.: Prentice-Hall, 1969.

Martin, James and Norman, Adrian R. D. *The Computerized Society*. Englewood Cliffs, N.J.: Prentice-Hall, 1970.

Martin, James *Design of Man-Computer Dialogues*. Englewood Cliffs, N.J.: Prentice-Hall, 1972.

Martin, James. *Future Developments in Telecommunications*. Englewood Cliffs, N.J.: Prentice-Hall, 1972.

CHAPTER 2

Trends in Switched Services

ARTHUR D. HALL, III

A communications system that can establish paths over which messages can be sent between any two (or more) end instruments in specified locations at desired times is, by definition, a *switched network*. The telephone system is the most ubiquitous example. If only permanent paths (or channels) exist between all subsets of end points, the communication system is said to be a *nonswitched network*. A local TV broadcasting system and a CATV system are good examples, even though each may have simple program switchers at the studio or head end, respectively.

If a network has the capability to store one or more complete messages at one or more points in the network and to forward them on to the next link at some later time, it is called a *message-switched* network. This contrasts with a *circuit-switched* network in which a path between the terminals is established by switching circuits together.

A mixture of forces including government regulatory decisions, user needs, system economics, and advancing technology is blurring the two dimensions just defined. Thus in television broadcasting, the need for sharing the enormous cost of programming among large audiences has long since required a switched network tying cities together, and the same force will, in part, cause the interconnection of CATV systems. Again, broadcasters have long employed message switching by means of recording programs for later transmission over the network to a different time zone. New, much less expensive storages, such as video tape cassettes will put message switching in the hands of the end user. In turn, very cheap digital storage and the future increased speed of switching may make switched networks for data service, or even for telephone, into message-switching networks. Thus the ARPA message-switching network has a round-trip delay of only 1 sec, which while too long for easy to-and-fro telephone conversation, indicates the trend of technology.

13

Another major dimension in the morphology of communications is the type of service given to the user. The major services along this dimension are:

Telephone.
Graphics (facsimile, telewriting, etc.).
Data and telemetry.
Television.
Videotelephone.
Audio.
Future (smell, touch, pain, etc.).

Any of these services can be differentiated further by the degree of mobility of the user(s) and by the bandwidth (or bit rate) needed to convey the messages in the time required by the user. Mobility is, of course, a vector with components that include the medium (land, air, space, underwater, etc.), speed, and other factors.

We have thus defined five dimensions in the morphology of communications systems:

1. Switched or nonswitched.
2. Circuit switched or message switched.
3. Service provided to the user.
4. Mobility of the user.
5. Bandwidth of the signal.

This chapter focuses mostly on new circuit-switched, nonmobile (fixed) services of a variety of bandwidths and concentrates on data and videotelephone services, because these appear to be the centers of major action in the coming decade.

TELEPHONE SERVICES

Our focus on new services does not mean the old ones are dormant. As the parent of most new services, telephone service in the United States is growing at the rate of about 6%/year on a base of 110 million telephones, and is still innovative.

The major producer of innovation in telephone service in recent years, and this portends to be even more so in the future, is the electronic stored program switching machine. This actually is a marriage of telephone switching and digital computer technologies. Starting with the Bell System's No. 101ESS (PBX) for customer-switched services (the first was installed in 1963), and with the No. 1 ESS for public central offices (1965), there is now a procession of systems coming from Bell, GTE, ITT, IBM and many other international companies. The following headings list just a few of the telephone services that are all available today, on some working system. Since the capital value of the existing plant in the United States alone is $60 billion, and the rate of capital formation is about $10 billion/year, it will be some years before these services can be enjoyed everywhere.

Pushbutton Dialing (or Touchtone®)

A pushbutton pad, currently with 12 keys and expandable to 16 keys, permits the user to key in the telephone address about twice as fast as with a familiar dial. Even more important, keying can continue after the wanted number is reached. This opens up a world of remote control and computer access by telephone. (A "zero-cost" computer terminal would be possible in every home.) About 9% of United States telephones are so equipped, and the rate of growth is about 10%/year (1971 data).

Call Transfer

Call transfer includes a variety of services, including call forwarding, which means that an incoming call to your line may be transferred to an attendant, or to someone else, if your line is busy or does not answer. Even during the conversation, the called station can transfer the call to another station.

Add-On

Found now in some PBXs, a station can add another party for a three-way conference.

Call Waiting

A busy line is notified that another call is waiting by a "beep" sound.

Speed (or Abbreviated) Addressing

A frequently dialed number can be called by dialing only two digits; the switching machine then translates these into the 10 or 11 digits needed. This service is also available by means of an automatic dialing device located at the telephone.

Wake Up (or Alarm Clock Service)

The user can dial a request for a call at a specified time that he also dials into the machine.

Many more specialized telephone services will undoubtedly be available in the future. For example, automatic answering service, available now by customer-owned equipment or from independent entrepreneurs, could much more cheaply be provided by a stored-program switching machine. Another is international direct distance dialing, analogous to today's DDD; this service has been planned, but its implementation calls for upgrading of existing plants around the world.

A major force for a great increase in the rate of telephone service innovation was the FCC Carterfone decision in 1968. Although there are still impediments to implementing this decision, it now seems assured that all manner of customer-owned peripheral equipment will be permitted to be interconnected with the public switched network, providing that such equipment does it no harm.

Although public telephone network service may seem almost infinitely extendable, it is not. Again, it has been the computer industry that has brought certain of its limitations into bold relief.

SWITCHED-DATA SERVICES

As computers have become larger in capacity, faster, and more expensive, the need has arisen and rapidly grown to interconnect remote terminals to them over data channels, and to interconnect computers for sharing of loads, programs, and data banks. At the present time there are about 300,000 computer terminals, but time-shared peripherals have caused the fastest-growing part of the communications market, and 5 million terminals are projected for 1980. Since the telephone network was the only one available, these requirements were superimposed on it. Yet the characteristics needed for "computer talk" are different from those for speech transmission for which the telephone network was and is designed. Below are some of the major contrasting requirements.

Rate of Transmission

The telephone network operates at only one bandwidth, which extends from about 200 to 3300 Hz. When used for switched data service, the maximum rate that can be transmitted over ordinary unconditioned lines is about 2000 bps. Data services today require a variety of speeds, and the number is growing. For example, environmental sensors may need very low speeds, say 1 bit/hour. Common teletypewriter speeds are 45, 50, 75, and 150 bps. For any of these services the telephone channel is wasteful. On the other end, high-speed computer printers and card readers operate at about 20,000 bps—too fast for connection over a telephone system.

Speed of Connection (Set-Up Time)

A person needs about 20 sec to dial a 10-digit telephone number; then a delay of as much as 30 sec may ensue while a string of switching machines sets up a number of links in tandem. While good enough for telephone service, this is much too slow for many computer applications. Airline reservation systems, for example, require a 3-sec response time.

Error Rate

The public-switched telephone network gives an average of about one error in every 100,000 bits. While adequate for voice, and also for videotelephone, it is too high for many data applications, which may need to be 100 times better.

Cost and Price

The cost of sending digital data over today's analog channels is high because of the cost of modems (*modulators* and *demodulators*) needed to convert signals at the interfaces, and other reasons. On long-distance service, rates are based on a 3-min initial period and a 1-min overtime period, and these intervals are badly matched to data messages, many of which may be less than 1 sec.

These and other limitations of the public-switched network have created great pressure for change among users. There has been and continues to be a variety of responses:

The common carriers have upgraded the quality of the network for data services.

Private line services, operating independently of the public-switched telephone network have been offered and bought by users able to justify full-time channels.

Very large users have established their own networks. The U.S. military AUTODIN and AUTOVON networks are major examples of this on a worldwide scale.

Private entrepreneurs in command of the technology have proposed separate networks for data and other "specialized" services, and the FCC has agreed, thus opening up a new world of competition.

The carriers and independents alike have pressed forward on digital technology adequate for the challenges.

As a result, in the 1970s we shall undoubtedly see new networks with dramatic user capabilities and benefits. As an example, here are some of the features of the system proposed and planned by Data Transmission Co. (Datran), a subsidiary of University Computing Corp. The Datran system is significant for proposing the most complete departure from the existing public network.

1. The system will be digital from end to end with no modems for conversion from analog to digital form or vice versa.

2. Addressing will be either automatic or manual, with abbreviated addressing available.

3. Connection (set-up) time will be less than 3 sec.

4. Transmission speeds initially offered will be 150, 4800, and 9600 bps, with 19,200 and 48,000 bps later, and speed and/or code conversion between incompatible terminals will be possible.

5. The average error rate will be less than one in 10 million bits.

6. The minimum charge time for a dial-up call will be only 6 sec, compared to 3 min for a telephone call. Rates may be independent of distance, in contrast to rates stepwise proportional to distance for telephone calls.

To realize these and other features within a few years, Datran is using essentially state-of-the-art technology, but the system is designed and engineered specifically for data transmission. Initially, it will feature a single transcontinental microwave system of 244 repeater stations, carrying 4000 channels each of 4800 bps. Initially there will be spur routes to provide service to 35 cities. The time-division switching system will employ centralized computer control, wherein the control computer may be in one place while the switches may be in another, thus economically sharing the services of the computer.

There are many market, financial, cost, and even some technical questions as to whether Datran will succeed in its very ambitious venture. If it does, the services offered will probably become less and less "specialized," since any analog signal such as telephone, facsimile, or videotelephone can be digitized. Whether it succeeds or not, Datran has spurred healthy competitive response. Bell has announced its own (nonswitched) Digital Data System.

SWITCHED VIDEOTELEPHONE SERVICE

The blending of telephone and television technologies was foreseen almost from the beginning of television. In 1930, the first (nonswitching) two-way video telephone was demonstrated by AT&T between New York and Washington. And from 1935 to 1938 the German Post Office operated a commercial appointment-type videotelephone service between four major cities over coaxial cable. When commercial television began after World War II, it was even more technically possible, but early studies based on this technology showed it would be too expensive. The time was not yet ripe. It was the invention of the transistor, the development of digital transmission, and the application of the systems approach that provided the keys to a complete demonstration of both economic and technical feasibility of video telephony as a mass-produced service.

The desirability of combining a television picture with a telephone for face-to-face conversation is still debated, and the arguments will not be over until all the contestants have actually experienced the service for a significant period in their daily lives. It seems almost that one has to be blind to appreciate the full importance of eyesight in communication, so self-satisfied has the world become with the telephone. More important, nearly all of the reaction obtained in field trials of the service and in market tests has been very favorable. These included "in-house" trials at AT&T and Bell Laboratories headquarters, at the New York World's Fair from 1964 to 1965, at the Union Carbide Co. in New York and Chicago from 1965 to 1966, and at Westinghouse Electric Co. in Pittsburgh and New York from 1969 to 1970. The service trial in 1965 of appointment-type service (similar to the German trial) between one location each in New York, Chicago, and Washington was not a commercial success because of the inconvenience of making an appointment and traveling to the centers.

Recently AT&T announced that Picturephone service was being withdrawn for "lack of demand." The reason given can hardly be taken at face value, since the "community of interest" and other requirements for successfully marketing the service were well known through years of research costing hundreds of millions of dollars.

The picture standards and many of the features of the Bell System's Picturephone® probably will be adopted as standard in most countries, although Bell's recent action may lose her the position of leadership in the choice of standards. A 1 MHz base band is used for the picture. The picture size of the current set is 5 1/2 in. wide by 5 in. high, and has 251 scanning lines/frame; thus the vertical resolution is 45 lines/inch—about the same as United States commercial television. Subjectively, the average viewer will not be able to tell the difference in picture quality as compared to his black and white home receiver when used for face-to-face viewing. The animation of the picture, 30 frames/second, is the same as on United States television.

Except for bandwidth, the limitations of the public-switched telephone network did not prove so limiting as it did with switched-data services. Thus the telephone set and the telephone network are used to set up, control, and provide the voice path for the Picturephone system. This is very important to the eventual success of the system. Also important is that, although the bandwidth is 250 times that of a telephone channel, the band can be carried for limited distances (6 miles) over ordinary telephone wires, when properly engineered. Thus only two pairs are needed for connecting the picture channel to the central office. At this point the video channels are switched in a network slaved to the telephone network, and the outgoing trunks are encoded into a 6.3 million bps (Mbps) signal for transmission over digital transmission facilities that may be used for any similarly encoded signal.

Until recently Picturephones were available in the Gateway Center District of Pittsburgh, Chicago, and Washington. The monthly charge quoted first was $160/month for the first 30 min of conversation, and 25 cents/minute after that. Lower rates were offered experimentally. The customer was permitted to own his own station set, and if he does, the charge for the service was $110/month. Bell expected to have 1 million sets in service in most major cities by 1980. It is clear that as the service develops the economies of the scale will set in to permit rate reductions. This factor will be more important than new technology in the short run of 10 years. The capital needed to develop the service is such that rates may be held up to control the demand. Development of the service may depend also on competitors entering the market. Indeed, several Japanese manufacturers already sell videotelephones to the Bell standards for less than the Bell cost, believed to be about $1200.

The Rural Electrification Administration, which subsidizes telephone systems in rural areas of the United States, believes that a target of $20/month is possible for service, based on its own analysis. While this seems to me to be too low for the next 10 years, it is indicative of what the longer range future may hold. Even within the broad range from 20 to $160/month, there will be a market for those who will trade transportation cost, time, and hazards for the convenience, speed, and safety of good communications; but even these benefits do not sell themselves, they have to be marketed expertly.

Although the major impact of the service is expected to be in person-to-person conversations—and all the uses that implies (buying, selling, interviews, diagnosis, etc.)—the implications of a national (later international) 1 MHz switched public network for all other communications services simply boggles the imagination. For instance, a 1 MHz analog Picturephone loop can now carry a 460 kbps data signal, and it will be able to carry 1.3 Mbps. For a business PBX, the digital line (T2 carrier) may be brought to the office, making a 6.3 Mbps data channel available. Such lines can be used for almost any conceivable form of computer access or interaction.

The current station sets will not permit a readable display of one full page of typewritten material (neither will a standard TV set). But a possible future slow-scan set (either as another larger screen set in parallel or in an integral "gear shift" version) will do this at the sacrifice of animation during the interval used for display. This opens the practical prospect of dial-up interrogation of microfilm or video tape libraries or files. For delivery of mail, or receiving hard copy of visually selected material, a photocopy device, similar to that already developed for CRT computer terminals, will undoubtedly be developed.

Facsimile service, hobbled by the band limitation of telephone channels, now delivers one page in 6 min, and about the most that can be expected in the future is 1 page/minute. If a machine were developed for use on the Picturephone network, it could deliver 150 pages/minute of the same quality (using

T2 carrier lines), or if good halftone copy were wanted (2000 lines/page), it could deliver 40 pages/minute. Such rates, compatible with present-day Xerox machines, will finally make practical the delivery of books by wire.

Video conference facilities also will be needed if communications are to become an effective substitute for travel. Good facilities already exist in a few places, using standard TV equipment on a nonswitched basis; hence they are inherently expensive. Dial-up Picturephone conferencing is still a few years away, but voice-switched systems, to point the camera to the person talking, are in experimental operation.

Color Picturephone with large screens, a gear shift for slow scan, and a printer for hard copy in color may be the ultimate in communications, but this is a long way off. The dream does come first.

THE RELATION OF PICTUREPHONE TO CATV

The parallel evolution of Picturephone and cable television (or CATV) invites questions about their relationship. First, the picture standards are different and hence incompatible (without expensive interface converters). This may produce the incentive to revise Picturephone standards to bring them into line with NTSC color, or some other compatible system. Second, all present CATV systems are nonswitched; essentially they are one-way mass party lines. There are experimental systems featuring two-way transmission, but in nearly all of these the "upstream" direction is narrowband while the "downstream" direction is wide band to allow for many TV channels. The two systems thus cannot be combined.

A significant exception to these observations is a system proposed by General Telephone and Electronics Corporation. In this system, each home or office is connected to a switch not more than 1000 yd away, using a small coaxial cable. Each home can dial the channel it wishes to view from a selection of 36 (or multiples thereof) TV channels brought to the switch from program sources, or other switches. The coaxial unit can be multiplexed for several channels of wide band transmission in each direction. The system appears to be two to four times more expensive than a conventional CATV system, but it does much more. By straightforward extension of these concepts, virtually any communication service could be provided over the same system, including Pictel (GTE's name for videotelephone), telephone, CATV, data, facsimile, and so on. It is clear that from a purely technical and economic point of view, a combined system can furnish these services much cheaper than they can be furnished independently. Whether the political, economic, and regulatory forces ever will allow such a combined system, only the future can tell.

It is conceivable that a common system could be built, perhaps by a consortium, and that the different services be provided by separate companies: for example, one for CATV, one for pay TV, and one for telephone. The technical and economic opportunity is bound eventually to present broad issues in public policy.

If the cable television services remain separate, as they are now, they will very likely evolve into a national switched network to provide for interconnection of program sources with their markets. But if Picturephone service grows, it will create a demand for the mass production of bandwidth on a scale unknown to date, far greater than that needed for national cable television. This will create an economic magnet that will tend to draw in all other services. The need will be satisfied by the millimeter waveguide and later by the laser technologies. Microwave and satellite radio will be completely inadequate for this task in terms both of cost and supply of channel capacity. The services thus will probably be combined on a national level if not on a local level.

The "field" of the economic magnet also will be felt by data transmission, where the repeatered line only cost advantage against independent data systems may be as high as 10 to 1. A way to maintain the viability of the independent or "specialized" carriers is to allow them interconnection at the system level, where if rates bear any semblance to costs, they can enjoy the benefits of mass production.

REFERENCES

Bell System Technical Journal (special issue on Picturephone), *50*, 2 (February 1971).

Hall, Arthur D. "Communication Systems," in *Handbook of Systems Engineering*, R. E. Machol (ed.). New York: McGraw Hill, 1965.

Martin, J. T. *Future Developments in Telecommunications*. New York: Prentice Hall, 1971.

CHAPTER 3

Broadcasting Technologies

JOHN W. WENTWORTH

Because we live in a technological age, it is important that analysts and decision makers in each sphere of human activity be conversant with the key concepts of science and technology that impinge on their areas of concern. It is appropriate, therefore, that we be given opportunities to review the highlights of the technologies on which modern broadcasting services are based. This paper provides *interpretive* comments on the various broadcasting media, stressing their technical limits, their interrelationships, and the current trends that provide the best basis for predicting future developments.

Broadcasting is defined as the branch of communications concerned with the dissemination of messages by electronic techniques from a few sources to many destinations. This definition is sufficiently broad to include audio and video recordings, as well as radio and television. The phrase "by electronic techniques" has been inserted to exclude consideration of ink-on-paper publishing, much of which would otherwise come under the "umbrella" of our broad definition of broadcasting.

RADIO BROADCASTING AND AUDIO RECORDING

It is appropriate that we consider the technologies of radio broadcasting and audio recording in association with each other, because their histories have always been interwoven. Although the phonograph preceded radio broadcasting by several decades, it was not until broadcasting stimulated the development of high-quality microphones and electronic amplifiers that true high-fidelity recording became possible. From the earliest days of broadcasting up to the present time, recorded materials have been used extensively as source materials for radio programs.

23

There have been relatively few "casualties" among the technologies associated with radio broadcasting and audio recording. New media have emerged at frequent intervals, but few of these have succeeded in displacing the longer-established media. Although it is true that the disc record eventually replaced the cylindrical record (at least in the consumer market) and the magnetic tape recorder effectively superseded the wire recorder, the net result of many innovations in audio systems has been to enrich the "mix" of products and services available to both producers and consumers.

Tutorial Comments on "High-Fidelity" Concepts

A unifying trend that can be discerned in the history of each of the broadcast media is the *quest for high fidelity*. Each new medium has been "launched" on the basis of a few truly basic inventions or discoveries coupled with suitable business organizations and (in the case of broadcast services) a regulatory framework. The services of engineers and technicians have always been required to maintain an appropriate *quantity* of service (that is, to conduct routine operations), but the most creative aspect of engineering support for each medium has been concerned with improvements in the *quality* of the product or service.

It is difficult to explain the concepts of high fidelity without making frequent use of the word *signal*. In audio systems, a *signal* is the electronic form of an auditory message, usually consisting of a fluctuating voltage or current whose *waveform* (or pattern of fluctuations as a function of time) is a faithful analog of the original air-pressure waves generated by the sound source. If such a signal could be generated, handled, and reproduced with perfect fidelity, it would produce sound waves at the receiving end of a system identical to those reaching the microphone at the transmitting end. There are three basic requirements for the high-fidelity handling of audio signals: (1) adequate *bandwidth*, (2) good *linearity* (or freedom from distortion), and (3) a high *signal-to-noise ratio* (or freedom from extraneous signals). Where recording is involved, a fourth requirement is *time-base stability* in the reproduced signal.

An audio system has adequate bandwidth when it provides uniform response for all significant frequencies in the audio signal. The normal range of human hearing varies somewhat from one person to another, but extends from approximately 15 to 15,000 Hz. (The *hertz*, often abbreviated to Hz, is the modern term for *cycles per second* (cps).) Thus, a true "high-fidelity" system should provide uniform response over the 15 Hz to 15 kHz range. In some applications, the bandwidth requirement can be compromised; there is general public acceptance of a bandwidth extending from about 50 to 5000 Hz as basically *suitable* (if not *optimum*) for both speech and music signals.

Good linearity in an audio system implies strict proportionality between the output and input signals of each device so that there is no alteration in the shape of the signal waveform. When a signal-handling device or circuit is nonlinear, it tends to distort the signal waveform in a characteristic way, typically by flattening either the peaks or the valleys; the reproduced sound then becomes "fuzzy" or "mushy." Mathematically speaking, a distorted signal differs from the original in that it includes unwanted *harmonics* of the input signal frequencies, so this type of departure from good fidelity is sometimes called *harmonic distortion*. Distortion specifications are usually expressed in terms of the maximum allowable amplitude of all unwanted harmonics as a percentage of the desired signal amplitude at a specified power level.

Noise or *extraneous signals* in an audio system can arise from a variety of sources, such as random currents in vacuum tubes or transistors (stimulated by heat energy), surface noise from phonograph records, atmospheric noises in radio links arising from lightning and other natural causes, man-made electrical noise arising from faulty automotive ignition systems and industrial machinery, and "cross-talk" signals arising from other communication channels operated in proximity to the desired channel. When a system has a high signal-to-noise ratio, the extraneous signals are so low in amplitude in comparison with the desired signal that they are of no practical significance. If the signal-to-noise ratio is low, however, the noise may be manifested to the listener as a hissing sound, as a series of crackles, clicks, pops, or whistles, or as faint speech or music signals heard in the "background" of the desired signal. A system with a poor signal-to-noise ratio also has a poor *dynamic range*; that is, it cannot properly handle both very soft passages and very loud passages with a single set of adjustments.

Time-base stability in audio systems involving recording devices is essential if one wishes to avoid annoying variations in the pitch and loudness of reproduced sounds. Distortions resulting from poor time-base stability (usually caused by insufficient precision in mechanical components) are usually called *wow* and *flutter*.

Standard Broadcasting

Amplitude-modulated (AM) broadcast services are now available throughout the world in the frequency band between 535 and 1605 kHz, and additional international broadcast services using similar modulation techniques are provided in several "short-wave" bands. From a technological point of view, standard broadcasting techniques are now very well refined. The major problems in studio acoustics, equipment design, and operating techniques have long since been solved, and standard radio broadcasting will probably continue indefinitely as a relatively low-cost medium for delivering informational services, entertainment programs, and advertising messages. AM transmitting equipment is now sufficiently stable that unattended operation is commonplace, and the extensive use of special recording and playback devices in AM broadcast studios helps to keep operating costs low.

With respect to fidelity, modern AM transmitters and receivers usually have very adequate *linearity*, but the *bandwidth* is limited in practice to about 5000 Hz, providing signal quality that falls somewhat short of true "high fidelity." It should be noted that this bandwidth limitation is not intrinsic in the amplitude-modulation technique per se, but results from the regulatory framework of standard broadcasting—standard broadcast channels are spaced only 10 kHz apart in the 535 to 1605 kHz band, and this spacing permits an audio bandwidth of only 5 kHz. Within the coverage area of each standard broadcasting station, the *signal-to-noise ratio* is kept up to a reasonable level by the expedient of using a very high-power level in the transmitted signal (which has the distinct economic advantage of permitting the use of relatively simple, low-cost receivers); the signal-to-noise ratio tends to become rather poor when AM stations are received at great distances.

Frequency-Modulation Broadcasting

Frequency-modulation (FM) broadcasting got off to a very slow start in the United States because of a combination of wartime conditions (the service was first authorized in January 1941, only a short time before America's direct involvement in World War II), a postwar reassignment to a totally different frequency band, the entrenched competition of standard broadcasting, and the general preoccupation of both broadcasters and the general public with the establishment of television broadcasting. FM broadcasting is now thoroughly established in the 88 to 108 MHz band, however, and will probably remain a popular medium for the indefinite future.

The chief advantages of FM broadcasting are its provisions for high-fidelity transmission, stereo broadcasting, and special auxiliary services. The high-fidelity capability results from a combination of wide audio bandwidth (a full 15 to 15,000 Hz range), good linearity, and the high signal-to-noise ratio made possible by use of the frequency modulation technique. Because audio information in conveyed in the form of *frequency* variations of the transmitted wave instead of in *amplitude* variations, it is possible to remove most extraneous signals from an FM wave by *limiters* or clippers before demodulation takes place, resulting in a very "clean" audio signal. It is unlikely that FM broadcasting will ever completely replace AM broadcasting, however, because the improvement in the *quality* of FM service comes at a rather high price in terms of reduced *quantity* of service. An FM signal requires a channel width of 200 kHz in the limited radio-frequency spectrum, in comparison with the 10 kHz allocated to a standard broadcasting station. In other words, a given frequency band could accommodate approximately 20 times as many AM stations as FM stations.

There are two additional technological factors that account for the "survival" of AM broadcasting in the face of FM competition: (1) FM receivers will always cost somewhat more than comparable AM receivers because of the need for greater bandwidth and more complex detector circuits, and (2) because of their wide channel requirements, FM broadcasting services have been allocated to the very high frequency (VHF) portion of the spectrum where propagation distances are more nearly limited to line-of-sight paths than is the case for the lower frequencies allocated to standard broadcasting. Thus, the coverage area for a station with a given power output is smaller for FM than for AM.

Stereophonic Broadcasting

One of the advantages of frequency modulation is that it is possible to add *subcarriers* to the audio signal (in the frequency range above 15 kHz) before it is applied to the frequency modulator, and to modulate each of these subcarriers with independent signals that can be transmitted through the basic FM channel without objectionable interference to the main audio signal. For the past decade, FM stations in the United States have been authorized to transmit *stereo* programs, involving two independent audio signals to feed the left and right loudspeakers of a stereophonic reproduction system. To provide high-quality service to monophonic receivers, the signal applied to the main channel is the *sum* of the left and right audio signals. Another signal consisting of the *difference* between the left and right audio signals is then applied, by amplitude-modulation, suppressed-carrier techniques, to a subcarrier at 38 kHz. This added subcarrier is not detected by ordinary monophonic receivers, but in stereo receivers suitable filters, detectors, and mixing circuits are provided to recover both the main signal and the subcarrier signal, and to combine them in a way that reconstructs the original left and right signals. Stereo broadcasting has become quite popular because of the greater "presence" and realism of the sound produced in both home and vehicular environments.

Variations of the same basic subcarrier technique used for FM stereo broadcasting are also used for so-called Subsidiary Communications Authorization (SCA) services offered by FM stations. Such services include background music, special programs for limited audiences, and facsimile transmissions. Further comments on such services are presented later in this Chapter.

Audio Recording

One of the most striking characteristics of the modern audio recording industry is the wide range of media that seem to be quite permanently established. There are disc records of various diameters, designed for operation at speeds of 45, 33 1/3, and 16 2/3 rpm. (The old standard at 78 rpm appears to be clearly on the way out.) There are 1/4 in. magnetic tapes on reels, employing a number of track formats and designed for playing speeds from 15 ips down to 15/16 ips (with 7.5 and 3.75 ips speeds being most popular). There are endless-loop tape cartridges in both four-track and eight-track formats, and there are miniaturized hub-to-hub tape cassettes in both stereo and monaural formats. Each of these media has certain characteristics making it particularly suitable for at least one class of applications.

The quest for high-fidelity performance has been apparent in each audio-recording medium. Professional quality magnetic tape recorders are now used almost universally for the production of original or "master" recordings; such machines, although quite expensive, provide extremely good fidelity and make possible the convenient editing of the master recordings. Many of the techniques originally developed for professional quality tape recorders have been adapted to the lower-cost units sold as consumer products.

A long series of improvements has been made in the art of disc recording; through modern materials and manufacturing methods, it is now possible to produce both monaural and stereophonic disc recordings with excellent fidelity.

Endless-loop tape cartridges have become firmly established as a preferred medium for the reproduction of stereo music in automobiles, where shock, vibration, and dust conditions make the operation of disc playback units very difficult. Miniature cassette units have become very popular in situations where consumer-controlled recording is desired or where the compact storage space required for prerecorded cassettes (in comparison with other media) is a definite advantage.

Until quite recently, miniaturized audio tape cassettes have been considered inferior to stereo discs in terms of fidelity, mostly because of somewhat limited bandwidth and a relatively poor signal-to-noise ratio attributable to the very narrow tape tracks. Special signal-processing techniques now being introduced may soon reduce or even eliminate the "quality gap" in audio cassettes relative to stereo disc recordings. In essence, these techniques involve precompensation of the audio signal before recording so that the high-frequency components are recorded at substantially higher levels than in ordinary recording so that these higher frequencies can override the "hissing" noise that is characteristic of the tape medium. In the playback process, the high-frequency components are restored to their normal levels. A specific system of this nature known as the "Dolby Process" has found recent commercial favor.

Even if the performance of tape cartridge and cassette systems can be made fully equivalent to that of disc recordings, it is probable that discs will remain a popular medium because of their cost advantages and the ease with which a given passage or selection within a disc recording may be reached.

TELEVISION BROADCASTING AND VIDEO RECORDING

Although television broadcasting has emerged as a significant industry only in the past 25 years, its impact on human civilization is already recognized as very substantial, probably equivalent to that of the great railroad systems developed during the last half of the nineteenth century or the automobile industry in the first half of the twentieth century.

In setting up practical standards for television broadcast systems, engineering compromises must be made between the *quality* and *quantity* aspects of the service—high resolution images that are free of flicker can be obtained only through the use of wide-bandwidth channels, but the wider the bandwidth the fewer the channels that can be accommodated in a given range of the frequency spectrum. The bandwidth requirements for a television signal are, at best, very great in comparison with those for either AM or FM sound broadcasting. In the United States, broadcast television services employ a 525-line scanning pattern, provide 30 images/second, and occupy a 6-MHz transmission channel. (The channel includes provision for a frequency-modulated sound carrier in addition to an amplitude-modulated picture carrier.) In other countries, scanning patterns range from 405 to 819 lines, image rates are either 25 or 30 images/second, and transmission channel widths range from 5 to 14 MHz. The great variety of technical standards creates serious problems for the internationally minded broadcaster, but broadcast engineers have devised a number of techniques for accomplishing conversions from one set of standards to another.

From a technological point of view, the two most significant innovations in television broadcasting since its period of initial growth are (*a*) the introduction of color, and (*b*) the development of video tape recording.

Color Television

The development of color television during the late 1940s and early 1950s was a particularly challenging problem for broadcast engineers and scientists, because the television industry was already quite well established and it was essential that the color system be *compatible* with the existing monochrome system. (That is, it was essential that the broadcaster who converted to color not lose that part of his audience possessing only monochrome receivers, and that the purchaser of a color receiver be assured of service from either monochrome or color transmissions.) A further constraint was imposed by the Federal Communications Commission (FCC) in its 1949 declaration that any color system to be considered for formal approval must operate within the same 6-MHz channel previously used for monochrome television.

It is now a matter of history that these challenges were very successfully met by an all-industry effort under the guidance of a National Television Systems

Committee (NTSC). A series of techniques was developed that enabled the red, green, and blue primary-color signals developed by a color television camera to be processed to yield a high-quality monochrome signal component (capable of rendering good service to monochrome receivers) plus a set of color-difference or *chrominance* signals that could be modulated on a video subcarrier and transmitted right along with the monochrome signal in a manner that avoided objectionable interference with monochrome sets. Meanwhile, a practical color picture tube was developed that made possible the reproduction of high-quality color television images in the form of closely intermingled red, green, and blue phosphor dots. Color television in the United States has now matured to the point where the great majority of all programs are now broadcast in color.

Many other countries have also adopted color television standards, some of them the same as in the United States. In much of Europe and Great Britain, the approved color system is known as the PAL system, which differs in concept from the American NTSC system only in the manner in which the chrominance signals are modulated on the subcarrier. (The numerical constants associated with the PAL system are also different because of the different scanning standards and video bandwidths used in Europe as compared to the United States.) In France and the Soviet Union, another type of color television, known as the SECAM system, is employed; this system uses a frequency-modulated subcarrier and a special delay-line storage technique instead of the amplitude-and-phase modulated subcarrier used in the NTSC and PAL systems.

Video Recording

The television broadcasting industry has been greatly influenced by the development of practical video tape recorders, first introduced in 1956 and subsequently refined through the efforts of several different companies. These machines are now capable of very high-fidelity performance in either monochrome or color, and permit the use of a variety of editing and animation techniques. Special types of recorders employing magnetic discs instead of tape are now widely used in special applications, such as "instant replays" of key bits of action in athletic contests.

Although the high-quality video recorders used by broadcast stations are typically priced in the range of 40,000 to $100,000, other machines using helical-scan techniques have been developed that make possible video recording through instruments that cost less than $1000. (The very high "writing speeds" needed to record wide-bandwidth video signals are usually obtained by moving the recording head itself, mounted on some type of scanning disc; tape moves through the machine at a rate comparable to that of an audio tape recorder.) Many of these low-cost video recorders are almost as easy to use as audio recorders, and have opened up a wide range of nonbroadcast applications for television.

At the present time, very intensive development work is in progress on several different types of "video cassette" systems, some of them based on magnetic tape (encased in suitable cartridges), others based on alternative recording media, such as holographic tape or optical film. Some very promising developments are also taking place in the area of video recording on discs; a practical system for recording audiovisual programs up to 5 min in length on a plastic disc has already been demonstrated. Comments on the significance of these developments are offered later in this Chapter.

CABLE TELEVISION SYSTEMS

Cable television (CATV), a medium that was once regarded only as a minor extension of television broadcasting, is rapidly growing to the point where it deserves attention as a significant medium in its own right, offering an interesting range of opportunities for new broadcast-type services. Historically, the technology of cable television was developed in response to the need for *community antenna systems* in areas where direct television reception is very poor. Pioneering CATV operators learned that satisfactory service at reasonable rates could be provided by mounting elaborate receiving antennas on a nearby hilltop or tall tower and stringing coaxial cables to each receiver location. In recent years, many systems have been installed in larger cities where residents have been found willing to pay reasonable service fees (typically of the order of 4 to $6/month) to obtain the advantages of improved signal quality and a wider range of program choices. It is estimated that approximately 10% of the households in the United States are now served by CATV systems, and this number is expected to grow rapidly during the coming decade.

Cable television deserves attention as a potential new medium because CATV systems can effectively multiply the quantity of broadcast services that can be provided, especially when the CATV industry grows to the point that it will become practical to link many cable systems to form regional or even national grids. The number of television broadcast stations that can be permitted in a particular geographic area is severely limited by the number of channels available in the electromagnetic spectrum. A modern cable system, however, provides an *additional* spectrum of anywhere from 120 to 300 MHz bandwidth (depending upon its technical sophistication). This additional spectrum space can theoretically provide from 20 to 50 television channels to every reception point, and it is possible to provide two, three, or even more separate cables if still more channels are needed.

In principle, the available spectrum within a cable could be used for *any* type of communication service without regard for the spectrum allocations that have been established for the "natural" or "open-circuit" electromagnetic spectrum. As a practical matter, however, most of today's cable systems use the available cable spectrum primarily for relaying the signals of television and FM broadcasting stations in the same VHF frequency bands used for such services in the "natural" broadcast spectrum. The opportunities implicit in the availability of a wide communications spectrum are gradually being recognized, however, and some cable system operators are beginning to develop these opportunities in at least a modest way. Many systems, for example, use one or more channels for the distribution of local programs developed through closed-circuit studios located at or near the "head end" of the CATV distribution system. One of the most intriguing opportunities is that of providing *two-way* communication so that entirely new services involving consumer interaction can be developed. The cable itself is intrinsically a two-way medium, but special filters and amplifiers must be installed in a CATV system before the two-way capability can be exploited.

POSSIBLE FUTURE DEVELOPMENTS

Quadrasonic Recording and Broadcasting

There is a great deal of current interest in four-channel audio recording, involving reproduction systems in which loudspeakers are placed in front of and behind the listener as well as on his left and right. Enthusiasts for these systems claim that the enhancement of the listening experience by this technique relative to conventional stereo systems is at least as great as the enhancement that stereo offers over monophonic listening. This author predicts that four-channel systems will shortly win acceptance in the recording industry, but not to the extent that they will supersede the existing two-channel stereo systems for at least two decades. Interest in three-channel or four-channel audio systems for FM broadcasting will probably grow in parallel with the interest in the recording industry. Basic technological tools required to accomplish four-channel recording and broadcasting (even within the existing FM channel structure) are already at hand.

Audio Cassettes as a "Broadcasting" Medium

Several pioneering "audio publications" involving the periodic distribution of audio program materials on cassettes to specialized audiences (including doctors, electronics engineers, and audiovisual specialists) suggest that we may even now be witnessing the birth of a significant new "broadcasting" industry. A number of these pioneering ventures are aimed at helping busy people make more efficient use of time spent in automobiles, waiting rooms, and even before shaving mirrors! Costs tend to be rather high per unit of information received in comparison with printed media, but these costs can be lowered when the ventures mature to the point where it becomes practical to use a "returnable bottle" concept, whereby each cassette is used repeatedly with only a small service charge for the actual program material. It is conceivable that in the future such public institutions as libraries or such commercial establishments as bookstores might provide fast duplicating machines into which the consumer inserts his own audio cassette to receive a fast copy of an audio program of his choice. The technology is available to support this type of service.

Video Cassettes or Discs

There is a great deal of current interest in developing the video equivalent of the audio recording industry, and very intensive development and design work is underway on related "hardware" systems. Many different approaches are being explored, and it is impossible at this time to predict which media will ultimately win acceptance in the marketplace. In my opinion, at least three different systems are likely to become established, corresponding roughly to the discs, tape cartridges, and tape cassettes now available in the audio recording field. It is probable that some type of video disc will offer the lowest-cost replication, but may be suitable only for relatively short prerecorded programs. There will probably eventually be a market (possibly on a rental basis instead of on an outright sale basis) for longer prerecorded program materials in some type of easy-to-handle cartridge or cassette. For some applications, there will probably be strong interest in a magnetic tape cartridge or cassette system that permits the consumer to do his own recording as well as playback.

The commercial aspects of the video cassette business are still largely unknown. Rather large-scale market tests will probably be necessary before anyone can predict with confidence what level of interest the public really has in prerecorded audiovisual programs and what prices the public will be willing to pay.

Communications Satellites for Broadcasting

The impact of communications satellites has already been felt in the broadcasting industry through the use of satellite relays for television coverage of special events of international significance. It is becoming increasingly clear that it is only a matter of time before satellites are used routinely for the interconnection of television stations in nationwide networks and for the transmission of broadcast services to outlying areas (such as Hawaii and Alaska in the United States).

It is possible that the establishment of one or more large-scale domestic satellite systems in the United States will lead to a rather significant shift in the character of the CATV industry, since it would become technically and economically feasible to link the "head end" facilities of CATV systems directly to the sources of television network programs and to other sources developing programs specifically for CATV audiences. Should this come about, CATV systems may no longer function primarily as "community antenna" systems limited to the relaying of signals picked up "off the air." There would be, of course, many nontechnological problems involving copyright fees, licensing arrangements, and regulatory arrangements to be worked out before CATV systems can take full advantage of the new opportunities that might become available through satellite interconnections.

Direct broadcasting from satellites to home receivers remains as a possibility for future development, but I believe that the very practical problems of language barriers, time zones, and administrative control (to say nothing of the technological problems related to required power levels and receiving antenna designs) will prevent rapid development of this possibility. It is quite probable, however, that direct satellite service to community viewing centers in sparsely populated areas (such as Alaska) will become established within the next decade.

Interactive CATV Systems

Some of the advanced CATV systems now in the planning stage or in the early stages of implementation include special two-way amplifiers that make possible the allocation of certain channels (usually narrow-band channels, suitable for data messages instead of voice or picture signals) for carrying signals from the consumer location back to the "head end" of the system. As such two-way CATV systems become more widespread, some very intriguing opportunities for new services are opened up, including:

1. Remote reading of utility meters.
2. Remote (and nearly instantaneous) public opinion polling.
3. A "shop-by-wire" service, whereby the consumer signifies his desire to buy specific items shown in a "video catalog" program.
4. Remote selection by the consumer of certain program materials transmitted on reserved channels.
5. Interactive instructional programs.
6. Remote access to a computer for "desk calculator" functions.
7. Remote banking by direct connection to a bank's computer.

Certain other technological developments can be utilized in conjunction with the two-way capability of advanced CATV systems, notably techniques for multiplexing a great many independent signals into conventional TV

What Can the New Technology Do? The Dimensions of Change

There is reason for some concern about problems of copyrights when con-
ner-controlled recorders are used in combination with broadcast channels,
. as long as the cost of a cartridge or cassette that permits recording by the
isumer is greater than the cost of the program on some other medium (such
audio or video discs), the risk of unreasonable "pirating" of program mate-
is would appear to be minimal.

channels. Through "freeze-frame" techniques, for example, i
to use a single TV channel for the transmission of the vid
many as 600 independent programs consisting of still pictur
10 sec; accompanying audio signals could be transmitted by
rable audio multiplexing techniques. Such a capability coul
basis of an exciting new approach to home study—the stude
choice of as many as 600 different illustrated lectures at any p

A prototype system having many features similar to those
currently under development in conjunction with Project
University of Illinois.

Facsimile Services

The technology to supply facsimile services directly to the
available for quite some time, but analyses of the commercial
yet shown sufficient promise to lead any major corporation t
investment that will be required to introduce such a service on
viable scale. It is conceivable that sometime in the next decade
copy printed materials delivered directly to the home will er
cient clarity to bring about such a service. Several alternativ
available with respect to transmission channels for facsimile s
special channels on CATV systems, the use of subcarriers on
missions, and the use of vertical blanking intervals in televisior

Broadcasting to Local Recorders

One of the technical possibilities that has not yet been signifi
as a consumer service is the combination of broadcast transi
and local recording devices to adjust schedules to suit the co
consumer. In many areas, a consumer has a wide range of
available to him through simple use of the tuning dials on his
sion receivers. His range of choice could be even greater, howe
easy-to-use, clock-operated recording device that could be turi
cally to record some program he has selected in advance (ever
sion occurs at some very inconvenient time, such as the "wee
the morning), and hold it for later use at his convenience.

Audio recording devices suitable for this type of service ar
ble, but little attention has been given to their "packaging" in
clock mechanisms. Certain helical-scan video tape recor
designed with clock timers, but these have been too expensive
impact in the consumer market. Some of the newer video ca
now under development might lend themselves to this type of s

CHAPTER 4

Communicating with the Man on the Move

JONA COHN

This chapter describes the "mobile communication" capabilities that are likely to be available in the coming one to two decades. In the process, there will be a review of the factors that will stimulate these developments, such as new technology or newly recognized communication needs, as well as comments on the factors that may impede these developments, such as limited availability of radio spectrum or channels.

MAJOR TECHNICAL DEVELOPMENTS

Consider the major technological developments that have created the technical environment within which these new communication developments will take place. There are two extraordinary developments and a third that is potentially so that will have major impacts on communications of all kinds, including mobile communication.

First is the development of microcircuits that are improving by orders of magnitude the size, cost, power consumption, and reliability of certain types of electronic circuits. The drawer full of electronics that I worked on early in my career can now be put on the head of a pin. These tremendous gains apply most readily to the manipulation of arithmetic or logic-type information. For proper perspective, it is important to note that microcircuits will have much less effect wherever power is involved, such as in generating radio signal power to permit transmission to distant points. As a result, micro-circuits (*a*) will provide some reduction in transmitter size, (*b*) will provide large reduction in receiver size, and most important, (c) will permit equipment of approximately current size to contain functions that are orders of magnitude more complex than for current systems. For example, microcircuits make it possible to design a coding device smaller than the size of a penny, which would provide a unique code for every person in the world.

To illustrate the use and impact of microcircuits, an example of their application in equipment used to manage a bus transportation system is discussed in a later section.

The second revolutionary factor is the continued development and application in software or programming capabilities of the computer. Simply stated, computers will augment our brain power as effectively as machines currently have augmented and exceeded our muscle power.

For systems consisting of units dispersed over a geographic area (as, for example, a metropolitan area), mobile communication can provide the link tying these units into an organized system. The computer will provide the key tool in taking advantage of this communication capability to organize and coordinate the system.

There is a second important way in which mobile communications and computers will be utilized. Mobile communications will permit the tremendous power of the computer to be available literally at our fingertips (and eventually by spoken word) as we move about. Computer terminals are now beginning to be available in vehicles and will be available in portable form.

The example on managing a bus system will illustrate this capability and application also.

A third development that may rank with the first two would be the installation of very high-capacity, two-way cable communication links throughout metropolitan areas. The very high communications capacity provided by cable not only frees the spectrum for mobile communication requirements, but it can also act as a "super communications highway," that is, a point-to-point interconnection segment for mobile communications systems.

FRAMEWORK FOR DISCUSSION

In considering what future communication developments will be, it is useful to consider four factors that play a role in having those developments occur. These are: (1) *need* (the driving force, the problem or opportunity that is the key to having the whole process gain momentum); (2) *technical barriers.* (From an engineering point of view, can we design the equipment to meet that need or are there natural laws of physics, barriers that we do not as yet know how to overcome?); (3) *economic barriers.* (In looking at the tradeoff between the cost of the solution and the benefits derived in meeting the need, is the solution being considered economically attractive?); and (4) *legal or organiza-*

tional barriers. (Assuming a need has been identified and technical solutions presented that are viable economically, is there something in the way we have organized our society which is a barrier? Are there man made barriers which prevent the particular development from coming to successful completion?).

It is in the latter area that the question of how the radio spectrum is allocated to potential users frequently occurs. This question will receive attention later in the paper.

CATEGORIES OF NEEDS

It may help the reader gain a better understanding of mobile communications to examine in greater detail the various categories of needs being met by mobile communications.

Safety

There are many important needs for mobile communications in combating a whole range of natural disasters due to fires, hurricanes, rain, snowstorms, floods, and earthquakes. There are communication needs in the medical field in providing medical assistance, and in biomedical telemetry from ambulance to hospital. There is, of course, the widely recognized need for communication by all of the police and other crime-combating activities, and in transportation by air, bus, train, and so on, in which mobile communication is necessary for the maintenance of safety.

Efficiency

Everyone is familiar with the use of two-way communications in the taxi field, which so obviously improved the efficiency of that type of operation. Less recognized but equally valid is the fact that the efficiency of almost every endeavor that involves a group of individuals who are spread out beyond a 20- to 30-ft voice range can be significantly improved by the use of mobile communications. (The term "mobile communications" refers to any "movable" form, including hand-carried equipment as well as vehicular-mounted equipment.) An examination of today's application of mobile communication verifies this in every conceivable type of endeavor, be it industrial, business, governmental, and so on. *It is estimated that by 1990, 1 in 10 of those employed will utilize two-way mobile communication.*

Convenience

In addition to the fundamental requirements of safety and efficiency, there are needs and uses that would be described as matters of convenience. Once the fundamental safety and efficiency needs are met, there are many circumstances in which individuals would like to communicate with each other. Their ability to do so may, in fact, have profound effects on the society. (What would our society be like if we could always reach our teen-agers wherever they drive in the car?)

It is important to recognize that in many situations the communications system can be used to meet all three needs, as in the following illustration.

Two other categories and needs are very significant, but are not covered in this chapter: namely, mobile communications for entertainment and education.

EXAMPLE APPLICATION

The discussion has so far been somewhat general. A specific example tying these concepts together may be helpful. The Chicago Transit Authority has in operation a demonstration system that is used for monitoring the adherence to schedule of 500 buses. This system involves the use of a *computer* to rapidly handle the mass of information involved, a radio system to link the computer to the buses, and *microcircuits* in the buses to permit the various functions to be performed there, with a resulting system that has important *safety, efficiency*, and *convenience* characteristics.

The "Monitor-CTA," through immediate identification and location of buses, assists in the maintenance of schedules with increased *efficiency* of equipment and personnel, and provision of up-to-date statistical data for analysis of systems performance. The system provides greater *safety* for passengers and bus operators alike who feel safer in the knowledge that assistance can be quickly forthcoming in emergencies either through transmitting a silent alarm signal or through two-way voice communications. The passengers enjoy the *convenience* of more dependable bus transportation with fewer and smaller gaps in service.

For the pilot program 500 motor buses are equipped with the "Monitor-CTA" equipment. The complete "Monitor-CTA" system consists of three basic elements: a computerized vehicle location and identification system that monitors bus schedules and allows "management by exception"; a silent alarm radio system that instantly relays the location and bus and run numbers of an operator in need of help; and a two-way radio system that provides voice communications between the bus operator and central dispatcher.

Located at various intersections along the bus routes are low-powered radio transmitters called "electronic signposts" that perform in the radio media a similar function to the familiar visual street identification signs. Each "signpost" is assigned an identification number. This coded number forms a digital pulse message that is constantly being transmitted by the "signpost" and is of sufficient power to reach any bus passing the "signpost" in any direction. On the bus is a location receiver unit that receives and stores the signal from the "signpost." Also in the unit is an elapsed time generator. As the bus leaves the vicinity of the "signpost" the elapsed time generator is turned on.

The buses are interrogated from the control center computer sequentially at the rate of *12 buses/second*. The buses respond with their last "signpost" number and elapsed time count.

The computer interprets the reply from each bus, calculates that bus deviation from schedule, and compares it to the tolerance limits. These limits allow the bus operator considerable leeway to contend with traffic and passenger delays, yet give the dispatcher rapid warning of a possible major interruption of service. If the deviation is excessive, a message is displayed at the dispatcher's console and a permanent record is made. The dispatcher can then act appropriately to restore scheduled service.

Another feature of this system is the emergency alarm. In the event of an emergency on the bus such as a disturbance, the bus operator can unobtrusively activate an alarm switch. Automatically, the equipment transmits the identification and location of the bus with no audible or visual indication to the riders. At the control center, the dispatcher immediately receives an audible alarm and the alarm message is visually displayed along with all necessary data to enable the dispatcher to direct the police to the location of the bus.

Modern technology, such as microcircuits (i.e., integrated) make the system possible, enabling the specialized coding, decoding, control, and alarm functions to be packaged compactly and operate reliably.

This is an excellent example of an important developing trend in which a computer combined with the mobile communications links can tie together, help organize, and assist in the management of a widespread complex system.

FUTURE MOBILE COMMUNICATIONS

We have examined in detail one example of the application of computers, microcircuits, and mobile radio. A large number of new capabilities are being developed for both voice and nonvoice communication functions including automatic vehicle location, vehicle status, record-form communications, repeater systems, alarming and signaling, telemetry, voice privacy, computer access, command and control, slow-scan TV, digital communications, paging systems, teleprinters, facsimile, and remote control.

Some Possible New Services

To look in detail at the new capabilities offered by each item in the list above would require volumes. This section looks at the new services those capabilities make possible in four areas: distribution of goods and services, public safety, medical emergency, motorist aid and information. These areas are selected because they are closely part of each of our day-to-day lives.

DISTRIBUTION OF GOODS AND SERVICES. Fast and efficient radio communications contributes directly to more timely and economical provision of all types of goods and services. Whether it be delivery of fuel or ready-mix concrete, furniture or groceries, or maintenance and repair services, the public benefits when these goods and services are provided in a timely and economical manner.

Effective operations in supplying goods and services are largely dependent on radio communications in order to direct and control personnel and equipment. This dependence on radio communications in many instances stems from the intrinsic high cost of providing a timely service in response to a public need. This high cost can be offset or reduced by application of effective radio communications.

To quantify the typical distribution savings, let us look at a fleet of 100 vehicles. Experience indicates that in general, three radio-equipped vehicles can do the work of four vehicles without radio. The cost of providing mobile radio for 100 vehicles can be compared to the cost of the alternative of adding 33 additional vehicles (Levin, 1971, and FCC, *2*, p. 398, 1967).

For the mobile radio costs assume:

Amoritization (over 10 years)	$8000
Maintenance	8000
Operating costs	9000

According to Casselberry and Gifford, the annual "cost of ownership" is then $25,000. And if the cost of operating 33 extra vehicles (in the absence of radio) is:

Per vehicle at 30,000 miles/year	$2,100
Cost of driver	5,000
Total cost per vehicle	7,100
Total cost for 33 vehicles	234,000

Net savings (after $25,000 annual ownership costs deducted) is then $209,300. The new savings is thus some eight times the ownership costs. Although costs have increased since these analyses had been made, the ratios or relative costs should be substantially the same.

For the entire United States the present savings from land mobile use is estimated at 8 to $13 billion/year.

PUBLIC SAFETY. Similarly, cost savings are currently estimated at $6 billion/year (not counting crime reduction) by utilizing mobile radio as an alternative to adding patrol cars to police departments.

Vehicle location systems constitute an excellent example of a technology that is in existence today, and is currently undergoing considerable development and marketing by manufacturers. These services will also have to be accommodated in the spectrum available for mobile communications.

Their value has already been cited for public safety agencies by the President's Commission on Law Enforcement and Administration of Justice. The U.S. Department of Transportation has issued several contracts for vehicle location research.

Related to vehicle location, and a function that could possibly be incorporated with it, is vehicle status reporting. By means of a computer that compiles information from each vehicle, a user is provided up-to-date information concerning each individual vehicle. He thus can maintain the status of his entire fleet and have on hand information that will assist him in being more effective with his resources.

The previously described systems can also be combined with a selective calling system, a computer, the regular two-way voice radio system, and a mobile teleprinter system to form a *computer-aided dispatch system*, a fully integrated command and control system. This system performs many of the manual, routine functions of the dispatcher, thereby producing greater efficiency and faster dispatch response. In addition to gathering and storing vehicle location and status information of police vehicles, the computer-aided system can select the nearest available patrol car to respond to a call; it can generate and display a situation map with all pertinent information regarding the fleet; and it can transmit information through the teleprinter to patrol vehicles. It can also provide automatic response information to requests from mobiles for the data on file for vehicles registration, wanted or missing persons, gun registration, and so on. Not only does the dispatcher have immediate information and control of the vehicle for which he is responsible, but he also is freed from many routine functions that he handled previously. Thus, he has improved capability for decision making, dispatch response, and flexibility.

Another new technique—now in limited use, but for which the demand is burgeoning—utilizes a group of satellite receivers to make a metropolitanwide portable transceiver system practical. Increasingly, police departments have utilized this approach to equip *all* of their foot patrolmen with portable radios in order to provide increased command and control functions, and to give policemen on the street access to immediate assistance. To insure coverage throughout an entire urbanized area, signals from the portable radios are received by the satellite receivers and retransmitted to the headquarters.

MEDICAL EMERGENCY. Hospital Emergency Administrative Radio (HEAR) systems provide communications channels for emergency assistance to and from hospitals during disaster conditions. It ties several area hospitals to a central or regional hospital; portable radio-equipped workers at the disaster scene are linked to the hospital, and ambulances, rescue squad trucks, and other emergency vehicles are linked to each other, to hospitals, and to workers on foot. These systems are available now. They rely on radio instead of wire line, because the same event that creates a hospital emergency (storm, fire, explosion, accident, etc.) may disrupt the wire-line connection. The Los Angeles HEAR system has been heralded for its performance during last year's earthquake in the Los Angeles area.

A Los Angeles ambulance service has tested a system that transmits the heartbeats of a cardiac patient over the ambulance radio to the dispatch office. The information is in turn relayed by a special telephone line to an EKG recorder at the hospital. Thus, a doctor can examine the electrocardiogram and make preparations for the reception and care of the patient while the patient is enroute to the hospital by ambulance. Another system transmits the data directly from the ambulance to a hospital receiver.

The availability of a number of channels should eventually make it possible to send information on other physiological functions such as blood pressure, pulse, respiratory volume, and skin temperature.

An improved version of the present coronary system has just been announced as a standard product. This Coronary Observation Radio system provides radio telemetry directly from the ambulance to the hospital Coronary Care Unit (CCU). At the CCU, a cardiologist interprets the EKG and advises the trained ambulance technician, through a standard two-way radio system, as to the proper action required to stabilize the victim for transport to the hospital.

MOTORIST AID SYSTEMS. Radio or telephone call boxes have been proposed and are being tested as aids to the motorist in distress, for the reporting of accidents, breakdowns, or sudden illness. The motorist can transmit an emergency call, identifying the problem and degree of seriousness. He can ask for a tow truck or an ambulance. Driving is made safer because assistance will be available immediately, and the motorist is reassured when a problem arises.

In today's motor vehicle, the driver is isolated from his surroundings. With windows up to keep out noise and air pollution, with airconditioner or heater blowing, and stereo tape deck or radio on, he cuts in and out of traffic among the maze of other vehicles, roads, turnoffs, merging traffic lanes, and road signs. All too frequently he misses a sign, warning, or siren and heads into disaster.

If local authorities had a means of communicating with the motorist in the privacy of his vehicle, many accidents and traffic jams could be avoided. Drivers have run into emergency vehicles, fire trucks, police cars, and ambulances because they could not hear the siren or their vision was blocked so that they did not see the flashing lights. Imagine the same situation with the emergency vehicle now able to transmit an audio warning into nearby vehicles. "Caution, emergency vehicle approaching . . . please pull over to the right and stop." The audio message could also trigger a flashing light within the vehicle as a visual sign. With this kind of warning system both the motorist and the emergency vehicle would stand a better chance of reaching their respective destinations safely.

There are many other applications for such an audio signing system, if only the motorist had a receiver in his vehicle. A portable transmitter with a taped message recorded at the scene of an accident could be left along the roadside to warn oncoming vehicles. Another application would be to warn motorists approaching fog-covered roads, ice-covered bridges, construction areas, and the like. Transmitters that broadcast these warnings could be installed in emergency vehicles, left by the roadside as portable units, or installed permanently at specific locations. Alternatively, information could be broadcast on emergency situations over large preselected areas.

SPECTRUM AVAILABILITY

The process whereby new technology is brought to bear to develop these new communication capabilities is continuing at a rapid pace. The technical barriers are being overcome, and products and services that meet the economic test of the marketplace are being created. We see this in the continuing spread in the application of two-way voice communication in vehicular and portable systems, and in one-way radio paging systems. The historical figures indicate the growth pattern shown in Table 1 along with the projections in the future.

Table 1   Growth of Land Mobile Radio

Year	Licensed Transmitters
1950	180,000[a]
1955	520,000[a]
1960	1,300,000[a]
1965	2,280,000[a]
1970	3,400,000[b]
1975	5,000,000[b]
1980	7,000,000[b]
1985	8,900,000[b]
1990	10,800,000[b]

[a] FCC records.
[b] Electronic Industries Association Projections in Exhibit 3 of Docket 18261 filing.

Table 2   Reductions of Channel Spacing for Mobile Radio Equipment in the High Band (150 to 174 MHz), 1940 to 1968

Year	Channel Spacing (kHz)	Relative Number of Channels
1940	240	1
1945	180	1.5
1948	120	2
1953	60	4
1958	30	8
1968	30/15	8/16

Source: Federal Communications Commission.

In addition to these current systems, a whole new array of digital communication systems is being developed to provide the multitude of services previously described. However, there is a barrier to these developments, and it is the limitation of radio spectrum. This causes crowding of current services and a lack of unused spectrum needed for the encouragement of new services.

The crowding of mobile radio channels has in the past been overcome by technology investments that have permitted a doubling in the number of channels available, three to four times since the initial channel spacing was set in 1940, as shown below. We have clearly reached the end of the road in this direction, as can be seen on Tables 2 and 3.

As the crisis has grown and the pressure has built up to break open the spectrum barrier, several proposals have been pursued. The first of these proposals is included in the considerations of FCC Docket 18261. This docket will permit some sharing by mobile radio services of spectrum in the 500 MHz band that had been allocated to television. Sharing means the use of a

Table 3   Reductions of Channel Spacing in Mobile Radio, 1953 to 1968

		Channel Spacing (kHz)		
		1953	1958	1968
Low	25–50 MHz	40	20	20
High	150–174 MHz	60	30	30/15
Ultra high	450–470 MHz	100	50	50/25

Source: Federal Communications Commission.

channel, unused by TV in a given city, by mobile radio systems, subject to technical standards that prevent interference with TV channels that are in use.

For the coming decade, it has been estimated that at least 42 MHz of additional spectrum would be required by land mobile radio services. An attempt to handle this problem for the first half of the decade has been made by proposing to make 12 MHz of spectrum available in the top 10 cities of the country. However, in some cities such as Chicago and Pittsburgh, only one TV channel (that is, 6 MHz), has been made available without severe technical restrictions; in other cities such as Detroit and Cleveland, the necessary agreements that would permit the designated channels to be utilized have not been reached with Canada. Thus, the TV sharing rulings that have been made in the FCC Docket 18261 have provided only a small portion of the essential short-term relief from spectrum crowding.

Significant spectrum in the 500 MHz band could be made available by going beyond sharing to reallocation of television channels to mobile radio use. The present TV channel allocation table was developed 20 years ago, when the FCC presassigned all of the channels throughout the United States. The channel assignments were made to provide for transmission of TV signals with a minimum of interference, based on the then available knowledge and technology for transmission and reception. As of this date, a large portion of the channel allocation remains unoccupied. These unused channels could provide a basis for reassignments of TV channels and reallocation of spectrum to mobile radio.

Our current, more accurate knowledge of TV interference shows that the initial guidelines to prevent interference were overly conservative. More accurate guidelines would result in making spectrum available for other uses. These guidelines do not require the use of new hardware technology. Use of new hardware technology would permit further reduction in the spectrum occupied by the television services.

There is a second area where the FCC is taking action to help break the long range spectrum barrier, and this is in the 900 MHz band. As part of the proceedings related to Docket 18262, spectrum is being made available in this new band for land mobile radio. While some technology and economic barriers are in the process of being overcome, this band will offer relief for some of the requirements starting in the mid-1970s. We will see a fully developed broad range of equipment for a broad range of requirements in the 1980s.

Because of the very modest relief afforded by TV sharing in the 500 MHz band as provided in Docket 18261, there a serious gap will remain during the 1970's. Spectrum limitations will continue to act as an inhibiting factor for the development of a number of new services previously described. Even meeting the basic crowding problems of the current services will be a problem until such time as use of the 900 MHz band is not merely started, but a broad range of equipment and services are available in that band.

Having reviewed this much of the spectrum situation, we come to what is possibly the most intriguing aspect, the long-term solution offered as a very important side benefit of the move of broadcast TV to broadband two-way cable communications. Two-way broadband cable communications has a number of very important characteristics to commend it. These include the improvement in the quality of the signals available and the expansion of the capacity of the communications offered not only for the traditional broadcast television, but for a whole host of new communications services that take advantage of the increased and two-way capabilities.

In developing national policy on the two-way broadband cable system, consideration must be given to the manner of utilization of the spectrum that will be made available by the transfer of broadcast TV to the cable system. Here it is important to recognize the fundamental fact that mobile communications of necessity must utilize radio spectrum as the only way to complete the link to the individual or vehicle in motion.

Putting broadcast TV, as a point-to-point service, on the cable system can free vast amounts of spectrum, which can act as a stimulus to the technology and economic investment required to truly tap the capability of mobile communications.

To conclude, the communications technology is ripe for a further harvest of products and services that are economically attractive for the meeting of a wide range of society's needs. Substantial contributions could be made to the safety, efficiency, and convenience with which our society functions. What remains are the organizational man-made barriers involved in our spectrum management policies. The contributions that this symposium makes toward laying the groundwork for appropriate policy decisions in this area could be a very significant contribution to the future well being of our society.

REFERENCES

Levin, Harvey J. *The Invisible Resource-Use and Regulation of the Radio Spectrum.* Baltimore: The Johns Hopkins University Press, 1971.

Joint Technical Advisory Committee. *Spectrum Engineering: The Key toProgress.* New York: Institute of Electrical and Electronics Engineers, 1968.

Report Of The Advisory Committee For Land Mobile Radio Services. (Federal Communications Commission). Washington: U.S. Government Printing Office, 1967.

Motorola, Inc. *Comments and Exhibits before the Federal Communications Commission on Docket No. 18261,* February 3, 1969.

CHAPTER 5

Graphics Systems

GEORGE R. WHITE

Mankind is in a new information revolution. Electronic computers, copying machines, and information theory all attest to the rapidly escalating capability and comprehension available for handling man's information. Graphics may well be crucial to future information systems, for a central perpetual truth is that man neither reads nor writes in digital bits.

The electronic information revolution started in 1832 with the invention of the telegraph by Morse. The telegraph, the telephone, radio, television, the electrostatic copier, and the electronic computer all depend intimately on electronic charges and currents to receive, manipulate, and present information. Specialists' myopia often leads one to overemphasize differences between analog and digital, audio and video, and graphic and alphanumeric; but these devices are strongly united with each other by their common electronic technology, and sharply divided in technique from any earlier devices. Electronic information technology revolutionizes society; *this fact* is historically established for the past and a commonplace for the future. The electronic revolution offers greatest benefit when smoothly merged with the highest previous means of expression, that of printing and associated arts, and yields graphics systems.

INFORMATION EXPANSION

It is interesting to sketch the expansion of man's information systems and crudely estimate some orders of magnitude involved. The parameter to be estimated is the total data base content accessible to man. If one believes that the data stored by a human is roughly equal to his vocabulary, then two Encyclopedia Americana estimates (Stone Age wrld population of 10 million, "pri-

Table 1   Human Cerebral Data Store

Era	Total Human Data (bits)
Prehistoric (3100 B.C.)	1.4×10^{12}
Preprinting (1450 A.D.)	5.6×10^{13}
Preelectronic (1830 A.D.)	3.8×10^{14}
Present (1971 A.D.)	1.3×10^{15}

mitive" language vocabulary of 10,000 words) times 14 bits/word is a rough measure of human cerebral data stored prior to the invention of writing. One obtains 1.4×10^{12} bits as an estimate of the human information system data content prior to the invention of writing.

The world population was about 400 million in 1450. Recognizing that the vast majority was still illiterate, one might still use a 10,000-word vocabulary of 14 bit words as the measure; the information system data content estimate is 5.6×10^{13} bits prior to the invention of printing.

The next significant date is 1830, the start of the electronic revolution. World population was 1 billion. Literacy expansion as a result of printing results in 25,000 words being taken as the vocabulary, with 15 bits required to encode, thus gives 3.8×10^{14} as the human cerebral data content estimate prior to the electronic revolution. Although these gross estimates are primarily a reflection of population growth, with only slight sensitivity to vocabulary expansion as a measure of each person's data base, they certainly could be interpreted as accurate to plus or minus an order of magnitude, as measures of human cerebral data store. The systematic trend exhibited is certainly significant.

The current estimate, from 1970 world population of 3.5 billion people, taking a 25,000 word vocabulary (although the Encyclopedia Americana ascribes 60,000 to 80,000 words to an educated English-speaking adult), of 15 bit words, is 1.3×10^{15} bits of cerebral data store. Thus, one may accept Table 1 as giving reasonable estimates for the human cerebral data store available to mankind.

A new data store of comparable significance arises from the printing revolution. In 1450 the world's document data store was clearly insignificant compared to cerebral store. Typical major libraries previous to printing's impact were the Sorbonne (1338 A.D.) with 1722 volumes or the Vatican Library (1484 A.D.) with 3650 volumes. If one estimates 250 pages/book, 30 lines/page, 10 words/line, 14 bits/word, there are 10^6 bits/book. Assuming the book data base to be 100 times that of the leading Vatican Library, one obtains 3.6×10^{11} bits as the estimated book data base at the onset of printing. The use of 100 times the leading library content as the estimated data base is arbitrary, but probably correct to the nearest order of magnitude.

By the onset of the electronic revolution in 1830 A.D., printing had swelled libraries tremendously. The imperial Library of Vienna had 200,000 volumes by 1812; in 1876 (the year of the invention of the telephone) Harvard University Library had 227,650 volumes. Retaining the estimate of 10^6 bits/book, times 100 equivalents of the Imperial Library, gives an estimate of 2×10^{13} bits as book data base at the onset of the electronic revolution.

Currently, the leading world library is the Library of Congress, with 14 million volumes. The assumption that today's book data base is 100 times greater, or 1.4 billion volumes, is roughly checked by noting that the holdings of the world's 144 largest libraries, from Number 1 (Library of Congress, 14 million volumes) to Number 2 (Lenin State Library of the USSR, 12.5 million volumes) to Number 144 (University of Oklahoma, 1 million volumes), total 366 million volumes. So 1.4 billion, or 4 times the holdings of these specific libraries, seems plausible. At 10^6 bits/book, the current estimate is thus 1.4×10^{15} bits as book data base.

Having used equally cavalier approaches to estimating the human cerebral data base and the book data base, one may wish to compare them, as in Table 2. These states of growth are plotted as successive growth curves in Figure 1.

It now becomes interesting to evaluate the current state of the electronic revolution in historical perspective. Standard and Poor state that the cumulative number of electronic computer installations by 1970 was 109,000. The memory store capacity (tape and disc drives, not fast core) is estimated as order of 10^8 bytes (8×10^8 bits)/computer, averaged over all installations by large and small. Thus, the United States computer memory capacity as of 1970 may be estimated as 0.9×10^{14} bits. The computers of the United States thus have on-line access to information capacity 1/10 that of the world's books. In 1970, the United States made 35 billion electrostatic copies. Ascribing an average of 300 words to each copy at 15 bits/word gives $1.6 \times$

Table 2   Comparison of Cerebral and Book Data Store

Era	Total Human Data (bits)	Total Book Data (bits)
Prehistoric (3100 B.C.)	1.4×10^{12}	—
Preprinting (1450 A.D.)	5.6×10^{13}	3.6×10^{11}
Preelectronic (1830 A.D.)	3.8×10^{14}	2.0×10^{13}
Present (1971 A.D.)	1.3×10^{15}	1.4×10^{15}

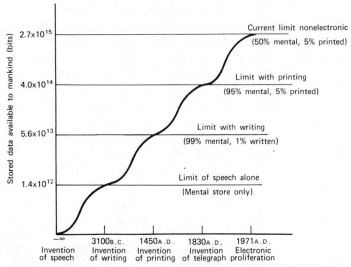

Figure 1   Growth of man's data base.

10^{14} bits annual copied information. These computer capacities have grown from zero in 20 years; the copier capacity has arisen in 10 years.

It is clear that a logarithmic acceleration on human information systems is underway. Speech was man's revolutionary information media for tens to hundreds of millenia; writing for a few millenia; printing for a few hundred years; and the recent electronic arts, computing and copying, for a few decades. As estimated in the preceding paragraph, capacity ascribed to these electronic arts is already within one order of magnitude to the total available human data base. But the differences in structure of the new systems are as important as their growth in size.

STRUCTURE OF INFORMATION SYSTEMS

It is useful to organize one's thinking about all information systems around three generalizations. The first generalization is that information systems are comprised of a "control channel" and a "content channel." The second generalization is that the information in a system may be classified as "historical" or "current." The third generalization is that information utilization may be categorized as "remote" or "local." These major dualities, control/content, historical/current, and remote/local, serve as bases for illuminating analyses of strengths and weaknesses for information systems.

A few examples clarify these concepts. In books, title page, index, chapter titles, page numbers, and call number all are elements of the control channel; preface, body text, and footnotes all are elements of the content channel. In computers, punch cards, paper tape, and ROM (read only memory) circuit chips are historical records; fast core memory, magnetic disc packs, and electronic circuit registers handle current information. In audio electronics, a telephone receiver utilizes remote information; a tape recorder utilizes local information. These duality characteristics can be considered as orthogonal or independent of each other. For example, whether information is control or content is not determined by whether it is historical or current, local or remote.

One final observation concludes these remarks on structure. Where the information-carrying modes can be common to all aspects of control, content, historical, current, remote, and local information systems usage, a homogeneous system results. Flexibility, power, and economy accrue to homogeneous systems such as the lecture (verbal elements), an AEC Document Repository (printed elements), or the NORAD Air Defense system (electronic elements). Where the information-carrying modes are not common to all aspects, a heterogeneous system results. Rigidity, weakness, and expense accrue to heterogeneous systems such as the dictating machine (electronic), secretary (verbal), typewriter (keystroke), signature (graphic), filing cabinet (alphabetic), and post office (pigeon hole) system for business letters.

FUNCTION OF INFORMATION SYSTEMS

It is desirable to construct a normative, not a descriptive, theory of the function and value of information systems. The dominant principle was perhaps stated originally by John Stuart Mill (Cherry, 1957, p. 62), and perhaps developed most completely by Tribus (1969, Chapter I), that inductive inference is the sole means of new knowledge arising. In this chapter, it is taken as true that man can "think" (perform this inductive inference) and that machines cannot. The well-known GIGO aphorism is taken as "Garbage In, Garbage Out" for machines, "Garbage In, Genius Out" for man.

The mathematical theory of communication (Shannon and Weaver, 1949) is essentially descriptive. The theory is an intellectual triumph, especially in identifying the entropy of physics with the uncertainty of communications; the theory itself and the resultant intellectual precision are bench marks of the current electronic information revolution. However, Weaver explicitly states that the theory applies in the first instance only to the technical problem of accuracy of transference of various types of signals from sender to receiver.

For this Chapter, the only purpose of communication systems is taken as conveyance of states of knowledge (both certainty and uncertainty) from the mind of one human to the mind of another. No matter how these minds are separated over time, by distance, by language or symbolism, by previous conditioning, or through mechanical intervention, this mind-to-mind transfer is the desired end. Physical experiments conveying data on states of nature are thus information systems but not communications systems. In human reference, they are soliloquies, not dialogues. The machine-to-machine transmission of information so thoroughly handled by today's science represents a means, a subsystem, which merely enables the end, the desired human mind-to-human mind communication of information.

This adaption of human-to-human transfer as the normative purpose of communications can only be supported, never derived. Similarly, this chapter is expected only to prophesy, not determine, the future course and value of communications systems development. If human taste, human choice, human participation, human thought, and human creativity do persevere as the primary desiderata, then the normative purpose of human-to-human communication will have been established.

It is now informative to analyze some existing systems. The first system of interest is that organized around human speech. Speech is effective for control signals and for content communication. Speech is weak for handling historical information, depending on inaccurate human memory, but is strong for handling current information. Speech is weak for remote information, but is strong for local information. The speech system is homogeneous, with vocal phonemes serving all purposes. Finally, speech has the readiest possible access to and from the human mind, with talking and listening being universal human skills.

Clearly speech is on balance a powerful and effective information system. It is interesting to note that the primary value of writing prior to 1450 A.D. (and universal literacy) could be ascribed to rectifying the first weakness noted for speech, that of historical records. Even today, we can ascribe the primary value of jet aircraft business travel as depending on the overwhelming strength of speech communication (the face-to-face meeting) by removing the second weakness, inability to communicate remote information.

The second system to be analyzed is that organized around the book. The book system uses print effectively for control and content. Books are extremely effective in handling historical information, which often survives centuries of loss or neglect; on the other hand, books are basically incapable of handling current information. The permanence and immutability of books is thus a great strength and a great weakness. Books, with but modest transportation delays, communicate remotely; they also serve as references for local information. The book system is homogeneous, with print serving all functions.

Finally, the book has strong access to and from the human mind, with reading and writing being broadly distributed human skills today.

The book also presents a powerful and effective information system. Magazines and newspapers can be considered special elements of the system, attempting to attack the greatest weakness, lack of timeliness. Also, magazines and newspapers specially address breadth of physical distribution to provide remote information.

The third system to be analyzed is considered the heart of the electronic revolution, the digital computer. Control, content, historical, and current information are all handled effectively by digital signals. With minor adaptations, our verbal analog electronic communications network can link remote and local information usage. The system is highly homogeneous; indeed the Von Neumann concept of the stored electronic program is a recognition of the powerful unity in similarly handling control, content, historical, and current information as homogeneous electronic bits. The digital computer per se has, however, the one crucial weakness, extremely weak access to and from the human mind.

Input to computers is by keystroke. The basic instrument, the key punch, actually antedates the electronic computer by many years. Even typing, the skill on which the key punch is based, is such an unusual competence that a special cadre of professional secretaries must be maintained to perform this service reliably and effectively. Output from computers comes primarily from chain printers. These devices, which also antedate the electronic computer, were performing at 600 lines/minute with the Univac, and are only up to 2000 lines/minute with System 370. Computer graphics quality, font selectivity, and format flexibility are miniscule compared to levels already shown to be significant for human ease by the achievements of printing. Although cathode ray tube displays, light pens, and interactive terminals already exist to address these weaknesses, it is clear that the interface between man and machine is a serious constraint on the potential capability of the man using the machines.

The points of analysis reviewed here are displayed in Table 3. The architecture of information advanced by this chapter is clearly displayed. Words are the central elements of human thought; the universal system organized around words is human speech. Coded data are the central elements of electronic information processing; the premier system organized around electronic coded data is the computer. Central to both systems are documents, graphics systems. The words of human thought are handled as alphabetically coded data. Thus, this chapter advances the central theme for graphics systems in the future; graphics systems play the central role, serving as the common meeting ground for the words of human minds and the codes of digital systems.

Table 3   Systems Comparison

	Speech	Books	Digital Computer
Control	Strong	Strong	Strong
Content	Strong	Strong	Strong
Historical	Weak	Strong	Strong
Current	Strong	Weak	Strong
Remote	Weak	Strong	Strong
Local	Strong	Strong	Strong
Human access	Universal	All but illiterates	Very poor

GRAPHICS SYSTEMS ESSENTIALS

Very little of graphics systems has as yet benefited from the electronic revolution. An insight especially poignant at my institution (Xerox Corp.) is that the memo copy explosion depends on a device of antiquity, the filing cabinet, an invention of the eighteenth century (by Ben Franklin—U.S. Post Office), the typewriter, an invention of the nineteenth century (by Christopher Sholes) and xerography, an invention of the twentieth century (by Chester Carlson). The mismatches and inefficiencies from archaic subsystems are the problem and are also the opportunity for graphics systems of the future.

Why bother? One point of view holds that man is foundering in obsolescent paper, that the only hope is a completely clean break with tradition for an all-electronic paperless future (Lindgren, 1971). Some current statistics are perhaps unappreciated by holders of such views. The current rate of book publication in the United States is 35,000 titles/year; at 250 pages/book this represents about 9 million pages of information. The current rate of electrostatic copying in the United States is 40 billion copies/year; since an average of slightly less than three copies are made per original, this represents about 15 billion pages of information. The electronic copier revolution is already processing about 10^3 times as much information as the book publishing system.

Another index of importance can be derived from corporate revenues. Money results here are taken as socially valuable indices of the importance of services rendered, not as stock tips. Table 4 compares revenues of the leading computer corporation with the leading copier corporation. Clearly the rate of growth at similar size shows electronic copying to have approximately the same utility to society as the computer.

The analysis of graphic systems might proceed by first identifying present weaknesses. The most serious weakness is the inflexibility for change and update. Essentially, documents handle historical information well and current information extremely weakly. There is no device more sophisticated than the pencil with eraser to present the human user with easy rewrite, addition, and deletion capability. It is largely the lack of flexibility for manipulating and processing current data that has led to the complicated, expensive, and heterogeneous system of dictation, secretary, and typewriter for business purposes.

The second major weakness of document systems is the lack of automated control capability. The systems control information is in human accessible form, but only in such form. Therefore, human clerical labor has been required as the central element for control purposes. This human clerical control

Table 4   Revenue Comparisons—Computers and Copiers

Large Computer Corporation			Large Copier Corporation		
Year	Revenues (Millions of Dollars)	Growth (%)	Year	Revenues (Millions of Dollars)	Growth (%)
1950	215		1960	37	
1951	267	24	1961	60	62
1952	334	25	1962	104	73
1953	410	23	1963	176	69
1954	461	12	1964	268	52
1955	564	22	1965	393	47
1956	734	30	1966	528	34
1957	1000	36	1967	701	33
1958	1172	17	1968	896	28
1959	1310	12	1969[a]	1483	66
1960	1436	10[b]	1970	1719	16[d]
1961	1694	18			
1962	1925	14			
1963	2060	7			
1964[a]	3239	57			
1965	3573	10			
1966	4248	19			
1967	5345	26			
1968	6889	29			
1969	7197	4			
1970	7504	4[c]			

[a] Worldwide revenues consolidated for the first time.
[b] Ten-year average, 21%.
[c] Ten-year average, 15% (less 1964).
[d] Ten-year average, 48% (less 1969).

labor surfaces as the mail-sorting problem for remote information use and as the file search and retrieval problem for historical information use. The fact that control signals are not readily handled by machines is the single biggest cause of the plaint, "we are choking on our paper."

The missing elements to rectify graphics systems weaknesses can be identified. Common document and digital data for addressing, indexing, and abstracting can provide both human and machines modes of control. Flexible data entry by automated typing, simultaneous to both document and digital records, readily changeable for correction, editing, and updating must provide an initial link between human thought and information system. Optical character recognition and facsimile scanning to convert graphics data to electronic signals must be provided. Automated machine sorting for remote document transmission and automated machine searching for historical document retrieval are required. High-quality (resolution and contrast) temporary display of full graphic presentation must be provided whether by electronic means or by throw-away printing. High-quality permanent printing in formats such as books, reports, film strip, or microfiche is required to be readily available at high speed and low cost. The typography and format of this printed material should be suitable and controlled for ease of electronic reentry by facsimile or optical character recognition. Table 5 summarizes these elements and their role in the merged document/digital system.

The fundamental strengths of this information systems architecture are digital signals for control and graphic documents for content. Processing, manipulation, automation, and update are apparently digital strengths and document weaknesses; the total system envisioned here would perform these functions electronically wherever direct human intervention is not involved. For the historical preservation of reference records, paper and microfilm documents for content with digital codes for access and retrieval present very low costs and high flexibility. For the remote use of information the physical transport of paper or microfilm documents, supported by automated sorting of coded addresses, provides transmission capacity much less costly than all-electronic channels, with only moderate delays. With human interaction based on broadly distributed literacy skills, continuing the use of graphic documents as a cultural norm, the system is expected to be as popular as the copier.

Emerging capabilities are already evolving in the directions derived as significant in this paper. Merged common alphanumeric/digital records are being investigated by the U.S. Air Force, among others. Input, by means of automatic typing, such as IBM MCST or Redactron Data Secretary, is available, and will certainly evolve to feed information not only to copiers but also to communications lines and to computers. Output by COM/COMP

(Computer Output Microfilmer/Computer Output Microfilm Printer) capability already exists to accept electronic data at speeds of 10^5 characters/second, to print at speeds much faster than 2000 lines/minute chain printers, and to render an eye-readable reference film record as well. Remote transmission of documents by electronic facsimile is perhaps just launched on a major growth cycle; optical character recognition and bandwidth compression processing of electronic facsimile data is probable. Historical record retrieval by digitally coded micrographics is expanding. This list merely identifies some current subsystems that are already performing as links in the expected architecture; these will be fruitful areas for invention and development of additional total systems capabilities.

It is not deemed foolhardy to prophesy specifically about the information system one decade hence. In 1982, it is expected that a major thrust of communications systems development will have become clearly the use of digital techniques for control and graphic techniques for content. Automated document creation devices will have displaced office typewriters in growth rate if not yet in total numbers. Automated document storage and retrieval will have become a major technique for large central files, even though fully electronic central files and manual local files will continue indefinitely for special purposes. Automatic document sorting against coded data will have become so

Table 5   Merged Document/Digital System Status and Expectations

	Document Status	Digital Status	Expected Links
Control	Readable by only humans	Not readable by all humans	Common document/ digital control data
Content	Typescript, line graphics, photos	Digital bits	OCR facsimile
Historical	Cheap, permanent	Expensive, volatile	High-quality printing of stored document/digital data
Current	Weak, costly	Flexible, inexpensive	Automated typing, display
Remote	Cheap, delays required	Expensive, instaneous	Mark-sense sorting
Local	Readily available to humans only	Readily available to machines only	Document retrieval by coded data, then display or print
Human access	Nearly universal	Highly restricted	Merged system access equal to that for documents

important that uncoded documents, requiring clerical intervention, will be prohibited or strongly discouraged from use as remote or historical records. Demand printing from electronic and graphic data bases will have replaced the hard copy preprinted inventory for all report and much book distribution. Finally, the recognition that machines accept, process, and deliver bits while men accept, process, and deliver words will have become the philosophical base line for analyzing and projecting communications systems.

CONCLUSION

Previous information revolutions have been most effective when the highest expression of the previous system has merged into the mainstream of the new system. The electronic information revolution has not yet accomplished the merging of graphic and digital techniques, but is about to do so. The human preference for words on paper will assure that this merged document/digital system prevails provided that the specific weaknesses of current document systems are rectified and the specific strengths of current digital systems are exploited. The new subsystem initiatives already underway should serve in the next decade to address all missing links and establish the dominance of the concept.

REFERENCES

Cherry, Colin. *On Human Communication*. New York: Wiley, 1957.

Lindgren, Nilo. "Our Paperless Future," *Innovation*, 24 (September 1971), p. 50.

Shannon, C. E., and Warren Weaver. *The Mathematical Theory of Communication*. Urbana: University of Illinois Press, 1949.

Tribus, Myron. *Rational Descriptions, Decisions, and Designs*. London: Pergamon, 1969.

CHAPTER 6

Communications Satellites

W. L. PRITCHARD AND P. L. BARGELLINI

COMMERCIAL COMMUNICATIONS SATELLITES

Communications satellites, the latest form of electrical communications, give speed, range, and coverage beyond those obtainable by other kinds of communications.

Two fundamental characteristics make satellites attractive: large bandwidth, and consequently, huge communications capacity, and the ability to serve points not specified *a priori*. Although future cables and waveguides operating at frequencies higher than those in current use may someday provide very high capacities, they can serve only fixed terminals. For instance, one could send all the pictures in the world from, say, New York to London, but what about the other potential terminal pairs? In this respect, satellites offer a flexibility that cannot otherwise be duplicated.

Although only about 30 years ago the radio spectrum above 30 MHz was almost empty of man-made signals, during and after World War II, new circuits involving waveguides and cavity resonators made possible radar and communications up to 100 GHz. Rocketry, lifted from the speculations of Tsiolkowsky and the works of Goddard and Oberth to the terrifying effectiveness of the German V-2's, has made artificial satellites and space exploration a reality.

Sputnik I shook the world in 1957, only 15 years after the first German V-2; but 12 years before, a British science fiction writer, Arthur C. Clarke, had written a prophetic article entitled, "Extraterrestrial Relays," suggesting the combination of rocketry and microwave engineering to provide relay stations in the sky. Today, with our successful experience with commercial communications satellites, a tribute to Clarke's vision is in order.

Curiously enough, his paper went almost totally unnoticed; thus, everything remained at the science fiction level until satellites became a reality with Sputnik (October 4, 1957) and Explorer I (January 1, 1958).

61

	INTELSAT I (EARLY BIRD)	INTELSAT II	INTELSAT III
● YEAR	1965	1967	1968
IN-ORBIT MASS (kg)	37	81	127
LAUNCH VEHICLE	DELTA	IMPROVED DELTA	LONG-TANK DELTA
PRIMARY POWER (WATTS)	40	75	120
● TRANSPONDERS	2	1	2
BANDWIDTH/TRANSPONDER (MHz)	25	130	225
ANTENNA TYPE	OMNISQUINTED	OMNI	MECH. DESPUN
CIRCUITS	240	240	1200
● DESIGN LIFETIME	1.5 YR	3 YR	5 YR

Figure 1   Three generations of satellites: INTELSAT I, II, and III.

The Experimental Era

Balloons launched by rockets to sufficient altitude and adequate ground transmitting and receiving terminals constitute a passive satellite communications system. In 1956, work in the United States by the National Advisory Committee for Aeronautics (later part of NASA) led to the development of aluminized plastic balloons of 30 m diameter, weighing less than 50 kg.

The balloon acts as a scatterer of electromagnetic energy coming from an earth transmitter, and part of this energy can be picked up at any point on earth from which the balloon is "visible."

The joint efforts of Bell Telephone Laboratories, NASA, and JPL resulted in the ECHO experiment with a balloon in an inclined orbit at 1500 km altitude. Communications across the United States were established in August 1960, at 960 and 2290 MHz.

Passive satellites are handicapped by the inefficient use of transmitter power, and as soon as space-flyable electronics became available, active satellites replaced them. The mathematics of the inverse distance square law that applies to active satellites versus the inverse distance fourth power law applicable to passive satellites are overwhelmingly in favor of the former.

The TELSTAR (1962 to 1963) and RELAY (1962 to 1964) experiments demonstrated the capability of active satellites in low- and medium-altitude orbits. The SYNCOM (1963 to 1964) project proved the feasibility of the synchronous orbit concept and A. C. Clarke's dream became reality.

The Commercial Era

The Communications Satellite Act of 1962 and the formation of the Communications Satellite Corporation in 1963 led two years later to the beginning of the INTELSAT worldwide communications satellite system.

Figure 1 illustrates the three generations of satellites, INTELSAT I, II, and III. From 1 satellite in orbit in 1965, with a capacity of 240 voice circuits, the system has expanded to 11 satellites in orbit at the end of 1971, with a combined potential capacity in excess of 10,000 circuits. The number of earth stations has grown from 5 in 1965 to 51 in 1971, with 62 antennas in 38 countries.

Figure 2 INTELSAT IV: exterior.

INTELSAT IV

● YEAR	1971
IN-ORBIT MASS (kg)	700
LAUNCH VEHICLE	ATLAS CENTAUR
PRIMARY POWER (WATTS)	400
● TRANSPONDERS	12
BANDWIDTH/TRANSPONDER (MHz)	36
ANTENNA TYPE	MULTIPLE, MECH. DESPUN
CIRCUITS	6000

Figure 3   Basic dimensions of INTELSAT IV.

Increased spacecraft size, weight, and power, increased antenna gain, increased bandwidth of the transponders, and improvements in the transponder design contributed to this spectacular growth.

Spin-stabilization, first demonstrated by SYNCOM and Early Bird (INTELSAT I), has been used in INTELSAT II, III, and IV. The first INTELSAT IV, launched on January 25, 1970 and now operational, epitomizes the state of satellite communications technology. The spacecraft is 530 cm high, 240 cm in diameter, and has a mess of 700 kg.

With 400 W of available prime electrical power, 6 antennas and 12 transponders receive signals at 6 GHz and retransmit them at 4 GHz. Depending on the distribution of power between the spot and the earth coverage antennas, and also on the type of modulation and the number of carriers per repeater, each satellite provides from 3000 to 9000 voice circuits. Twelve TV channels could be carried at one time. Figure 2 shows the exterior appearance of an INTELSAT IV; Figure 3 shows some of its characteristics, and Figure 4 shows a block diagram of the transponder.

INTELSAT IV will enhance the use of demand assignment multiple access techniques. The term "multiple access" designates the ability of two or more earth stations to communicate with each other through the same satellite. Earlier satellites made use of either full-time dedicated carriers between two points, or multidestination carriers; that is, channels are preassigned. Although preassignment of channels provides efficient operation in heavy traffic links, the utilization of satellite transponders becomes less efficient when the number of circuits per link is small. A solution to the problem of lightly loaded links is to share satellite channels among earth stations; the channels are then assigned on demand, forming a temporary link between any two earth stations. At the end of the communication, the channels are released and are available for the next call.

A common misunderstanding about telephone communications by satellite needs some clarification. Since it takes about 1/8 sec for signals to travel between earth and satellites in geosynchronous orbit, the channel delay (one way) is 1/4 sec and the circuit delay (round trip) is twice as long. Certain fears expressed in the beginning about possible adverse effects of these delays on conversational flow have failed to materialize. The experience of seven years of commercial satellite circuits has amply demonstrated that once echoes are under control, delays, per se, are not objectionable. In effect, two-hop satellite circuits involving delays twice as long are in current use. The echoes must, of course, be taken care of by improved termination of radio and wired circuits on the ground with the additional protection obtainable from antiecho devices such as echo suppressors and echo cancellers.

Future Trends

System studies have identified trends in spacecraft characteristics and technologies most likely to be employed in the future. In order to achieve higher communications capacity, brute force solutions that would require bigger and heavier spacecraft can be avoided. Although the future will probably witness the use of very powerful boosters for certain forms of communications satellites, for example, for direct broadcasting, for most commercial applications, satellites of size and weight not too different from those of an INTELSAT IV spacecraft appear as adequate.

Higher communications capacity will be provided by a combination of the following techniques: (a) active body stabilization, (b) higher efficiency solar cells, (c) increased efficiency of the energy storage devices required during eclipses, (d) electrical propulsion for stationkeeping and other functions, (e) use of additional frequencies higher than 10 GHz, (f) high-precision pointing of narrow antenna beams, (g) spectrum reuse by means of orthogonal polari-

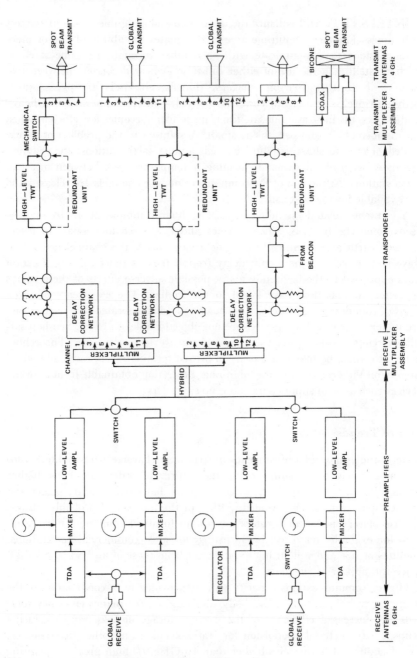

Figure 4 Block diagram of INTELSAT IV transponder.

zation and antenna directivity, *(h)* Time Division Multiple Access (TDMA) techniques, *(i)* combination of *f*, *g*, and *h* with Space Division Multiple Access (SDMA) techniques, *(j)* improved methods of modulation and coding for error control and reduction of source redundancy.

A spacecraft with active stabilization along its three axes permits the use of oriented panels of solar cells that yield a threefold increase of the available electrical power from an equal area occupied by solar cells in a spin-stabilized satellite. Although various thermal and mechanical problems will have to be solved, the technology is available to make the more efficient solution acceptable.

More efficient conversion of energy will be made possible by using deployable light arrays of solar cells. Figure 5 outlines the experience of the recent past and expected future trends.

Replacing the present heavy NiCd batteries with light rechargeable H_2-O_2 fuel cells will considerably ease the energy storage. Space-flyable units of these cells do not yet exist, but laboratory specimens have already been developed.

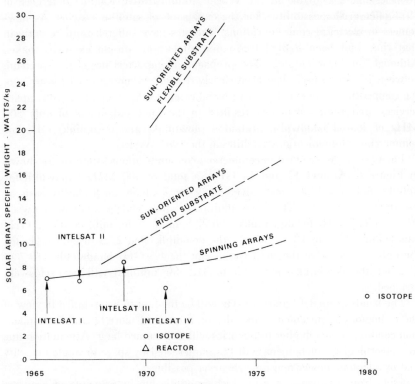

Figure 5   **Trends in solar energy conversion.**

North-south and east-west stationkeeping required to maintain a satellite in the desired position—or to move it whenever necessary from one longitude to another—has been obtained until now by jets from thrusters using hydrogen peroxide or hydrazine. Advances in electrical propulsion have resulted in cesium or mercury ion thrusters characterized by values of the exhaust velocity, and of the specific impulse higher than those obtainable with chemical propellants, with reasonably good efficiency.

These advantages, combined with those obtained in the field of solar cells, yield very low values of the thrusters' specific weight, that is, the ratio of the weight to the electrical power. Another advantage of electrical propulsion is its availability as soon as the solar cell panels are deployed.

North-south stationkeeping by electrical propulsion is feasible with the existing technology. East-west repositioning, and even the raising of satellites from low to synchronous orbits by electrical propulsion, are conceivable as future developments.

Improved spectrum utilization at 4 and 6 GHz and the use of the new frequencies made available at the World Administrative Radio Conference of 1971 offer great possibilities for the expansion of satellite systems. As newcomers in electrical communications, satellites have suffered until recently in that they had been denied frequency allocations on an exclusive basis. Although the favorable noise and propagation characteristics of the spectrum between 1.5 and 6 GHz had been clearly recognized more than 10 years ago, the competition from existing earth-based communication systems, other space services, and radio astronomy resulted in the final assignment of only 500 MHz of shared bandwidth around 4 (downlink) and 6 (uplink) GHz for commercial communications satellites at the 1961 WARC.

The new bands assigned to commercial communications satellites are shown in Figure 6. Around 12 and 14 GHz, a total of 500 MHz bandwidth for uplinks and downlinks, respectively, will be available on a shared basis as practiced at 4 and 6 GHz. Two additional bands will be available between 17.7 and 21.2 GHz for downlinks, and 27.5 and 31.0 for uplinks. The 2 GHz bandwidths (17.7 to 19.7 GHz for the downlink, and 27.5 to 29.5 GHz for the uplink) will be of the shared type, but the 1.5 GHz bandwidths (19.7 to 21.2 for the downlinks and 29.5 to 31.0 for uplinks) will be exclusively assigned.

The bands around 12 and 14 GHz will be the first to be occupied in view of the technology requirements, and also on account of more favorable propagation conditions. The higher frequencies will be occupied later. Attenuation due to meteorological phenomena will be counteracted by space or angle diversity and by adequate power margins whenever possible.

Aside from frequency reuse, which is applicable in principle at any frequency, the recent WARC agreements will permit an expansion of future communications satellites by a factor of eight, vis-a-vis the present system. In addition, the benefits resulting from the exclusive frequency assignments to satellites are considerable since satellite power can be increased, thus permitting earth stations with antennas of modest size in "downtown" locations.

Orthogonal polarization makes it possible to introduce an additional factor of two in the balance. At 4 and 6 GHz, this approach is immediately available

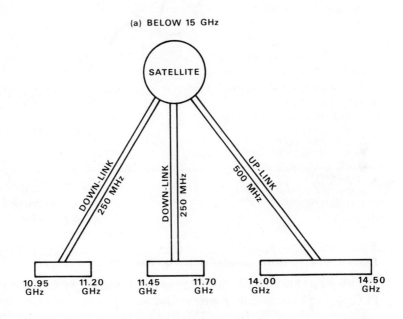

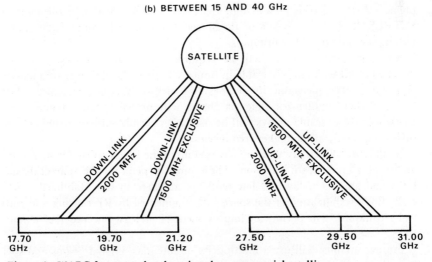

Figure 6 WARC frequency bands assigned to commercial satellites.

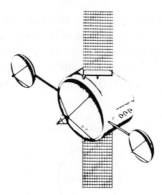

● YEAR   ~1978
   IN-ORBIT MASS (kg)   800
   LAUNCH VEHICLE   ATLAS CENTAUR/TITAN III (?)
   PRIMARY POWER (WATTS)   1000

● TRANSPONDERS   ~ 50
   BANDWIDTH/TRANSPONDER (MHz)   VARIABLE
   ANTENNA TYPE   MULTIBEAM
   CIRCUITS   ~100,000

● DESIGN LIFETIME   10 YR

Figure 7   Possible future communications satellite.

and practical. For instance, in the proposals submitted by COMSAT to the FCC for the United States domestic satellites, spacecraft of nearly the same size and weight of the current INTELSAT IV would provide 14,400 telephone circuits, that is, more than twice the number available through an INTELSAT IV as a result of the introduction of orthogonal polarization techniques and other circumstances.

Although earth coverage will be required for certain services, increased directivity will be highly desirable in future satellites. At the lower frequencies of 4 and 6 GHz, mechanically deployable structures will be developed, while at the higher frequencies rigid parabolas will probably remain dominant. Electronically steerable arrays will be used after certain difficult problems are solved, especially at the higher frequencies.

In all cases, directivity will be increased up to the limits set by the maintainability of a given beam direction. These limits depend on the spacecraft stability and amount today to a few tenths of a degree in spin-stabilized spacecraft. Body stabilization of the spacecraft, augmented by RF sensing and control techniques, will permit an improvement of at least one order of magnitude.

Future satellites will use a multiplicity of antennas. Global beams and spot beams a few degrees wide will be used at 4 and 6 GHz; in addition, beams less than 1 degree wide will be used at the higher frequencies. This concept is illustrated in Figure 7.

The choice of a particular method of multiple access or a combination of these methods constitutes a complex problem. Ultimately, solutions can be evaluated in terms of the number of channels which can be provided by a given set of satellites and earth stations. All the studies conducted so far have confirmed the great potential of SDMA. In order to take full advantage of the SDMA concept, on-board switching is necessary. Hence, systems of this kind are designated as Satellite Switched/Space Division Multiple Access, or SS/SDMA Systems. SS/SDMA can be combined with either TDMA or Frequency Division Multiple Access (FDMA). Figure 8 shows a possible system configuration combining TDMA with SS/SDMA. The fundamental concept is the frequency sharing and reuse by narrow antenna beams. As each beam occupies the fully available bandwidths, the desired capacity is achieved. A spaceborn distribution center, consisting of a switching matrix and a control unit, provides the interconnection of different transponders. Information about traffic flow is stored in on-board memory circuits that control the switching matrix. Command signals from earth will rearrange the connections in the matrix whenever necessary.

Three important advantages result from this scheme. With a single carrier present at any given time, the traveling wave tubes operate at saturation with

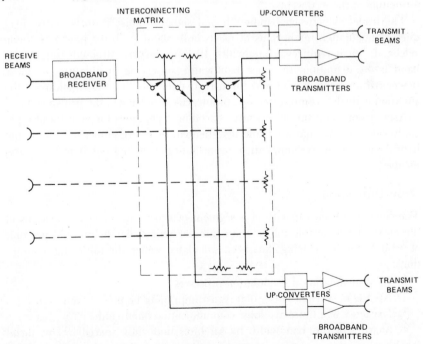

Figure 8   Block diagram illustrating TDMA-SS/SDMA concept.

maximum conversion efficiency, no intermodulation noise, and without the weight penalty of the multiplexing filters required in FDMA. The overall result is a weight reduction by a factor between three and four. An experimental TDMA system, TDMA-1, operating at 50 Mbps, has been tested successfully over the Pacific Ocean via the INTELSAT III F-4 satellite. Operation at rates from 500 Mbps to 1 Gbps is deemed feasible in the late 1970s.

For the sake of completion, FDMA-SS/SDMA systems should also be mentioned. In lieu of switches, filters are cross connected with the arrangements of the connections being commanded by ground signals. The weight of the filters and the backing off of the TWT's necessary in any FDMA system make this approach less attractive than the previously mentioned TDMA-SS/SDMA system.

Finally, another form of SS/SDMA would involve a single, wide band transponder, with a receive spot beam antenna sequentially scanning the earth station transmitters, while an on-board transient spot beam antenna sequentially scans the earth station receivers. Another form of SDMA, which involves scanning receive and transmit antennas, needs an on-board buffer memory. Scanning speed and capacity of buffer memory are, in general, complementary. Additional work is necessary in these areas, as well as in those of the required corresponding earth systems, in order to formulate a comparative estimation of these concepts.

The trade off between bandwidth and signal-to-noise ratio is costly. Frequency modulation, as now used, needs only about 1/20 the power of single sideband, but it requires a bandwidth 10 times larger. Although single sideband is, by definition, the most bandwidth-conserving modulation, its large power requirements do not make it attractive for satellite use. In addition, the situation is further complicated by nonlinearity problems in the transponder.

Aside from the promise of source encoding techniques for voice, and especially video signals that are being investigated, there are numerous digital and hybrid modulation systems with intermediate power and bandwidth requirements.

Orbit Utilization

The communication capacity of a segment of geostationary orbit, in terms of the maximum allowable information rate per unit of bandwidth and per unit of orbital angular spacing, has been calculated under the following assumptions:

1. Ideal modulation-demodulation processes.
2. Neglect of thermal noise, that is, assumption of no power limitations.
3. Existence of a finite available communication bandwidth.
4. Sharing of this bandwidth by satellites uniformly spaced in equatorial synchronous orbit.

5. Ground stations' antennas as uniformly illuminated rectangular aper-
tures with a given diameter-to-wavelength ratio.

With these assumptions, the noise in each circuit is produced by the radia-
tion spillover. Even for an infinite number of satellites, the noise power, as
defined, is constrained by an upper bound that is a function of a single geomet-
rical variable, that is, the satellite spacing. Neglecting differences in slant
range and choosing an optimum spacing $\Delta\Theta = \lambda/D$, the communication
capacity per unit angle is

$$C = 2\frac{D}{\lambda}\,\text{bps}/\text{Hz}/\text{rad} \tag{1}$$

Consequently, the information rate (bps), which can be handled by a segment
of synchronous orbit spanning Θ rad with an available (common) bandwidth
B, is

$$R = 2\frac{D}{\lambda}\cdot B \;\cdot\; \theta\,\text{bps} \tag{2}$$

A bandwidth of 500 MHz and a dish of 30 m at 6 GHz yields, then, a global
capacity in the neighborhood of 3.75×10^{12} bps, or roughly 10^8 telephone
channels.

In a practical engineering sense, these theoretical results must be corrected
to take into account: *(a)* mechanical problems of tight orbital spacing, *(b)* real
modulation-demodulation processes, *(c)* effects of thermal noise, and *(d)* actual
antenna configuration. While the first three items would lead to lower communication capacities, the last item leads to an increase of communication
capacity.

Other possibilities of augmenting the communications capacity are *(a)* inter-
satellite relaying, *(b)* increase in the allowable interference ratio, *(c)* channel
interleaving, *(d)* reversed use of frequencies, and *(e)* pseudostationary satellites
and two-dimensional orbit space.

Studies in depth of all these possibilities are underway; since the two problems of spectrum and orbit utilization are inseparable, compromises are necessary.

Earth Stations

Figures 9 to 11 illustrate the development of earth stations from the early
1960s to the late 1970s.

The trend is toward lower cost, fixed or partially steerable antennas
designed for unattended operation, and large bandwidths. The use of higher
frequencies will increase the amount of usable spectrum and provide greater
channel capacity. The present concepts will have to be expanded to include
small terminals. Multiple-beam operation will be made possible with toroidal
antennas and movable feeds.

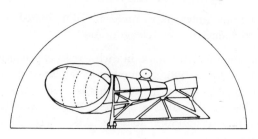

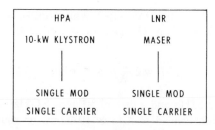

HPA	LNR
10-kW KLYSTRON	MASER
SINGLE MOD	SINGLE MOD
SINGLE CARRIER	SINGLE CARRIER

SINGLE CARRIER/XMTR
FM/FDMA

ANTENNA	
TYPE	FULLY STEERABLE
APERTURE	27 m
G/T	39 dB/°K
BW	30 MHz
XMTR POWER	10 kW

Figure 9   Earth station development 1965 to 1973: phase I.

BROADCAST SATELLITES

Preliminary Remarks

In broadcasting, although the use of high power increases somewhat the coverage, the service radius of medium wave (550 to 1500 kHz) stations rarely exceeds 100 km in the daytime. At night, although sky-wave propagation permits distances to be reached up to 1500 to 2000 km, fading and distortion are frequent. Receivers are simple; solid-state technology has made it possible to produce inexpensive (10 to $15) receivers of excellent quality.

From 1962 to 1965, low-definition (30 to 60 scanning lines) TV broadcast experiments were carried out at MF and at low HF, but the need for much higher carrier frequencies to accommodate high-definition TV prompted the move to VHF in the years immediately preceding World War II. Above 30 MHz, electromagnetic waves are not reflected by the ionosphere and the range is limited to the line of sight. Thus, the only way to increase range is to raise the transmitting antenna. As antenna tower heights in excess of a few hundred meters are impractical, TV service is by necessity "local" unless stations are connected via cables or microwave links.

Towers cannot, of course, be used over the oceans, and since submarine cables are inadequate in bandwidth, the oceans remained insurmountable barriers to TV until the advent of communications satellites.

Possible Forms of Satellite Broadcasting

A satellite in geosynchronous orbit covers about 1/3 of the earth's surface; an area of approximately 2×10^8 km^2, compared with the coverage of an earth-based transmitter that is of the order of 2×10^4 km^2 (radius of 80 km), that is, 10,000 times smaller. Thus, satellites offer the possibility to reach the entire world simultaneously and directly with voice and TV programs.

Three categories of television broadcasting by means of satellites should be recognized: *(a)* distribution, *(b)* community, and *(c)* direct to the home. In the

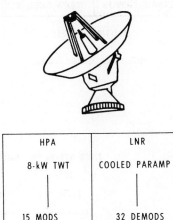

HPA	LNR
8-kW TWT	COOLED PARAMP
15 MODS	32 DEMODS
15 CARRIERS	32 CARRIERS

MULTIPLE ACCESS
FM/FDMA SPADE

ANTENNA	
TYPE	FULLY STEERABLE
APERTURE	32 m
G/T	41 dB/°K
BW	500 MHz
XMTR POWER	8 kW

Figure 10   Earth station development 1965 to 1973: phase II.

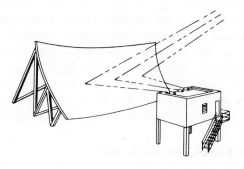

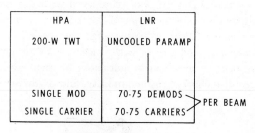

HPA	LNR
200-W TWT	UNCOOLED PARAMP
SINGLE MOD	70-75 DEMODS
SINGLE CARRIER	70-75 CARRIERS

PER BEAM

MULTIPLE ACCESS

MOD-FDMA, SPADE, TDMA

ANTENNA		
TYPE	FIXED REFLECTOR, MULTIPLE BEAMS	
APERTURE	12 m	32 m
G/T	32 db/°K	42 db/°K
BW	40-80 MHz	40-80 MHz
XMTR POWER	100-200 W	100-200 W

Figure 11   Earth station development 1965 to 1973: phase III.

first case, the program is relayed by a powerful station to the satellite and retransmitted to special earth stations that are linked to local TV stations. This service would introduce substantial savings and increased flexibility. Because it relies on TV stations, and to a lesser extent on microwave relays, this service seems better suited to highly developed nations. It can also serve areas not provided with conventional relay systems.

The Soviet Union has developed a system of this sort well matched to its geography. The area of coverage is in the northern hemisphere, and part of it lies beyond the latitudes easily covered by an equatorial satellite. The Molniya satellites in highly elliptical orbits with apogee over the northern hemisphere remain a large percentage of the time in view of the Soviet Union.

COMSAT has proposed a pilot program to demonstrate the operation of a distribution system in the United States. A satellite weighing about 800 kg, launchable by Titan II or Titan III-class vehicles, would provide 12 channels of color TV over the entire United States to receiving stations with 30-ft antennas costing around $75,000 each. Such a satellite could easily handle toll telephone calls as well.

Canada is planning similar dual-purpose satellites for telephone and TV coverage to its far-flung areas. The Germans and French have a joint program (Symphonie) for the development and launch of a satellite for TV distribution and long-haul telephone service.

In the second category, designated as "community" service, a satellite broadcasts to moderately expensive receivers for community viewing or possibly local redistribution by cables to schools, post offices, city halls, and so on. In the case of emerging nations, there is a justification for this kind of service. India, with a large area and a widely dispersed population greatly in need of educational material, appears to be a suitable place for the installation of such a system.

The third, and perhaps most farfetched service, is direct broadcasting from a satellite to the home. In this case, program material originated in metropolitan centers is transmitted to the satellite by large earth terminals and broadcast by satellite directly to ordinary home receivers.

Estimates of the size and cost of a satellite capable of direct broadcasting vary widely. An extreme case would be the simultaneous direct broadcast of a dozen color TV channels to unmodified portable TV sets using indoor antennas in downtown urban areas. Such a solution, although technically conceivable, would require hundreds of kilowatts of satellite power with a weight that could be placed in orbit only by a Saturn V rocket at an estimated cost of $200 million/launch. Fortunately, more elegant and less costly solutions are available. Modifications to the standard home receivers will make it possible to broadcast color TV programs from lighter satellites requiring less powerful rockets. Specially designed outdoor antennas and preamplifiers reduce the satellite weight requirement by two orders of magnitude. The cost of such modifications would be about $100/set, an amount easily acceptable by the average set owner, especially if good programs were not available otherwise.

In present television systems using VSB/AM, the transmitter power depends critically on picture quality. This power determines the weight of the spacecraft. In urban environments, man-made noise predominates at UHF frequencies over the receiver and the galactic noises. Consequently, for direct

Table 1   Summary of Satellite TV Systems

Type of System	Distribution	Community	Direct
Number of users	10^1 to 10^2 (USA)	10^2 to 10^3	Up to millions
Receiver antenna diameter (m)	8	3 to 8	1 to 3
System noise temperature (°K)	200	200 to 400	600 to 10,000
Approximate terminal cost ($)	75 to 200 K	0.5 to 75 K	100 to 300
Possible users	Broadcasters	Small communities (villages) schools CATV professional groups	General public Government

broadcast to urban homes, this rather poorly defined parameter has a significant effect on the final spacecraft weight.

Calculations show that a satellite weighing about 1000 kg/channel could supply adequate color TV service to a single United States time zone. A satellite large enough to supply the entire United States would have to be at least three times bigger, and if more than one channel were desired, corresponding increases would be necessary.

The numbers just quoted are based on use of the most advanced solar cell technology, that is, lightweight, thin film, sun-oriented CdS solar panels, and also on the assumption of not using storage batteries in the satellite. Clearly, while 24-hour availability of a satellite is essential for telephone service, it is not necessary for broadcasting. Eclipses always occur near local midnight; their maximum duration is a little over an hour, and the eclipse "season" lasts only a few weeks at the spring and autumn equinoxes. Thus, the total time during which service would not be available is short, is not prime time, and by doing without it, one saves enormously in spacecraft weight.

Two factors that noticeably distinguish the three types of service are the number of users and the complexity of the ground receiving station. Typical values for these are shown in Table 1.

Power and weight requirements for broadcast satellites can be determined in a straightforward fashion. With antenna gain and receiving system noise temperatures consistent with expected levels of environmental noise, a figure of merit expressed by the ratio of the antenna gain to the system noise temperature for the receiver installation can be chosen.

Picture quality has been the object of numerous studies. A certain coverage area on the ground defines the antenna beam width, and thereby its gain and

its size. From the previously determined effective radiated power requirements, the transmitter power is obtained. The antenna size sets the attitude
control requirements for the spacecraft and the transmitter power dictates the
primary power requirements. Estimates of the spacecraft weight are subject to
wide excursions, depending on the technology assumed for attitude control and
primary power.

The direct TV broadcast with vestigial side band/AM (VSB/AM) is possible, but the penalty compared to FM is of the order of 15 to 20 dB; hence, the
transmitter power and/or the antenna gain must be correspondingly adjusted.
With narrow-beam antennas on the spacecraft and improved home-receiving
installations (e.g., 10 m^2 aperture and high-gain, low-noise preamplifier at the
antenna), it would be possible to provide direct TV service to the home with
RF power on the spacecraft not exceeding 40 kW and values of the spacecraft
power in the range of 60 to 80 dBW. These values are between two and four
orders of magnitude higher than those used in current communications satellites. A clear distinction should be made, however, between what is already
available and what needs additional development. It should be reemphasized
that rockets are already available to place satellites in synchronous equatorial
orbit in the previously mentioned power range with corresponding payloads in
the range from 500 to 2000 kg.

The technologies that must still be developed for direct TV broadcast concern high electrical power (primary and RF) space systems. Major efforts are
needed in the following areas:

1. Development of primary power systems up to 50 kW; that is, large solar
arrays with areas in excess of 100 m^2, or nuclear generators in the same power
range.

2. Development of flyable high RF power equipment.

3. Development of large, deployable antennas with high pointing accuracy
(spacecraft stabilization and stationkeeping improved).

4. Development of thermal control systems for large amounts of power dis-
sipation.

In summary, high-power satellites for direct broadcasting to conventional
home receivers do not require major technical breakthrough; they require,
however, substantial engineering advances in the just-mentioned areas. If FM
were used in place of vestigial sideband/AM, the reduction of satellite power
(between 16 and 20 dB) would make it possible to adopt available RF and
spacecraft technology at the expense of more complex and costly receivers.
Frequencies in the UHF and SHF regions of the spectrum are preferable on
account of the low external noise (natural and manmade).

Finally, the cost of a space broadcast system would actually be lower than
the cost of terrestrial systems potentially capable of providing the same service.

Forthcoming Experiments

A wealth of technical information is available, especially in the following frequency bands: the 470 to 890 MHz UHF band now used for terrestrial TV broadcasting; the 2.5 GHz band for educational TV; and the 12 MHz SHF band that has not been used yet for broadcasting services. Frequencies lower than 800 MHz could be used, but they must be ruled out, simply because they are already used for other purposes. Frequencies higher than 12 GHz have been considered for possible use in the future. Experimental broadcast will first occur at around 900 MHz with FM transmission and community-type service, which allows the use of a limited number of medium cost ($300 to $1000) earth stations.

For large countries not yet provided with TV service, or with one or two transmitters only near big cities, the above-mentioned approach has considerable appeal. Around 1973, India will receive experimental TV broadcasts through the ATS-F NASA satellite. This satellite will have a beam-edge e.i.r.p. of 53 dBW at 890 MHz. The use of FM will allow good quality TV programs to be received with inexpensive ground terminals using parabolic antennas with diameters of 3 m. From the present number of TV receivers in India (less than 10,000), it is planned to have 30,000 receivers in 1972, via satellite broadcasting. Later, the program could be expanded by using an operational system with a bigger satellite capable of providing at least three TV channels.

Other countries actively interested in similar programs are Brazil, Canada, and Europe itself with its Symphonie program. In addition to Russia, which already has an operational broadcast distribution satellite system (as many as 15 Molniya-type spacecraft have been orbited since 1964), other countries will have TV via satellites in the next few years.

FUTURE OF SATELLITE BROADCASTING IN THE UNITED STATES

Curiously enough, TV broadcast via satellites will not come easily to countries such as the United States, where an extensive service is already provided by a complex network of stations connected by an elaborate system of radio/relay links and cables. Possibly the introduction and eventual success of satellite broadcast in the United States might come for somewhat different reasons. In addition to sparsely and widely distributed populations of emerging nations of vast and as yet unsettled areas, there are other examples of "sparse populations." Physicians, attorneys, engineers, scientists, and other professionals in the United States form, in effect, a population of somewhat around 1 1/2 million, with specific educational needs. These people are widely distributed and a large number of them live within the range of either no or only one television station (e.g., certain areas of the Rocky Mountains).

The educational needs of these people are self-evident. For instance, several states have passed laws requiring a minimum number of hours of professional training for physicians every year if they are to maintain their licenses, and especially their certification in specialties. Professionals living in large urban centers may go to local schools—but for someone living out of easy commuting distance, the situation is different. In order to maintain professional competence, some continuous educational program is necessary. Membership of professional societies in the United States amounts to well over 1 million people. If, for example, each one of this million were to spend $100 a year in support of a dedicated educational TV system, we would have $100 million; or perhaps more realistically, only 1/2 million of the members would all spend $200 a year, or even $300 or $400. It is therefore conceivable that the membership of the leading professional societies in the United States could together raise between 100 and $200 million a year to support a dedicated educational TV system. The obvious way to implement this system would be by the use of a direct broadcast satellite.

Coverage of the entire United States implies launching a number of satellites in the 5000 kg category using the vehicles mentioned before. The cost of the spacecraft and the launches for such a dedicated satellite system would be in the vicinity of $200 million, plus the cost of earth terminals to transmit to the satellite, and the cost of bringing the program material to the earth stations from the points of origin. By reasonably assuming the lifetime of satellites and earth stations at 10 years, operating costs, including capitalization, would amount to somewhere around 40 to $50 million a year. In addition to this, there will be the cost of generating the program material. Nevertheless, the total annual cost of such a system may well be regarded as economically viable by the people for whom it is intended.

CONCLUSIONS

Satellites are capable of providing the huge channel capacity needed in modern point-to-point communications systems. The success of commercial communications satellites will ultimately be measured by their effectiveness in reducing the cost of communications.

In 1927, when the first transatlantic radio-telephone service was inaugurated (on frequencies around 60 kHz), the cost of a New York to London telephone call was up to $60. The introduction and expansion of HF circuits in the 1930s and 1940s led to a fourfold cost reduction for the same call. Although the quality of the connection was greatly improved with the availability of submarine cables in the Atlantic, the costs did not change substantially until quite recently. Today, the cost of United States to Europe 3-min telephone calls ranges from about 5.00 to $12.00, clearly quite a decrease from the rate of 1927, especially when inflation is taken into account.

The impact of satellite circuits' availability on this reduction is self-evident. Yet, there is ample reason to believe that once appropriate competitive international arrangements are made, the cost should come down, possibly in the neighborhood of 1.00 to $2.00 for 3 min, between any pairs of points on earth. The expansion of the satellite system will permit worldwide dialing.

In the field of domestic communications, satellite systems that the industry has been ready to provide for quite some time will make economically attractive, novel, and versatile services in addition to contributing to the expansion of conventional ones.

Satellite broadcasting will be implemented in the next few years, and it is bound to eventually become a major industry.

REFERENCES

Bargellini, P. L. "Satellite Communications," *Telephone Engineer and Management, 75*, 24 (December 15, 1971), 44–49.

Clarke, Arthur C. "Extraterrestrial Relays," *Wireless World, 51*, 10 (October 1945), 305–328.

Communication Satellite Systems Technology, (AIAA Progress in Astronautics and Aeronautics, Vol. 19), R. M. Marsten (ed.). New York: Academic Press, 1966.

Metzger, S. "The Commercial Communications Satellite System, 1963–68," *Aeronautics and Astronautics, 6*, 4 (April 1968), 42–51.

Pierce, John R. *The Beginnings of Satellite Communications*. San Francisco: San Francisco Press, Inc., 1968.

Pritchard, W. L. "Trends in Communications Satellites," Paper 1 A.2. IEEE International Convention, New York, N.Y., March 22–25, 1971.

Puente, J. G., W. G. Schmidt, and A. M. Werth. "Multiple Access Techniques for Communications Satellites," *Proceedings of the IEEE, 59*, 2 (February 1971), 218–229.

Radio Spectrum Utilization in Space. A Report of the Joint Technical Advisory Council of the IEEE and the EIA. *XXXIII* (September 1970).

Useful Applications of Earth-Oriented Satellites, 9: Point-to-Point Communications. Prepared by Panel 9 of the Summer Study on Space Applications, Division of Engineering, National Research Council for the National Aeronautics and Space Administration. Washington, D.C.: National Academy of Sciences, 1969.

CHAPTER 7

Social Control
Through Communications

DENNIS GABOR

"How can we anticipate and shape—rather than only absorb—the human consequences of communications?"

If it were true that "the medium is the message," instead of arrant nonsense, communication technologists would be the masters of the human consequences. But in fact, we have hardly any power over them. We produce the loudspeaker; we cannot prevent Goebbels getting hold of it. What can we do against "Gresham's Law of Communications" that "bad truth drives out good truth," that violence and pornography, and much worse, hate messages, drive out knowledge dissemination and ethical messages? As technologists, we can do precious little. The most the medium can do is to preselect, to some extent, the conditions of reception. Books are for the individual, films for the theatre, television for the home. But we have no power whatever for directing, for instance, television features of violence to those to whom it does least harm, and enlightenment to those who need it most.

But even the technologists are not only technicians, but human beings with a sense of social responsibility. Nor are communication scientists interested only in technology. As this is the case, technologists and scientists must come to grips with certain concrete problems, including:

1. Can communications contribute to world peace?
2. Can communications help in countering the danger of overpopulation?
3. Can communications arrest the dangerous tendency toward overurbanization?
4. What can communications do against crime and drug addiction?

5. What can we do against the bottleneck that is developing between the almost unlimited information capacity of modern communication systems and the limited reception and assimilation rate of the individual?

6. What can we do against "information pollution"?

7. What can we do to educate children, adolescents, and adults to be fit citizens in a materially stationary civilization on a high cultural level?

In a general way the answer to all these questions is that communication techniques can do *some* good, but the answer is seldom "the more the better." Also, in every case, communication techniques are *tools* that can be used for good ends or for bad ends. Social control is achieved *through* and *not by* communications.

CAN COMMUNICATIONS CONTRIBUTE TO WORLD PEACE?

Wars *can* start by lack of communication. Colin Cherry, in *World Communication: Threat or Promise*, quotes a famous historical example of this, the War of 1812. Early in 1812, an "Order in Council" was issued in England, which legalized setting up a blockade to restrict American exports. When the English Government became aware of the strong reaction that this evoked in the United States, the order was revoked on June 23, 1812; but it was too late. News traveled too slowly across the Atlantic. The American Congress declared war two days *later*, on June 25, 1812.

Such crass cases of delayed information can hardly occur in our times, but a more sophisticated danger, insufficient "signal-to-noise" ratio, has played a dramatic part in the catastrophy of Pearl Harbor, as ably analyzed by Roberta Wohlstetter.

The supreme danger to peace in our time is the all-out nuclear war between the United States and the Soviet Union, and this, we must note with satisfaction, has now a very small probability, thanks to good information systems: the hot line between the White House and the Kremlin and mutual supervision systems by satellites, which have miraculously succeeded when all inspection plans have failed. The launching sites of ICBMs are now known to both sides, and a surprise attack has become so difficult that it will almost certainly not be attempted. So far instant information has served us well. But, as I have said before, one can seldom say that "the more, the better." At present, the deterrent effect of the almost complete knowledge of the launching sites is powerfully enforced by the complete ignorance of the position of the rocket-carrying submarines. The "dead men's revenge" would completely cripple the aggressor, even in the most unlikely case that he would destroy all land-based missiles in a first strike. But if means were found for tracking the nuclear submarines, the "ultimate deterrent" would be weakened, and this would again improve the chances of a determined aggressor. I base my hopes on the fact that such an information system is most unlikely, and even in that case, a nuclear Pearl Harbor would require a degree of madness several orders of magnitude worse than that which led to the historic Pearl Harbor.

So far I have mentioned only the United States and the Soviet Union. The case of China is somewhat different. Until very recently, China was in an information vacuum almost without a parallel since the days of Marco Polo. Now the veil has slightly lifted, and I think it has revealed a country not at all inclined to provoking the Soviet Union to the point of nuclear annihilation. Her launching sites must be very completely known to the Russians. So, thanks to superior communication techniques, I do not think that there is any serious danger of an all-out war between any of the superpowers.

I wish that I could be as confident in the matter of small wars. A better information system might have perhaps deterred Egypt from provoking Israel if they had known of the strength and readiness of the Israeli forces. But it would not have prevented the recent war between India and Pakistan; it would have made India even more confident of her victory. In the matter of quarrels between developing countries, in a state of adolescent nationalism, communications can help only slowly and indirectly.

CAN COMMUNICATIONS HELP IN COUNTERING THE DANGER OF OVERPOPULATION?

We cannot expect the developing nations to have a strong belief in the sanctity of human life so long as overpopulation threatens them with starvation. Here communications can fight successfully with two weapons, by education in better agricultural methods and by propaganda for family planning. By itself, better agriculture would fight a losing battle; no amount of improvements could feed an exponentially growing population. All the wonderful new crops that agricultural science has developed require more fertilizers, and these (particularly the phosphates) will run out at some time. Optimists (if this is the right word?) may be able to visualize a world population of 15 billion in 2050 A.D. but hardly in 2150 A.D. In the long run, the earth will hardly be able to sustain a population very much larger than the present, at a reasonable level of consumption.

The most important task for communications in the less developed countries is therefore birth control propaganda. Needless to say that we, too, must practice what we are teaching, all the more as a child born in the industrialized countries will consume about 12 times more in natural resources, and pollute about 50 times more than one in the less developed countries. But these countries must also realize that they are running into a catastrophy if they do not cut down their birthrate drastically. This requires recognition of an exceptional case in which "the more the better." If as many transistors and television sets as possible are distributed to the illiterate peasantry of countries whose governments have awakened to their responsibilities, the effectiveness of birth control propaganda will be increased. Let everybody realize that the situation is very precarious. Two bad harvests in succession could have frightful consequences in some countries at a time when the surplus-producing countries like the United States and Canada will not have more than a full harvest in their stocks, as they had in 1964!

CAN COMMUNICATIONS ARREST THE DANGEROUS TENDENCY TOWARDS OVERURBANIZATION?

All countries, developed and underdeveloped, are suffering from a tendency toward overurbanization, but the problem is perhaps more urgent in the United States than anywhere else. The United States are not overpopulated; one could put the whole population of the world into it and the density would still be only that of England. But 90% of the people are now living in 10% of the area, and the tendency is, unmistakably, toward even stronger concentration. Herman Kahn foresees three giant cities: "Boswash" (Boston to Washington), "Chicpitts" (Chicago to Pittsburgh), and "Sansan" (San Francisco to San Diego), each with more than 50 million inhabitants.

This is a trend against all reason, although certainly not without its causes. It is against reason, because the megalopolis has long become an anachronism. The old, walled city of the Middle Ages was a protection against the enemy, in a modern war the large city is a death trap. The working man had to live within a short distance of his working place so long as he had to walk; now 2 miles in the city take more time than 20 miles in the country. Once, but this was a very long time ago, all business was transacted face to face. This has become an anachronism with the invention of the mail, let alone with the invention of the telephone, telex, facsimile transmission, and the Picturephone. Yet millions of clerks are working in huge city skyscrapers, from the in-tray into the out-tray, not talking the whole day to any living person except through the telephone.

High-rise office buildings are still sprouting up in our cities, simply because they are somewhat cheaper to construct than human habitations and yet command a higher rent. By the time the British government stopped it, there were 10 million ft^2 of vacant office space in London. Perhaps this ailment of our times is now nearing its end now, when Centre Point, the finest new building in London, has been vacant for seven years, and the New York World Trade Center is half vacant.

While business is still increasingly concentrating in the cities, human habitations are crowding not in them, but *around* them, in suburbs. This has led to the decay of the city proper in the United States. Many of the city centers in this country are changing into slums, inhabited by racial minorities, where it does not pay even to demolish derelict buildings, let alone construct new habitations. From these poverty and crime-infested centers the taxpayers flee into another sort of hell; into vast, soulless "housing developments." John Aldridge, in *In the Country of the Young*, writing about the confusion and aimlessness of American youth asks: "What better could be expected of people who have grown up in housing developments?" Suburbs are certainly better than slums, but it would be vain to expect that they will ever grow into centers of culture.

What are the forces that drive people into these vast conurbations? I wish I knew of a satisfactory survey of this important question, instead of having to make guesses. Only a small part of the population can be attracted by the cultural amenities of cities, such as live theatre. I believe that their most powerful attraction is *job security*. If one loses a job in a large city, there is a better chance of finding another than in a small place with only a few local employers. Of course, this is often only an illusion. Many thousands of commuters are crossing the center of London twice a day, because their new job is at the other end of the town.

It is clear, therefore, that communications alone cannot be expected to solve this enormously important social problem *completely*, but they can make a very important contribution.

The technological means for spreading industrial and commercial activities into rural areas are available in abundance. Cherry has shown how radically the cost of communications per circuit mile has decreased by modern inventions. It costs only 10 times more to install 1000 times more telephone circuits! This is a one third power law of costs versus circuits, and a British expert has assured me than in the case of cables it may be even a one quarter power law. There is no difficulty in providing the office worker in a small town, hundreds of miles from a city, with several megacycles of waveband. The saving in commuter miles would, by itself, compensate for the additional circuit miles. But will this be sufficient for breaking what has become a bad, anachronistic social habit?

There are other, powerful arguments for small towns, for a new "rural America." The prospect of bringing up one's children in a neat, clean, small town, without slums, adequately policed, and made safe against drug pushers ought to be sufficient to persuade many a responsible family man for moving away from large conurbations. Small cities, unlike housing developments, dormitories, can also develop a worthwhile city culture. The argument for small towns could be made perfect if one could settle them with *basic* manufacturing and servicing industries, instead of with the "most advanced" ones, which are, unfortunately, most affected by economic cycles. This, too, I believe, could be achieved by an enlightened legislation.

WHAT CAN COMMUNICATIONS DO AGAINST CRIME AND DRUG ADDICTION?

I have already touched on this problem, which is one of the most painful and worrying of our times. A reasonable dispersion of the population is certainly the most powerful means for fighting crime and drug addiction, because it is well known how strongly these are concentrated in big cities. It has been estimated (I do not know how reliably), that one half of the heroin addicts of the United States are in New York City. But the dispersion of this population will take a long time, hardly less than one generation, and the crime and drug problem is a very urgent one. Crime could be fought at the roots by breaking up the slums, but this is a slow and expensive process. The cheaper alternative is law enforcement, and here communications can give very powerful aid.

The central problem is crime detection. If crime were certain to be detected and punished, only the insane would become criminals. But unfortunately, crime can be good business. The statistics collected by a committee of the National Academy of Engineering (Chairman, Dr. P. C. Goldmark) have shown that a robber has less probability of going to jail than a businessman of going bankrupt—especially a businessman who has founded a science-based industry!

At present the best instrument science can offer against crime committed in the streets is the electron camera. It is now possible, in principle, for a central authority to see *everything* that is going on in the streets, even in very darkly lit streets, at starlight level. For the time being this has been realized only on a modest scale. In Stockholm, one observer in the police department watches 24 television tubes that show the locations where traffic jams are most common. The observer follows also the locations of all the police cars and in an emergency can direct the nearest to the critical spot. Another observer watches all the underground stations and can shout warnings through loudspeakers. This is, of course, still a far cry from watching all the streets where muggings can occur, but it proves plainly that it is cheaper to multiply electron cameras than policemen. There is no need to watch every yard of the hundreds of miles of streets in a large town. The knowledge that you may be watched has a salutary effect on criminals.

Of course, street supervision, if it could be made perfect, would have only the effect of driving crime from the streets into the houses, or into dark doorways. Here, the best science can offer at the present time are burglar alarms and *personal alarms* for people who are attacked.

More radical technical means, such as wire tapping and bugging of slums, difficult in themselves because they require an enormous supervising personnel, come up against strong and well-justified objections in a free, open society. It is indeed an open and painful question how the freedom of the good citizen can be reconciled with constraints on the criminal. I think the good citizen will have to give up the not very important freedom of paying with cash, and will have to pay with checks and vouchers, under the vigilant eyes of the tax office. Drug pushers are always paid in cash; they do not accept checks. If the circulation of banknotes is radically reduced, we may be able to solve a great part of the crime and drug problem without wire tapping and bugging.

Who defines what crime is? An all-controlling dictator?

WHAT CAN WE DO AGAINST THE BOTTLENECK THAT IS DEVELOPING BETWEEN THE ALMOST UNLIMITED INFORMATION CAPACITY OF MODERN COMMUNICATION SYSTEMS AND THE LIMITED RECEPTION AND ASSIMILATION RATE OF THE INDIVIDUAL?

We have already come up against this bottleneck in the previous section. In principle it would be possible to watch all the streets with electronic cameras and tap all the telephones, but from where will the enormous monitoring personnel come? The answer of the technologist is, of course, almost automatically, more machines! But the intelligence of machines is limited. At a stretch one can imagine a computer that watches a radar screen in an airport tower and gives a warning when two planes are on a collision course. But we cannot even dream of a computer that will spot a mugging on the street or make a note of a suspicious conversation between heroin smugglers. Only human intelligence can achieve this.

Apart from arithmetic, it is the human memory that can expect serious support from the computer. *If* we have a perfect filing system, the computer can quickly retrieve any piece of information by pushing a few buttons. It still falls short of the performance of a good secretary, who can operate with an imperfect filing system, eked out by her memory. (Can you remember . . . we had some correspondence with this chap three or four years ago, . . . what was his name? . . . what was it about?)

Computers that will retrieve information from a few remembered features are not out of the question; Van Heerden made a first suggestion for such holographic memories 10 years ago. But at present, if one gives the computer imperfect data, it is likely to flood you with information. It is therefore indispensable to have a perfect taxonomy, and beyond this, a data bank of carefully prepared abstracts. Only one science possesses both, to my knowledge, and this is chemistry. The magnitude of this task can be judged from the fact that one year of the Chemical Abstracts (plus pro rata of the Quinquennial) costs $3000.

If we do not want to drown in information and reach the point at which only very narrow specialists can understand each other, we have an enormous preparatory task to perform. About 100 years ago all cultured people had read more or less the same books and could understand each other as if they had been members of the same family. By now the books on the shelves of the national libraries run into hundreds of miles. Must we give up the hope of a culture shared by all educated men? Can we have nothing in common except having read the bestseller? There may be some gold also in the tens of thousands of titles published every year, but how to get at it?

The problem is at its worst of course not in the natural sciences, which have mostly reasonably good classifications and abstracting services, but in the social sciences. Of the many thousand words written in this book, much will be repetition; how does one pick out the few seminal ideas that may be hidden in it?

I want to emphasize, therefore, that if we want to make good use of the information capacity of modern machine memories, we shall have to create a class of *expositors*, of people who can abstract in one paragraph the novelties in books of hundreds of pages, and put this into some system of classification of all human ideas, that does not yet exist.

Once we shall have such a system, there will be no need to fill the shelves of our scientific libraries with miles of journals of which only a small fraction will ever be read. We now have methods for making any number of copies, small or large of the original articles, in original size or in microfiles, and if the mail is too slow, we can also send them through teleprinters or facsimile machines. It has been said often enough that the whole Library of Congress could be accommodated in a corner of a room. The bottleneck is not in the size of the file, but in the mind of the man who is to use it.

WHAT CAN WE DO AGAINST "INFORMATION POLLUTION"?

"Information pollution" (scandal, violence, and pornography through the mass media) is a price that we must pay for a free society if we interpret freedom as the right of selling anything that sells. There is no such problem in the communist countries. They have no scandal and very little crime in their newspapers, consequently people do not read their newspapers. Of course, they watch their equally sterilized television, because there is nothing else.

There is no need, however, to adopt one extreme or the other. We cannot suddenly change human nature, which asks for scandal, sex, and violence, but we can deprive it of its favorite nourishment, I believe, without too strong "withdrawal symptoms." In Britain, the BBC has kept an ethical and cultural level far above that of the British press, and the independent television companies have been forced to conform to it. The license of one company has been withdrawn because it supplied the public with too much of what they liked. In Sweden, films are censored, but it is not the sex that is cut in American films (this they supply themselves in abundance), but the violence. Nevertheless, I would consider Britain and Sweden free countries.

Cable television and cassette television will, in time, enormously increase the *choice* of programs. But this great achievement of technology will probably produce a cleavage in the public instead of an improvement in taste. The educated will choose more and more cultural features, the culturally deprived even more pornography and violence. The technologist could no more than enlarge the choice. Directing the choice is the job of the legislator, who can make it pay for private interests to educate the public instead of corrupting it.

WHAT CAN WE DO TO EDUCATE CHILDREN, ADOLESCENTS, AND ADULTS TO BE FIT CITIZENS FOR A MATERIALLY STATIONARY CIVILIZATION ON A HIGH CULTURAL LEVEL?

It might appear that I am begging the question by assuming that the society of the future will be materially stationary. Our whole optimism is so tied up with continual material growth that most people are very unwilling to admit that it must have an end. And yet, who would dare to dispute that, for example, the manufacture of motorcars cannot forever rise exponentially, or that there is a limit beyond which the earth cannot sustain a population on the American standard? Of course, growth need not stop and will not stop today or tomorrow, but even if we reject the scare forecasts of some extreme ecologists, it is very likely to happen in one generation or two. And education for this new, more or less stationary economy ought to start at least one generation in advance. There are enough signs around us of the unwillingness to produce more and to consume more, for all but the willfully blind to read. I think that the preparation for the new education ought to start *right now*.

A society stationary in numbers and in material consumption need not and must not be, of course, stationary in civilization. This does not mean that we must emphasize *new* values. We must instead return to old values, much older than the consumer society.

Such statements will, of course, evoke a contemptuous smile from the side of "no nonsense," "practical" people. They will automatically assume that I am advocating a return to an era that never existed. I may be conceited, but I think I am a realist. Just let us look around us. In the United States, as in Britain, we now have millions of unemployed. In Britain, in the recent recession, industry has "shaken out" a million workers—but the production (and consumption), has not dropped back at all; it has even increased a little. Several more million could be "shaken out" with a little streamlining of production. Is this a reason for ringing our hands in despair? Does this not mean that we have grown potentially *richer*? Could we not use the millions of hands that are now idle for rebuilding our ugly towns, for building new ones of which we could be proud instead of making them only *just* sufficient? Could we not use the more intelligent part of the redundant office population for a great extension of education? Could we not afford not only longer holidays but also sabbatical years for everybody who can profit from them? And, in brief, a little of all the luxuries that individuals can afford when they get richer.

All this is *materially* possible, but there are formidable psychological and institutional obstacles in the way. Our world is so exclusively *growth oriented* that it has no preparation at all for slowing down. The economic mechanism has no brake; it is slowed down only by somebody throwing a spanner into it, such as strikes or financial crises. Any such stoppage produces a wave of downheartedness, which shows itself in a decrease of productive investments, which in turn make the crisis worse. There exist no institutional mechanisms or sound economic habits for directing investments that have become redundant in production toward investments in services. Such cultural improvements, are considered "luxuries" that we cannot afford in times of recession— although in my opinion we can afford them just by the fact that we have unemployment at undiminished production.

I do not want to discuss institutional reforms that would make a smooth transition possible because I have already transgressed somewhat beyond our subject, which is control through communications. We come closer to our subject by stating that once the *intelligentsia* of a country is penetrated by the belief that such a transition is both desirable and possible, the obstacles will get out of the way because they are only anachronistic institutions and bad habits of thought.

Where there is a will, there is a way, so let the new education foster the *will* toward climbing up to a higher plateau of civilization. There is a fine field here for the new methods of audiovisual education. I believe that we have enough gifted architects for producing seductive pictures of the beautiful new towns and parks that could be created in America with the spare labor that is now being demoralized in idleness. I believe that it is not the artistic talent that is lacking, but the patrons. Why could our rich world not create beauty and diversity to equal that of Renaissance Italy?

The men and women who will fit into the society of the future must be of the *rational-altruistic* type. Responsible people must fight with all their power the wave of *irrationality* that has been spreading like wildfire in the industrial countries of the world (*Horizon*, Spring 1972). It is born of boredom and disillusion, not so much with the failings as with the successes of the consumer society. Though milder, and sometimes verging on the ridiculous, this irrationalism has some of the features of the "escape into darkness," into madness, so familiar to psychiatrists.

I do not claim to have a cure for this ailment of our times. In its extreme form, as it manifests itself in the "Weathermen," it is outside the reach of any rational discourse. But perhaps I can make a suggestion for fighting its milder form, which shows itself in many young people who *hate our times because they do not know the past*. I believe, with H. G. Wells, that cultural history is the basis of all education for civilization. But education by books goes only into the cerebral cortex, not into the thalamus! We must teach the long, painful, heroic history of Man from the cave to the consumer society and the welfare state with such emotional intensity that the boy or girl shall feel *as if he had lived through it*. We can do this with three-dimensional films that surround the viewer from all sides, but it is an expensive method of education that might appear to be a luxury. I believe that in time it will be seen as a necessity.

Although I have often come up against the impotence of technology as such for solving social problems, I wish to finish on a more optimistic note. We are before a great transition, and for this we need a new education. Whenever such a problem has turned up in the past, its solution took a long time, because first one had to educate an army of educators. Now, with television, and in particular with cable and cassette television, if there is only *one* man in an area of one language who can give a course, his teaching can be multiplied many thousandfold. So perhaps the great problems of retraining and reeducation may not take as long as one might think. History, as H. G Wells said, is a race between education and destruction. Perhaps in our time technology has given education the chance to win the race.

PART II

Institutional Powers and Controls; The Direction of Change

CONTENTS

INTRODUCTION

WILLIAM H. MELODY

Institutions in society provide a cohesiveness and stability to its functioning. Institutional powers and interrelationships constitute the matrix through which major changes are absorbed, massaged, promoted, redirected, or resisted. At any point in time, a particular network of institutional arrangements may or may not be responsive to the needs of society. Institutions may be adaptable to changing circumstances or, like vintage technology, become obsolete for changing public needs. Furthermore, some institutions may be vigorous, active, and aggressive in exercising their power while others remain timid and passive, responding only to the initiative of others. In fact, many institutions appear to follow a marked life cycle from vigorous youth to retirement.

There has been very little study of the interrelationships between institutional structure and the direction and control of communications technology. Technological change is too frequently viewed as some autonomous development that suddenly appears on the scene and that is going to have its deterministic influence regardless of our institutional arrangements. The chapters in this section demonstrate otherwise.

The commission system of regulating communications, as represented by the Federal Communications Commission, the Canadian Radio and Television Commission, and other agencies at federal, state, and local levels, is unique to America. Most other countries have opted for direct public ownership of such vital industries as telecommunications and broadcasting. However, the United States in particular has pursued a course of development in these industries under privately owned and operated, but publicly regulated enterprise. The regulatory commission has been created as a government agency for influencing the course of development of these industries through the policies and standards that it implements. In the structure of government, the commission has been separated from the three main branches and relates to each of them in different ways. Indeed, the conscious distribution and balancing of government power was a central concern in the design of the regulatory structure.

Unfortunately, there has been considerably less concern about the distribution of economic and political power in the communications industrial structures, where economic concentration is extremely high. This unusually high concentration is defended as being required by the state of technology and the objective of economic efficiency. As a result, the established firms are provided with a substantial incentive to ensure that new technology does not destroy that rationale, render older technologies obsolete, and reduce their market power.

But what happens when new technological opportunities appear on the horizon? We might expect that this would call for a reexamination of inherited institutional arrangements as part of a larger process of planning the rate and direction of application of the new technology. But it does not. Indeed, new technology appears to be viewed as a threat to both the economic institutions that may lose a share of their markets and the governmental institutions to which they relate. New technology creates great uncertainties for all institutions by threatening to alter or destroy established, stable, and understood relationships and to introduce new institutions and new relationships of an unknown character. Rarely is any institution, whether it is the American Telephone and Telegraph Company, the French Post-Telephone-Telegraph, the Federal Communications Commission, or the telecommunications department of the Province of Saskatchewan prepared to consider the possibility of a decline in its present power position as being in the public interest.

The following chapters examine particular institutional issues associated with major technological change. Richard Gabel describes the evolution of telecommunications interconnection policies. The structure of the United States domestic telecommunications industry has been heavily influenced by policies of refusal to interconnect among systems and to attach privately owned terminal equipment to the network. In a landmark 1968 decision (Carterfone), the FCC declared such interconnection restrictions unlawful unless necessary for protection from technical harm to the communications network. This prompted a shift in the continuing debate over the terms and conditions of interconnection from economic to technical grounds. Gabel analyzes the changing nature of the interconnection issue, its consequences for telecommunications services, and its potential impact on the structure of the industry.

Ralph Lee Smith addresses the institutional issues surrounding the introduction of cable television in the United States. After examining the characteristics of the technological and institutional conflicts, he evaluates the long-awaited FCC Cable Television Rules (1972) as guidelines directing the future development of the CATV and broadcasting industries.

Nicholas Johnson focuses his analysis on the changing circumstances surrounding the FCC's Fairness Doctrine during a period of significant technological change. He observes that the technology of cable television could provide an opportunity for solving or alleviating crucial regulatory problems that relate to the Fairness Doctrine and the use of the radio frequency spectrum.

But such an opportunity would require substantial realignments of regulatory policies and institutional relationships. Hence, he questions whether such a proposal could be seriously considered within the present institutional framework of broadcasting.

Dieter Kimbel's chapter provides an assessment of the implications of yet another major change in communications technology. He specifically evaluates consequences arising from the marriage of computer and telecommunication technologies in the provision of specialized data and teleprocessing communication services, comparing developments in Europe, Japan, and North America. His analysis demonstrates the lack of effective international integration and coordination in this area and points to the need for international, as well as national, policy planning.

In his chapter, William Melody directs his attention to the technological determinism rationale that supports both the institutional arrangements that are found in the United States telecommunications and broadcasting industries and the direction of governmental regulatory policies. He examines the public policy responses to the introduction of television transmission service, satellites, specialized data communications, and CATV as a basis for outlining a model of regulatory response to new technology. Noting the absence of policy planning and advocacy of public interest positions, he suggests a re-evaluation of regulatory objectives and activities that would, among other things, require examination of industry structures and justification of concentrations of economic power as a major part of continuing regulatory responsibility.

CHAPTER 8

Telecommunications Interconnection: Wherefrom and Whitherto?*

RICHARD GABEL

Contemporary textbooks on public utility regulation explain the development of communication regulation as stimulated by public need to obtain the
economies obtained from an industry of large increasing returns to scale. Of
course, the development of regulation of communications did not occur in this
fashion. Most current arguments for monopoly organization of communication services are post hoc rationalization instead of valid historic descrip-
tors. The establishment of communications as a public utility with monopoly
status was encouraged by energetic and farsighted businessmen who perferred
the incursion of state regulation to the risks and instability of competition.
Among the most powerful tools that were developed to insure the primacy
of monopoly organization was the prohibition against "foreign attachments,"
the refusal to interconnect the public message network with customer-owned
terminal devices.

Since its inception, the Bell system has refused to connect its local lines
with customer-furnished equipment. This prohibition has been enforced by
state regulatory commissions. The Carterfone decision represents the end of
an era and the demise of public acceptance of the classical arguments against
"foreign attachments."

By the Carterfone decision of June 27, 1968, the Federal Communications Commission opened the way for private, noncarrier suppliers of terminal devices to interconnect with the Bell system, consistent with the

* The views expressed herein are the personal views of the author and are not to be construed in
any way as reflecting the views of the Department of Commerce or other agency of the federal
government. Research assistance was provided by Shirley Holton.

integrity of that system. If the decision is implemented in the spirit in which it was intended, it promises sharp acceleration in the pace of technical and service innovation in the telecommunications industry and serious inroads in the prevailing monopoly domination of that industry. But there are strong indications that the FCC and Bell are dragging their feet in terms of implementation, and that full implementation will require several years of negotiation and procedural development.

The Carterfone decision did not spring full blown like Athena from the head of Zeus. Interconnection restrictions and prohibitions date to the earliest days of telephony. In the early years extensive interconnection restrictions were used to prevent competitive inroads by independent telephone companies. The development of the Kingsbury Commitment (AT&T, 1913, pp. 24–26) and enactment of the Willis-Graham Act in 1921 ended Bell refusal to interconnect with nonduplicative independent telephone companies. The removal of this prohibition coincided with the intensified enforcement of the "foreign attachment" provision in Bell system tariffs forbidding the use by a customer of any device or appurtenance as attachment to the message network, on pain of removal of service.

Bell's partial motivation in opposing interconnection with independent telephone compaines in the pre-Kingsbury period and reenforcing the "foreign attachment" rule throughout its history has been to impose obstacles to the growth of competing suppliers and to establish its own domination of the field. Originally, this motivation was candid and open, but more recently, and particularly throughout the Carterfone controversy, the Bell system has recognized that public relations need a more subtle and sophisticated approach. Private profit motives have been submerged and interconnection opposed on "technical" grounds. A major part of this chapter reviews the issues that surrounded the interconnection controversy throughout its history. Perhaps if we know where we came from, we will be able to better predict where we can go.

INTERCONNECTION RESTRICTIONS AFFECTING INDEPENDENT TELEPHONE COMPANIES, 1893 to 1920

The expiration of the basic Bell patents in 1893 marked the end of its monopoly over the telephone field and the beginning of the independent telephone industry.

From that time until about 1913, there was limited interconnection between Bell and independent exchanges (FCC [b], pp. 134–66) since ability to interconnect would have strengthened the position of the independent companies, and Bell was not interested at that time in helping the independent companies grow.

Inability of the independent exchanges to connect with Bell for long-distance service meant that they could provide service only within the geographic limits of the particular territories they served. The independent companies early recognized this obstacle to their growth and in 1899 attempted to form an independent long-distance network. Construction of the network required substantial outside financing, and the independents tried to arrange the necessary financing through a private financial consortium. It collapsed, however, with the withdrawal of its principal member at the request of J. P. Morgan of the banking firm (Commercial & Financial Chronicle, 1899).

Bell's President Theodore Vail clearly set out the company objections to interconnection in 1909 before the N.Y. State Joint Legislative Committee on Telephone and Telegraph Companies.

VICE-CHAIRMAN MERRITT.   You have said that inter-communicating is possible. What do you say as whether it is feasible?

MR. VAIL.   Well, it is not feasible.

VICE-CHAIRMAN MERRITT.   Yes, what are those objections, and what are the business objections, if any?

MR. VAIL.   Well, our principal objection, of course, is that it would give a competing exchange advantage of our big system, without any corresponding advantage to us, that we built up and maintained—often times at a loss to ourselves. We have a large system extending all over the United States and perfectly connected, and if a competing exchange in the town should have all the advantages we have, there would be no corresponding advantage to us. [N.Y.S. Joint Committee, 1909, p. 458]

While there were also objections to interconnection on technical grounds, they were clearly subordinated to economic considerations, as is apparent from the following dialogue between Mr. Vail and Committee Counsel.

Q.   What do you say as to the possibility or practicability of one telephone company giving interconnection with the subscribers of another telephone company when they are equipped with different styles of apparatus?

A.   Why, it would be unsatisfactory—there is no question about that at all.

Q.   Would it be possible to communicate?

A.   Most anything is possible when people undertake to do it.

Q.   Has it ever been done?

A.   Yes. It is being done every day. [N.Y.S. Joint Committee, 1909, p. 459]

Nor was Bell alone in opposing interconnection. The independent telephone companies also opposed interconnection and took active measures to block state legislation that would compel physical ties between competing telephone companies. Burt Hubbell, President of the Federal Telephone

Company (serving Buffalo, N.Y.) and a leader in independent telephony in New York state stated his reasons against interconnection before the N.Y. State Joint Legislature Committee (1909, p. 457):

> . . . A business man using a complete telephone service must have our service. . . . If the Bell Company had access to our exclusive list, it would be a very easy matter for it, with its large income in New York City, to divert its unusual profits to Buffalo to care for the extreme loads that it would afford to take, and underbid for telephone service. There would not be any incentive to take the service of the independent company . . . it would ultimately create a monopoly.

Had state policy makers pressed the issue of interconnection between Bell and the independents at that time, a markedly different industrial structure would probably have evolved. However, since both Bell and the independents opposed interconnection at the time, the state legislators did not do so.

The issue was ultimately joined at the federal level. Bell's rapid acquisition of independent properties led the independents in 1913 to file an antitrust complaint with Attorney General Wickersham. The Kingsbury Commitment resulted from the intervention of the U.S. Justice Department. Under this agreement, the Bell system agreed not to acquire control over any competing company, and to connect its system with those owned by independents if the latter met the Bell system equipment standards.

This commitment to interconnect with independent telephone companies was actually the culmination of a series of steps leading in the same direction. In the first place, the bitter price and service rivalry with the independents had resulted in an overextension of financial resources by Bell. Between 1895 and 1905, the Bell system assets nearly quadrupled, increasing from $120 million to $453 million (FCC [b], p. 52). Competition, however, had caused Bell's revenues to be decreased from $88/station in 1895, the first year of competition, to $43 in 1907 (FCC, 1958, pp. 243–250). The sharp reduction in Bell's earnings was the source of repeated lament to stockholders (AT&T, 1904, p. 10 and AT&T, 1906, p. 12). When financial stability is at stake, the financiers intervene to obtain stability.

> In January 1911 . . . an effort was made to effect a working agreement between J. P. Morgan, the independents and Bell. . . . Bell proposed that both sides seek to furnish an efficient service to the public and suggested that in cities where both operated, the stronger exchanges absorb the other on an equitable basis. . . . Bell was finding difficulty in obtaining sufficient capital for development. Morgan, it was rumored, having told the Bell officials that the telephone business was so great that the financial interests could not furnish capital for continued warfare. [MacNeal, 1934, p. 184]

As noted here, the independents had also been intransigent over the subject of interconnection, but they, too, now recognized the business desirability of eliminating uneconomic competition. For example, Theodore Gary, founder of the Independent Telephone Holding Company, urged: "An effort to make a fair deal with Bell in the interest of the telephone business" (*Telephony*, 1911). The previous year Frank H. Woods, founder of the Lincoln (Nebraska) Tel. & Tel. had worked out a division of territory with the local Bell Company, including interconnection of his exchanges with Bell's long-distance network. At the December 1910 convention of the National Independent Telephone Association, Woods proposed: " . . . the Independents ask Congress and the state legislature for laws compelling interchange" (MacNeal, 1934, p. 183).

Interconnection required the cooperation of both the Bell and the independents and cooperation was financially less risky than continued price warfare and competition for markets. In this sense, the Kingsbury Commitment's agreement to interconnect was the capstone to internal financial needs of the industry.

Problems after Interconnection with Independents

The tail on Bell's obligation to interconnect with independent systems under The Kingsbury Commitment soon produced new problems. The independent telephone industry had largely grown in small towns and rural areas of the country, ignored by the Bell system. Concurrent with this growth, a large number of independent telephone manufacturing firms developed, showing considerable versatility in equipment and systems design for their own markets.

With acquisition by Bell of several hundred thousand independent stations, interconnection problems arose in three different levels.

First, in an effort to achieve technical uniformity, Bell frequently replaced the equipment in those exchanges with Western Electric-type equipment. Thus, for example, the entire south Los Angeles central offices were reconverted from automatic dial service instituted by its previous independent owners to manual common battery service (California, 1948). Since Bell drew its entire switching supply from Western Electric, this type of changeover foreclosed the growth and replacement markets to the independent manufacturing industry. The acquired exchanges were now interconnected with Bell as part of the latter's system, and in this sense, the interconnection problem was "solved" for these exchanges.

A second form of interconnection problem resulted from the efforts of the remaining independent exchanges to obtain interconnection with Bell's long-distance network. While the signal supervisory techniques developed by independent manufacturers sometimes surpassed those generally available in Bell plants, there were real questions of compatibility as well as market supremacy underlying the resulting disputes.

A third interconnection problem arose with customers who had been furnishing their own terminal apparatus, including private branch exchange equipment. For financial and other reasons, the independents were very tolerant in permitting the use of subscriber terminals on their lines.

On occasion, the subscribers objected to the change in interconnection policy. These matters were submitted for resolution to the state regulatory commission. It is interesting to review the arguments presented by Bell, arguments that state commissions almost unanimously adopted as their own. These arguments were basically of three types.

First was Bell's claim that public regulators must be able to rely on some responsible source for providing adequate service. Private individuals who have as their primary objective their own economic and service advancement cannot be relied on to shoulder the burden of "total service responsibility" PUR 1915A, 928; PUR 1915A, 1032; PUR 1915C, 106; PUR 1920E, PUR 1925A).

Second was the challenge to commission authority to require interconnection. This argument in turn had three major subdivisions: *(a)* since a significant portion of customer traffic was interstate in nature, regulation of interconnection was beyond the province of state commissions (PUR 1926C). It is interesting to note that this argument does not appear in any of the state cases after 1934, when the Federal Communications Commission came into existence; *(b)* the proper selection of facilities on the network is a managerial judgment beyond regulatory control (PUR 1928B); and *(c)* the state commission lacked the power to order interconnection to equipment not owned by the telephone company (PUR 1920D).

These two sets of arguments were presented up to 1925. The first argument has assumed modified form, the second has been abandoned, and the third, which still survives, is to the effect that if noncarrier ownership of end instruments is permitted, the telephone company would lose control of the service. Purportedly supporting this argument was detailed engineering evidence allegedly demonstrating the immense technical complexity of providing voice telephony. The network is a nationwide system: "It is only as good as its weakest link." Only the telephone company is effectively concerned with the operations of the entire system and only it can effectively determine how the constituent parts mesh (PUR 1921C; 2PUR(NS) 1934; 3PUR(NS) 1934; 6PUR(NS) 1935).

In affirming Bell tariff restrictions against interconnection with customer devices, state regulatory commissions saw no conflict between the telephone companies' professed concern over public service responsibilities and its private profit motive. A summary of two of these cases may help illustrate the problem.

A lumber company in the state of Washington operated and maintained its own PBX and instruments at the mill in addition to providing grounded circuits from the mill to numerous logging operations scattered throughout its operating area. Without proposing to serve the remote logging operations, the local telephone company sought an order from the Commission to require the lumber mill to replace its PBX with the telephone company PBX. In approving this division in responsibility for providing service, the Washington Commission solemnly stated: "Dividing responsibility and authority in the provision of instruments is apt to result in confusion and trouble and poor service" (PUR 1925A).

Frank King wanted to purchase his extension sets rather than pay monthly rental. He submitted testimony showing that annual rental of a dial business extension was $15 "while the instrument cost $13.89 and estimated cost of wiring $1.50." The Oregon Commissioner, however, denied King's application: "The (protestant) proceeds on the erroneous theory that the company is renting its equipment rather than furnishing a service . . . no important relationship appears between the cost of the instrumentalities . . . and the rates for service." And finally, customer-owned equipment "would result in injury and impairment to the telephone service" (16 PUR(NS) 1937).

The report of the Joint Committee established by the New York legislature in 1915 to investigate the matter is very illuminating on these problems. Despite the 57-year interval between that report and the deliberations of the FCC PBX Advisory Committee in 1971, the similarity between the problems is clear.

Connection With Privately Owned and Installed Equipment:

In the development of the telephone systems in the state a very serious and difficult problem has arisen by reason of the diversity of instruments in use. Many of the independent systems, some of which have been purchased by the New York Telephone Company and its subsidiaries, use in the construction and operation of their lines different styles and types of apparatus, manufactured by various concerns.

It is the claim of the Bell System that it should not be compelled to connect its lines with a privately owned installation because of the inferiority in many instances of the instruments used. . . . It is the claim of the Bell system that no amount of supervision, no matter how thorough, can guarantee satisfactory service over dissimilar instruments. . . .

With this contention the Committee cannot and does not concur . . . Not the telephone company, but the Public Service Commission should have complete supervision and control (a) over the design of installation; (b) over its maintenance; (c) over relocations and extensions; (d) over operation. Perhaps such an extension of the supervisory powers of the Public Service Commission over a privately owned and installed telephone system might render practicable the enforced connection by the telephone company with such privately owned and installed telephone system . . . at the present time the Public Service Commission has not adopted any definite standard for construction, maintenance and operation of telephones, and, therefore, the New York Telephone Company, through its monopoly, fixes the only standard now existing. [N.Y.S. Joint Committee, 1915, pp. 17–19]

INTERCONNECTION RESTRICTIONS BARRING CUSTOMER ATTACHMENTS—STATE REGULATORY ACTIVITY

As nearly as can be determined, the Bell system has always had a policy of restricting customer devices from the message network. The earliest tariffs of public record (1915) contain the classical phrase: " . . . no apparatus or appliance not furnished by the company shall be attached to or used in connection therewith" (PUR 1915C, p. 137).

The acquisition of thousands of independent telephone exchanges and farmer mutual lines had introduced a variety of unstandardized terminal equipment. "Of the thousands of men who went into the business, almost none had any engineering knowledge or any practical experience in the telephone field" (MacNeal, 1934, p. 79). Faced with recurrent capital shortage, many of the independents and almost all the mutuals encouraged customer ownership of station equipment. In acquiring many of these systems, Bell had inherited these former practices.

Although many user facilities had been furnishing adequate services while operating under independent aegis, Bell proceeded systematically to weed out nonconforming equipment. Most users accepted the change. In some instances, customer station apparatus was "sold" back to the telephone company through remission of several months exchange rental, the same apparatus remaining in place while the carrier assumed responsibility for maintenance. In other locations, the equipment was replaced by Western Electric-type instruments.

INTERCONNECTION RESTRICTIONS—PRE-CARTERFONE CASES ON FEDERAL LEVEL

The creation of the Federal Communications Commission in 1934 added a new forum for controversy over interconnection of noncarrier devices. Prior to the Commission's Carterfone decision, there were at least five major contested proceedings involving interconnection: *(a)* Recording Devices (Docket 6787); *(b)* Hush-A-Phone (Docket 9189); *(c)* Jordaphone (Dockets 9383 and 9700);

(d) Railroad and Right-of-Way Interconnection with Telephone Companies (Docket 12940); and *(e)* The Telegraph Investigation (Docket 14650).

As is apparent from the following summary of these cases, Bell's hostility to interconnection was as vigorous as it was before state regulatory bodies. However, the nature of arguments used by Bell became more complicated and sophisticated in line with the shift in the character of the approach. Whereas the earlier state proceedings usually involved irate individuals lacking technical resources with but relatively small issues at stake, the pleas to FCC frequently involved larger corporate bodies with extensive knowledge of communication engineering and large communications budgets.

In the proceeding involving recording devices, the Commission conditioned its authorization permitting interconnection with customer-furnished recorders on a requirement that each unit be equipped with a jack and plug arrangement "so as not to impair state freedom of action."

The Commission's denial of interconnection in the Hush-A-Phone case was overridden by the U.S. Court of Appeals. The Court, in an oft-quoted phrase, decided a "subscriber has the right to use his telephone in ways which are privately beneficial without being publicly detrimental."

The Association of State Regulatory Officials (NARUC) interceded in the Jordaphone case involving connection with electronic answering services. Yielding to these jurisdictional concerns, the FCC questioned the need for authorizing the service on interstate calls, but expressed no objection if the state commission found a local need for answering services. State commissions, more concerned with who exercises authority than with the wisdom of its use, have played a persistent role in interconnection policy.

In the late 1950s the railroad industry sought reestablishment of its historic rights of interconnection with the public message telephone network. Again the Commission avoided making a decision by accepting a compromise negotiated between the railroads and Bell.

Interconnection was precipitated as an issue in the Telegraph Investigation of the FCC in 1961 as Western Union Telegraph Company threatened to enter the market for voice services. When the telegraph company retreated, the Commission declined to rule on the matter.

Six basic arguments have been offered by Bell to the FCC in explanation of its refusal to interconnect with "foreign" attachments or facilities. Those submitted in the railroad interconnection proceeding and the responses thereto are sufficiently representative to summarize briefly.

Service Quality

Probably the most important reason urged by Bell is "the desire to furnish high quality service." Bell claims that it must have full control over the design, installation, operation, and maintenance of all facilities. Division of responsibility would be detrimental to telephone service.

In response the railroads pointed out that because of the dependence of railroad operations on railroad communication, their standards of service were at least equal to, if not higher than, normal Bell standards. Responsibility within the telephone industry has always been divided between thousands of independent telephone companies and within the Bell system.

The Need for Integration

Since telephone service is provided through a system made up of thousands of complicated and delicately balanced parts, these parts must be coordinated by a responsible entity, namely, the telephone company, in order to function effectively.

The railroad answered that coordination had been effectively maintained through 70 years of Bell's interconnection with the railroad communication systems. While Bell's witnesses testified at length concerning the potential harm from interconnection, they failed to furnish a single instance where such harm actually occurred from railroad connections.

Quality of Maintenance

Bell reactivated its long-time criticism of customer-provided equipment that the quality of maintenance of customer-furnished equipment once installed would depend on the customer's own ideas of what is needed and on his willingness and ability to obtain and pay for maintenance service. There would be no assurance that the customer equipment would be kept in satisfactory condition.

The railroads undercut the argument as applied to them by showing in great detail the lengths to which they went to train their staffs and their overriding need for rapid and dependable communication service. Interconnection was essential to railroad operations, for safety of life, public safety, and national defense. The telephone industry, on the other hand, applies higher standards of maintenance in urban areas than in small towns and less lucrative operating territory. The concern expressed by Bell over the need for high maintenance standards did not apply to service in rural areas where farmer mutuals provided the major source of communication.

Equipment Obsolescence

Bell contended that customers providing portions of the facilities used in their telephone service would tend to oppose Bell's technological innovations if this obsoleted their facilities, thereby retarding the introduction of improved service facilities.

The railroad answered that they were ahead of Bell in adopting improved signaling methods. It was Bell, not the railroads, who had done what they could to stop the replacement of open wire plant with microwave relay by refusing to connect the newer facilities, and "grandfathering" the old. The most conservative organization with regard to rate of equipment obsolescence was the telephone company. They normally improvise "appliques" on older plants to prevent early plant retirement.

Company Responsibility

The fifth reason was that where customer-furnished equipment gave trouble, the public would be disposed to blame the telephone company. But the railroads pointed out that over 90% of interconnected calls are made between railroad personnel and practically all calls are made at acceptable transmission standard.

Effect on Charges to the Public

Finally, Bell argued that interconnection generally will mean increased costs and a need to charge higher rates. Bell's reasoning in this regard ran about as follows. Bell has the responsibility of serving all customers at uniform rates. Customers would elect to furnish their own facilities only when their costs would be lower than Bell's uniform charges. On the other hand, when their costs would be higher than Bell's charges, the customer would take that portion of his communication service from Bell. Since Bell would be left with the more costly portions of the service, the result would mean higher overall cost of service and therefore higher rates to the general public, a burden that would fall largely on the smaller users who are unable to provide facilities of their own.

The railroads' answer briefly was that railroad communication services were initially established because of the unavailability of service and because the telephone companies had concentrated their attention in the more lucrative urban communities. It was only in recent years that the telephone companies extended their plant to points that would enable the provision of communications to the railroads service areas. If there had been "cream skimming," the telephone companies had been the earliest practitioner.

THE CARTERFONE DECISION

The upshot of the FCC decision in the Jordaphone case and the court ruling in the Hush-A-Phone case was a relaxation of the flat prohibitions against "foreign attachments" in the Bell tariffs. Practice was still a different matter. Bell undertook on the one hand to interpret the exceptions as narrowly as possible and on the other hand to refuse generally to interconnect with customer-furnished terminal devices or systems. The Carterfone case brought this inconsistency to a head.

Carterfone is a device that provides an acoustic coupling to enable two-way communications between the public message network and mobile radio units. Between 1959 and 1966, the Carter Electronics Corporation marketed some 3500 Carterfones. To stop the marketing of these devices, Bell unleashed its standard technical and economic arguments.

In addition to these standard arguments, Bell came up with a few new ones. For example: "If a call (to a Carterfone user) had been placed person-to-person the operator as well as the calling party would have to wait until the desired party was reached . . . we would have circuits connected for an additional period which would increase the cost of service." The examiner, however, rejected these arguments and found

> . . . no reason to anticipate that the Carterfone will have an adverse effect on the telephone system or any part thereof. It takes nothing from the system other than the inductive force of the electrical field in the earpiece of the handset. . . . It puts nothing into the system except the sound of a human voice into the mouthpiece of the handset, and that is the precise purpose for which that portion of the system is engineered.

The FCC, upon its review of the Examiner's opinion, went much farther. In a landmark decision it found the tariff provision against "foreign attachments" unreasonable, as well as discriminatorily applied against Carterfone (FCC Reports, 1937).

> Our conclusion here is that a customer desiring to use an interconnecting device to improve the utility to him of both the telephone system and a private radio system should be able to do so, so long as the interconnection does not adversely affect the telephone company's operation. . . . A tariff which prevents this is unreasonable; it is also unduly discriminatory. . . . No one entity need provide all interconnection equipment for our telephone system any more than a single source is needed to supply the parts for a space probe. [FCC Release, 1960, p. 7]

The Carterfone decision compelled the Bell system to undertake a fundamental reexamination of its interconnection policies. In December 1968 the FCC permitted Bell to put into effect modified tariffs that permitted customer provision of various terminal devices subject only to two primary reservations: (1) Bell provides (at a charge) an interface device between the customer facilities and the telephone message network, and (2) Bell furnishes the address signaling device (rotary dial or Touchtone pad) with the customer equipment. Over the strong objections of independent manufacturers and users, but without expressing approval, the Commission permitted the new tariffs to go into effect January 1, 1969 (FCC Release, 1968).

IMPLEMENTATION OF CARTERFONE

The Carterfone decision emphasized the desirability of developing publicly useful and innovative devices and systems not heretofore provided by the sole common carrier source while at the same time recognizing the need for continued protection of the network and limiting harmful exposure.

Bell has stressed both these objectives but with differing emphasis in speech and in practice. Bell's board chairman has voiced enthusiasm toward the new interconnection provisions:

We believe that new regulations will open up new communications potentialities for our customers and afford new opportunities for the many fine customers that are making and marketing information-handling devices. . . . We want to make the connection of such equipment as easy as possible . . . we want to find more ways to say "yes" to our customers—to approach these things imaginatively and flexibly. [AT&T, 1968]

The NAS Panel

Bell took full advantage of the opportunity to emphasize the potential "harm" to the network occasioned by the appointment of a special panel of the National Academy of Sciences (1970) to "make an assessment of the technical factors affecting the common carrier/user interconnection area of public communication" without regard to economic and cost implications. Specifically, the Chief of the Commission's Common Carrier Bureau, in a letter dated September 25, 1969, asked the panel to comment on the following matters:

1. The *propriety* of the telephone company—provided network control signalling requirements and *various alternatives* to the provisions thereof by the telephone company.

2. The *necessity* and *characteristic* of telephone company-provided connecting arrangements and various alternatives to the provisions thereof by the telephone company, and

3. Basic standards and specifications for interconnection and the appropriate method to administer them. [Author's emphasis]

In response, the panel came forward with six primary conclusions.

The first conclusion was that "Uncontrolled interconnection can cause harm to personnel, network performance and property" (National Academy of Science, 1970). Of the four-score suggestions offered by intervenors to the FCC tariff inquiry to implement the Carterfone decision, virtually none had suggested "uncontrolled interconnection."

The second observation by the panel was that "the signal criteria . . . relating to signal amplitude, waveform and spectrum are technically based and valid and if exceeded can cause harm by interfering with service to other users." This second conclusion is not very instructive since there is no question that signal amplitude, waveform, and spectrum are technically based. It is virtually a truism that communication equipment standards must be compatible with one another. It is equally true that existing common carrier facilities are characterized by wide statistical distribution of the signal and spectrum features. The important problems in this regard, to which the panel provided no solution, are what are the consequences of infractions of these standards, with what tolerances, and what is the impact in terms of relative costs and benefits of variation in the Bell standards for signal amplitude, waveform, and spectrum. The panel conclusion stated an absolute, but did not provide an operational tool.

The panel's third conclusion was that "present tariff criteria together with carrier provided connecting arrangements are an acceptable basis of assuring protection." Elsewhere the panel reported four potential "harms" that can arise from interconnection: (a) voltages dangerous to human life; (b) signals of excessive amplitude or improper spectrum; (c) improper line balance; and (d) improper control signals. The protection afforded by the connecting arrangements, referred to in the panel's third conclusion, has limited deterrent effect on customer devices that generate improper spectrum or improper control signals. The deterrent is actually contained in the Bell tariff, which specifies frequency spread and a requirement that the rotary dial address network be furnished by Bell. Improper off-hook and supervisory signals can be generated by customer devices without contravening the tariff, or can be impeded by the coupling device. In fact, impairment of performance and reliability can be introduced by the connecting arrangement required by Bell. The panel was asked whether any connecting arrangements were necessary; if they were, a description of the character of the connecting arrangements and feasible alternatives was requested, but the panel did not respond to any of these questions.

The panel's fourth conclusion was that "present tariff criteria, together with a properly authorized and enforced program of standards development, equipment certification, and controlled installation and maintenance are an acceptable basis of achieving direct user interconnection." This is a fairly broad conclusion containing within itself all the difficult questions for resolution. To be fair to the panel it should be observed that under FCC directive the panel was circumscribed from examining the economic aspects of interconnection. Development of standards may appear to be wholly a technical problem, when, in fact, they are closely interrelated with economic questions. Limiting consideration solely to technical aspects, and excluding economic aspects, may appear to provide a conclusion when, in fact, it may just be preliminary to consideration of many questions.

In its fifth conclusion, the panel noted: "Innovation by carriers need not be significantly impeded by a certification program. Opportunities for innovation by users would be increased." The successful improvisations undertaken in many allied industries that have been subject to certification procedures is strong evidence that innovation has not, in fact, been blocked by certification procedure, but how much such innovation has been held up is another unanswered question. In contrast, there can be no question but that the opportunities for innovation would be enhanced by reduced strictures on attempts to innovate.

The panel's final conclusion was that "mechanisms are needed to promote the exchange of information among carriers, users and suppliers." This observation was another step in the right direction. The full exchange of information among the affected interests could obviously serve to foster the resolution of the remaining problems with regard to interconnection.

INTERCONNECTION—WHITHERTO?

The PBX Advisory Committee

In March 1971, the FCC established a PBX Advisory Committee (Telecommunications Report, 1971). Although there are hundreds of terminal devices that can provide a useful communication function, the FCC established the advisory unit for private branch equipment presumably as a mechanism to set the field for other terminal devices and attachments. At the opening meeting of the PBX Advisory Committee, the FCC staff chairman provided two salient directives to the industry membership (FCC PBX, 1971). The group was requested to use the report of the Advisory Panel of the National Academy of Science as a starting point for its deliberations. Economic issues would be treated separately and be developed subsequent to generation of the material assembled by the Advisory Group.

The FCC Advisory Committee divided into two groups. The first group, the Technical Standards Committee, was requested to define the electrical parameters beyond which "harm" to the network or personnel could be identified. A second group, the Procedure and Enforcement Committee, was established to develop recommendations for certification, installation, and maintenance of PBX equipment.

Following the lead suggested by the National Academy of Sciences Panel, the Technical Standards Committee broke into four working parties, each of which was requested to address the "harms" potentially to be encountered with interconnection with customer-furnished PBX equipment, such as hazardous voltages, network control signaling, longitudinal balance, and signal amplitude and signal spectra. A recurrent difficulty of the Technical Standards Committee was the inability of any of the common carriers to furnish *any* statistical evidence of "harms" resultant from past interconnections or to indicate their relative significance with carrier-supplied equipment.

The Procedures and Enforcement Committee divided its work among six working parties covering equipment certification procedures, installation procedures, maintenance procedures, personnel licensing, quality control requirements, and laboratory facility requirements for testing. The precedence-making role of the Procedures Committee involves service risks; in developing, testing, licensing, and procedural requirements for terminal equipment of relatively high cost, it may invite prohibitive economic penalties against equipment of much lower unit cost.

The PBX Advisory Committee includes in its membership some of the outstanding professional engineering talent in this country in the field of private branch exchange equipment. The separate committees have met continuously on the average of once a month since their creation, while the working groups have met more frequently with a constant exchange of technical papers.

Representation on the committees includes strong participation from the regulated common carrier industry and their manufacturing subsidiaries, and the independent manufacturing firms, as well as the installation and maintenance organizations. Consumer representation is led by the National Association of Manufacturers.

The Opportunities Ahead

The program that is developing out of implementation of the Carterfone decision represents an area of vast opportunity for increasing customer service features and for innovation in the provision of communications. The FCC PBX Advisory Committee is composed of men of high technical calibre who have conscientiously been seeking to provide a worthwhile series of recommendations to the FCC. Interconnection is fundamentally an economic problem, however, based on technical consideration. To be publicly useful a program must be simple, flexible, and inexpensive to administer so that the potential benefits from interconnection may be realized. There appears to be a tendency to become engrossed in the technical problems and minimize economic solutions. Procedural bottlenecks may be made so cumbersome and costly that the latent benefits of competition will have limited play. The FCC must be hospitable toward interconnection by viewing it as an opportunity to make positive contributions to user service instead of taking a negative approach of accentuating the "harms" that continue to remain statistically undefined. Carterfone has already made a positive contribution as the carriers have accelerated their own equipment innovations. The increased emphasis by the communication utilities on "cost" versus "value" pricing will precipitate new kinds of problems.

The potential opened up by interconnection has its counterpart in the fact that less than a decade ago the large computer main frame manufacturers asserted the same concerns over customer use of other manufacturer's equipment as the common carriers exhibit today. The spawning of independent peripherial device businesses handling high-speed printers, tape, and disc handlers raised parallel questions by main frame manufacturers over "system integrity," quality of maintenance, and concern over divided responsibility. The greater speeds and complexity of computer peripheral devices make the interconnection problems in telecommunications simple by comparison. Users took the risk. "Foreign attachments" today are a way of life for the computer industry. Thirty years ago the telephone industry was surrounded by users who did not know or care what their terminals consisted of. Today, the telephone industry serves great numbers of highly sophisticated customers who not only are on a level with it technologically, but are anxious and capable to contribute solutions to customers' needs. The impediments cannot endure. As in the computer field, customer attachments will be a way of life.

REFERENCES

AT&T Annual Report to Stockholders, 1904.

AT&T Annual Report to Stockholders, 1906.

AT&T Annual Report to Stockholders, 1909.

AT&T Annual Report to Stockholders, 1913.

AT&T Annual Report to Stockholders, 1968.

California P.U.C. Decision No. 41416, "Investigation of Rates and Practices, Pacific Tel. & Tel. Co.", April 6, 1948.

Commercial and Financial Chronicle, 69, p. 1151, 1899.

FCC(a) "Investigation of the Telephone Industry", Staff Exhibit 2096-D, "Control of Independent Telephone Companies."

FCC(b) "Investigation of the Telephone Industry", Staff Exhibit 1360-A "AT&T Co.—Corporate and Financial History."

FCC-PBX Standards Advisory Committee, Minutes of Meeting, July 20, 1971.

FCC "Proposed Report—Telephone Investigation", 1938 (Pursuant to Pub. Res. No. 8, 74th Congress) referred to as Walker Report.

Gabel, Richard. "The Early Competitive Era in Telephone Communications, 1893–1920", Law and Contemporary Problems, Spring, 1969.

MacNeal, Harry B. *The Story of Independent Telephony*. Chicago: Indep. Pioneer Tel. Assn., 1934.

195 Magazine, September 16, 1960.

National Academy of Sciences, Computer Science and Engineering Board, "Report on Technical Analyses of Common Carrier/User Interconnection", June 10, 1970.

N.Y.S. Final Report of Joint Committee of the Senate and Assembly on Telephone and Telegraph Companies, February 26, 1915.

N.Y.S. Joint Committee of the Senate and Assembly on Telephone and Telegraph Companies, Report No. 58, December 8, 1909.

Sears, Vinton A. *Telephone Development—Scope and Effect of Competition.* Boston: Barta Press, 1905.

Telecommunications Report, March 8, 1971, pp. 1–5, November 1, 1971, p. 15.

Telephony Magazine, March 4, 1911.

CASES

California PUC Rec. No. 78894, July 13, 1971, *Phonetile, Inc., v. General Tel. Co. of California.*

FCC Docket 9383, *Jordaphone Corp. of America and Mohawk Business Machines v. AT&T Co.,* May 5, 1954.

FCC Docket 12940, *RR Interconnection,* January 17, 1962.

FCC Docket 14650, Staff Outline, March 6, 1963; *Testimony of George Best, W. M. Davidson and Jere Cave,* AT&T Exhibits 3 and 31.

FCC Docket 16942, *Use of the Carterfone Device in Message Toll Telephone Service, Testimony of Hubert Kertz,* Bell Exhibit 1, April 3, 1967, *Testimony of C. R. Williamson,* Bell Exhibit 3.

FCC Release 68–661, June 27, 1960, *In the Matter of Thomas F. Carter v. A.T.&T. Co.*

FCC Release 68–1234, December 26, 1968, *AT&T Foreign Attachment Tariff Revision, Memorandum Opinion and Order.*

FCC Reports, *11,* 1036, *12,* 1005, *In the Matter of Use of Recording Devices in Connection With Telephone Service.*

FCC Reports, *13,* 430 (1967), *In the Matter of the Carterfone Device in Message Toll Telephone Service.*

FCC Reports, *14,* 282, Docket 9189, November 1, 1949, *Hush-A-Phone Corp. v. AT&T Co.* Cf Complainant Exhibit 36—A record maintained by N.Y. Tel. of different "foreign attachments" showed: 1916–30:10; 1931–35:49; 1936–40:24; 1941–45:7; 1946–49:34. N.Y. Tel. attempted to match the equipment made by Hush-A-Phone but withdrew it for lack of success.

FCC Reports, *Jordaphone Corp., 18,* pp. 644–678.

FCC Reports, *20,* 391, 415, 424 (1955).

FCC Reports, *29,* 825 (1960) *In the Matter of Allocation of Frequencies in the Bands Above 890 MC.*

FCC Reports, *32,* 337 *Railroad and Right-of-Way Interconnection with Telephone Companies.*

FCC Reports, *37,* 111 (1964) *Telpak Rates and Changes.*

PUR 1915A, 928, Ill. PUC in re *Bluffs v. Winchester Tel. Co.*

PUR 1915A, 1032, in re *Investigation of Rates, Changes and Practice of Telephone Companies.*

PUR 1915C, 106, S.D. RR Com. in re *Rates and Practices of Social Tel. Co.*

PUR 1915C, 776 *Hotel Sherman et. al. v. Chicago Tel. Co.*

PUR 1920D, 137 *Quick Action Collection Co. v. N.Y. Tel. Co.*

PUR 1920E, 633, Calif. RR Com. re *Robert L. Swanson.*

PUR 1921C, 72 Calif. RR Com. re *Tognini, Gerezzi & Dalidio Tel. Co.*

PUR 1923C, 374, Wisc. RR Com. re *Peoples Tel. Co.*

PUR 1925A, 676 Wash. Dept. *Public Works v. Montesone Tel. Co.*

PUR 1926C, 367, Indiana P.S.C. re *Steuben Tel. Co.*

PUR 1928B 396, Mass. Sup. Ind. Ct., *N. England T&T* v. *Dept. P.U. and Hotel Statler, Inc.*

PUR 1929A, 224, N.Y. PSC re *Gibsam Realty* v. *N.Y. Tel. Co.*

2 PUR(NS)1934, 247 *City of Los Angeles* v. *Southern Calif. Tel. Co.*

3PUR(NS)1934, re *Customers of Concondia Tel. Co.*

6PUR(NS) 1935, 27 *Archie Newton* v. *Jamestown Tel. Co.*

16PUR(NS)1937, 348 Oregon PUC, *Frank Knight* v. *P.T.&T. Co.*

72PUR3d, 76 re *P.T.&T. Co.*

CHAPTER 9

CATV: FCC Rules
and the Public Interest

RALPH LEE SMITH

Cable television is more and more referred to as broadband communications, reflecting a rising national awareness of cable's diverse capabilities. These capabilities will inevitably have an impact on existing technology and institutions, and, in fact, the conflict is already well advanced. Many of its facets were reflected in the Federal Communications Commission's proceedings leading to the adoption of the Commission's new cable rules on February 3, 1972, and it is these proceedings and their consequences that will be discussed in this chapter.

Interest and participation in the three years of FCC activity that preceded the issuance of the rules, were on a scale without precedent in the Commission's rule-making history. Several hundred parties, many of them public interest groups, community groups, and even individual citizens, filed written opinions and comments with the Commission. The proceedings included five days of panel discussions unique in communications rule making, held before a large audience in the auditorium of the National Academy of Sciences. On these panels, several dozen leaders in communications made oral and written statements and exchanged views with each other and with the Commissioners.

Issuance of the rules occasioned a sharp exchange between Commissioner Nicholas Johnson and FCC Chairman Dean Burch. Johnson charged that in framing key provisions of the rules, FCC and the White House Office of Telecommunications Policy had acted neither as spokesmen for, nor as protectors of, the public interest. Instead, they had played the role of Helpful Harry to the three industries that have a stake in the growth of cable—the broadcasters, the cable entrepreneurs, and the owners of program material—assisting them to come to a private, mutually beneficial economic accomodation among themselves, and then obligingly freezing the result in federal bronze. The resulting regulations, Johnson charged, "carved up the action among the three industries at the expense of the viewing public."

Chairman Burch, who had fostered the interindustry compromises that had become law through the rules, slugged it out with Johnson toe to toe. "In the manner of demagogues," Burch said, "he [Commissioner Johnson] elevates gross oversimplification to the level of a moral imperative. For him all differences are by definition *dis*-honest. Accommodation and compromise equal 'sellouts.' Any desire to preserve what we have—warts and all—can only be motivated by 'greed.' Commissioner Johnson's world is peopled wholly by white hats and black hats, and every role is type-cast in advance. I almost envy him the simplicity of his perspective. But I cannot wallow with him in his irresponsibility."

The exchange between the two Commissioners makes some of the best reading to come out of a federal regulatory agency in recent times. In addition to conveying the anger and immediacy of battle, it reflects profound differences of viewpoint between the two men on the role that the federal government should play in the regulation of communications. The shape that cable and other new technologies will take in America and the extent to which they will serve public and social goals will be greatly affected by the extent to which one or the other of the two philosophies is pursued.

At the heart of the conflicts over cable lie certain inherent capabilities of the cable itself.

CHARACTERISTICS OF THE CABLE

First, cable can deliver many more TV channels than it is technologically or economically feasible to broadcast over the air.

Three factors conspire against the delivery of large numbers of channels of TV over the air. They are the inherent limitations on the amount of good transmission space on the over-the-air spectrum, the large amount of spectrum space required to transmit each TV signal, and the economics of over-the-air broadcasting, which limit the number of individual transmitters that can be profitably built and operated in any given community. This combination of limitations mean that no more than five or six TV channels can usually be seen, even in large metropolitan centers, and most United States TV viewers receive less.

By contrast, a single coaxial cable can deliver between 28 and 35 channels of TV with full clarity, plus the entire AM and FM radio broadcast bands. With a new generation of more sophisticated input equipment and better amplifiers, this capacity could probably be increased. In practice, this may not prove to be necessary, at least in the near future, since more capacity can be added by laying a second or third cable. Uniting more cables with more advanced technology can result in a true "channel explosion." Nathaniel Feldman, a cable technology expert on the staff of The RAND Corporation, predicts that by the year 1990 a four-cable system will be able to accommodate 400 channels of TV.

Second, cable systems offer low cost of transmission. In addition to rebroadcasting regular TV signals, it is possible to transmit material directly over the cable to the subscribers. With the cable system installed and in operation, the cost of opening up unused channels in the system is very small—a tiny fraction of the cost of building and operating a new over-the-air TV transmitter.

Third, cable achieves exact coverage. An over-the-air TV station broadcasts a signal to a large, vaguely defined area, where it is received with varying degrees of clarity. By comparison, a cable net covers an exact area, and even an exactly known group of households. Through the use of switching and filtering arrangements, subscribers can be broken down into geographical subgroups such as school districts or election districts. Subscribers can also be grouped by interests or specialties—for example, all the doctors, all those wishing to take college courses by TV, or all the Mah-Jongg players. When regional and national interconnections among cable systems are established, it will be possible to assemble audiences for all kinds of specialized programming that would not appeal to enough people in any given area or locality to be worth broadcasting.

Fourth, cable will carry signals in both directions. In a literal sense, an individual in his home can talk back to his TV set. The exact kind of "talking" will depend on the complexity of the equipment, and the economic feasibility of things that are feasible in a technical sense.

In its simplest form, two-way communications would enable the set owner to send back electronic responses to a set of options described in the televised programming. One could indicate one's choice of products, or could listen to the mayor describe a new bond issue, and then indicate whether one favored it or not. More complex forms of interaction include voice conversations with the source of the programming, or even with any other subscriber; data interchange between and among subscribers; and ultimately, direct interaction with high-capacity computers, enabling one to receive on demand immense amounts of informational material in the home, either visually on the screen or as hardcopy printout.

These capabilities of cable lend themselves to uses that cannot be performed as well, or that cannot be performed at all, by existing media. Individual communities can have their own TV broadcasting facility, overcoming the limitations on local and community service that are imposed by the economics and technology of over-the-air broadcasting. Educational material, library service, and easy access to information on governmental services could all be provided by cable. Individual dependence on inflexible time schedules of over-the-air broadcasting could be lessened by frequent rebroadcasting at different times, of educational, informational, cultural, and entertainment materials, over the numerous available channels.

The ending of channel scarcity could result in the transmission of everything from first-run movies to the Metropolitan Opera. In some instances, such specialized programming could be financed through payments made by those who wish to see it—that is, through designating certain channels as pay channels—instead of through the willingness of advertisers to sponsor, which now largely governs what televised material reaches the home screen.

CONFLICT OF TECHNOLOGIES

The capabilities of cable that make it welcome to some parties have caused other parties to view it with puzzlement and even with alarm. Among these latter parties are

1. The over-the-air TV broadcasting industry.
2. The telephone companies.
3. The print media, including newspapers, magazines, and book publishers.

Cable poses potential competitive threats to all three, and each industry has given much thought to its present and future relationship to the new medium. The group that feels most immediately threatened is the over-the-air broadcasting industry. This industry's interest in the problem places it significantly at odds with the interests of the public.

From a business point of view, the over-the-air broadcaster's activity consists of delivering an audience to advertisers. Programming is a means to that end. The larger the audience that a broadcaster can deliver, the more advertisers will pay to reach that audience. The broadcaster's interest therefore lies in restricting as much as possible the public's selection of programming to watch, so that as many people as possible will be watching the broadcaster's own offering. The opening up of numerous channels and the airing of numerous competing offerings, could fragment broadcaster's present audiences and reduce revenues.

Cable TV began in the United States in the late 1940s. Before the industry was 10 years old, importation of distant signals by cable systems into towns and cities whose residents were once the captive audience of local stations had become a major controversy, and broadcasters were pounding on the doors of the federal government for protection. In 1958, 12 TV station owners who testified at Senate Commerce Committee hearings on the problems of bringing TV service to smaller communities begged Congress to prohibit cable systems from importing distant signals because it could interfere with local stations' revenues.

This cry of alarm from broadcasters has since become familiar everywhere, and it has decisively molded the fortunes of the cable industry. Interestingly, however, it has never been substantiated by any significant amount of real evidence. For example, none of the 12 stations that in 1958 asked Congress for protection has gone out of business. Despite the existence of cable systems in all of their communities, all have fared well and all have increased their advertising rates. In the cities in which two of the stations are located, additional TV stations have gone on the air. In two other cities, the owners who subjected Congress to such a woeful picture of their predicament have since sold their stations at a handsome profit.

In the larger picture, cable TV has more than quadrupled since 1958, both in number of systems and in audience size. In addition, most cable systems that offered only five channels in 1958, offer 12 or more channels now, thereby greatly increasing the alleged "competition." Nevertheless, 173 new commercial and 120 educational TV stations have gone on the air in this period, 44 of them in communities in which cable systems were already operating. In 1971 FCC Chairman Dean Burch told the Senate Communications Subcommittee that thus far the Commission possessed no convincing evidence of damage to broadcasting interests by the advent of cable systems.

CABLE AND THE FCC

The danger of serious competition nevertheless obviously exists, and broadcasters have pressed FCC for protection. The FCC has responded, issuing a series of rulings to restrict the expansion of the cable industry. The Commission has afforded special protection to broadcasters in the nation's top markets. These markets contain the largest and strongest broadcasting stations, which a number of studies have shown to be best able to withstand the projected competition of cable. But these large stations also possess the most economic and political power, and they have displayed it impressively in the FCC's cable proceedings.

In 1966 the FCC prohibited cable systems from importing distant signals into the nation's top 100 markets unless the cable system could prove to the Commission's satisfaction that such importation would not do economic harm to over-the-air TV stations in the area. Such "proof" was of course nearly impossible to adduce.

In 1968 the Commission clamped on the lid even tighter. It ruled that cable operators wishing to import distant signals into the top 100 markets would have to secure permission on a program-to-program basis from the originator of the distant signals. Even if the stations originating the signals were inclined to grant such permission—which, of course, they were not—their copyright and exclusivity agreements with the owners of the programming material would prohibit them from doing so.

Since distant signals are, at this stage in the cable industry's growth, the principal thing that cable systems have to sell, the effect of these rulings was to virtually exclude cable from the nation's largest urban and metropolitan areas. The purpose of the rulings was not disguised by the Commission. "I think you should remember the fundamental hypothesis of the Second Report and Order adopting our CATV rules in March of 1966," FCC Commissioner Kenneth A. Cox said to a National Association of Broadcasters audience in 1968. "It concluded that rules were necessary in order to keep CATV a supplemental service, rather than risk its displacing our basic over-the-air system." The FCC's goal, Commission Chairman Rosel H. Hyde told members of the House in 1969, has been "to integrate the CATV operation into the national television structure in a manner which does not undermine the television broadcast service."

THE NEW CABLE RULES

At the time that Hyde spoke, the FCC had launched a new cable rule-making proceeding. The Commission found itself under increasing pressure from civic and community groups and from city governments to permit cable systems to import at least some signals into the top 100 markets so that cable would be economically viable in cities and its benefits could be brought to these cities.

Other sources of pressure were the cable industry itself, which was getting bigger and stronger, and the electronic equipment manufacturing industry, which was beginning to realize that there might be more new business in wiring the nation's cities for cable than in updating and replacement of equipment for the over-the-air broadcasting industry, which was past its period of rapid growth in physical plant. The equipment manufacturers broke ranks with the broadcasters in an important filing made with the FCC in 1969 by the Industrial Electronics Division of the Electrical Industries Association, calling for wiring the entire nation for broadband communications.

These new FCC proceedings attracted the great interest described earlier in this paper. The rule making covered many problems and issues, but as always the problem of distant signals proved to be the hard nut to crack.

On August 5, 1971, the FCC issued its proposed new rules in the form of a 55-page letter addressed to the chairmen of the House and Senate Communications subcommittees. "Our basic objective," said the FCC, "is to get cable moving so that the public may receive its benefits and do so *without, at the same time, jeopardizing the basic structure of over-the-air television.* The fundamental question is the number of signals that cable should be permitted to carry to meet that objective. . . . We propose to act in a conservative, pragmatic fashion—*in the sense of protecting the present system* and adding to it in a significant way, taking a sound and realistic first step, and then evaluating our experience," (Emphasis supplied).

The formula that the FCC proposed divided the nation's TV audience into three groups—those living in the top 50 markets, those living in markets 51 through 100, and those living in markets below 100. The formula ran as follows:

1. In the top 50 markets, cable would be allowed to carry three network-affiliated stations and three independents.

2. In markets 51 through 100, the quota would be three network-affiliated stations and two independents.

3. In markets below the top 100, carriage would be limited to three network-affiliated stations and one independent.

4. Cable systems in the top 100 markets would be permitted to carry two distant signals. If importation was required to fill the system's basic quota, this would be subtracted from the two that they could carry.

5. In addition, cable systems could carry any number of distant educational TV station signals, provided the ETV stations in question did not object.

THE EXCLUSIVITY PROBLEM

A big joker in this deck was the issue of exclusivity. At present FCC permits networks and broadcasting stations to buy exclusive rights to films and other programming and to hold those exclusive rights for long periods of time, preventing others from using or showing the material. Thus, a network or broadcasting station may buy the right to show a film once or twice within a two- or three-year period, and to prevent any other party within the network's or station's area of coverage from showing the film during the entire period. Some exclusivity contracts run for one year, some for two years, many for seven years, and some for indefinite periods.

Exclusivity contracts are used by networks and major broadcasting stations to control markets and limit competition. Potential competitors, such as independent UHF stations and cable operators, cannot afford to bid competitively for the purchase of exclusive rights at the price levels that the giants have established, nor can they get the programming once the giants have swallowed it up.

FCC has been interested in the problem for some time. In 1971 it opened a rule-making docket on "the exclusivity practices of broadcast stations in terms of both time and geography, and the impact of these practices on the ability of UHF broadcasters and cable operators to obtain programming." As with so many issues that have an important bearing on the coming communications revolution, public knowledge of, and interest in this issue is nil. Most of the filings in the FCC's docket have been by parties with an economic stake in its outcome, and the FCC has little constituency for any move that it might wish to make that would offend them.

To understand how the exclusivity problem relates to the distant signal controversy, let us assume that a New York TV station has bought exclusive rights in the New York area, for two years, to a movie—let us say, *The Dawn Patrol*. Let us also assume that a station in Philadelphia has bought exclusive rights to *The Dawn Patrol* in Philadelphia. Now, obviously, a cable system in New York could import the Philadelphia signal into New York when the Philadelphia station was showing *The Dawn Patrol*, thereby undermining the exclusivity for which the New York station had paid. This could lessen the value to broadcasters of buying exclusive rights. And for the owners of films like *The Dawn Patrol*, it could lower the amounts that broadcasters would be willing to pay for "exclusive" rights to material that had become less than exclusive because of signal importation.

For broadcasters there is a further issue. Loss of the power to buy and enforce exclusivity would mean diminished control over markets and diminished ability to dampen competition. If broadcasters can limit cable's access to appealing programming, they can limit cable's popularity and thereby its growth. Conversely, inability to control cable's growth will mean that broadcasters will soon have to reckon with a competitor big enough and strong enough to enter the program-buying game on equal terms with the broadcasters. Genuine competition, the eventuality that every free enterpriser dreads most, will come to pass.

EXCLUSIVITY AND THE NEW RULES

In its proposed rules that it issued on August 5, 1971, the FCC tried to duck the exclusivity issue. The Commission already has a rule prohibiting cable systems from importing and showing on the same day network programming being shown by local TV stations. The new rules continued this requirement in effect, but took no steps to protect nonnetwork programming to which stations had bought exclusive rights from importation of the same material on distant signals. The Commission noted that it was involved in a rule-making proceeding on the exclusivity problem, and perhaps hoped that this would calm any ruffled feathers.

But it was not to be. When the proposed rules were issued, the broadcasters stated as usual that they would be destroyed by the audience fragmentation that the importation of distant signals by cable companies into the top markets would create. But on the exclusivity issue, they did more than complain. They went straight to the White House and to its Office of Telecommunications Policy.

A series of closed-door conferences ensued. Persons present were FCC Chairman Dean Burch, officials of the Office of Telecommunications Policy, and representatives of the National Association of Broadcasters, the National Cable Television Association, and the motion picture industry.

In November the closed doors opened and the parties emerged with the announcement that they had come to an agreement. According to this agreement, broadcasters in the top 100 markets would be allowed to control nonnetwork programming that had been sold in more than one market and would be able to prohibit cable systems from importing such material as a distant signal. In the top 50 markets the broadcasters' exclusivity protection would run for 1 year on programming that had never been shown on TV before, and for "the run of the contract"—that is, as long as the broadcaster wished to pay for exclusivity—on other materials. In markets 51 to 100, broadcasters would also be granted exclusivity, but a complex set of arrangements would allow for greater accessibility of programs and shorter terms of exclusivity. In markets below 100, exclusivity could not be enforced.

Another section of the rules further restricting signal importation is called the leapfrogging provision. Logically, a cable system in a given city that is looking for distant programming that would be attractive to its subscribers would import a signal from a larger city, perhaps a major independent station carrying a substantial number of offerings not available locally. Thus, for example, a cable system in Washington, D.C., might logically want to import a distant signal from an independent station in New York. But the rules say no. The system cannot "leapfrog" intermediate markets such as Pittsburgh or Philadelphia, but must fill its distant signal quotas with stations in these markets. Furthermore, it must give first priority to UHF stations in these markets whose programming is almost invariably weaker than that of the VHF outlets. Such distant signals will be of almost no interest to residents of Washington, and of little or no value in attracting subscribers.*

At the conclusion of the private conferences, the agreements reached by the interested parties were incorporated into the rules, which were then rewritten and promulgated on February 3, 1972. They went into effect on March 31, 1972. It is generally agreed that the restrictive provisions of the rules substantially nullify the signal importation advantages in the nation's top 50 markets.

---

* A provision in the labyrinthine rules states that if a distant signal must be blacked out because of exclusivity, the cable system can then import a signal from anywhere in the country while the regularly imported signal is blacked out. In practice, the complexity of finding a nonexclusive imported signal to fill the exact time slot and the high microwaving cost that would be required to bring it in from any really great distance make this provision of little or no value to cable operators.

According to reliable sources, a study commissioned by the Office of Telecommunications Policy, but not made public, indicates that even under the most optimistic assumptions, no more than 11 of the top 50 markets will be economically viable for cable under the rules.

The motivation of the cable industry in accepting the compromise agreement appears to be twofold. First, markets 51 through 100, where the exclusivity restrictions are less stringent, are less costly to wire, and are for the most part easier markets in which to sell conventional cable TV, because many or most of them have inferior over-the-air service. Second, it was made abundantly clear to cable interests by the FCC and OTP that the compromise represented the most that they would be able to get in view of the pressures of the broadcasters and the program owners. Faced with the choice of the compromise or nothing, the cable industry accepted the compromise, hoping for a better break in the future.

CABLE, THE GOVERNMENT, AND THE PUBLIC INTEREST

What is one to say of these events, and of the regulatory philosophy that underlies them?

One can say that no legitimate public interest is served by limiting or barring the entry of cable into the nation's largest cities and that, in fact, the public interest is harmed. A federal policy directed to achieving the public interest in this field would not expend its energies and ingenuity on limiting the growth of cable in order to protect the economic interests of broadcasters and program owners. Instead, it would foster the competition, the increased variety, and the new public and social services that cable can bring, and would direct its attention to the complex problems of making sure that maximum public benefits accrue from a new medium that has such great and important potential.

None of the Commissioners, in the statements that they issued when the rules were released, maintained that the rules reflected an assertion of the public's interest in cable. Dean Burch said that he believed that the rules were "consistent with the public interest," but did not aver that public interest criteria had shaped them. Others appeared to see no function for the Commission beyond the discovery of formulas to which competing interests will agree and the promulgation of these formulas as rules. "While I do not find myself in complete accord with each and every item set forth in the new Rules," said Commissioner Charlotte Reid, "the fact that these rules reflect the consensus agreement reached by the principal parties (cable television system owners, broadcasters and copyright owners) are [sic] far better than no rules at all. It, therefore, seems clearly in the public interest to give implementation to the compromise agreement and for that reason, I concur with the results of the Commission's action."

What remains somewhat mysterious is why the Commission feels so powerless to base its rule making not on what the economically interested parties would like, but on what will promote the broadest interest of the public, and to promulgate rules whose direct purpose and intention is to achieve and enforce public interest goals. Of what, after all, is the Commission really afraid? And if the Commission will not do it, then who will?

If our present institutions are in fact incapable of implementing any concept of the public interest other than the accommodation of economic interests, then the question of devising new and better institutional arrangements that are capable of giving better expression to a larger concept of public interest is a matter of national priority that should not be delayed. The new communications technologies are coming fast, and time is short.

REFERENCES

Federal Register, 37, 30, Part II, (Saturday, February 12, 1972), pp. 3252–3341. This issue contains the FCC's new cable television rules, discussed in this chapter.

Price, Monroe and John Wicklein. *Cable Television: A Guide for Citizen Action*. Philadelphia: Pilgrim Press, 1972. 160 pp.

Sloan Commission on Cable Communications, *On the Cable: The Television of Abundance*. New York: McGraw-Hill, 1971. 256 pp.

Smith, Ralph Lee. *The Wired Nation*. New York: Harper & Row, 1972.

Tate, Charles (ed.). *Cable Television in the Cities: Community Control, Public Access, and Minority Ownership*. Washington: Urban Institute, 1971. 184 pp.

CHAPTER 10

Institutional Pressures and Response at the FCC: Cable and the Fairness Doctrine as a Case Study

NICHOLAS JOHNSON

During my association with the Federal Communications Commission a large part of the major controversial activities affecting broadcasting can be thought of as falling into four general categories:

1. A reanalysis and rethinking of the patterns of ownership and economic organization of the broadcast industry—multiple and cross-media ownership patterns; conglomerate ownership; the function of competition; program production and network relationships.

2. The accommodation of newly represented interests in the license renewal and approval of sale of stations process—minimum standards of performance; participation by consumer interests; service to minorities; enforcement of equal employment and nondiscrimination; evaluation of new entrants.

3. The integration of new technologies into the broadcasting system (often under the inconsistent encouragement of innovation and protection from competitive injury)—cable television; domestic satellites; individualized broadcast production and reception systems (cassettes); and some UHF television station improvement in commercial and public broadcasting.

4. The arbitration of claims to access of broadcast facilities—fairness in the coverage of controversial issues; reply time for those personally attacked; enforcement of "equal time" in political campaigns; emerging theories of First Amendment rights of access in broadcasting; the application of fairness to

commercial product messages; allocation of broadcast time between the executive and legislative branches; allocation of time between the political party in power and opposition parties during noncampaign periods.

This leaves out other activities such as the Commission's affirmative decision not to regulate the presentation of deceptive advertising, and the Commission's inquiry on children's television programming. But these four categories include major Commission concerns.

The institutional process in each of these major activity areas suggests a number of broad conclusions.

First, the broadcast industry as a whole is in retreat wherever it is touched by Commission activity. The industry seems to be fighting a four-front war just in these areas—not to mention its difficulties in the use of campaign funds for political broadcasting, reallocation of broadcast spectrum for other uses, and increases in rates for broadcast network interconnection. In almost every situation the position of the broadcast industry is not an effort to gain better regulatory treatment, but a defensive struggle to maintain as much of the status quo as possible. In every area the industry's sand foundation has been washed away, and further erosion is a very strong possibility. It attacked the fairness doctrine as unconstitutional, and lost (Cases, *Red Lion*). It is now supporting the doctrine as effective in its present form and opposing its expansion (Comments, FCC Dkt. No. 18260). Proposals for longer license terms, immunity from renewal challenge, or relaxing of present FCC ownership rules have all the characteristics of unrealistic anachronisms totally removed from the likely future course of events.

Second, much of the impetus for change comes from agencies outside the FCC. The courts have played a central role, confirming and strengthening the FCC position when it has tried to reform and change, and reversing the Commission sharply in numerous cases where the Commission has resisted public interest arguments (Cases, *Citizens, United Church*, 1972; *United Church*, 1966; *United Church*, 1969). The executive branch has often played a crucial role. The Department of Justice Antitrust Division strongly supported diversification in ownership patterns, and has acted to undo the Commission's actions on several occasions. The Division's support on prime time access was important, and it has filed a recent separate suit against the networks. Equal employment rules received a boost from the Department when they were in a crucial stage, as well as from the Equal Employment Opportunity Commission. The Federal Trade Commission, to which the FCC has graciously deferred on matters of advertising regulation, made specific proposals to the FCC on "counteradvertising." Of increasing importance is the role being assumed by the White House Office of Telecommunications Policy. The interests of this office seem to encompass the full range of broadcast regulation—its dominant role in the cable compromise is a case in point (FCC

Cable Report). The result of these increased outside pressures is that the FCC —swamped by work, burdened with a primitive management system, and conditioned to a role of reaction—frantically struggles to sort out the arguments and proposals before it. There is only a limited inclination or ability within the agency to generate policy analyses and proposals on its own. And whatever path it chooses, the agency knows it will be buffeted from one side and then another.

Third, significant stimulation has come from outside private parties representing groups of consumers or others presenting "public interest" arguments. The importance of this participation can be seen in its effects. It presents the Commission with an alternative set of backup data and analyses to compete with the constant stream of special interest representation to which any regulatory body is subjected. It serves as an antidote for the often one-sided and self-serving arguments made by special interest groups, and decisions made by the agency. Public participation insures constraints on the discretion of the regulatory body. A decision may be reversed by court appeal. Even the threat of court appeal may set constraints on agency behavior. Special interest groups may feel constrained by the potentiality of adverse public opinion generated by outside groups' public complaints, outrage, or ridicule. The agency may also act to avoid adverse public reaction. It is difficult to measure the impact over the past 5 years of increased nonindustry participation before the FCC, but it is clear that the impact has been substantial.

Finally, the technological possibilities in broadcasting—cable, enhanced network interconnection alternatives, and personalized production/reception facilities—have profound implications for each of the Commission's regulatory thrusts. Each of these technological alternatives offer increased consumer choice and market diversification. Not only is the supply of consumer alternatives increased, but the power of present institutions could be eroded, both in terms of oligopoly control of markets and the power to control programming.

Without trying to outline the types of responses that different institutional groups make to technological innovation in broadcasting, certain general descriptions can be made. Underlying much of the response of institutions is a facade of rational intellectual inquiry. By this I mean that the arguments made by interested institutional groups attempt to persuade the Commission and other decision makers on the basis of rationality—making claims that reflect accepted tenets of equity (this is what is fair), public interest (the nation will be best off if the decision is thus and so), or traditional free, competitive, private enterprise (the efficient "competitive" solution is this). In reality, the intellectual quality of the arguments is generally subsumed in the overpowering effect of political power relationships. The FCC becomes an arbitrator and consensus-finder of those power relationships. And when the FCC begins to act in ways that are inconsistent with them, the issues are appealed—to the

White House or the Congress—so that the alignment of power relationships and decision effects can be made more congruent.

The bargaining responses of various institutional groups are not difficult to predict on the question of technological innovation. Existing suppliers, fearing increased competition, argue for total prohibition or substantial inhibition on the technological threat using a combination of accepted tenets of rationality. Sometimes the vigor of their opposition is blunted by their hedge positions; for example, a number of broadcasters who argue that they may be harmed by cable simultaneously have acquired substantial cable ownership interests. New entrants attempt to achieve the most favorable entry conditions and appeal to the accepted tenets of rationality in doing so.

Public participants presenting nonindustry alternatives have a number of goals in their responses. They are primarily concerned with molding new technological developments to achieve what they perceive as appropriate social outcomes. Some public participants may speak for neglected minority or ethnic interests. Others seek to enhance a particular value, such as First Amendment freedoms. A third group may assert their evaluation of a combination of social goals, including "economic efficiency."

The FCC's reaction to these types of pressures and participation is a combination of subreactions. The agency tries, with limited resources and lack of experience, to formulate its own independent evaluation of what is equitable, or in the public interest, or most nearly an approximation of an "efficient competitive solution." But usually this initial effort at "rational decisionmaking" is quickly overwhelmed. By habit, necessity, and inclination the agency becomes the adjudicator of the special interests before it—seeking the path of consensus and least resistance. Its decision-making biases are conservative (known states of reality with their attendant flaws and evils are preferred to unknown states) and fragmented (issues are considered so as to achieve accommodation and consensus instead of taking account of their interrelationships).

These general conclusions and observations could be demonstrated with numerous case studies of specific issues and controversies at the FCC. Cable is a particularly inviting and timely example, but I will leave it for another time. Equally as useful is the present controversy surrounding the Commission's implementation of the Fairness Doctrine.

Without overcomplicating the study or unduly extending the length of this paper, a brief examination of the present posture of the Fairness Doctrine is useful.

Under the Fairness Doctrine (and related doctrines) a Commission licensee must

1. Air controversial issues of public importance to its community of license.
2. Provide for the presentation of opposing viewpoints when a controversial issue of public importance is discussed.

3. Provide reply time to persons or groups personally attacked on the station.

4. Provide reply time to editorials supporting political candidates.

5. Evaluate responsibly whether it has incurred obligations under the Fairness Doctrine when requests for time are initially presented to the station (following which there may be appeals to the FCC). [FCC Fairness Primer, FCC Personal Attack Rules]

This is all fairly straightforward, although sometimes the Fairness Doctrine (which is primarily directed toward issues) is confused with the "equal time" provisions that apply to election campaigns. Many people believe there is a right of "equal time" or personal access under the Fairness Doctrine, when, in general, there is not.

The rationale of the Fairness Doctrine has been expressed in several ways.

1. Fairness is required because access to broadcasting is limited by the limits on availability of spectrum.

2. Fairness is required because of the unique power of broadcasting in a democratic society.

3. Fairness is required because the government chooses the operators of the system and is therefore implicated in any exclusions from that system.

4. Because of the tendency to seek mass audiences, the medium will inherently exclude minority or unpopular views unless there are fairness requirements. [Westen/Johnson 1971]

The Doctrine originally grew up as an enunciation of public interest standards for the operation of broadcast stations. Its growth, limits, and operating principles were developed almost entirely in case-by-case decisions by the Commission and the courts.

The fullest Commission explanation was given in the 1949 Report on Editorializing, the last time a thorough review of the issue was undertaken. In 1964 the Commission summarized the existing law in a Primer, hoping to make the subject area clearer and easier to enforce, and in 1967 issued rules on personal attacks and editorializing in campaigns (FCC Fairness Primer, FCC Personal Attack Rules). The Doctrine's constitutionality was validated in the only Supreme Court case to consider it—the 1969 decision in *Red Lion* (Cases, *Red Lion*). Accelerating developments in the Commission and courts since *Red Lion* have now led the Commission to undertake a formal proceeding reviewing all Fairness Doctrine issues (FCC Dkt. No. 18260). The renewed interest has also spawned a spiraling number of speeches, studies, law review articles, and Congressional hearings—now including this one as well.

The development of the Fairness Doctrine proceeds almost entirely as a result of public participation. A person or group may seek access or relief from the station, and, failing to receive satisfaction, then appeals to the FCC. It is also possible for a licensee to seek a declaratory ruling. FCC decisions may then be appealed to court. One need only examine the list of plaintiffs in recent court cases to see the importance of public activity—David Green, Alan Neckritz, Friends of the Earth, John Banzhaf, Business Executives' Move for Vietnam Peace, Retail Store Employees Union, the Republican and Democratic National Committees, United Church of Christ (Cases: *Green; Neckritz; Friends; Banzhaf; BEM; Retail Store; DNC; CBS; Healey*).

Extension and modification of the Doctrine begins with a public complaint. It is up to public groups that feel their rights have been violated to seek redress —the Commission does not go looking for Fairness Doctrine violations. With the new technology of cable, the FCC has imposed fairness requirements on cable operators when they present their own programs (FCC Cable Fairness Rules). But in the Commission's most recent decision on cable, the Commission concluded that fairness requirements were not required for those channels accorded quasicommon carrier status when the cable operator does not control programming, and makes access available on a nondiscriminatory basis (FCC Cable Fairness Rules). This is an important modification in response to a changed technology.

The position of the broadcast industry has changed markedly on issues concerning the Fairness Doctrine. I do not mean to suggest that a single "position" can be identified, but certain general trends can be stated. Up to 3 years ago, the main effort was to attack the constitutionality and general wisdom of the Doctrine. That was before *Red Lion* (Cases, *Red Lion*), sometimes called the Waterloo of the industry. Now, with the Doctrine and the Commission's implementation of it under attack by public groups, broadcasters praise its past operation and suggest that, in fact, no change is needed except to limit the effect of court cases where the Commission has been reversed. Broadcasters simply want to enhance and confirm their range of discretion under the Doctrine. This shift in institutional position on the part of the industry occurred because of the loss of its constitutional argument and the public attacks on the status quo in specific cases and general policy proceedings. As a result, there is now a proceeding questioning the basic philosophical and operating premises of the Doctrine.

Along the way a new ingredient has been added. In the past, the role of the executive branch has been limited. When the Commission goes to court on a Fairness Doctrine matter its briefs in the courts of appeal must be reviewed and approved by the Antitrust Division, Department of Justice. In the process of preparing briefs there is opportunity for a discussion of the legal and policy grounds for a particular decision, and there has been the possibility of negotiation between the Commission and the Division. Matters going to the Supreme Court go to the Solicitor General's office, with the same opportunity. In fact, on the *Red Lion* case there were varying theories about how the arguments should be made. And there is always the possibility that a Commission decision will be opposed by the Department of Justice (as in *United States* v. *The Federal Communications Commission*—the ITT-ABC merger approval), which substantially undercuts the Commission's position.

Recently the degree of executive branch intervention in Commission matters has increased. First, the President has himself become a substantial issue under the Fairness Doctrine, particularly in his use of broadcast facilities for the presentation of administration views on controversial issues. Opposition parties and spokesmen, competing with the President and the incumbent party, have attempted to assert claims for broadcast time under the Fairness Doctrine—as have members of the legislative branch competing with the executive branch. Second, the Office of Telecommunications Policy (OTP) in the executive office of the President has intervened in an increasing number of issues that were once the sole preserve of the FCC acting as an "arm of Congress." On the Fairness Doctrine, OTP has proposed that the doctrine be replaced with a system of paid access based apparently on a quasicommon carrier status for broadcast stations (Whitehead, 1971). Depending on the issue and the environment, the rise of the Office of Telecommunications Policy as a force in communications policy decision making at the FCC is somewhat analogous to the rise of nonindustry private groups and their participation in FCC matters.

The Commission presently has a comprehensive review of the Fairness Doctrine underway. The issues in the proceeding go to the heart of the Commission's present implementation of the doctrine. Some of the major questions being considered include

1. How do you tell when an issue is "controversial"?
2. When is a "controversial" issue one of "public importance"?
3. What constitutes a "personal attack"?
4. Since the Commission relies on the public for complaints, what should the public have to show before it can get Commission action in support of its claims?
5. What should be the operational mechanics of Commission consideration of Fairness Doctrine matters? Should broadcast licensees be examined only at the triennial renewal time, rather than in response to individual complaints?
6. What should be the limits of licensee discretion to decide when fairness claims are valid and by what criteria should the Commission decide that licensee discretion has been abused?
7. How and to what extent should "minority viewpoints" be accommodated and given access for expression?
8. What would be the financial impact on broadcasters of alternative policy conclusions?

9. How do product commercials and other paid speech interrelate with the Fairness Doctrine?

10. What should be the relationship of the Fairness Doctrine to "equal time" in political campaigns? How should ballot issues be treated where campaign time is purchased? If a station only offers to sell time for political campaigns, and does so on an equal basis, but only one candidate can "afford" to buy time, has the station discharged its obligation to be "fair" on the controversial issue of which candidate should be elected to political office?

11. What are the limits of the Commission's discretion under "the law" as enunciated in the basic Commission statute and subsequent court decisions?

With the issues specified here, along with the others mentioned by the Commission and by parties commenting in the proceeding, it is difficult to imagine any facet of the doctrine that will not receive attention.

It is important to note the impact of the Fairness Doctrine on certain fundamental legal principles. There is a tension within the concepts of freedom of press and freedom of speech. The broadcaster, as journalist, believes he has a right to decide what he wants to broadcast totally without interference by government. A newspaper need not now—legally—provide opportunity for the presentation of opposing views on a controversial issue of public importance nor equal opportunities to political candidates (Cases, *Chicago*). But freedom of speech suggests the right of the individual to hear viewpoints of his choice, and to present his own views in suitable forums—such as the "soapboxes" of the mass media. Access under fairness and equal time theories tends to enhance the exercise of this freedom by individuals. There is another interesting tension developing. In the past, governmental controls on commercial speech—the regulation of advertising, for example—have been premised on the belief that this speech occupies an inferior position to other speech for First Amendment purposes. However, as commercial speech, for example, product commercials, becomes the subject of fairness complaints because it contains controversial viewpoints on issues of public importance, the rationale for leaving commercial speech unprotected from governmental interference is undercut.

Within the framework of the present FCC proceeding on fairness, and given the present "sensitized" state of the various institutions participating in that consideration, it is difficult to say that the FCC proceeding lacks comprehensiveness in its deliberations. But it is a particular type of institutional "comprehensiveness"—it addresses all the issues anyone can think of within a context that holds all other variables in broadcast regulation constant. Thus, while a few participants in the fairness proceeding have mentioned cable television as a way to meet some of the policy goals of the Fairness Doctrine, it is likely the proceeding will be conducted midst the lore of past decisions, court cases, and present problems. The relationship of the Fairness Doctrine to other Commission activities and possible policy alternatives probably will not occupy much of our time.

As an example, consider this alternative policy approach, and particularly notice its potential impact on the Fairness Doctrine as well as other FCC broadcasting issues.

1. Cable television of advanced technological design (40 channels) to be installed in every home that has a telephone.

2. Cable service to be free, or of nominal cost, to the subscriber. Operators would receive most of their revenue from leasees using the system to distribute programs.

3. Homeowners would have "slave" television sets for cable of much lower cost than present sets since only "reception" capability from the cable would be required.

4. Cable systems across the country would be interconnected.

5. The system would be designed to provide full shielding against spectrum interference.

6. All over-the-air television stations would receive a free, dedicated use of two cable channels on which to program—a programming right that could be transferred or sold—and the stations would cease broadcasting over the air. The free right to channels would be a form of compensation to offset in part the effects of technological change brought by cable.

7. All over-the-air spectrum presently used for television would be reallocated for land mobile and other uses, for fees payable to the general treasury, or used to carry out internal subsidies in the cable system (cable in rural areas.)

8. All other cable channel space would be available for lease on a common carrier basis for any use—except that certain "public policy" allocations to free public access, educational, and local government use could be made.

It is safe to say that this approach will never seriously be considered in the present institutional framework of broadcasting. It would be silly to attempt to assess it; there is no possibility of its adoption as national policy, no matter how favorable its evaluation might be. The implications of such a policy alternative are simply too devastating for existing institutions and interest groups. And existing policy-making organizations appear unable to examine the full range of policy alternatives, nor to relate policy choices in one subarena to policy implications in another subarena. The potential relationships between Fairness Doctrine issues and innovation through cable broadband mass communications will never be fully explored.

There are alternative institutional frameworks within which such a proposal might be considered. In a totally planned economy the government might be able to arrange the institutional realignments and manage the bargaining process so that all who would be affected by the change would be compensated with some degree of certainty. I do not suggest that planned economies necessarily handle technological innovation more successfully, but it is at least conceivable that the full range of policy alternatives and implications would be considered in the context of a government-directed economy. And the problems of accommodating those affected by the change in terms of personnel changes, resource shifts, and so on, might be handled internally. Where the impetus for change would come from is another question. A purely competitive model, with no governmental involvement whatsoever, might also "consider" this proposal in the context of decentralized and individualized decision making. If a different technological system could economically replace an existing one, it would. Even an institutional environment with no FCC regulatory limitations on cable, no copyright liability, and a reasonably free policy of granting building rights to local cable systems might have resulted in massive cable building—with the only limitation being the marginal cost/marginal revenue relationships for relatively unattractive potential subscribers. Unserved areas might then be filled in by mechanisms such as the rural electrification and telephone programs that the federal government used to provide those two fundamental services. But a "mixed economy," in which government and special interest groups work out accommodations, will not lead to consideration of technological changes that make anything more than gradual changes in the institutional framework.

At the beginning of this chapter, I outlined four major FCC policy thrusts in broadcast regulation. The Fairness Doctrine review I have described is a part of one of these—the FCC's arbitration of claims for access to broadcast facilities. This proceeding is viewed as a comprehensive proceeding—looking at all the issues. But in posing a hypothetical policy alternative—a national cable system—that apparently can never be considered in the present institutional framework, I am suggesting that comprehensiveness for the FCC is defined in terms of rather narrowly drawn boxes of specific issues drawn on the basis of historical developments.

Let me now turn to the effects the national cable system hypothetical alternative might have on the issues in the Fairness Doctrine proceeding, and indeed on many of the Commission's major broadcast regulation concerns. One effect of a national cable system would be to undercut substantially the need for any FCC activity whatsoever in many of these areas. For example, the arbitration of access would be unnecessary with a new medium open to all, including those who cannot pay. "Fairness" could be achieved by any group seeking and obtaining access—and access will be available with the elimination of channel scarcity. There would be access to respond to personal attacks or to present counterinformation on commercial products. Political broadcasts would be no different than any other access question. The ability of the President of the United States to monopolize television would be lessened—he would not be able to control every cable channel.

There would be little need for ownership restrictions outside of traditional antitrust principles—for example, owners of other media could not own cable systems and there would be limits on multiple ownership. The present oligopoly situation in television networking would be broken. A common carrier access policy with the elimination of the scarcity of channels would go a long way to transforming broadcast/cable communications into a market more approaching the paperback book, recording, or magazine publishing industries in terms of ease of access, entry, and exercise of market power.

Concern over license renewal and sale of stations would be gone, since many of the groups seeking stations or seeking to improve performance would have easy access to channels. Monopoly profits in broadcasting—a major incentive in the acquisition of stations—would be eroded by the new competitive environment. Concern by government for "performance" of stations would end with the end of government licenses; performance would be regulated by consumer choice and direct government subsidy, for example, public broadcasting. The present controversial and relatively ineffective FCC efforts to induce broadcast licensees to finance "good" programming, news and public affairs, for example, from monopoly profits would end.

The concern over accommodation of new technologies would also end since government would no longer have an external public or bureaucratic interest to protect—cable would compete on its own against alternative technologies such as cassette/video recorder systems, information utilities, or any other new technology that might develop. There would be no public benefit externalities other than the end result of competition to justify restrictions on the competitive impact of new technologies.

In short, the proposal outlined in this chapter could produce a total "deregulation" of the television industry. This would not come about because any of the present policy goals had been abandoned. This is not to say that a national cable system would not need to be structured to yield the policy outcomes we seek today, only that such a system could be structured to meet those policy goals without detailed and continuing regulation.

The ability of institutions to respond creatively is under question from every group in our society. The web of broadcasting and governmental relationships is no exception. Technological innovation has the capacity of making broadcasting more responsive. But the limitations on the institutional framework—its ability to respond and incorporate technological innovation—severely limit not only the uses of new technology, but even the process of critical assessment of its potential.

It is easy to look at the incredible record of the FCC and conclude that the problem must be that the Commissioners are either stupid or on the take or both. Indeed, my colleagues often react to my criticism of FCC process and decisions as if I had made such charges. In fact, I never have, and I have often gone out of my way to make that clear. The reasons why the FCC does not work are much more complicated than that. The consequences, however, are no less serious. I hope this chapter may have contributed a little more thorough understanding of the problem.

REFERENCES

CASES

Office of Communication of United Church of Christ v. *FCC,* 359 F. 2d 994 (D.C. Cir. 1966).

Banzhaf v. *FCC*, 405 F.2d 1082 (D.C. Cir. 1968).

Red Lion Broadcasting Co. v. *FCC*, 394 U.S. 367 (1969).

Office of Communication of United Church v. *FCC*, 425 F.2d 543 (D.C. Cir. 1969).

Retail Store Employees Union, Local 880, R.C.I.A. v. *FCC*, 436 F.2d 248 (D.C. Cir. 1970).

Citizens Communication Center v. *FCC*, 447 F.2d 1201 (D.C. Cir. 1971).

Green v. *FCC*, 447 F.2d 323 (D.C. Cir. 1971).

Neckritz v. *FCC*, 446 F.2d 501 (9 Cir. 1971).

Business Executives' Move for Vietnam Peace v. *FCC*, 450 F.2d 642 (D.C. Cir. 1971) appeal pending in the Supreme Court.

Friends of the Earth v. *FCC*, 449 F.2d 1164 (D.C. Cir. 1971).

Columbia Broadcasting System, Inc. v. *FCC*, 454 F.2d 1018 (D.C. Cir. 1971).

Healey v. *FCC* (D.C. Cir. 1972).

Office of Communication of the United Church of Christ v. *FCC* (D.C. Cir. 1972).

Democratic National Committee v. *FCC* (D.C. Cir. 1972).

Chicago Joint Board Amalgamated Clothing Workers of America v. *Chicago Tribune Co.*, 435 F.2d 470 (7 Cir. 1970).

FCC MATERIALS

FCC Personal Attack Rules—47 Code of Federal Regulations §§ 73.123, 73.300, 73.598, 73.679 (1971).

FCC Cable Fairness Rules—47 Code of Federal Regulations § 74.1115 (1971).

FCC 1972 Cable Opinions—   Federal Communications Commission Reports 2d series 24 Pike & Fischer Radio Regulation, 2d series, 1501 (1972).

FCC Fairness Doctrine Review—Study of Fairness Doctrine, 33 Federal Communications Commission Reports 2d series 554 (1972) and materials filed in that docket at the FCC (Dkt. No. 19260).

FCC Fairness Doctrine Primer, 29, *Federal Register, 10,* 415 (1964).

Editorialization by Broadcast Licensees, 13 Federal Communications Commission Reports 1246 (1949).

OTHER SOURCES

Johnson, N. and T. Westen. A Twentieth-Century Soapbox, *Virginia Law Review, 57,* 574, 1971.

Whitehead, Clay T. *Remarks at the International Radio and Television Society Newsmaker Lucheon,* October 6, 1971. But see Whitehead, *Address Before Sigma Delta Chi, Indianapolis Chapter,* Indianapolis, Indiana, December 18, 1972 for an apparent reversal of Administration position.

CHAPTER 11

An Assessment of the Computer-Telecommunications Complex in Europe, Japan, and North America

DIETER KIMBEL

During the 1960s the formerly disparate technologies of computers and telecommunications merged to create a new class of computer-telecommunication systems. Although these systems are well developed, they are not yet defined with respect to possible applications. Such systems use telecommunication links such as the telex-network, telegraph networks, public-switched-telephone-networks, networks using wideband circuits, and a varity of terminal equipment and time-sharing techniques, which bring the capabilities of computers and the information in data banks to millions of locations. The system(s) overhead costs are shared between many users who are widely dispersed. Ideally, such a utility would provide each user with a private computer capability as powerful as the current technology permitted whenever he needed it, but at a small fraction of the cost of an individually owned system (Parkhill, 1966).

The application of such systems extends far beyond the field of computation. In addition to making computer power available in a convenient economical form, telecommunication-based computer systems can serve any function related to the collection, processing, storage, manipulation, or distribution of information. Such systems cut across all sectors of society both as a productive resource in the economic infrastructure and as an incentive to further industrial economic growth. Many authorities predict that the computer-tele-communication industries will be key industries in advanced economies by 1980.

The widespread distribution and utilization of computer and information power will result in social changes and opportunities for human development, promising to make the next few decades among the most critical that man has ever faced. As a result, governments are now confronted with many funda-mental problems of public policy of vital importance to the future of their countries and the ways of life and their citizens.

Among these issues are the implications held by computer-telecommuni-cation technologies for the computer industry, the telecommunication industry, and the general public, the kinds of systems and services that should be built, and the institutions and policies needed to encourage and guide their fast and furious growth. Developments in computer-telecommunications systems are likely to confront policy makers with the problem of achieving a balanced approach toward certain targets; that is, quantitative and qualitative economic growth, the infrastructure of various services and activities, and social implica-tions of the introduction of such technologies.

With respect to economic growth potential of computer-telecommunication systems, American "experts" estimate that by 1980 about $260 billion will be invested, $160 billion for computer systems, and about $100 billion for the expansion and improvement of telecommunication facilities (*Business Week*, December 6, 1969 and November 6, 1971).

A British market analysis evaluated that the total expenditure of computing in the United Kingdom will approach 4% of the GNP by 1980 (Hoskyns Group Limited, 1969). Similar results of studies are reported from Canada and France. In Canada it is estimated that the total value of the industry—including telecommunications—will account for possibly 5% of the Canadian GNP in 1979 (Science Council of Canada, 1971).

In France, the computer industry is expected to overtake the automobile industry in dollar volume by 1976 (Le Marché de l'Informatique, 1969). The total net investment in computer-telecommunication systems in Japan will count for 6% of the GNP in 1985.

The benefits of the application of such systems are more debatable. How-ever, there are good reasons to believe that computer-telecommunication sys-tems will lead to the "second industrial revolution," particularly in the service sector. Through the integrated use of such systems, tremendous savings and productivity increases could be achieved in this sector. In governmental plan-ning and operations, it is "guesstimated" that such systems could cut the annual budget by about 10% without reducing the quality or quantity of serv-ices now provided. This could mean either a tax-free year every 10 years or a freeing of resources for urgent socially relevant projects that governments cannot tackle at present.

The possible and economically feasible developments in computer utilities cannot be assessed directly in terms of dollars saved alone. Much higher values are at stake.

Universal access to knowledge, where the total accumulated knowledge and experience of all mankind is continuously distilled and made available in any desired concentration to everyone through personal information terminals, becomes a possibility.

Individualized computer-assisted instruction, making education independent of age and geographical location and transforming it into a continuing process in which anyone who has access to an information terminal is provided with the equivalent of a private tutor precisely tailored to his personal needs, can be realized.

As the computer is fully integrated into every aspect of our lives, computer power will become as abundant and as effortless to employ as electric power is today.

An automated economy in which the myriad industrial, distribution, and business functions become, in effect, a single distribution machine controlled and linked together by the national computer networks may result.

Finally and perhaps of greatest long-term significance in a "communications affluent" society where techniques of television, computer graphics, computerized data bases, data processing, and normal telecommunications are combined and their services made universally available through "fireside" terminals, many of the pressures for urbanization may be reversed. If people can access and manipulate any piece of information without leaving their homes and simultaneously interact with other people and machines as easily as if they were sitting in the same room with them, then there would seem to be little reason for concentrating workers in large office buildings. They might better conduct their routine business activities from the comfort of their homes and gather together only for formal affairs, laboratory work, and social occasions. The total effect might be one of decentralization.

The list of possibilities could be continued indefinitely and include the cashless society, true participatory democracy, automatic publishing, and other innovations just over the horizon. The precise details of the list, however, are unimportant. The promise of computer-telecommunications systems lies not in any particular function or service, but instead in the possibility of making information services in whatever forms they may evolve available to every human being. The computer's enormous manipulative storage and processing power can be used to multiply the capabilities of every man and lift all mankind to unprecedented levels of achievement. Indeed, some have predicted that in the post-industrial society, per capita information processing power may well be the principal indicator of a nation's wealth.

Questions arise as to how a nation might avoid the myriad dangers and false paths inherent in the creation of the kind of society suggested by such possibilities:

> What are the boundaries between communications and data (information) processing?
>
> Should the common carriers be allowed to enter the data processing business and vice versa?
>
> Are the industrial structures and relationship between common carriers and common carrier suppliers economically and technically optimal for timely and effective responses to user needs?
>
> What rates should be set?
>
> What is the proper balance between social objectives and private gain?

These questions are fundamental in making the potentially revolutionary promises of computer power available to the largest number of people while providing safeguards against its misuse.

A MANAGERIAL OPPORTUNITY

As the technical trends and the urgent need for computer-telecommunication systems as a working tool become more evident, the need to make technical, economic, and political plans to deal with the challenge of a revolutionary society becomes apparent.

The preparation of such plans would be a major milestone in the history of the human species; for the first time the distribution of a major technological change could be planned and directed in the public interest with the probable social consequences taken into account. The alternative is to proceed as in the past, to be the unwilling victims of unplanned technological development resulting in a multitude of noncompatible, individual objectives and unnecessary expense.

During the first industrial revolution there may have been a general awareness that momentous changes were occurring, but there was certainly no realization on the part of the public of the extent to which the process could have been subjected to rational control. As a result the brutal doctrine of laissez-faire with its disadvantages of human degradation and unnecessary suffering, pollution, and uncontrolled urban development was pursued.

Current information allows us to project without exaggeration the proliferation of computer-telecommunication development exclusively toward the future needs of commercial industries. The needs of governments and the individual are for the most part not even considered. In the absence of any long-range public strategy, this might lead to a wide range of independent private networks within each country, causing interconnection difficulties similar to those currently on the international scene. This could not occur without incurring enormous opportunity costs, both political and economic.

Within the social context, the development of independent private networks for a clientele indifferent to the prices and costs of such systems may lead to yet another concentration of information and power over those who do not have access to information and the capabilities to manipulate it according to individual needs.

THE INTERDEPENDENCE OF COMPUTERS AND TELECOMMUNICATIONS

In elaborating policy issues aimed at realizing the promises of computer-tele-communications, a number of basic considerations must be kept in mind. The technical, economic, and institutional interdependence of computer-telecom-munication systems is of prime importance.

The Technical Interdependence

Computer-telecommunication has required the merger of digital computer and telecommunications technologies. Consequently, and because these systems depend entirely on telecommunication lines, the normal boundaries between data processing and communications are becoming blurred.

The basic design of the existing telecommunication network goes back 30 or 40 years and was exclusively tailored for the transmission of the human voice. Consequently, this network, that is, its main segments, the circuits (lines) for transmission and the switching centers to select the path from numerous possible combinations of circuits for the signal to follow, is designed to transmit the frequency range of 300 to 3400 cycles in analog wave forms. The digital computer, however, works on binary coded strings of bits that cannot be transmitted directly over the public network. Thus, to introduce a computer system into this public network is to couple two different techniques and two different user characteristics. Consequently, the use of the existing telephone network for computer-telecommunication has certain negative cost effects. These include inappropriate pricing philosophy, the need for costly modems to transfer the digital signals being sent and received by computers to make it suitable for telecommunication's analogue network; the low transmission speeds if compared to the data flow rate capabilities and needs of the central

Table 1   Common Carriers Plans for Digital Data Terminals and/or Modems[a]

Country	1970 to 1971	1975	1980
Belgium	400	5,000	50,000
Denmark	700	16,000	56,000
France	4,000	50,000	—
Germany	4,300	8,000	68,000
Italy	2,200	13,000	—
Japan	27,000	120,000	—
Netherlands	600	—	—
Norway	300	8,000	—
Sweden	1,100	20–30,000	—
Switzerland	500	5,000	10,000
United Kingdom	12,000[b]	57,000[b]	234,000[b]
			434,000
United States	185,000[c]	820,000[c]	2,425,000[c]

[a] The figures for France, Germany, Japan, the United Kingdom, and the United States represent data terminals, that is, a connection or interface with common carrier transmission services irrespective of the actual number of connections. Correspondingly the quantification in terms of modems (modulator-demodulator) or of so-called data-sets identifies data transmission users on the analog network but not the number of terminals connected to one data set or modem.
[b] The figures for the United Kingdom refer to the years 1973, 1978, and 1983.
[c] Datran Report.

processing unit; the saturation of the existing telephone network; and the disturbance of traditional telecommunication services.

Thus, if the promises offered by computer-telecommunication are to be realized, adequate telecommunication facilities meeting the quality, quantity, and bandwidth characteristics of these applications must be provided.

The Economic Interdependence

Besides the technical deficiencies of the existing telephone network for data transmission, the tariffs for data transmission do not seem to be responsive or adequate for the characteristics of these new telecommunications services. They are generally considered prohibitive for most telecommunication-supported computer systems.

Because of the technical deficiencies and pricing philosophy of the telecommunication network, some negative trends are becoming evident, throwing doubt on the economic viability of the computer utility. Unfortunately, as many considerations such as desired terminal equipment, required bandwidth, distance of the remotely sited user, expected usage, mode of operation, and the myriad applications are involved in determining the system costs, it is impossible at present to establish a typical cost-performance calculation. However, it seems possible to identify certain trends within the structure of system costs.

Depending on its components the computer-telecommunication system costs might be expressed as the sum of the costs of CPU and storage device, costs of terminals and modems, and the cost of data transmission and switching. The costs of these three components have followed quite different trends. The computer and the terminal industries have realized massive gains in productivity because of rapid technological changes and the highly competitive environment. Consequently, the cost of raw computing power has declined by an order of magnitude every four years, an apparently continuing trend.

The communication costs, that is, the charges of telephone channels to purvey raw computer power and services, have not followed CPU cost trends. Tables 1 and 2 provide a survey of national tariffs and an idea as to which country offers the best climate for a widespread use of the computer utility.

As computation costs decrease faster (50% every two years) than communications costs (2%/year, as predicted for the United States and Canada) these latter become a more and more significant factor in large telecommunication computer systems. Some systems already equally divide costs between commu-

Table 2   Data Terminals per Civil Employee in Selected Countries[a]

Country	Number per 100 Civil Employees			Number per 100 in the Tertiary Sector		Civil Employees
	1970 to 1971	1975	1980	1970 to 1971	1975	1980
Belgium	0.01	0.1	1.3	0.2	3	27
Denmark	0.03	0.7	2	0.6	14	49
France	0.01	0.3	—	0.02	0.6	—
Germany	0.006	0.03	0.3	0.01	0.07	0.6
Italy	0.01	0.07	—	0.03	0.2	—
Japan	0.2	1	—	0.3	3	—
Netherlands	0.01	—	—	0.03	—	—
Norway	0.02	0.5	—	0.04	1.1	—
Sweden	0.02	0.5–0.7	—	0.06	1–1.5	—
Switzerland	0.02	0.2	0.4	0.04	0.4	—
United Kingdom	0.05	0.2	1–2	0.09	0.5	1.8–3.5
United States	0.2	1	3	0.4	2	5

[a] The reported data-set and terminals have been divided by the civilian employment figures for 1969 and the figures of civil employees in the (service) tertiary sector.

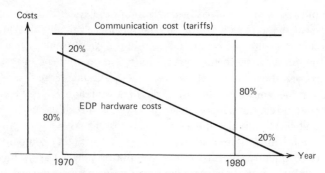

Figure 1   Development of computer and telecommunication costs.

nications and data processing. From an interactive banking computer system in Germany, communication costs even represented 75% of the total system costs.

Consequently, fears have been expressed that the costs of telecommunication services may prove to be the limiting factor in the future growth of the industry, and thus the strategic parameter deciding which services will be computerized and which groups of society will enjoy the tangible as well as intangible benefits of these systems.

Figure 1 illustrates the anticipated effect of new technology on computer and telecommunication costs.

Institutional Interdependence

The EDP segment of computer-telecommunication services is experiencing rapid technological change and considerable cost decreases.

By contrast, the telecommunication segment is characterized by relatively slow innovation and the absence of price reductions.

As a consequence, communication costs may prove to be the limiting, dominant economic factor in computer (information) utilities and thus the strategic variable affecting the speed of its development and the structure of this escalating industry. Therefore, the growth rate of computer-telecommunications and the quantitative and qualitative economies through its application in the service sector will depend on the rate at which new telecommunication services are offered and telecommunication costs fall. Thus, the governmental administration that operates telecommunications as a monopoly, and governmental agencies, which have to ensure that the franchised common carriers respond in good time to users' needs, face a tremendous responsibility. It is evident that the effective application of the computer art to meet individual, commercial, scientific, and governmental requirements is becoming more dependent on the availability of adequate telecommunication facilities at reasonable cost (infrastructure approach). As the technologies are still in an initial stage of future growth, the U.S. Federal Communications Commission, the Canadian Department of Communications, and postal and telecommunication administrations in Europe and Japan have a unique opportunity to guide this possibly key industry in desired directions.

There seems to be general understanding that these institutions could, through an early announcement of their planning activities, stimulate considerable technical assistance for the developments in the various components of the computer segment, such as minicomputers, video-recording devices, terminals, and so forth, with respect to new telecommunication networks. This is further underlined by the report that "there is no major technology which will be used on a large scale in public telecommunication systems by 1980 which is not already under development or at least explored at this date" (President's Task Force, 1969, p. 1).

THE WIRED CITY CONCEPT

It is generally agreed that the telephone system, although highly developed and allowing some 300 million people around the world to communicate, suffers from the fact that it utilizes pairs of copper wires as its local distribution facility and is thus only suitable for handling signals of the telephone or low-speed data type.

To overcome these bandwidth limitations, a recent seminar discussed the possibilities of a "wired city" being characterized as having the capacity "for total communications" (Telecommission Study, 1970).

According to this study it seems possible, at least conceptually, to increase significantly the telephone system's capacity by replacing its copper pairs with coaxial cables. The resulting switched coaxial cable system would also allow the reallocation of the overcrowded electromagnetic spectrum.

It is reported that the British Post Office is convinced of the viability of this concept and have an experimental system operating in their laboratories at Wembley. It is considered as the forerunner of multiservice coaxial cable systems to be installed throughout Britain during the next 20 years. However, it was reported at the same seminar that the Dutch Post Office authorities have come to the conclusion that it is impossible to integrate their television system with other telecommunication services and have abandoned the idea (Telecommission Study, 1970).

Distinct from existing CATV networks, which employ coaxial cable providing 300 times the potentiality of copper pair, this configuration is two-way oriented and switched (Parker, 1970, p. 55). Experts believe that such a broadband system might be possible within 10 to 15 years. "A switched coaxial cable system would have the same philosophy of operation as the existing telephone system," and it could accommodate such services as advertising, pictorial consumer information, alarms, banking, facsimile production, emergency communication, communication between subscribers and computers, meter reading, radio programming, shopping, television, audiovisual playback, educational television, computer teaching, voting, and so on.

Perhaps a harbinger of greater concern with the public interest is the report of the New York City Mayor's Advisory Task Force on CATV and Telecommunications (1968), which proposed making cable television service available to every home in New York wishing to subscribe within two or three years.

It recommended that in addition to carrying local television signals, the cable should be used for program organization and should reserve three channels for municipal purposes. The report also noted the potential future development of CATV into a new urban telecommunications system, perhaps becoming "the transmission belt for all mass information."

In Japan, the concept of the wired city will become reality in Tama New Town—a new satellite town of Tokyo. In a $19 million project—to be started in 1972—some 300 apartments will be interconnected through a coaxial cable network with access to schools, hospitals, and other public services (Ministry for International Trade and Industry, 1972).

Similarly, a recent statement from the FCC has publicly noted for the first time that "the expanding multichannel capacity of cable systems could be utilized to provide a variety of new communications services to homes and businesses in the community."

It listed, among the possibilities of a "wired city" concept, such information utility services as "facsimile reproduction of newspapers, magazines, documents, etc.; electronic mail delivery; merchandising; business concern links to branch offices, primary customers or suppliers; access to computers . . . information retrieval (library and other reference material, etc.) and computer to computer communications." It referred to the possibility of CATV's developing a "capability for two-way and switched services," and, through high capacity intercity communications and computer technology, to become an element in "new nationwide or regional services of various kinds" (FCC, 1968). More recently, in a proceeding involving telephone carrier-CATV relationships, the Commission again referred to the variety of potential services involving data transmission that could be provided over the broadband cable in addition to CATV, and to the "real potential that such services will be furnished over regional and national networks consisting of local broadband cable systems interconnected by intercity microwave, coaxial cable and communications satellite systems" (FCC, 1970).

STATE-OF-THE-ART AND PLANNED ACTIVITIES

With this background, it may now be asked whether and how the relevant authorities in the communication field take up their options in these economic and socially vital spheres. The question at stake is how far individual, industrial-commercial, and government needs for computer-telecommunication have progressed toward a permanent communication policy process that would allow a continuous adjustment to new developments of systems. Ideally, such a comparative analysis should also reveal whether the organizational structure of the traditional common carriers and their relationship to the suppliers are appropriate or insufficient to access these technologies.

Unfortunately, and this is particularly true for most European countries, there is little information available about what common carriers intend to do in the years ahead to meet the complex telecommunication requirements. What is available are some terminal/modem forecasts and planning parameters. (See Table 1.)

Consequently, these forecasts must be interpreted as the quantified demand for computer-telecommunication as the respective telecommunication administrations are considering it. These forecasts show the existing demand and reveal the expected growth of these services expressed in digital terminals and/or modems in some selected countries.

Table 3   Data Terminals and Telephones and the Population in Selected Countries

	Data Terminals per 100 Inhabitants			Telephones per 100 Inhabitants[a]
	1970	1975	1980	1970
Belgium	0.004	0.05	0.5	20
Denmark	0.01	0.3	1	33
France	0.004	0.1	—	16
Germany	0.002	0.01	0.1	20
Italy	0.004	0.02	—	16
Japan	0.03	0.1	—	22
Netherlands	0.005	—	—	24
Norway	0.008	—	—	28
Sweden	0.02	0.3–0.4	—	54
Switzerland	0.008	0.08	0.2	45
United Kingdom	0.02	0.1	0.4–0.8	25
United States	0.09	0.4	1.1	56

[a] See The World's Telephones at 1st January, 1970, American Telephone and Telegraph Company.

The absolute figures of data terminals included in Table 1 look quite promising. They look less so, however, in conjunction with the number of people for whom they are planned.

Tables 2 and 3 show how many inhabitants and civil employees will have to share a terminal by 1980. The figures in the tables—owing to lack of other data—are compared with the number of telephone sets per 100 inhabitants and civil employees. If expert opinion is correct in its prediction that by 1980 data terminals connected to computers and data bases will be as common as the telephone is today, then the market estimates of the common carriers are very conservative.

Other factors mark the common carrier approach as overly conservative. These include considerations of the potential applications and users of tele-communication-based computer systems; the social context if information and its manipulation capabilities were reserved for a privileged class of people only, a potentially explosive situation; and industrial costs of terminal production on large scale.

Another criticism is connected with the speed philosophy that underlies the planned telecommunication networks. The United States' definitions of speed have developed in different ways than most European countries, as Table 4 shows in an oversimplified way.

As the transmission speed (bps) that the telecommunications link allows is a key question for on-line systems applications, shortcomings here may be disastrous. It is mostly decided within the telecommunication network which remote operating applications are possible and which are not (or only with tremendous additional costs).

Unfortunately, a doctrinal status has developed as to the speeds necessary in a public-switched network for the next 10 to 15 years. There are no sophisticated applications described to be considered for planning but the respective telecommunications authorities believe and use as planning parameters "that 85 percent to 97 percent of all data transmission applications will only need low or medium transmission speeds, i.e. 2,400 bits and below" (Fernmeldewesen, 1969) by 1985.

Correspondingly, a British consultant's report on which the British Post Office is said to rely heavily for its present planning has estimated that by

Table 4   Transmission Speeds of Planned Telecommunications Networks

Country	Low Speed	Medium Speed	High Speed
Europe	200 bps and below	201 to 10,000 bps	Above 10,000 bps
United States	50,000 bps	7 bps	50 bps

1983, 99.9% of all data terminals likely to be installed will operate at slow and medium transmission speeds, slow and medium being defined as up to 10 kbps (Data Transmission).

With a switched network designed for these limited possibilities, most of the applications using visual displays, that is, computer-aided instruction, would be excluded again by 1985. Only "less sophisticated" terminals such as printers would be possible. Unfortunately, some European countries regard the data network requirements as merely an upgrading of their present telex or other switched telegraph networks.

This seems to be all the more regrettable as in most European countries all means of telecommunications such as radio, TV, and telephone activities are more or less under the same governmental jurisdiction and thus could be much better and more "easily" integrated than, for example, in the United States, where these high capacity techniques of telecommunications are in private hands and some, such as CATV, are even not subject to FCC regulation. Consequently, the conditions for integrated use of these facilities meet many more "natural" constraints in the United States than in Europe or Japan.

It was mentioned earlier that these systems might help to ease the time-consuming and paper-consuming activities in the service sector. Thus, in order to realize the trend toward paperless or less paper "instant" societies, the terminals then in use cannot be teleprinter devices but will be ultrafast facsimile and visual display terminals allowing instant interaction, such as dynamic line drawing for CAI. To do so, however, and abstracted from resolution techniques, which still have to be developed, hundreds of thousands of bits have to be transported economically over the network.

This refers particularly to the local loop-network. Consequently, the FITCE-Experts (1969, p. 35) recommend that a switched telecommunications network with capacities *of some million bps* should be envisaged.

A NATIONAL PROGRAM

The preceding discussions emphasize the multifaceted nature of the growing interdependence of computers and telecommunication technologies and its possible importance to most countries. The potential of these technologies, the payoff from their proper exploitation, is expected to be greater than the payoff from the peaceful uses of atomic energy, the moonshot or the supersonic aircraft.

The giant strides made in the technology of the various computer-hardware components and the impending revolution in telecommunication technologies indicate that the trend toward the "total computer telecommunication" utility concept is clear and unmistakable (Dunlop, p. 45).

Table 5   Organization Structure of the Industrial Informatics Promotion Program for Japan

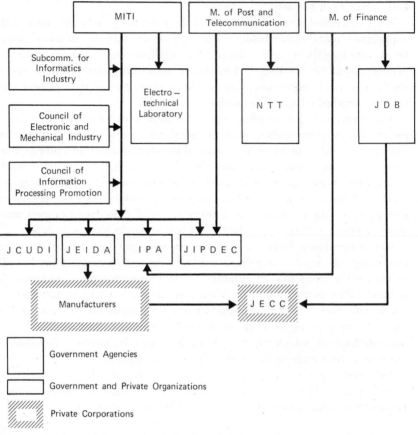

Source.   Ministry of International Trade and Industry, Electronics Policy Division, February 1972.
Legend.   MITI, Ministry of International Trade and Industry. NTT, Nippon Telegraph and Telephone Corporation. JDB, Japan Development Bank, (a new department will be created to finance and support software and consultancy industry). JEIDA, Japan Electronics Industry Development Association (Standardisation is achieved together with MITI). JCUDI, Japan Computer Utilisation Development Institute (Association of large EDP users). JIPDEC, Japan Information Processing Development Centre. JECC, Japan Electronic Computer Company (leased equipment promotion). IPA, Information Technology Promotion Agency.

To prevent politics, lobbying, monopolistic sloth, or destructive competition from diminishing part of the riches that the technology could bring, many experts are calling for a detailed plan to facilitate the introduction and guide the shape of this new infrastructure; to reassess existing telecommunication services and the structure of the computer-telecommunication common carriers industry; and to guide these industries. If most of the promises are to be translated into action and the most serious dangers avoided, something more imaginative than total laissez-faire and shareholders' interests of common carriers and the PTT concept, which leave most decisions to "the consideration" of the administration, will be needed.

Leaving the optimal assessment to the existing carriers and administrations within the present environment is a course that is not welcomed because of the built-in stifling effect with regard to innovation and the lack of imagination and courage to invest sufficiently in this vital future resource. But the creation of national systems of the scale and the potential described in the report appears economically and politically impossible without a concerted approach including computer manufacturers, common carriers and broadcasters, software houses, and present or potential users for the broadest possible base.

As this is the case, there is need for a "major program approach" to try, like any other national undertaking, such as the Moon-Apollo Project, to identify major objectives of social and economic needs likely to be fulfilled by the computer-telecommunication technology, and to embed it in national science and technology policy as well as economic policy.

At the moment, I am only aware of two nations where such a program seems to find an echo on a high policy level: Canada and Japan. A look at Table 5, the Japanese program, seems to be worthwhile, although the social means and ends seem to be omitted.

With respect to the institutions that could coordinate the activities and incentives involved in such a national undertaking to assure the integrated and balanced development in technical, economic, and social terms of computer-telecommunication systems, a new approach seems to be necessary. There have been loud voices for a "new institutional environment" that could be characterized by the strict separation of planning and policy making from operational functions of telecommunication and computing services respectively.

Physically, the network would integrate perfectly the various subnetworks such as telephone, telex and facsimile data and video, cable TV, and radio broadcasting, and thus correspond to the systems integrity in technical terms.

To do so, it can no longer be beyond the scope of any regulatory policy to make a thorough examination of the relationship between common carriers and the equipment supplier(s). It would thus have to go into such issues as the degree of economic advantage to be gained by horizontal or vertical diversification of this relationship.

Unfortunately, however, this new look at the structure of the telecommunication industry is famous only in North America. Particularly in Europe, the governments' monopoly in telecommunications has remained the "holy cow."

REFERENCES

American Telephone and Telegraph Company, New York, Annual Reports 1960 and 1970.

Armer, Paul. The Individual: His Privacy, Self-Image, and Obsolescence. The Management of Information and Knowledge. Committee on Science and Astronautics, United States House of Representatives, 1970.

Auerbach, I. L., Technological Forecast 1971, IFIP-Congress 71, Ljubljana, August 1971.

Business Equipment Manufacturers Association. Comments on FCC Docket No. 16, 979, Washington, D.C., 1968.

Business Week, December 6, 1969 and November 6, 1971.

COPEP, Comité du 6e Plan. Groupe "Transmission des Données," Paris, 1970.

Data Transmission, Post Office Telecommunications, United Kingdom.

"Datran Report." In *A Preliminary Survey of Data Communications in the United States*, John M. Richardson and Robert Gary (eds.). O.E.C.D. Document DAS/SPR/70.66, November 4, 1970.

Duggan, M. A., E. F. McCartan, and M. R. Irwin (eds.). *The Computer Utility: Implications for Higher Education*. Lexington, Mass.: Heath, 1970.

Dunlop, Robert A. *The Emerging Technology of Information Utilities*.

The Emerging Technology. Santa Monica, Calif.: Rand Corporation, September 1970. (R-503 CCOM/NSF RC.)

FCC Docket No. 18397. Notice of Proposed Rule Making and Notice of Inquiry, adopted December 12, 1968.

FCC Docket No. 18509. Computer Inquiry, January 28, 1970.

Fernmeldewesen: Weiterentwicklung und Forschung. Prospektive Studie für den Zeithorizont 1985. Bericht abgefasst von der FITCE (Föderation der Hoch-Schulingenieure des Fernmeldewesens der Europäischen Gemeinschaft). Brüssel, Kommission der Europäischen Gemeinschaften, November 1969.

Gruenberger, Fred. *Computers and Communications: Towards a Computer Utility*. Englewood Cliffs, N.J.: Prentice Hall, 1968.

Hoskyns Group Limited, "United Kingdom Computer Industry Trends 1970–1980," October 1969.

"How Bell Labs Answers Calls for Help." *Business Week* (January 23, 1971).

"Interdependence of Computers and Communications Services and Facilities." FCC Docket No. 16, 976, November 10, 1966. (FCC 66–1004–90954.)

Layton, Christopher. *European Advanced Technology: A Programme for Integration*. London: Allen & Unwin, 1969.

McKay, K. G. "The Network." *Science and Technology* (1968).

"Le Marché de l'Informatique." *L'Expansion*. 21 (July—August 1969).

Martin, James. *Future Developments in Telecommunications*. Englewood Cliffs, N.J.: Prentice Hall, 1971.

Martin, James. *Telecommunications and the Computer*. Englewood Cliffs, N.J.: Prentice Hall, 1969.

Mayor's Advisory Task Force on CATV and Telecommunication. A report on cable television and telecommunications in New York City, September 14, 1968.

Melody, William H. "Market Structure and Public Policy in Communications." Annual Meeting of the American Economics Association, New York, December 28, 1969.

Moonman, Eric (ed.). *British Computers and Industrial Innovation: Implications of the Parliamentary Select Committee*. London: Allen & Unwin, 1971.

Parker, Edwin B. "Information Utilities and Mass Communication." In *The Information Utility and Social Choice*, H. Sackmann and Norman Nie (eds.). Montvale, N.J.: AFIPS Press, 1970.

Parkhill, D. F. *The Challenge of the Computer Utility*. London: Addison-Wesley, 1966.

Policy Considerations with Respect to Computer Utilities. Telecommission, Studies 5(a), (c), (d), (e). Canada: Department of Communications, 1971.

Postal Services and Telecommunications. Telecommission, Study 7 (i). Canada: Department of Communications, 1971.

President's Task Force on Communications Policy, Staff Paper 1, Part 2, Washington: Clearinghouse, June 1969.

"The Revolution of Communications." *Science and Technology* (April 1968).

Richardson, John M. and Robert Gary (eds.). *A Preliminary Survey of Data Communications in the United States*. O.E.C.D. Document DAS/SPR/70.66, Novemeber 4, 1970, Pans.

Schéma général d'aménagement de la France. Eléments pour un schéma directeur de l'informatique, Paris, La Documentation Française, January 1971. (Travaux et Recherches de Prospective 13.)

Science Council of Canada. A Trans-Canada Computer Communications Network, August 1971. (Report No. 13.)

Telecommission Study 6(a). Reports on the Seminar on the Wired City. Ottawa: Department of Communications, 1970.

United Kingdom Computer Industry Trends 1970–1980. Market Study. Hoskyns Group, October 1969.

Whitehead, Clay T. Electronics Industries Association Conference, Washington, March 9, 1971.

Whitehead, Clay T. Meeting of the Society of Civil Engineers of France. Paris, June 9, 1971.

Whitehead, Clay T. "International Communication, An American View." American Bar Association Meeting, London, July 14, 1971.

The World's Telephones at 1st January, 1970. New York: American Telephone and Telegraph Co.

Zeidler, H. M., et al. *Patterns of Technology in Data Processing and Data Communications*. Menlo Park, California: Stanford Research Institute, February 1969. (S.R.I. Report No. 7379 B-4.)

CHAPTER 12

The Role of Advocacy in Public Policy Planning

WILLIAM H. MELODY

Technological change in communications portends enormous advances in the potential for information processing and distribution by making information accessible in kinds and quantities not dreamed of heretofore. In addition, it offers the potential for completely altering the environment of mass communications both in terms of the audience networks to be reached and the conditions of access for communication to them. By transmitting the information necessary for improved selection from alternative courses of action, expanded communications and additional information can enable the alternative performance of particular activities at reduced social costs. They can enhance the awareness of decision makers to alternatives that would not otherwise have been considered. They can permit planners to free themselves from the myopic frame of reference that traditionally has confined them and to extend their planning horizon to encompass consequences that otherwise would have been ignored.

The expansion of communication and information processing capacities may have consequences of enormous social benefit or detriment to different groups in a society depending on the particular direction of application of the technology, the institutional structure of controls over the technology, and the particular environment in which it is introduced. This makes the dual tasks of planning the direction of application of communications technology and devising an institutional structure in which its development will be responsive to the publics' interests crucially important.

In most fields, new technology is conceived primarily in terms of its consequences for economic efficiency in expanded production. But the direct economic effects of communications technology, although substantial, may be one of the less significant areas of impact on society. We have learned only belatedly that an ever-expanding gross national product may bring with it enormous social costs. In communications, the "externalities"—the consequences that occur outside the narrow analytical framework of the private decision maker—are likely to dwarf the effects considered in making specific economic decisions. Hence, effective planning in communications must go beyond the aggregative forecasting of direct and quantifiable economic effects. It must address the more fundamental issues relating to the particular structure of technological applications and evaluate the consequences for society as a total system.

To state that the stakes are high in new communications technology is an understatement. The potential for acquiring economic and political power by means of controlling communications technology is enormous. As the communications industries are the cornerstone of the infrastructure of economic systems, their impact and influence is spread and multiplied throughout the economy. In addition, new technologies can provide improved circumstances for the development of communication systems as tools of overt manipulation of behavior. In technologically advanced economies, timely information and an effective communications network can be assets of far greater value than those recorded on balance sheets.

TECHNOLOGICAL DETERMINISM AND INSTITUTIONAL STRUCTURE

Technological conditions and institutional arrangements have always been interrelated and have frequently been in conflict. Will new technology be directed and controlled by established institutional arrangements and the distribution of economic and political power that they reflect? Or will institutional arrangements be determined by the new technological conditions? History provides illustrations of each case. Although one would hope that neither technological nor institutional determinism would arbitrarily prevail, it is clearly desirable that technology be directed toward socially productive ends within a framework of feasible and acceptable institutional arrangements that reflect the values of the society. In the United States, and in other countries as well, this does not appear to be the basis for the development and introduction of new communications technologies.

In most instances, today's technologies are being forced to accommodate almost completely to the power structure inherent in our established institutional arrangements. But these arrangements do not represent a socially desirable institutional structure that reflects the ideals of United States democracy. They represent deviations from the framework of generally acceptable arrangements, believed to be required by the very special characteristics of technologies introduced at various times in the past. Hence, the direction of modern communications technology is determined by neither the unique characteristics of the technology nor a requirement to preserve a socially desirable institutional structure. It is determined by a requirement to accommodate a distorted institutional structure supposedly created because of the particular characteristics of past technologies. As a result, the social costs of technological opportunities delayed, neglected, misdirected, and foregone will be tremendous.

The Heritage of Business Affected with a Public Interest

Jacques Ellul and others have described how technology shapes and conditions the environment of society. However, in communications, technology has provided an additional kind of technological determinism that has directly influenced the structure of our communications institutions and the direction of our public policies. Both the telecommunications common carrier and the television broadcast industries have been recognized as falling within a special category of "business affected with a public interest."

As such, these industries have a unique relationship with governmental authorities which provides both special rights and imposes special obligations on the firms. The concept of business affected with a public interest is a legal one, developed in English common law but generally traced to the landmark decision in *Munn* v. *Illinois* (1877) where the United States Supreme Court stated:

> When, therefore, one devotes his property to a use in which the public has an interest, he, in effect grants to the public an interest in that use, and must submit to be controlled by the public for the common good, to the extent of the interest he has thus created. [94 U.S. 113, 126]

Special franchises or licenses define the firm's rights and public responsibilities. As an instrument of public policy conferring special privileges on its recipients, the franchise or license has always had an economic value directly related to the extent and significance of the privileges granted and the effectiveness of the public regulations imposed.

The regulation of private industry has been attempted by governments since earliest times. Most attempts at regulation reflect the ancient ideal of social justice perhaps best illustrated by the doctrine of the "just price." We find its modern counterpart in virtually all legislation proscribing the regulation of public utilities, which requires that prices be fair, reasonable, and not unduly discriminatory.

The basis of this special classification of business affected with a public interest is economic and technological. Where the supply of a particular good or service is monopolized, or control over the supply is so concentrated that competition is ineffective in protecting the public's interests, the economic power of the supplier(s) must be limited so that he will deal reasonably with those who are dependent on him for supply. The status of being required to deal reasonably classifies a firm as business affected with a public interest. The earliest businesses so classified were transport and public utility companies, including telecommunications common carriers.

Since the beginning of regulation of business in the United States, it has been recognized, at least by the regulated companies, that one of the most important determinants of the license or franchise value is the degree of monopoly power conferred on the firm. From the firm's perspective, the ultimate franchise is exclusive and in perpetuity. The terms and conditions of the franchise have been generally negotiated between the firm's management and the governmental authority. In most industries franchises were granted initially during an era of promotion where the primary concern was to provide sufficient inducements to attract capital for the creation of initial services. Hence, regulatory authorities were frequently quite magnanimous in granting special privileges, including that of exclusive monopoly or highly restricted oligopoly.

For purposes of analyzing changing technology, the most significant characteristic of these regulated industries is that they are extremely difficult for a new firm to enter. Like the established firms, a potential new firm must be certified by the regulatory authority before it can enter the market. Direct regulatory policies restricting entry exist because unrestricted entry into these industries would not be in the public interest. In other words, the competitive force of market regulation has been determined to be inadequate as a means of protecting the public interest. Government regulation acts as a surrogate for market regulation. Although the institution of government regulation provides a vehicle for going beyond the traditional economic criteria of free markets by establishing policy based on public and social objectives, it has rarely done so.

The Technologically Based Rationale

Technological conditions are the fundamental determinant on which industries are classified as business affected with a public interest and segregated from the norm of competitive capitalism. In the interest of economic efficiency within a particular static state of technology, corporate institutions of substantial economic power have been permitted and encouraged to develop. As the countervailing force, regulatory institutions have been created to apply political power as a restriction of the private economic power of the firms in regulated industries.

The rationale for regulating telecommunications common carriers is that the technology creates a condition of "natural monopoly." Natural monopoly exists when the technological alternatives are such that competition must inevitably lead to monopoly because large-scale supply is believed to be more efficient than smaller scales of competitive supply. The monopoly is natural because it is determined by the state of the technological alternatives available at a particular point in time.

The telecommunications industry is the most monopolized industry in the United States economy. The industry has been dominated by the Bell System since long before the federal commission was created in 1934. Through its many operating companies Bell provides local service in the most densely populated areas of the country to about 85% of the nation's telephones and supplies nearly all interstate long-distance service. Its manufacturing affiliate, Western Electric, provides more than 80% of the sales in the domestic communications equipment market. Bell Laboratories dominates the communications research and development field. When the Federal Communications Commission was created in 1934, there was no public interest determination that such a monopolistic structure of the industry was either required by the technology or in the public interest. The institutional structure of the industry was simply inherited. There is a case to be made that communications technology for local distribution facilities may require a local monopoly for any specific geographic area. But the creation of a $50 billion company with overwhelming nationwide monopoly power extending from research and development through the provision of all types of communication services in order to prevent duplicative wiring of a community is a clear case of institutional overkill. It represents the myth of technological determinism rationalized to a ridiculous extreme in order to justify the inherited concentration of monopoly power.

The rationale for regulating television broadcasting is not natural monopoly, but rather "natural oligopoly." It, too, is founded on technological determinism that creates a substantial concentration of economic and political power in the industry. The relative scarcity of the radio frequency spectrum resource limits the number of signals that can be broadcast in any given geographical area. But this limitation is as much a function of the evolution of regulations for allocating the radio frequency spectrum among broadcast and other communications uses as of the technology itself.

Here also the technological determinism rationale neither requires nor supports the concentration of economic power prevailing in the industry. That power is wielded principally by the broadcast networks, which are not regulated. The networks stand between the regulatory agency and the broadcast stations and can exercise their power in either direction. Ironically, the conventional wisdom that technological determinism has required the existing structure of the communications industries has come to be used as the rationale for preventing the rapid introduction of new communications technologies which would alter that structure.

IMPLICATIONS FOR THE APPROACH TO REGULATION

Acceptance of the concept of technological determinism, which established communications firms have carefully nurtured, has had a devastating effect on the approach to regulation adopted by regulatory authorities at federal, state, and local levels. It has narrowed the scope of analysis of public policy makers to encompass only changes within the established institutional structure. Since technology dictates the structure of the industries, the prevailing view has been that this is not a matter for continuing review by the regulatory agencies. The regulated communication companies are treated in the main as entities to whom the public interest has been entrusted.

Regulatory agencies apply periodic, or sporadic, reviews of performance which, in the absence of implementable standards, result in considerable exhortation, wrist-slapping, and hand-wringing. Telecommunications common carrier regulation focuses almost entirely on past profit levels, while broadcast regulation is addressed to a somewhat nebulous evaluation of performance in the public interest at license renewal time.

The substantive areas of decision making that affect the direction of the industries, the services they provide, and the application of new technologies have been generally left to the expertise of the companies. These decisions typically fall into the category of managerial prerogatives and are made within the inner sanctums of management. They are not visibly justified to the public and there is no accountability in terms of alternative possibilities that were foregone. As a result, public policy remains narrowly focused, static, and passive.

This traditional philosophy of communications regulation is undoubtedly a product of the environment in which it was spawned. The Federal Communications Commission was created in an era of depression when the general economic philosophy of government was directed toward discouraging active competition (in fear that it might be destructive) and promoting stable institutional arrangements. This philosophy apparently set the framework for the traditional approach to communications regulation.

However, when new technology comes from outside the established institutional arrangements, it threatens the established structure and regulation is forced to come to grips with it. Under these circumstances, the established industry seeks protection of its inherited position of economic power, and the regulatory agency seeks a compromise of interests reflecting the relative economic and political power of the opposing interests. Whether or not the resulting arrangement is responsive to interests of the broad public is incidental to the basic market division among technologies. Once the new rules, official and tacit, are established, regulation returns to normal.

During the past decade, substantial technological change affecting communications has come from outside the established industry. This has prompted the FCC to begin to address issues of policy planning and to reexamine some of its traditional policies based on regulated monopoly.

PUBLIC POLICY RESPONSE TO CHANGING MARKET CIRCUMSTANCES

Television Transmission Service

Over the past 25 years, FCC response to new technologies and new communications demands has varied substantially. When a demand for the transmission of television broadcast signals developed during the late 1940s, the FCC followed the traditional policy of reserving the market for the established telecommunications monopoly, the Bell System, even though others, including the television broadcasters themselves, wanted to supply television transmission links. Because of a shortage of Bell facilities over its existing intercity routes at the time and an expected shortage of facilities for some time into the future, available channel time had to be allocated among applicants for service. After many complaints, the FCC permitted others to build television relay links where Bell could not supply that service, but warned them that as soon as Bell could provide the service, they would have to abandon their own supply and move to Bell's.

During the 1950s Bell gradually developed its television transmission network throughout the major cities of the United States. However, Bell did not extend the network to the smaller towns and rural areas. As a result, in 1958 the FCC concluded that Bell was not interested in developing any more of the market than it had already developed. Therefore, the Commission changed its policy to permit broadcasters to use independent television intercity relay facilities, and allow other communication carriers to pick up the television signals from Bell and deliver them to the small towns and rural areas. These new telecommunications carriers were aptly named miscellaneous common carriers because they could do nothing more than pick up television signals and deliver them to unserved areas. They could not put signals into Bell's television transmission network.

During the 1960s there was rapid growth in the supply of television transmission service by miscellaneous common carriers. It was stimulated by developments in the cable antenna television field (CATV), which enabled these carriers to deliver broadcast signals for both over-the-air broadcasters and CATV customers. The existence of sizable networks of miscellaneous carriers permitted the rapid growth of cable television in small towns and rural areas. This in turn brought an awareness of the potential threat of CATV to the established broadcasting structure, prompting the FCC to retard CATV growth by requiring the broadcast signal to be sent over a higher frequency (11 Ghz instead of 6 Ghz) if it was being delivered to CATV customers. FCC restrictions on CATV followed in 1966 and 1968, which froze CATV development in the top 100 broadcast markets.

Although the transmission of television signals did not involve an explicit application of a new technology, it did require the special adaptation of communications technology to the demands of a newly created specialized service. The public policy response was to entrust and protect the market for development within the established institutional structure. Only after the economic interest of the established monopolist was fully satisfied did policy permit market development by others. The policy was not based on an evaluation of the consequences of alternative possible policies, but on the presumption of a chosen entity in whom the public interest was entrusted and to whom new communications markets and technological applications were reserved.

Satellite Technology

Because of its unique characteristics, satellite technology offers enormous potential for providing new and different telecommunication services. Satellite technology permits the distribution of a single transmission to multiple reception points; it eliminates transmission costs as a function of either distance or the number of receiving points; it eliminates a major part of the hierachical switching system required by landline transmission systems; it can provide a flexibility for providing many communications services that cannot be matched by other technologies; it can provide communications services that cannot be supplied at all by other technologies, especially to geographically dispersed points and remote areas; and it can permit a specialization in services that can be provided more efficiently by satellite than by landline systems.

Although commercial applications of satellite technology have been in existence for nearly a decade the technology really has never been fully developed. The application of the technology has been so compromised by the established institutional structure of the communications industry and its vested interest in cable technology that the satellite technology has been treated as if it were just another point to point communications facility.

The design of a particular satellite system determines the direction in which the technology will be applied. Thus, satellite systems can be designed to provide conventional point to point service, like a cable system, in which case expensive ground stations are likely to be employed. On the other hand, if the system is designed for maximum communication to many geographically dispersed locations, each of which has small communications requirements, the system will be designed with inexpensive ground stations and differently designed satellites. Similarly, a general purpose system will be designed differently than a specialized system. Thus, satellites can be designed to maximize their similarity to existing technologies or to maximize their unique characteristics. Unfortunately, to date, the former approach has prevailed.

The introduction of satellites for international telecommunications produced a classic institutional struggle during the early 1960s. On one side we had the established institutions, led by AT&T and including the FCC, trying to preserve the technology for the established institutions. On the other side we had the aerospace manufacturers who feared that if AT&T controlled satellite technology, they would be frozen out of the satellite equipment and facilities market. The cause of this large-scale institutional conflict was AT&T's vertical integration into the telecommunications equipment market through its Western Electric subsidiary. A faint voice in the wilderness was muttering something about a public dividend because the taxpayer had paid for most of the research and development.

From this struggle was born a new corporation, Comsat, a unique corporation designed to do little else than accommodate the established economic interests. The satellite manufacturers were taken care of by a requirement that Comsat purchase its equipment and facilities by competitive bidding. The established United States international telecommunications industry was given 50% ownership in Comsat and therefore effective control over its policies (the other 50% was to be held by the public). The FCC soon followed with policies limiting the purchase of satellite channels from Comsat to "authorized users," the established international cable carriers. This almost entirely removed Comsat from the user market and turned it into a carrier's carrier. In addition, Comsat was to share ownership of satellite earth stations with the established cable carriers.

At present, the international cable carriers offer the same communications services regardless of the technology used to supply the service. They charge "composite rates" supposedly based on an averaging of the costs of the international cable and satellite systems. To make matters even worse, since the international cable carriers own the cables but only lease satellite channels, the traditional method of profit regulation provides an incentive for the cable carriers to maximize their use and ownership of cables regardless of the greater efficiency of satellites. At the time of this writing (1972) AT&T has an application before the FCC to add another cable in the Pacific even though Comsat is operating its Pacific satellite system at considerably less than 50% of capacity and will have idle capacity for many years.

Clearly Comsat is simply a convenient institutional mechanism for preserving traditional institutional relationships. The company does not manufacture equipment, launch satellites, or sell services. It serves principally as a managerial intermediary for the established communications industry and the United States government on international satellite policy.

International satellite communication developments illustrate a clear case of almost absolute institutional determinism. The application of the satellite technology has been completely subservient to the preservation of established power relationships. As a result, the unique communications opportunities of the satellite technology remain to be developed.

How the satellite technology will be applied to domestic telecommunications is now under consideration at the FCC. However, judging from the FCC policy statement of March 1970, the Commission is not going to follow the torturous satellite path that developed in international satellites. The Commission specifically did not opt for the chosen entity approach despite the pressures from AT&T and Comsat. Instead it followed in the path of its recent decisions in domestic communications opening new and specialized communications markets to new potential suppliers as well as the established industry. The Commission stated:

The most important value of domestic satellites at the present time appears to lie in their potential for opening new communications markets, for expanding the beneficial role of competition in the existing markets for specialized communication services, and for developing new and differentiated services that reflect the special characteristics of the satellite technology. [Domestic Satellite Facilities, 1970, p. 14]

The policy statement also emphasized that the authorized user policy of international communications would not be applied to domestic satellites. Moreover, satellite planners could assume that there would be opportunities for direct access to the markets they desired to serve and that there would be no arbitrary restrictions on interconnection or through-route arrangements. Finally, the Commission raised the institutional problem of the Bell System monopoly and specifically recognized the constraining effects of AT&T's existing terrestrial facilities and services on the company's own planning horizons.

In response to the commission's policy statement inviting domestic satellite system proposals from any qualified applicant, eight major applications were filed. AT&T and Comsat filed a proposal that would simply extend the cartelized international satellite institutional structure to domestic satellites. Apparently Comsat is content to extend its redundant role as a carrier's carrier and be, in essence, AT&T's domestic satellite subsidiarly. AT&T has stated that satellites will enable it to supply no new services that it is not supplying now and that it could provide the same communications more efficiently with terrestrial facilities. Nevertheless, it would still like to own and control the application of the technology.

Comsat submitted a second satellite proposal to supplement its system with AT&T. Under this proposal Comsat would be the carrier's carrier for any other potential users of satellite channels. It includes a plea for an exclusive satellite monopoly, arguing that in the interest of single-system economies of scale and technological determinism the FCC should reserve the technology for Comsat and its two satellite systems.

The AT&T and Comsat domestic satellite proposals simply ignore the FCC's domestic satellite policy statement. Instead they have responded to the existing institutional arrangements in the international satellite field. Their proposals are diametrically opposed to the spirit of the stated policy. The FCC is now confronted with the classic dilemma of established institutional power versus stated public policy in domestic satellites. It is entirely possible that the full potential of satellite technology may never be developed in the United States.

Data Communications and Teleprocessing

Growing demands for data communication and teleprocessing services and changing technology in the computer industries have prompted the FCC to reexamine some of its fundamental monopoly-based policies in domestic telecommunication. The result has been a liberalization of policy in the areas of interconnection and attachment to the nationwide telecommunications system and in the area of new entry of specialized telecommunications carriers.

The response to these changes in the direction of FCC policy has been enormous. Both equipment and transmission interests are rushing into markets previously closed to them by the policies and practices of the FCC and the Bell System.

Policy developments in the area of data communications and other nonvoice specialized services come closest to reflecting policy planning by governmental authority. But the planning was more a recognition that the existing institutional arrangements are not likely to foster an effective response to either the new technology or the new and changing demands for telecommunications services than an attempt to guide technological applications and institutional arrangements in any specific direction. The policy direction tends more to correct an unnecessarily distorted institutional structure than point a path to the future. Nevertheless, such a step may be a prerequisite to the establishment of an active and continuing program of policy planning.

At present there is much to be said for a series of regulatory policies that maximizes opportunities for entry into data and teleprocessing markets. Many diverse consumer demands are growing rapidly; the technology is changing rapidly and is subject to many different adaptations. The entire future market environment is quite uncertain for all planners and forecasters. There is a premium on innovation, imagination, new ideas, and willingness to assume the risks of the potential of future markets. At present there seems little justification for excluding any entrant from these markets except perhaps those firms that already have too much power and that would monopolize the market. But should special circumstances arise, the FCC still has the power to reject any proposed entrant if the Commission believes that it is in the public interest. Over the next several years the most serious problem confronting the FCC in this area will be preventing AT&T from using its enormous monopoly power to engage in predatory competitive activities, and most specifically use the monopoly voice telephone service to subsidize its competitive efforts.

Cable Television

The FCC's recent policy decision on CATV reflects the culmination of yet another illustration of a new technology being completely warped to the fundamental requirement of preserving the established institutional arrangements. Aside from the procedure employed to arrive at its decision that raises questions about whether this policy reflects the deliberations of public policy makers or private policy makers, the rationale on the basic substantive issue has never made much sense. If one were interested in protecting over-the-air broadcasters with marginal profits, he would not put the most rigid restrictions against entry by cable operations on that part of the existing broadcast market realizing the greatest profit. The policy statement is responsive to the large profitable broadcast interests, not the marginal broadcaster.

However, there is merit in some parts of the policy statement that reflect consideration of long-range planning and at least some public interest considerations. There is at least room for local interests to influence significantly developments in their areas.

What is most disturbing is that the basic issue of the relationship between the technology and the industrial structure of CATV has not been addressed. The industry is developing along much the same lines as the telephone and electric power industries. Conglomerate holding companies are being formed to extend the limits of centralized management control, provide expanded financial leverage, and diversify short-term risks. Of course, this also gives the additional advantage to management of shielding more and more internal transactions from external visibility and the requirement of meeting an arm's length market test on internal sales and exchanges. The companies are integrating horizontally and vertically and the rate of mergers can be expected to continue. Indeed, if Teleprompter's plans are fulfilled, it will eventually become a resident East India Company. Will we once again fail to acknowledge that cable is common carriage until the industry has been structured, the economic power has been established, and the dominant firm(s) would like the security of some legal barriers to entry? Now is the time for implementation of policy guidelines that permit responsive and efficient development of the industry without the accumulation of unnecessary monopoly power.

PUBLIC POLICY: PLANNING AND ADVOCACY

The theoretical efficiency in resource allocation within free competitive markets is derived from the absence of economic power. The invisible hand of economic democracy is guided by the market mechanism to allocate resources in the direction of the public's demands. The inefficiency of monopoly is derived from an imbalance in economic power that enables the monopolist to exploit his position. It is a structure of private government of the market and its terms of exchange. In the major communications industries we have institutional structures that are not only heavily concentrated, but regulated by government authority. However, regulation does not neutralize the economic power of the regulated firms. It simply prevents that power from being exercised in certain traditional ways. Telecommunication carriers have their accounting profits regulated; broadcasters are required to perform some public interest functions that prevent them from maximizing profits. But there are many other ways that this economic power can be exercized, not the least of which is ready conversion to political power.

The full economic and political power of the communications industries is clearly used to insulate, preserve, and extend their markets, as well as to control the development and application of new technologies. But at present, public regulatory authorities do not focus on the direction and application of technology as a matter of continuing responsibility. These decisions have fallen within the sanctity of managerial prerogatives where monopoly managers make decisions that are not visible outside the firm and for which there is no public accountability. These have been the key areas where the chosen entity performs its government-assigned role in making decisions that are supposed to represent the public interest, but that can always be better explained in terms of the firm's private interest.

In this decision-making environment the regulatory agency generally can be overwhelmed by the power of the established economic interests, even when the issue involving new technology is raised from outside the established industries. Most of the time the regulators simply have no real alternatives. They are subject to continuous and overwhelming advocacy from established interests, occasional advocacy from special outside interests, and no advocacy from the general public. Sometimes the only apparent choice for regulatory agencies is to select an entity to entrust with the public interest. And it is always safer to go with an old friend, or even an old adversary, than a relatively unknown newcomer with whom your relationship has not yet stabilized.

On issues involving public policy toward new technology or new communications markets, policy is always very responsive to the breadth and depth of advocacy of various special interests. If there is only the advocacy of established firms, policy will entrust the technology to them. If there is countervailing advocacy from an adverse interest that exceeds a given threshhold level, then the adverse interest will get a piece of the action within a policy framework that protects the vital interests of the established industry. If the adverse interest is advocating the removal of an arbitrary restriction used by the established industry to preserve its power at the expense of the adverse interest, it has a fighting chance of modifying past practices. Thus, the satellite equipment and facilities market can be kept open for the aerospace industry, restrictions against interconnection and attachment to the telecommunications network can be liberalized, some specialized data communications systems finally can be permitted to enter new data communications markets, and ultimately CATV can win access to the market. Moreover, there is a chance that the result will coincide in some way with some interests of the public.

However, if the adverse interest advocates policy that would restrict activities of the established firms, such as to protect themselves from predatory competitive practices or to restrict the participation of the established firms in new technologies, it has a negligible chance even with a strong case. Despite AT&T's stated views on domestic satellites, the possibility of the FCC even considering seriously rejecting AT&T's participation at the present time is remote. Yet, it would seem that with a full-fledged counteradvocacy to the AT&T satellite proposal, the desirability of such a policy could be readily demonstrated to anyone.

The deficiencies of public policy in communications are due in substantial part to the fact that neither policy planning nor advocacy have been viewed as part of the permanent and continuing regulatory responsibility. Hence, there is a need to initiate permanent policy planning and advocacy responsibilities into the regulatory process. Hopefully, the institutionalization of planning and advocacy would change regulatory philosophy in a very fundamental way. It would put an end to the separate sovereignty of major managerial decisions relating to the direction and application of the technology, the development of communication services, and major investment programs. Rather than select the private interest that will be entrusted to make major public interest decisions in an environment where there is no effective public accountability, the regulatory agency would select the plan that can be best justified as serving the public's interests.

This means that we cannot rely on a single planner, whether it be a large corporation or a single government agency, for an optimum plan. There really is no such thing as an optimum plan for the public. There are several optimum plans depending on the interests of the groups doing the planning. Hence, we cannot place our hopes in the Office of Telecommunications Policy (OTP) to come up with policy plans that necessarily reflect the public interest of all publics. Like the FCC, OTP is subject to the pressures of economic and political power. But since the framework of analysis for its planning will be quite different from that of the communications industries, OTP has the potential to contribute immeasurably in policy planning. And its planning will be even better if it knows that alternatives are being prepared and that it will be publicly accountable for its recommended plans.

Plans and forecasts, once exposed to the light of critical examination, are the very essence of advocacy. They are determined by the approach taken, the assumptions employed, the data selected, and the concepts adopted. They contain a myriad of hidden valuations that reflect the biases (conscious and unconscious) of the planners. Alternative and competitive plans derived from the viewpoint of different interests will tend to force the hidden valuations into the open. This will provide fundamental information on the range of perceptions and interests that make up the public's interests and establish a relevant framework for the development of public policy.

Two primary perspectives require representation in planning approaches. One is the perspective of the unrepresented users of monopoly services who are presently treated as the residual claimants to service and guarantors of common carrier revenue requirements. The other is the perspective of the total system benefits, costs, and consequences for society. This approach would focus specifically on those consequences that lie outside the normal realm of the private decision maker and would include an evaluation of economic externality, social and cultural consequences of alternative possible plans.

Advocacy in the planning process would focus continuing regulatory debate on such issues as the allocation of research and development expenditures, the institution of major investment programs, and planned changes in the complement of services. Continuing study in the trends of technology and potential services should be undertaken by the agency not only because it is necessary in order to participate in the regulatory debate but also to provide periodic public assessments of the state of communications technology and service. On these matters the regulatory agency's activities should be just as visible as those of the regulated companies.

In addition, the periodic assessment should include an evaluation of the existing and developing institutional arrangements in the industry as part of a continuing evaluation of the existence of any undue concentration of economic power. This analysis should draw clear distinctions between the concepts of technological determinism, which may require a certain degree of monopoly power in a market, and technological permissiveness, which simply makes it possible to acquire additional monopoly power. It appears that the range is very wide and that a small amount of technological determinism provides the rationale for a large amount of technological permissiveness.

The policy planning approach suggested here could be implemented by starting with the most fundamental decision areas and the periodic threshhold decisions that significantly affect the course of development of the firm or industry. There should be no intention to get involved with the details of the myriad of continuing specific investment decisions. It should be confined to policy direction for the firm's major programs and decisions. Moreover, the policy planning alternatives derived from different perspectives need not be exhaustive detailed studies. Their purpose is to provide guidelines and direction to decisions, not to make every decision. The objective should be to create a decision-making environment that constrains the private decision maker from pursuing alternatives that are contrary to the public's interests and that makes him aware of public interest requirements that he would otherwise ignore. With advocacy of alternative policy plans, visible and justified decisions, and continuing accountability, public policy formation at least has the potential to be fully responsive to the public's interests.

REFERENCES

Churchman, C. West. *Challenge to Reason*. New York: McGraw-Hill, 1968.

Ellul, Jacques. *The Technological Society*. New York: Knopf, 1964.

Galbraith, J. K. *Economics and the Art of Controversy*. New York: Random House, 1955.

———. *The New Industrial State*. Boston: Houghton Mifflin, 1967.

Irwin, M. R., *The Telecommunications Industry: Integration* v. *Competition*. New York: Praeger, 1971.

Melody, W. H., "Regulation and Competition in Data Communications in the United States." Paris: OECD Publications, 1971.

———. "Economics and Regulation of Domestic Satellite Communications." New York: IEEE, 1971.

———. "Technological Determinism and Monopoly Power in Communications." annual meetings of the American Economic Association, New Orleans, La., December 1971.

Myrdal, G. *Objectivity in Social Research*. New York: Random House, 1969.

Telecommission Studies, Ottawa, Canada: Department of Communications, 1971.

CASES AND POLICY DECISIONS

Munn v. *Illinois* 94 U.S. 113 (1877).

Changes and Regulations for Television Transmission Services and Facilities, 5 RR 639. See also 17 FCC 152 (1952).

Allocation of Frequencies in the Bands Above 890 Mc., 27 FCC 359 (1959), 29 FCC 825 (1960).

Use of the Carterfone Device, 13 FCC 2d (1968).

Microwave Communications Inc., 18 FCC 2d (1969).

FCC Computer Inquiry, Docket 16979. Notice of Inquiry (1966). Supplemental Notice of Inquiry (1967). Report and Further Notice of Inquiry (1969). Report and Order (1971).

FCC Policy Statement re Establishment of Domestic Satellite Facilities, Docket 16495, March 1971.

FCC Report and Order re Specialized Common Carriers, Docket 18,920, May, 1971

FCC Cable Television Policies, Dockets 18, 397 et al., February, 1972.

PART III

Communication and Education for the Full Employment of Human Potential

CONTENTS

INTRODUCTION

LARRY P. GROSS

How can we move our children to a cultural level that we ourselves have not attained? How can we teach things that we probably have not learned, through media that we are not yet fully capable of understanding? This is a question with which we have just begun to struggle.

"Education," wrote Maslow (1971), "is learning to grow, learning what to grow toward, learning what is good and bad, learning what is desirable and undesirable, learning what to choose and what not to choose." In a time of rapid cultural and technological change it is important that we give serious thought and attention to the question of what we choose to have our children learn, how we choose to teach them, and why. We may have little to say about the overall nature of the culture of which our children become members, but the aspects that are part of the formally institutionalized processes of education can and must come under severe scrutiny. Schooling is a mechanism by which societies direct much of the culturation of their members, and we must ask whether our present modes of education are producing a citizenry that is willing and capable of striving for a just and full employment of its human potential.

This question is particularly relevant at a time in which the technological media of communications have radically extended the range and power of relatively small numbers of message producers to reach and influence large, heterogeneous, and widespread audiences. The growing powers of communications technology can either greatly aid or powerfully hinder the movement toward a truly just and equalitarian educational process.

We have inherited a concept and a system of education that grew out of the need of an industrializing and urbanizing society for a populace that possessed a minimal set of literacy and arithmetic skills. This has remained the essential ingredient and core of our school system. " . . . The basic structure of American education had been fixed by about 1880 and . . . it has not altered fundamentally since that time. . . . (Its) purpose has been, basically, the inculcation of attitudes that reflect dominant social and industrial values . . . .The result has been school systems that treat children as units to be processed into particular shapes and dropped into slots roughly congruent with the status of their parents" (Katz, 1971).

The fact that the nature of our society is changing and that this limited focus of educational objectives is no longer acceptable to a significant proportion of parents, students, and teachers is what gives us hope that we can utilize the new possibilities provided by social unrest and technological progress in order to create the conditions for a humanizing educational system.

In order to achieve such a grand ambition we must recognize that our current definitions of human potential and intelligence need to be reexamined and recast in the light of more just and humanistic concerns. We can no longer settle for a system in which " . . . it seems very likely that the intelligence which the schools reward most highly is not of the highest type, that it is a matter of incomplete but docile assimilation and glib repetition rather than of fertile or rebellious creation" (Waller, 1965).

We must realize that the narrow and restricted view of intelligence and the exclusion of vast realms of human ability and feeling from the educational environment is a central aspect of the process of social control and domination by a system that is endangered instead of enriched by the full realization of human potential. "The repression of aesthetic and moral needs is a vehicle of domination" (Marcuse, 1972). Restructuring our concepts and modes of education is a requirement for social liberation on a meaningful scale. "To the degree to which liberation presupposes the development of a radically different consciousness (a veritable *counterconsciousness*) capable of breaking through the fetishism of the consumer society, it presupposes a knowledge and sensibility which the established order, through its class system of education, *blocks* for the majority of the people" (Marcuse, 1972).

We must attempt to capture the opportunities offered by our present state of change and transition because, if we do not, these same opportunities will be turned against the tide of humanization and merely increase the power of our system to limit and confine the potential of a majority of its members. It is hard to view the prospect for beneficial and creative change with undaunted optimism. Societies are rarely willing to support modes of education that threaten their basic values or their definitions of what it means to be a member of their culture. The only hope for such a movement is through the growing realization that there is much that needs to be changed, that change is possible, and that the potential for improvement outweighs the frustrations and defeats that stand in the way.

The chapters in this section are beginning steps on this road. They lay out basic assumptions and viewpoints that can guide and inform those who wish to understand the nature of the challenge to our present modes of thinking, and to begin the effort of recasting and redirecting our educational processes. The rapidly expanding technologies of communications do not, in and of themselves, contain a key to educational reform or revolution. They do provide us, however, with an incentive to pause and reflect on current policies and future directions, and with the means to modify some of these policies and chart some of these directions. We must guard, however, against the temptation to let the imperatives of technology dictate the course of educational innovations.

The chapters by Gross and Olson and Bruner deal with fundamental issues of the very conceptions of the nature of knowledge, intelligence, and competence, and of the means by which these can be fostered and cultivated. These chapters present an outlook that is richer and more humanistic in scope than the spirit that dominates the current educational establishment. They also attempt to lay a theoretical groundwork on the basis of which the role of the technological media can be seen in proper perspective in terms of their very real capabilities and their equally real limitations. Technology must be the servant, not the master of our educational fate.

The chapters by Palmer and Joyce represent the experience of those who are developing ways to put the new technologies to work in the service of more just and truly equalitarian educational objectives. Palmer, writing from the vantage point of the most large-scale and successful effort in the history of educational television, gives us an insight into the unique marriage of creative production technique and research-grounded theoretical insight that makes the Children's Television Workshop a particularly valuable example of technological potential. It is a tribute to the wisdom and integrity of its producers that even skeptics (like myself) must admire and applaud the many achievements of "Sesame Street" and "The Electric Company" even while cautioning against those who are overeager to see in them a panacea for all educational ills.

Joyce presents us with an argument against the role and nature of our current school system (one with which I obviously agree), and then outlines a bold scheme for a technologically based, but not dominated system of learning centers that would combine to provide a meaningful and lifelong range of educational modes and objectives. Whether this is a hopelessly utopian and impractical proposal or a realistic and feasible alternative will depend on the strength and courage of our determination to raise new questions and seek new answers to our educational dilemmas.

The optimism inherent in the first four chapters is offset somewhat by Gerbner's analysis of the role of teachers and educational institutions as they have been formulated and promulgated in the mass media of several countries. Gerbner shows that Western industrial societies have systematically portrayed teachers and schools through images that "humiliate and depress them." Whether a profession that has been "despised and rejected" by the popular culture of its society can seize the opportunities inherent in our technological revolution and become the cornerstone of a radically new social and cultural movement is a very important and difficult question to answer. The chapters in this section help to create a climate of opinion in which the answer would be clearly affirmative.

REFERENCES

Katz, M. B. *Class, Bureaucracy, and Schools*. New York: Praeger, 1971.

Marcuse, H. *Counterrevolution and Revolt*. Boston: Beacon, 1972.

Maslow, A. H. *The Farther Reaches of Human Nature*. New York: Viking, 1971.

Waller, W. *The Sociology of Teaching*. New York: Wiley, 1965.

CHAPTER 13

Modes of Communication and the Acquisition of Symbolic Competence*

LARRY P. GROSS

The aims of education include the transmission of knowledge, the instillation of values, and the development of intellectual, physical, social, and artistic skills and competencies. This chapter will concentrate on the nature of these skills and competencies and the conditions governing their acquisition. This emphasis, shaped by certain normative assumptions, is specially conditioned by the way in which I formulate the concepts of intelligence and ability.

Symbolic thought and communication is a uniquely human attainment and can be thought of as the source and substance of culture and civilization. Only through competence in the modes of symbolic behavior does man transcend private experience and achieve even a modicum of creative mastery over his environment. To the extent that, as educators, we accept the responsibility of contributing to a citizenry that is capable of thinking in the fullest sense of the term and of fully realizing their particular creative potentialities, we must devise educational systems that permit and encourage the acquisition of the widest possible range of symbolic competence.

I will argue that thought and knowledge are always active processes and that they exist in a variety of distinct modes. These modes are systems of symbolic thought and action that (dependent upon the nature of our biological structures and physical environment) determine the kinds of information we can perceive, manipulate, and communicate.

* I would like to thank Ray Birdwhistell, Charles Hoban, Susan Schwartz, and Sol Worth for their useful criticisms of earlier drafts of this chapter.

189

Meaning can only be understood or purposively communicated within a symbolic mode, and some minimal level of competence is the basic precondition for the creation or comprehension of symbolic meaning within such a mode. The central goal of education, therefore, must be the acquisition of competence in the modes of symbolic behavior. I will concentrate in this chapter on the early stages of education because it is there that the emergence of these competencies seems to be best facilitated or tragically discouraged.

The concept that symbolic thought is created and communicated not merely by one but through a variety of modes will be seen to have fundamental implications for the determination of educational techniques and priorities. Among these implications is an emphasis on direct experience and active exploration and manipulation. And if we accept the importance of direct, immediate, and active participation as fundamental learning processes, we must view the growth of the technological media of communication with guarded enthusiasm.

The view of thinking "in the fullest sense of the term," basic to the arguments to be presented, is derived largely from Piaget's formulations of intelligence and learning:

> The essential functions of intelligence consist in understanding and inventing, in other words in building up structures by structuring reality. . . . The essential fact . . . which has revolutionized our concepts of intelligence, is that knowledge is derived from action. . . . To know an object is to act upon it and to transform it, in order to grasp the mechanisms of that transformation as they function in connection with the transformative actions themselves. [1970, p. 27 ff]

Thinking is an *activity* embracing the perception and the cognitive processing, storage, and retrieval of structured information. Structured and meaningful information can be received, stored, transformed, and communicated through a variety of symbolic modes that are variously amenable to formulation in symbolic code systems. These modes are partially but not totally susceptible to translation into other modes. Thus, they are basically only learned through actions appropriate to the particular mode.

To comprehend a concept of "symbolic modes" as related to thinking demands that the common Western assumption (increasingly questioned, but still powerful), that thinking is above all a verbal activity be rejected along with the attendant almost exclusive emphasis on verbal skills in education. The primary flaw in our educational system is a fixation on reading, writing, and arithmetic, which has resulted in a blindness to the richness of symbolic thought in other modes. A pioneer in the use of visual methods in education has recently testified (Hoban, 1970) that educators "have tended to narrow the spectrum of symbol systems to (1) the verbal and (2) the mathematical."

Accepting the assumption that thinking is a multimodal process requires that we focus on a wide range of physical, perceptual, and cognitive skills that are appropriate and necessary for the full development of intellectual and social potentialities. These skills can and must be acquired fairly early in the educational process. Without them the child is largely dependent for encouragement and direction on the one primary symbolic system that we all attain —verbal language—and will be exposed to other modes almost entirely through impoverished and inherently limited translations into the verbal mode.

This is tragic for two reasons. It guarantees that the child will never attain the ability to be creative (or truly appreciative) in the nonverbal modes, because he will lack the generative ability to go beyond the given. To either produce or comprehend new forms requires skill in manipulating objects and/ or symbols within the terms of these modes. Perhaps more important, he will not be motivated to develop and extend his skill in these modes because he cannot experience or share the pleasures that derive only through the exercise and appreciation of these skills and competencies. While it is true that learning a symbolic system by translation may provide the child with many extrinsic rewards of social approval, mere translation does not offer the child (and his teacher) the more powerful and meaningful satisfactions of creative performance. And it is these satisfactions that potentiate his continued growth and development as a multimodal creator and thinker.

THE MULTIMODAL NATURE OF SYMBOLIC THOUGHT AND ACTION

A mode of symbolic behavior is a system of potential actions and operations (external and internal) in terms of which objects and events can be perceived, coded cognitively for long-term storage and retrieval, subjected to transformations and orderings, and organized into forms that can elicit meaningful inferences (of whatever level of consciousness) by the creator and/or others who possess competence in the same mode.

Modes of symbolic behavior are not identified with specific sensory systems. The fundamental system of symbolic communication, the lexical mode, requires at the minimum both the auditory and the visual senses. Symbolic modes may be largely organized within a single sensory system, but they may also blend and overlap. Present evidence indicates that the same sensory system is capable of being utilized for performance in various distinct modes of perceptual organization and symbolic communication. I am only dealing with culturally determined modes of symbolic behavior. While there may be innately determined, universal communicational codes, this is far from demon-·

strated and, even if assumed, of little relevance to the comprehension of the vast range of culturally specific codes of symbolic knowledge and action.

Modes, Codes, and Media

A code may be defined as an organized subset of the total range of elements, operations, and ordering principles that are possible in a given mode. In the simplest sense, then, any single language is a code existing within the verbal mode. We invariably encounter symbolic modes in terms of a particular "native" code that will shape our perceptions, memories, and cognitive processes in that mode. Most human beings (throughout history and even today) need never be aware that the code they know is not coextensive with the symbolic mode. Phenomenologically, then, the code—not the mode—is the primary level of analysis.

Most adult Americans, for example, are probably unaware that our base-10 number system is but one of many possible mathematical codes for the symbolic expression of quantities. While there can be no doubt that we never really transcend our native codes, the precise extent to which one code determines our ability to operate creatively in related codes is not known.

Many of our symbolic codes have been formalized through the development of notational systems such as the alphabet, numerical and mathematical signs, and musical notation. These are remarkably powerful cultural tools for the storage and transmission of symbolic messages. Skill in decoding a notational system and retrieving the stored information in the mode in which it was formulated frees one from dependence on one's immediate experience and allows the widest possible access to the heritage of our culture.

The development of technological media for the storage and transmission of performances and messages also permits the simultaneous exposure of symbolic communications to vast and widely separated audiences. The role and importance of these systems should be neither underestimated nor overrated.

It is clear that our civilization would not exist in its present form had it not been for the development of technologies to record symbolic codes. Very likely, as McLuhan argues (1964), these have operated, as well, to shape the very nature of the symbolic behavior that they were created to serve. However, the modes of symbolic communication precede the development of whatever technological media may be used to store and transmit their coded products. Moreover, the fact that means exist to make available to the learner the performances of others more competent than himself in a symbolic mode does not in any way obviate the necessity for active performance by the learner if he is to acquire such competence himself. Such performances, whether observed directly or through a technological medium, may provide inspiration and guidance, but they will substitute only to a limited extent for active learning.

Primary, Derived, and Technical Modes

The *primary* modes of symbolic behavior that roughly characterize a culture are (1) the linguistic, (2) the social-gestural, (3) the iconic, (4) the logico-mathematical, and (5) the musical. While each of these possesses the basic communicational characteristics previously outlined, it should be noted immediately that they are not all equally elaborated and formalized, that they have many areas of interpenetration, and that they do not exhaust the total range of human symbolic activity.

Hence it is necessary to recognize the existence of *derived* modes that seem to be built on one or more of the primary modes. Among these I would include poetry, dance, and film, as well as technical modes.

Modes of expression are not, of course, restricted to the symbolic. *Technical* modes of knowledge and action involve the application of competence in the primary modes to the understanding of physical and biological systems and structures; these modes of expression function as the basis for skills that are not primarily symbolic in nature. Such skills are involved in the production of material goods and the execution of complex nonsymbolic performances. These practical modes would include the various sciences, engineering and technologies, architecture, and so on. All of these utilize verbal, social, logico-mathematical, and visual skills (at least) and are therefore dependent on the prior acquisition of competence in the primary symbolic modes.

The criteria that determine whether a mode will be considered primary are those of independence and self-sufficiency. A primary mode is one that can be identified with (1) a range of objects and events, (2) a distinctive memory-storage capacity, (3) a set of operations and transformations, and (4) specific principles of ordering that govern the formulation and communication of meaning.

They are also, in an important sense, nontranslatable. This means that information that is coded within one mode will not be capable of being fully recoded in terms of another. The "essence" of a specific symbolic message will only be appreciated within the code in which it was created. There is no adequate verbal translation of an equation in differential calculus, or of an elaborate physical gesture, or a Bach fugue, or the smile of *La Gioconda*. All of these convey specific meanings with great precision, but only within the terms of the proper mode. It is this criterion of ultimate nontranslatability of meanings (and, to a lesser extent, of skills) that I am utilizing in generating the list of primary modes.

The derived modes, while not fully translatable into any one primary mode, are each dependent on at least one of these for the formulation and communication of symbolic meaning. Scientific knowledge is verified by logico-mathematical operations as well as by empirical observation. Poetry is understood in the context of our verbal competence:

First of all, a poem cannot be regarded as totally independent of the poet's and reader's extrinsic experiences—not if we recognize that our experiences include *language itself,* and that it is upon our past linguistic experiences that poetry depends for its most characteristic effects. Moreover, a poem does not, like the proposition systems of mathematical logic, make its own rules; it adopts and adapts the rules (i.e. conventions) of nonliterary discourse so that the principles which generate and conclude the one are conspicuously reflected in those of the other. [Smith, 1968]

The most important implication of the nontranslatability of the symbolic modes is that in the absence of sufficient competence in a mode, one will be unable to fully appreciate, much less be creative in that mode. The vast riches of our culture testify to the almost infinite range of creative and productive performance, and the equally unlimited range of sophisticated appreciation and connoisseurship available to those who have achieved the necessary competencies.

There may be a sense, however, in which symbolic communication is always an impoverished translation. It is not clear whether or not creative thought is ever carried on entirely within any symbolic mode. Possibly, the ultimate nature of thought is solipsistic and fluid to the extent that we can never fully communicate the nature of our internal thoughts and feelings. Jakobson, writing to Hadamard about the use of signs in thought, notes:

For socialized thought (the stage of communication) and for the thought which is being socialized (the stage of formulation), the most usual system of signs is language properly called; but internal thought, especially when creative, willingly uses other systems of signs which are more flexible, less standardized than language and leave more liberty, more dynamism to creative thought. [Hadamard, 1954]

The Tacitness of Skill and Competence

Competence in a symbolic mode involves the ability to perceive and/or manipulate symbols. The basic skill is that of receiving and comprehending an organized symbolic message. The process of reception and decoding is not passive and, it will be argued, the process of acquiring even a minimal level of skill cannot be passive.

Symbolic competence minimally involves (1) knowing the range of symbols and the range of referents to which they apply, (2) some wareness of the operations and transformations involved in coding such messages and activities, (3) the ability to store and retrieve information coded in the proper mode, and often, if not always (4) some awareness of the results of prior performances (by oneself or others) that may serve as the basis for evaluating the quality of the encoded behavior/message.

Beyond these requirements for the proper reception and comprehension (and, possibly, evaluation) of symbolic messages, one may see the development of two complex, distinct, but not mutually exclusive levels of skill in a symbolic mode. These are the levels of creative production and sophisticated appreciation.

Creative activity in a symbolic mode is deeply rooted in the process of reception and comprehension in that same mode. In the most fundamental sense, appreciation is a constant aspect of the exercise of any symbolic skill. One of the most important emphases of the generative grammarians has been that in order to understand verbal behavior, one must deal with the fact that any member of a linguistic community is capable of, and constantly involved in creating and comprehending sentences and sentence combinations that are completely novel to that individual (Chomsky, 1965). In a higher and more complex sense of the term, we tend to call an individual creative if he can regularly produce organized symbolic objects, events, or messages that are novel (exactly in what sense is a very complicated question), and that satisfy certain criteria (rarely made consciously explicit) of beauty, scientific and/or practical utility, expressive meaning, and so on. The sine qua non of creative performance is competence in the proper mode. The act of performing creatively is one in which appreciative skill (reception and comprehension) is constantly being exercised, however tacitly and unconsciously, at a high level of competence.

Sophisticated appreciation of organized symbolic events requires competence in perceiving and attending to skillful aspects of the performance, remembering previous performances and comparing them with present ones, understanding the levels of decision making involved, and evolving and applying criteria for the evaluation of the beauty, utility, expressiveness, and integrity of performances (Gross, 1972).

Symbolic competence, whether at the level of basic decoding or at more complex levels of appreciation and creativity, is characterized by a further set of psychological properties. The existence of competence is dependent on extensive and continual action. Skillful action in a mode is intelligence and knowledge itself (Piaget, 1970, 1971; Olson, 1970), and at the same time it is the only way in which such knowledge can be acquired, maintained, extended, and utilized in creative and productive activity. "Intelligence is skill in a medium, or, more precisely, skill in a cultural medium" (Olson, 1970). All such skills are largely *tacit, transparent*, serve to *involuntarily structure* perception, memory, and cognition, and are *generative*.

In discussing the nature of science, Polanyi (1958) insists on the centrality of the notion of skill, and on its tacitness: "The aim of a skillful performance is achieved by the observance of a set of rules which are not known as such to the person following them."

Basic motor and perceptual-cognitive skills become increasingly less explicit and conscious as they become better "known" through practice. In this sense it seems that skillful activity can only be carried on efficiently when we need not (cannot) consciously and explicitly attend to the ongoing physical, perceptual, and cognitive operations.

At least one of the reasons for this often-noted tacit property of skillful action is that, while we tend to conceive of conscious and explicit attention in terms of verbally coded information and thought, in fact, much if not most of the physical, perceptual, and cognitive elements of skillful performance are not amenable to being coded and comprehended verbally.

It is indeed true, as Polanyi claims, that "we know more than we can tell" (1966). We know much that we cannot tell in words but that we tell in other symbolic forms that express and communicate that knowledge. This second sense of the tacit nature of skill is the crucial nontranslatability of knowledge in one symbolic mode into the verbal lingua franca of individual consciousness and social communication. I would hold that this fact is of utmost importance for the understanding of the proper method of instruction, which aims to develop skill and competence in a symbolic mode.

Symbolic codes are also involuntary and transparent structurings of thought and action. In the sense defined by Piaget (and implied by Whorf and Sapir), symbolic codes govern our structuring of reality. We assimilate the world through perceptual-cognitive schemata that, although dependent on innate structure, are developed, modified, and extended through interactions with, and accommodations to our environment. Knowledge is acquired and expressed through performance in a medium and the use of that medium becomes automatic and transparent. As we become competent "native speakers," words and sentences in our native language become carriers of meanings and we no longer need pay (or, in fact, can pay) conscious attention to their actual auditory or visual characteristics. We begin to hear meanings instead of sounds, and we do so involuntarily.

As evidence of this aspect of lexical transparency, think of the difficulty encountered in training students of linguistics to attend to the actual phonetic aspects of speech and ignore the phonemically carried meaning. This kind of perceptual transparency involves an involuntary structuring and organization of symbolic information. One cannot voluntarily fail to understand words spoken or written in one's native tongue or voluntarily not hear musical dissonances in a style one has learned (or become familiar with). Beyond a certain level of competence in any symbolic mode it is only with great effort (actually, with the acquisition of a rather specialized secondary skill) that one is able to avoid structuring a symbolic message in terms of its organized meaning. While it is not at all clear what the "perceptual units" are in the various modes, it is certainly the case that most of us comprehend symbolic messages in organized chunks before we are able to deliberately attend to and evaluate their constituent elements.

Only when one can perceive, select, store, recall, transform, and order objects, events, and information without constant recourse to nontacit levels of attention and consciousness is one capable of generating or comprehending novel and aesthetically pleasing symbolic performances. In our culture, such competence is found almost exclusively in the lexical and social modes.

The acquisition of this fundamental level of appreciative-generative skill, however, requires learning and practice in performance *in the mode itself.* Although much of what we learn and what we hope to teach our children can only, or most efficiently, be learned through a linguistic code, skill and competence in the full range of symbolic modes cannot be acquired in translation.

VARIETIES OF THE SYMBOLIC EXPERIENCE

The Social-Gestural Mode

One of the major achievements of modern cultural anthropology has been the increasingly explicit insight that most human behavior, particularly in the presence of others, is determined by learned, culturally specific modes and patterns that communicate to fellow members of that culture a great deal of precise information about the stable characteristics and situational intentions of the actor (Birdwhistell, 1970; Hall, 1959, 1966; Goffman, 1959, 1963). Along with verbal language, in fact, the social-gestural mode is the only mode of thought and action that every member of a culture will acquire. It is also, along with verbal language, one that is never really taught by formal instruction, but is acquired primarily through observation, imitation, and trial and error.

Usually one will only become consciously aware of the fact that actions and gestures carry decodable information about one's background, intentions, and so on, when one is (1) placed in a foreign culture, (2) being trained to deliberately observe such processes (e.g., in ethnographic or psychiatric training), or (3) attempting to consciously convey misleading information. In the latter instance, for example, one can avail oneself of books on etiquette if one wishes to "pass" successfully in social circles in which one was not originally acculturated. Such code dictionaries are not written for "native speakers" of a culture, who often equate the use of such devices as a sign that the user "does not belong."

A child who cannot acquire the basic verbal and social codes of his culture will not be able to function as a normal human being:

A system exists into which [the child] must be assimilated if the society is to sustain itself. If his behavior cannot, afer a period of time, become predictable to a degree expected in that society, he must be specially treated. . . . This special treatment can range from deification to incarceration. But ultimately the goal is the same: to make *that child's* behavior sufficiently predictable that the society can go about the rest of its business. [Birdwhistell, 1970]

It is an open question whether the other cultural modes have not attained this sort of priority in our culture because we do not yet know how to ensure the successful achievement of competence in them; or, conversely, we have not learned how to cultivate them universally because they are not crucial to the business of our society.

The Iconic Mode

Arnheim (1969) has recently discussed the relation of thought and visual perception. He establishes in great detail the nature, extent, and importance of perceptual-cognitive symbolic behavior, which is organized visually in the iconic mode. Minimally, it must be accepted that visual images and symbols are capable of communicating and expressing meaningful information that cannot be formulated in the lexical or, indeed, any other mode. Ivins (1953) points out that words, being in essence "conventional symbols for similarities," are incapable of communicating the unique and singular aspects of objects and events that can be depicted visually.

Iconic symbols are highly suitable for the purpose of organizing and communicating information about the spatial, topological nature of objects, about relations between objects in space, and for expressing and evoking emotional responses. To see visual images as merely a peculiar way to tell a story is, as Ivins (1953), Gombrich (1960, 1965a, 1965b), Arnheim (1969), and many others have shown, to misunderstand totally the nature of the iconic mode.

A topical issue here is the status of film as a mode of communication. More and more we are being told that our children are becoming primarily oriented toward visually communicated information through film and television. A corollary argument has been that they are thus learning a (new?) "visual language" and that educators should accept the conclusion that the "language" used by these media be taught in schools.

Film is considerably more complex than "simple" visual communication in that it is organized in temporally sequential images that do, indeed, require the viewer to exercise skill in order to comprehend and appreciate the intended meaning (cf. Panofsky, 1947).

However, it is not at all clear exactly to what extent this organization is culturally specific or, if so, whether it is a function of the linguistic code of the culture (Worth and Adair, 1972).

Second, it does not seem to be the case that competence in appreciating the meaning conveyed in a film is dependent on the same degree of performatory skill as has been claimed to be true for the primary modes. Finally, it is highly debatable whether there is any heuristic benefit to be gained in the understanding of how film communicates by starting from the assumption that film is a "language" (cf. Worth, 1970, for a detailed discussion of this point).

The Logico-Mathematical Mode

One can conceptualize the mode that I am calling logico-mathematical at two rather distinct levels. In his recent book, *Biology and Knowledge* (1971), Piaget describes logico-mathematical knowledge as "one of the three main categories of knowledge, coming between innate structures and knowledge based on physical or external experience." He also notes, however, that it "takes on a differentiated form only in the higher ranges of human intelligence." This is an important distinction because it allows us to see the aspects of action that are logical in nature as well as the particular culturally elaborated mode of logical and mathematical thought. In describing the basic category of logico-mathematical knowledge, Piaget further clarifies the independence of knowledge from verbal language:

> Logic . . . is not to be reduced, as some people would have it, to a system of notations inherent in speech or in any sort of language. It also consists of a system of operations (classifying, making series, making connections, making use of combinative or "transformation groups," etc.) and the source of these operations is to be found beyond language, in the general coordinations of action. [1971]

In this sense, then, I would maintain that all of the modes of symbolic thought are logical in nature. However, the higher, "differentiated" form of logico-mathematical thinking can most certainly be seen as an organized system of operations that permits those who have acquired the requisite competence to manipulate, store, retrieve, and organize symbolic information in a rather complex and specific code.

The French mathematician Jacques Hadamard (1954) obtained statements from many eminent mathematicians about their conceptions of the nature of logico-mathematical thought. The most consistent aspect of these reports was the claim that mathematical thought is not performed in the linguistic mode: "I insist that words are totally absent from my mind when I really think."

The Musical Mode

In every known human culture there exists a musical code of formally organized communication. At the level of a culturally determined symbolic code, music, like mathematics, expresses and communicates specific but verbally ineffable meanings.

> It seems to me quite clear that music, far from being in any sense vague or imprecise, is within its own sphere the most precise possible language. I have tried to imply this by saying that music embodies a certain type of movement rather than that it expresses it. All of the elements of this movement—rhythm, pitch, accent, dynamic shading, tone quality and others sometimes even more subtle—are, in competent hands, kept under the most exquisite control, by composer and performer alike; the movement that is the

stuff of music is given the most precise possible shape. It was for just this reason that both the ancients and the teachers of the Middle Ages accorded to music such high place in educational discipline. By these means, a musical gesture gains what we sometimes call "musical sense." It achieves a meaning which can be achieved in no other way. [Sessions, 1968]

The creation or appreciation of musical meaning depends, as in the other symbolic modes, on the same order of tacit fluency. As Meyer (1956) explains:

. . . we perceive and think in terms of a specific musical language just as we think in terms of a specific vocabulary and grammar; and the possibilities presented to us by a particular musical vocabulary and grammar condition the operation of our mental processes and hence of the expectations which are entertained on the basis of those processes.

Meyer shows that Western music is architectonic and hierarchic in nature, and that music appreciation involves the continual arousal of expectations, which are confirmed or modified as the listener "recreates" the structural organization of the piece. While this is not the only possible mode of musical organization (cf. Keil, 1966), it does seem to be the primary mode of Western "classical" music. The ability to decode the structure of musical organization is dependent on the competence of the listener in the particular cultural code or style in which a piece has been formed. Musical meaning exists only in the perceptions of those who have acquired specific culturally determined habits and dispositions.

These dispositions and habits are learned by constant practice in listening and performing, practice which should, and usually does, begin in early childhood. Objective knowledge and conceptual understanding [verbally coded information] do not provide the automatic, instinctive perceptions and responses which will enable the listener to understand the swift, subtle, changeable course of the musical stream.[Meyer, 1956]

This is a point of crucial importance for the understanding of symbolic communication. Only on the basis of the competence to appreciate meaning presented in a symbolic mode can one hope to achieve the realization of creative potential in that mode. Early learning, through action and performance in a cultural mode, is necessary for the achievement of appreciative skill, and creative skill can emerge only if the individual has already attained this fundamental competence. The process of creation in a symbolic communicational mode presupposes and constantly involves the process of appreciation. "It is because the composer is also a listener that he is able to control his inspiration with reference to the listener" (Meyer, 1956).

THE ACQUISITION OF SKILLS AND COMPETENCIES

Competence Is Its Own Reward

The kinds of activities that lead to the acquisition of skill and competence in the modes of symbolic behavior seem to be intrinsically rewarding. This fact is implicit in the conclusions drawn by White (1959) on the basis of evidence from a wide range of research on animal learning, child development and psychoanalytic processes. White describes a class of behaviors that seem to share a common biological and psychological significance:

. . . they all form part of the process whereby the animal or child learns to interact effectively with his environment. The word *competence* is chosen as suitable to indicate this common property. Such activities in the ultimate service of competence must . . . be conceived to be motivated in their own right.

Among the behaviors that White claims are motivated by their role in the development of competence are the activities that are crucial for the development of competence in interacting effectively with the symbolically organized aspects of the environment. The intrinsic satisfactions of such activities are related to the playful character of many of the basic experiences in symbolic communication.

[Play] in its two essential forms of sensorimotor exercise and symbolism, is an assimilation of reality into activity proper, providing the latter with its necessary sustenance and transforming reality in accordance with the self's complex of needs. [Piaget, 1970]

The pleasures of effective interaction with the environment are not at all limited to the young child's development of competence. At all ages and at all levels the exercise of competence in the skillful manipulation of the physical and symbolic environment provides continual intrinsic satisfaction. In fact, I would claim that the most quintessentially human form of pleasure derives from the exercise of creative and appreciative skills. Moreover, these are open-ended skills. One cannot speak of a point at which competence in a perceptual-performance skill has been completely achieved. Instead, I would say that competence is continually being acquired and extended through performance, and therefore it is continually satisfying.

Learning Through Action

All competence in a skillful mode is acquired on the basis of constant practice and repetition (as well as observation and appreciation). While at the higher levels of competence, practice may get you to Carnegie Hall, at the initial stages of learning it is the only way to get anywhere. One achieves competence in a medium by slowly building on routines that have been performed over and over until they have become tacit and habitual. This basic repetitious activity can be easily seen in children who derive enormous satisfaction from performing over and over some action that results in a predictable effect. The feeling of efficacy, as White terms it, is the basic and initial form of satisfaction in competence. It is on the basis of a repertoire of often-repeated actions that the child can begin to introduce and perceive slight variations and thus extend the range of his perceptual-intellectual competence to more complex forms of organized behavior.

The acquisition of competence in modes of symbolic communication entails the learning of the "vocabulary" for representing objects and events proper to a particular mode, and the "grammatical" and "syntactical" operations, transformations, and organizational principles that are used to structure these into conveyers of meaning and intention. In our culture some of the basic modes of symbolic knowledge have been formalized in terms of notational systems and technological media. As a result, an important aspect of the acquisition of communicative competence in these modes is the development of skill in manipulating these notational and technological forms.

The most important of the notational systems is the visual recording of phonetic symbols in written language. It does not necessarily follow, however, that skill in reading and writing should be taken as a paradigm for learning and instruction in other modes that also possess formal notational systems.

The Verbal Fallacy

A basic corollary of the assumption that thinking is primarily linguistic is the view that skill in reading and writing is a precondition for all meaningful learning. The enormous cultural value of reading skill is undeniable, as it is largely through reading that one can come into contact with the vast realms of stored knowledge that are the key to our highly complex civilization. However, knowledge—in the sense of competence in the linguistic mode—is required in order to learn to read and write in the first place. In fact, based on the work of Piaget, Furth argues that, "A school that in the earliest grades focuses primarily on reading cannot also focus on thinking" (1970).

While I do not agree with Furth that reading cannot offer the intrinsic satisfactions of competence-extending activities, it does seem clear that an overemphasis on reading skills is detrimental to the fullest development of competence in the nonverbal modes of symbolic knowledge. This approach tends to commit the errors of introducing the child to the nonverbal modes almost exclusively through impoverished verbal translation (which largely consists of applying verbal labels to objects and events coded in these symbolic modes);

and, hence, of not allowing the child to engage in these forms of active behavior within the modes themselves, which are the only route to the acquisition of competence.

The mistaken identification of verbal assimilation of information with the acquisition of knowledge about the referential world is the source of many unfortunate educational practices. Once one has achieved a symbolic skill, it can be highly efficient to use verbally coded information in order to record or convey certain aspects of the knowledge embodied in that skill. But one would not have been able to acquire that skill, in most instances, on the basis of such verbal information. "Verbal description appears never to be equivalent to a motor performatory act" (Olson, 1970).

The realization that we cannot convey most information about the important aspects of reality to children through the linguistic mode is, unfortunately, still relatively rare. One of the most difficult intellectual tasks is that of being able to understand what it means not to know something that one has already assimilated at the level of tacit and transparent knowledge.

> . . . whenever it is a question of speech or verbal instruction, we tend to start off from the implicit postulate that this educational transmission supplies the child with the instruments of assimilation as such simultaneously with the knowledge to be assimilated, forgetting that such instruments cannot be acquired except by means of internal activity, and that all assimilation is a restructuration or a reinvention. [Piaget, 1970]

We also tend to mistakenly treat the child's ability to repeat verbally coded information as evidence that he has the ability to use that information constructively. The performatory ability of a child, for example, to recite the alphabet is not at all the same as competence in the verbal mode, let alone the same as being able to read. In fact, it is not at all clear whether this ability is a useful or necessary precondition for the development of skill in reading and writing (cf. Rozin et al., 1971; Makita, 1968; Smith, 1970).

At the point at which they begin their "formal education," most children are already quite fluent in the linguistic and social-gestural codes of their culture. If they are to achieve even this not inconsiderable degree of competence in the modes of logico-mathematical, iconic and musical communication (or in other, derived modes), they must be given the same kind of ample opportunity for active exploration and the same kind of responsive and appropriately coded feedback that were necessary preconditions for the emergence of their verbal and social competence. It is only possible to attempt such "instruction" once the initial level of linguistic and social culturation has occurred, but it is impossible to achieve it if one relies primarily on verbally mediated communication and social approval.

The Seductiveness of the Visual

As psychological research and pedagogic experience have exposed the dangers of the "linguistic fallacy" in education, it has become more and more tempting to turn to the increasingly available and pervasive technologies of visual communications. While it is true that still photography, film, and television are capable of yielding enormous educational benefits and offer valuable opportunities that could not otherwise be realized, they are a mixed blessing. These technologies of communication are not equally appropriate and useful at all levels of instruction and learning.

At the higher levels of education, when basic competence in symbolic modes of communication can be taken for granted, the availability of iconic images is a fundamental requirement for the attainment of skill and knowledge in a great number of specialized fields of intellectual activity. As Ivins (1953) has clearly demonstrated, the development of many scientific fields became possible only when it was technologically feasible to produce exactly repeatable pictorial images. The later development of exactly repeatable moving visual images has further extended the range of such specialized intellectual activities through our ability to capture, store, and review movement and action, and even permits us, by slowing down or speeding up the rate of viewing, to observe aspects of the events that could never be directly perceived.

Many of the potentials of visual communications are of value also in earlier stages of education. As McLuhan has most clearly indicated, it is more and more the case that we come to know the world as it exists beyond our immediate horizon primarily through the media of electronic telecommunications. We should avail ourselves of these technologies for the instructionally guided exposure of our children to the world they are living in. Several dangers and dysfunctions inevitably arise, however, if we fail to realize the important limitations of these methods and to take into account, once again, the nature of the audience.

One potential dysfunction of improperly conceptualized enthusiasm for the educational use of pictures and film results from the failure to realize that children may not perceive or interpret these images in the way we intend them to. Just as with the verbal transmission of information, we tend to forget that children often lack the assimilative structures necessary to comprehend properly the intended meaning.

There is a level of sophistication necessary for the proper understanding of visual images. Pictures and films often convey misleading impressions of scale, distance, time, and relationship. By overcoming the limitations of space and time they may also fail to communicate the reality and importance of these dimensions. More importantly, perhaps, the images conveyed through these media may be deliberately or inadvertently false. The potential for misleading and dissembling, for confusing fiction and reality, is at least as great with photographs and films as with words and actions, and quite possibly much greater.

On the one hand, this would imply that an important task for modern education is the development of a level of sophistication sufficient to permit our children to be aware of this dangerous potential of communications technologies. On the other hand, however, it is clearly incumbent on those who utilize these media for instructional aims to be aware themselves of these problems, even of the extent to which they themselves are susceptible to them.

There is, however, a more important dysfunction in the reliance on these media of visual communications at the early stages of learning. It has been argued that verbally coded information is not sufficient for the acquisition of competence in the nonverbal modes of symbolic thought. Similarly, it is a mistake to think that by showing the child visually presented images of the actions that constitute performance in a symbolic mode one will instill competence in him. One will not learn to compose music or even to play the piano by watching a film of a virtuoso pianist performing.

The observation and imitation of sounds and actions are crucial to the development of verbal communicational competence. They are also a basic part of the process of acquiring any mode of symbolic skill. But the observation is important only insofar as it stimulates the child to perform on his own and insofar as it makes him perform the kind of actions that will evoke meaningful feedback.

> . . . if we compare the memories that distinct groups of children retain of a grouping of cubes, according to whether the grouping has been *(a)* simply looked at or perceived, *(b)* reconstructed by the child itself, or *(c)* constructed by an adult while the child watches, we find that the memories produced by case *(b)* are clearly superior. The demonstration by an adult *(c)* produces no better results than simple perception *(a)*, which shows once again that by carrying out experiments in the child's presence instead of making the child carry them out, one loses the entire informational value offered by action proper as such. [Piaget, 1970]

It is quite possible that one might be able to avoid substituting passive observation for meaningful and instructive action but there is no reason to assume that this is more easily achieved through pictures and films.

Olson (1970) found that a particular kind of nonverbal modeling, developed by the Montessori school, seemed effective in evoking skillful performance in the viewer. The important aspect of this method was that instead of simply performing the task correctly and smoothly, the model went through a simulated process of trial and error, committing and rectifying errors as well as making the right moves. Thus, Olson concludes, he "indicated to the child what the choice points were, what his alternatives were, and how he was to choose among them."

The medium of live or animated film could quite possibly be used to present performance sequences organized in this fashion. While this would not be a substitute for actual performatory activities on the part of the child, they would be a useful way of introducing and stimulating a wide range of learning behavior.

As new media emerge they evoke new forms in which symbolic knowledge and action can be organized and communicated. What we must always keep in mind, however, is that the emergence of new forms of symbolic skill and knowledge at the higher levels of complexity and sophistication does not in any way reduce the vital importance of competence in the basic modes of cultural intelligence and communication. It is folly to assume that new media of technological communication will obviate the necessity for these competencies or provide shortcuts that permit their acquisition without the basic experiences of learning through active performance within the domain of the modes themselves.

EDUCATIONAL POLICIES AND TECHNOLOGICAL IMPERATIVES

The implications of the arguments presented are not entirely in accord with many of the current technological and economic considerations that influence our educational institutions. The position taken with regard to the usefulness of the new forms of communications technology for early education in particular is fundamentally conservative and skeptical.

I believe that the necessary conditions for the acquisition of the kinds of skill and competence described above are:

1. That the child be exposed to physical and social settings that permit and encourage the initial exploratory behavior that can lead to the point where he begins to enjoy the pleasures intrinsic to activities in the symbolic modes that develop and extend his competence in effectively interacting with his environment.

2. That this exploratory activity, and the increasingly competent behavior that it leads to, be carried on with the assistance of adults who are themselves skilled and competent in the various modes of symbolic communication, and who are also trained to provide appropriate feedback and reinforcement.

3. That, as the child attains higher and more complex levels of competence, he be encouraged and guided in becoming familiar with the existing bodies of stored knowledge and meaning in these modes, and thereby develop the ability to apply criteria for judging the aesthetic, scientific, and practical success of his own and others' performances.

While this may be a somewhat utopian set of requirements, it is also the only way to evolve a society in which all can meaningfully appreciate the highest achievements of the culture and in which those with creative potential (in any mode) will be most likely to develop and express that potential.

Even at a more modest level, however, these arguments suggest that the focal points for improvement in educational practice are the provision of proper learning environments and the training of teachers who are competent in the various modes of communication and also sensitive to the particular requirements of early education.

It is obvious, however, that the centralized production and mass distribution of television programs that are viewed by millions of children are both faster, easier, and cheaper than the recruitment, training, and distribution of teachers, or the accomplishment of fundamental changes in the prejudices and policies of those who already staff our educational institutions. Given the current economic limitations on our school systems, and given the need to improve the educational opportunities of children whose earliest learning environment has not allowed them to attain the level of initial performance skills we expect of children when they enter school, it seems inevitable that we will see more and more attention and resources devoted to centrally produced, mass media-carried education programs.

It will be necessary, therefore, to find means whereby these media can be utilized more successfully to evoke and direct activities that will facilitate the acquisition of skills and competencies in the fullest possible range of symbolic communication modes. At the same time, it is even more important that we work toward the provision of the kinds of settings and the kinds of teachers that will remain the fundamental preconditions for the full acquisition of these skills. "If you want to gain knowledge you must participate in the practice of changing reality. If you want to know the taste of a pear you must change the pear by eating it. . . . All genuine knowledge originates in direct experience" (Mao Tse-Tung, *On Practice*).

REFERENCES

Arnheim, R. *Visual Thinking*. Berkeley: University of California Press, 1969.

Birdwhistell, R. *Kinesics and Context*. Philadelphia: University of Pennsylvania Press, 1970.

Chomsky, N. *Aspects of the Theory of Syntax*. Cambridge: M.I.T. Press, 1965.

Furth, M. *Piaget for Teachers*. Englewood Cliffs, N.J.: Prentice-Hall, 1970.

Goffman, E. *The Presentation of Self in Everyday Life*. New York: Doubleday, 1959.

———. *Behavior in Public Places*. Glencoe, Ill.: The Free Press, 1963.

Gombrich, E. H. *Art and Illusion*. New York: Bollingen Series, 1960.

―――. "The Use of Art for the Study of Symbols," *American Psychologist, 20* (1965a), 34–50.

―――. "Visual Discovery Through Art," *Arts Magazine*, November, 1965b.

Gross, L. "Art as the Communication of Competence," paper presented at the Symposium on Communication and the Individual in Contemporary Society, under the auspices of the International Social Science Council, the International Council for Philosophy and Humanistic ‚Studies and UNESCO, Rome, 1972.

Hadamard, J. *The Psychology of Invention in the Mathematical Field*. New York: Dover, 1954.

Hall, E. *The Silent Language*. New York: Doubleday, 1959.

―――. *The Hidden Dimension*. New York: Doubleday, 1966.

Hoban, C. F. "Communication in Education in a Revolutionary Age," *AV Communication Review, 18* (1970).

Ivins, W. *Prints and Visual Communication*. Cambridge: M.I.T. Press, 1953.

Keil, C. M. "Motion and Feeling Through Music," *Journal of Aesthetics and Art Criticism, 34* (1966).

Makita, K. "The Rarity of Reading Disability in Japanese Children," *American Journal of Orthopsychiatry, 38* (1968), 599–614.

McLuhan, M. *Understanding Media: The Extensions of Man*. New York: Signet Books, 1964.

Meyer, L. B. *Emotion and Meaning in Music*. Chicago: The University of Chicago Press, 1956.

Olson, D. R. *Cognitive Development: The Child's Acquisition of Diagonality*. New York: Academic Press, 1970.

Panofsky, E. "Style and Medium in the Motion Pictures," *Critique*, 1947.

Piaget, J. *Science of Education and the Psychology of the Child*. New York: The Viking Press, 1970.

―――. *Biology and Knowledge*. Chicago: The University of Chicago Press, 1971.

Polanyi, M. *Personal Knowledge*. Chicago: The University of Chicago Press, 1958.

―――. *The Tacit Dimension*. New York: Doubleday, 1966.

Rozin, P., S. Poritsky and R. Sotsky. "American Children with Reading Problems Can Easily Learn to Read English Represented by Chinese Characters," *Science, 171* (1971), 1264–1267.

Sessions, R. *The Musical Experience*. New York: Atheneum, 1968.

Smith, B. H. *Poetic Closure: A Study of How Poems End*. Chicago: The University of Chicago Press, 1968.

Smith, F. *Understanding Reading*. New York: Holt, 1970.

White, R. W. "Motivation Reconsidered: The Concept of Competence," *Psychological Review, 66* (1959).

Worth, S. "The Development of a Semiotic of Film," *Semiotica*, 1970.

Worth, S., and J. Adair, *Through Navajo Eyes: An Exploration in Film Communication and Anthropology*. Bloomington: Indiana University Press, 1972.

CHAPTER 14

Learning Through Experience and Learning Through Media

JEROME S. BRUNER AND DAVID R. OLSON

This chapter is concerned broadly with the consequences of two types of experience that may be designated as direct experience and mediated experience, their partial equivalence and substitutability, and their differing potential roles in the intellectual development and acculturation of children. Our analysis will begin with the problem of the nature of direct experience and its effect on development. A clearer conception of the processes involved in direct experience will permit us to examine better the manner and extent to which mediate experience may complement, elaborate, and substitute for that direct experience.

Much of a child's experience is formalized through schooling. Whether for reasons of economy or effectiveness, schools have settled on learning out of context through media that are primarily symbolic. Schooling generally reflects the naive psychology that has been made explicit by Fritz Heider (Baldwin, 1967). The general assumption of such a naive psychology is that the effects of experience can be considered as knowledge, that knowledge is conscious, and that knowledge can be translated into words. Symmetrically, words can be translated into knowledge, hence, one can learn, that is, acquire knowledge, from being told.

Congruent with this is the belief that what differentiates child from adult is also knowledge and that the chief mission of school is to impart it by the formal mode of pedagogy. Concern for "character" or "virtue" centers not on the school, but on the home and the child's more intimate surrounding, the sources that provide models.

The assumptions that knowledge was central to the educational enterprise and that it was independent of both the form of experience from which it derived and the goals for which it was used had several important and persisting effects on educational thought. First, it led to a certain blindness to the effects of the *medium* of instruction as opposed to the content, a blindness that McLuhan (1964) has diagnosed well; this led, second, to a deemphasis of and a restricted conception of the nature and development of *ability*. As the effects of experience were increasingly equated to the accumulation of knowledge, experience was considered less and less often as the source of ability. Since knowledge was all, ability could be taken for granted—simply, one *had* abilities that could be used to acquire knowledge. Abilities were then projected rather directly into the mind in the form of genetic traits (Jensen, 1969). Culture and experience were both ignored as possible candidates to account for their development. The result of this strange turn has been to downgrade the task of cultivating abilities in students, often thereby making schooling a poor instrument for the attainment of those important effects.

Education critics have, of course, long attacked educational goals formulated in terms of the simple acquisition of knowledge. Dewey's (1916) concerns with the relationship between knowledge and experience have much in common with contemporary reanalysis. Genuine experience involved the initiation of some activity and a recognition of the consequences that ensued. Experiences of this sort would result, Dewey argued, in the natural and integrated development of knowledge, skills, and thinking. Schooling, on the other hand, attempted to develop the three independently of each other and with little regard for the experience of which they are products. No surprise then, that schools frequently failed to achieve any of them. Dewey's revised conception of the relationship between experience and knowledge reappears in the current attempts at educational reform, which emphasize the role of process instead of content, or more specifically, of activity, participation, and experience instead of the acquisition of factual information (Bruner, 1960; *Learning and Living*, 1969). The contemporary critic and Dewey alike would attack the assumption that knowledge is acquired independently of the means of instruction and independently of the intended uses to which knowledge is to be put.

That knowledge is dependent on or is in some ways limited by the purpose for which it was acquired has been illustrated in experiments by Dunker (1945), by Maier (1931), and by many other students of thinking and problem solving. The conventional use of a pliers as a gripping instrument makes them difficult to perceive as a pendulum bob. Knowledge per se does not make it possible to solve problems. The same appears to be true of verbally coded information. Maier, Thurber, and Janzen (1969) showed that information coded appropriately for purposes of recall was thereby inappropriate for purposes of solving a problem. Information picked up from experience is limited in important ways to the purpose for which it is acquired—unless special means are arranged to free it from its context. But this conclusion is at odds with the naive view that one can substitute "instruction" for "learning through experience."

We must, then, reexamine the nature of direct experience and its relationship to both knowledge *and* skills or abilities. Of course, the term "direct" experience is somewhat misleading in that all knowledge is mediated through activity; the resulting knowledge is not independent of the nature of those activities. But if we consider both the knowledge of objects and events that result from experience and the structure of activities involved in experiencing, we may come closer to an adequate conception of "direct experience." We will then be in a better position to contrast it with mediated, or more accurately, with the symbolically encoded and vicarious experience that is so important in acculturation.

DIRECT EXPERIENCE

Psychology, mirroring an earlier physics, often begins an account of the nature of experience with the concept of the "stimulus." What occurs in behavior is thought to be a reflection of the stimulus acting on the organism. At a more abstract level of analysis, the shape of the effective stimulus is seen as the result of certain physical filterings or transformations of the input given by the nature of the nervous system and its transducers. This conception is much too passive and nonselective with respect to what affects organisms. Living systems have an integrity of their own; they have commerce with the environment on their own terms, selecting from the environment and building representations of this environment as required for the survival and fulfillment of the individual and the species. It follows that our conception of physical reality is itself achieved by selective mediation (Tolman, 1932; Pribram, 1971; Sokolov, 1969; von Holst 1937). The search for a psychological account of behavior must begin with the organism's activities and then determine the nature of the "reality" sustained by that type of activity. It is a point that is explicit and central to Piaget's (1971) conception of adaptive behavior in general and intelligence in particular: objects and events are not passively recorded or copied but rather acted upon and perceived in terms of action performed.

What does this imply about the nature and consequence of experience? As we have said, we have a picture of reality that is biased by or coded in terms of our actions on it; knowledge is always mediated or specified through some form of human activity. Any knowledge acquired through any such activity, however, has two facets: information about the *world* and information about the *activity* used in gaining knowledge. In an aphorism: from sitting on chairs one learns both about "chairs" and about "sitting." This distinction is reflected in ordinary language in the terms knowledge and skill or ability. There are therefore, two types of invariants that are specified through experience. The set of features that are more or less invariant across different activities may be considered as the structural or invariant features of *objects* and *events*. Similarly the set of operations or constituent acts held invariant when performed across different objects and events may be considered as the structural basis of the activities themselves—what we call *skills and abilities*. It is our hypothesis that "knowledge" reflects the invariants in the natural and social environment while "skills or abilities" reflect the structure of the medium or performatory domain in which various activities are carried out (see Figure 1). Obviously, major value must be attributed to *both* facets of experience.

Consider more specifically how both facets are realized in practice. The performance of any act may be considered a sequence of decision points, each involving a set of alternatives. These decision points are specified jointly by the intention motivating the act and the structure of the medium or environment in which the act occurs. An effective performance requires that the actor have information available that permits him to choose between these alternatives. In making choices, he must "analyze the task" in the sense of keeping in mind not only the end state he is seeking, but also where he is with respect to that end state. He must assess the means while keeping the end criteria in mind. It is a universal routine—in love, in war, in writing a paragraph or solving an

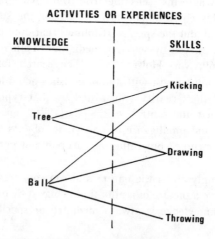

Figure 1 The relationship of knowledge and skills to physical activity.

equation, or indeed, in managing to get hold of objects during the initial phases of infant mastery of reaching.

From this point of view, mastery depends on the acquisition of information required for choosing between alternative courses of action that could lead to a sought-after end. The most obvious way to acquire such information is through active attempts to achieve various goals in a variety of performatory domains. This is surely what is meant by learning through one's own direct contingent experience. In virtually all accounts of learning the consequences of one's acts are postulated as the major source of skill and/or knowledge.

MEDIATE EXPERIENCE

But there are other ways to acquire information. From seeing a man struggle with his load, one can make some estimate of its weight. That is, one can experience vicariously or mediately. Psychological studies have repeatedly shown that learning can occur when neither of the primary conditions for learning through contingent experience—self-initiated action or direct knowledge of its results—is fulfilled. An illustrative experiment was performed long ago by Herbert and Harsh (1944) that reveals something particularly interesting about observational learning. Two groups of cats learned to pull strings and open doors by observing other cats. One group of cats saw only the final errorless performances of cat models while the other group saw the early error-filled performances as well as the correct performances. Both groups of cats learned to solve the problems more quickly than the control cats, who learned only from their own contingent experience. But the cats that saw the error-filled performances learned more readily than those who saw only the error-free performances. Might this finding point to an important difference between good and poor demonstrations? We shall return to the question shortly.

A second alternative to learning directly is through symbolically coded information—transmitted through the media—such as the spoken or printed word, film, or diagram. It is *learning through the media* that most readily substitutes for direct experience in formal schooling. Vygotsky (1962) and Bruner (1966) have emphasized the extent to which language per se provides the means *par excellence* for teaching and learning out of context, out of a situation in which action is in process and contingent consequences are clearest. Language, then, provides the best opportunity for acquiring knowledge by criteria other than its usefulness in a particular line of action. It is the medium that most directly lends itself to ordering of knowledge in terms of the rules of abstract thought.

We have, therefore, three modes of experience that map roughly on to the three forms of representation discussed earlier (Bruner, 1966) as enactive, iconic, and symbolic: the first related to direct action, the second to models, and the third to symbolic systems.

More important to our purpose is the fact that these three modes of experience map on to evolutionary development (Bruner, 1971). While all animals learn from contingent experience, primates are distinctive in their capacity for learning by observation—there is an enormous amount of observation of adult behavior by the young with incorporation of what has been learned into a pattern of play. The human species is, of course, marked by its reliance on symbolically coded experience—so much so that the development of language is taken as the distinctive characteristic of the human species and the development of literacy in various symbolic codes is the primary concern of formal schooling. It follows that these three forms of experience differ greatly in the assumptions they make about the organism, that is, they differ primarily in terms of the skills they both assume and develop.

ON THE ACQUISITION OF KNOWLEDGE

To this point, our argument has shown that one can learn from three very different forms of experience and that these forms of experience, whether mediate or direct, qualify what is learned. This section of the chapter is directed to an examination of what is common to different forms of experience; the final section is concerned with what is distinctive about them. Our conclusion will be that different forms of experience converge as to the knowledge they specify, but they diverge as to the skills they develop.

The assertion to be examined here is that different forms of experiencing an object or an event can be mapped on to a common underlying structure—a coherent and generalized conception of reality. Information about a particular event, regardless of the activity or medium through which it is obtained, has in common the property that it permits the differentiation of that event from the alternatives with which it is in danger of being confused. Consider the experience of actually seeing a zebra with that of hearing the instructional statement, "A zebra is a striped, horselike animal." The same features detected in the act of discrimination are specified in the statement, hence, they are informationally equivalent and they can both be mapped on to an underlying conception of zebras, horses, animals, and so on. This is not to deny that each mode has a residual uniqueness but only to insist that they share a common structure as well. The range of topographically dissimilar forms of experience, including various forms of instruction, may be considered as various "surface structures" that relate in their special ways to a common underlying structure of knowledge. Indeed, it is the very fact that information relevant for action can be acquired through means *other* than direct action itself that makes instruction possible. Thus, one can learn to sail, perhaps only to a limited extent, through watching films and reading books. There is considerable evidence from controlled laboratory experiments to show that common learning results from different types of experience. A child can learn to construct a diagonal either through activity coupled with reinforcement, through an appropriate demonstration, or through instruction in language (Olson, 1970a). Others have shown that, although difficult to teach, children can learn Piagetian conservation tasks through quite different training techniques (Bryant, 1971; Halford, 1970; Gelman, 1969). And it is well known that there is almost an infinite number of ways to teach reading (Chall, 1965). The problem is to specify as far as possible the structure of information in these various instructional forms or surface structures and to see how they each relate to the underlying structure described above. Once these forms of instruction have been specified it may be possible to indicate how each of them relates to the various technologies involved in their production and distribution. These relationships are laid out in a preliminary way in Figure 2. Figure 2 indicates that there are three basic forms of instruction, through arranged contingent experience, observational learning, and symbolic systems, all of which have their effect by providing information relevant to the knowledge and skills discussed earlier.

All three forms of *instruction* can only be extensions of basic forms of interaction with the world. They may be characterized as "instructional" only when their use is marked by the intent of another person who, for some reason, usually institutionally derived, accepts responsibility for the learner. The first, learning from one's own contingent experience, can be considered as instruction only in special circumstances such as when the environment is intentionally prearranged by another person. The learner's role in this process is readily described as *"learning by doing"* and the instructor's role is primarily that of selecting, simplifying, or otherwise ordering the environment and buffering its consequences. The second form of instruction may be designated *observational learning*. The learner's role may be described as "learning by seeing," or perhaps more appropriately, as "learning by matching," and the instructor's role is primarily that of providing a demonstration or model and perhaps some feedback. The third form of instruction involves the use of various symbolic systems, including a natural language. The learner's role is primarily that of "learning by being told" and the instructor's role is that of telling—providing facts, descriptions, explanations, and much more if he can.

Consider briefly how each of these surface structures relates to the common underlying knowledge structure. Specifically the problem is one of determining what information is invariant to all instruction, and more specifically, how information is coded in the instructional programs we have examined to this point: reinforcement, modeling, and verbal instruction.

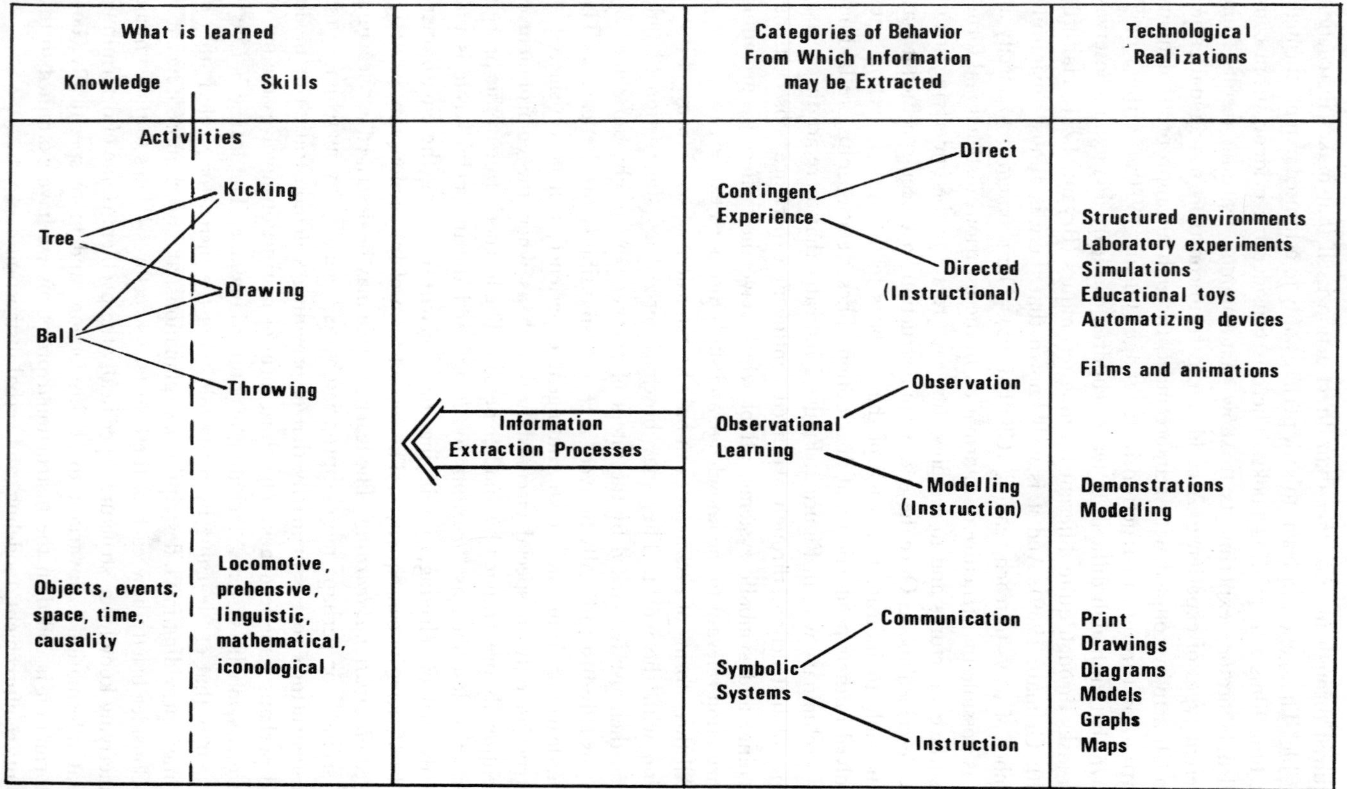

Figure 2 The acquisition of knowledge and skills and the forms of experience from which they are derived.

REINFORCEMENT

First consider the instructional effects of reinforcement. Reinforcement broadly conceived is knowledge of the consequences of an act (Glaser, 1971). Reinforcement assures a means of determining when an appropriate choice among alternatives has been made. Obviously, reinforcement mediates much interacting with the inanimate environment. But while the discovery of new knowledge may be dependent on our direct contingent experiences with nature and with other organisms, reinforcement has the limitation of being ambiguous in outcome. When a teacher reinforces a child for asking a question, the child may not know if it was the question-asking that she approved or the merits of that specific question. Reinforcement can rarely indicate the critical alternatives but only the consequences of the final performance. And given the fact that other human beings are often the principal source of reinforcement—with their notorious variability—the ambiguity is confounded. It is very surprising how uncritically many people accept the idea of control of behavior by reinforcement in view of the very constrained circumstances necessary for it to be effective at all. And more important for instruction, a child obtains no relevant information from a reinforcement if he happens not to be considering the critical alternatives.

Three devices are widely used to render reinforcement less ambiguous. One is by immediacy—tagging the reinforcement directly to the act. The second is by disambiguating the feature of the stimulus to be attended to by placing it in a context that differentiates it from an alternative (Garner, 1966; Bryant and Trabasso, 1971). The third is through "scientific method," by assigning unambiguously certain sequalae to certain prior events so that the necessity of the conceptual link cannot be overlooked. This is typically the way of "guided discovery" that, as with the other two techniques, acts heavily on control of attention. In time, one who must learn by direct encounter comes to control his own attention in one of the three ways suggested: by keeping an eye peeled for immediate results, by being selective in his scanning of features, and by attending to necessity and regularity of relationship. Obviously, there is a technology and a form of materials that must go with learning such "reinforcement skills." It would be foolish to assume that such learning is not crucially subject to instruction. If such were not the case, there would be far more learning from direct experience than there seems to be.

MODELING

One of the more transparent instructional approaches is that of modeling or providing demonstrations, an approach that makes up an important part of

Montessori programs. How is information conveyed through modeling? Complex acts cannot be simply imitated unless the performer already knows how to carry out the act. That is, modeling may initiate or instigate known behavior, but not in any simple manner produce learning.

How can information be conveyed appropriately through modeling? In line with the general theory advanced above, information permits the choice between alternatives. That is, consciousness of the alternatives is a necessary prerequisite for the pickup or acquisition of new information. In another context, it was reported how a Montessori teacher successfully modeled for a 3-year-old child the process of reconstructing a diagonal pattern, a task that is normally solved only by 4- or 5-year olds. (Olson, 1970a). The demonstration consisted of showing the child each of the choice points, that is, the critical alternatives and then indicating how to choose between them. The demonstration of where *not* to go or what *not* to do will be important to the extent that those alternatives are likely to mislead the child. In this light, it is possible to understand the finding of Herbert and Harsh (1944) mentioned earlier, that the cats who saw the error-filled performance learned more than those who saw the error-free performance.

Good instruction through modeling, then, depends on the sensitivity of the instructor to the alternatives likely to be entertained by the child. Modeling or providing demonstrations is a skill to which most pedagogic theories are blind (Richards, 1968).

Just as providing clear demonstrations involves skill, it seems possible *that learning from demonstrations itself demands a skill*; depending on its generality and utility, it may be a skill worth including in our educational aims (aside from the knowledge conveyed by that means).

Any skilled performance, be it doing, saying, or making something requires perceptual information for the guidance of each component of the act, that is, for selecting between all possible alternatives at each choice point in the performance. Modeling as an instructional technique is successful to the extent that it creates an awareness both of the critical alternatives and how to choose between them. To this extent a good demonstration is different from a skilled performance.

VERBAL INSTRUCTION

Finally, consider language as an instructional medium. It is an instructional device par excellence by virtue of the fact that a word indicates not only a perceived referent, but also, in the nature of things, an excluded set of alternatives. Words function contrastively—they differentiate alternatives. The ordinary claim that "words name things" overlooks the fact that words indicate or point to objects or events *in the implied context of the excluded alternatives* (Olson, 1970b). This point may be grasped by noting that the name or the description of an event is determined by the contrasting alternatives. Thus, a large white block in the context of a small white block is called "the large one" while the same block in the context of a large black block is called "the white one." Reciprocally, hearing such a sentence, or any other instructional sentence, the listener knows about both the intended referent *and the likely alternatives*. That is, language is structured precisely in the way that is required for instructional purposes in general. For this reason, the training of literate subjects almost always involves language; when experimentally tested such instruction competes favorably with that of reinforcement and ordinary demonstrations; language coding is less ambiguous, that is, it conveys more information than other instruction media.

But there are many ways in which language can specify an intended referent and these ways provide a microcosm for examining the major premises of the instructional model presented in this paper. The point is that very different sentences convey the same information and hence are generally called paraphrases of each other or synonymous sentences. Consider these simple examples:

1. (*a*) George is here.
   (*b*) My father's brother is here.
   (*c*) My uncle is here.
2. (*a*) The stick is too short.
   (*b*) The stick is not long enough.

The sentences in statement 1 all designate the same intended referent and in some contexts are informationally equivalent. The specific sentences in each case differ, however, in the way the information is coded and in the specific mental processes involved in arriving at that meaning. They also differ in the assumptions they make about the listener; the first could be used only if the listener already knew who George was, and so on. This picture is complicated by the fact that different sentences frequently appear to arrive at a common effect without having a common meaning.

Given the choice between two equivalent instructional sentences, one would choose between them on the same basis that one would choose between two instructional forms in general, that is, in terms of the complexity of the demands they make on the learner and their generalizability to new but related problems. It is interesting to note that the greater the generalizability of an instructional sentence (roughly, its instructional value), the greater the demands it places upon the learner (roughly, its ease of comprehension).

The teaching of rules and strategies falls into a similar position; they are very difficult to comprehend but they have very wide generality. There is always a tradeoff between these two factors, a tradeoff that is reflected in an instructional rule of thumb coined by Bereiter (personal communication) to the effect that if the rules are easily stated and have few exceptions, teach the rule and let the learners practice applying it to various problems; if the rules are not easily stated or have many exceptions, simply give practice on the problems and let them extract what rules they care to for themselves.

The major limitation of language as an instructional medium, along with all cultural media such as graphs, diagrams, numbers, mime, and so on, is that the information is conveyed through a symbolic system that places high demands on literacy in that medium. Furthermore, the meaning extracted from those symbolic systems will be limited to the meaning acquired by the use of that symbol in the referential or experiential world. Thus, the meaning of the word "square" develops along with perceptual knowledge of squares and as the set of perceptual features increases; this in turn is accomplished by differentiating squares from an increasingly large set of alternatives, as shown in Figure 3. Stated generally, this limitation of language implies an ancient point that no new information can be conveyed through language. If the information intended by the speaker falls outside the listener's "competence," the listener will interpret that sentence in terms of the knowledge he already possesses. It follows that instruction through language is limited to rearranging, ordering, and differentiating knowledge or information that the listener already has available from other sources such as modeling or through his own direct expe-

	Utterance	Event	Alternative	Meaning
Case 1	This is a square	☐	—	Ambiguous
Case 2	This is a square	☐	△	4-sided
Case 3	This is a square	☐	○	Straight edged
Case 4	This is a square	☐	○ △ ☐	Straight edged 4-sided Equilateral

Figure 3 The learning of word meanings as a function of alternatives differentiated.

riences. In spite of this dependence of language on perception, perception does come to be shaped in a way to permit easier comment.

ON THE ACQUISITION OF SKILLS AND ABILITIES

Having said that knowledge from different forms of experience can map on to a common deep structure, we must now make it plain that there are also differences. The most important nonequivalence between experiences of events in the three forms is manifest not so much in the knowledge acquired, but in the skills involved in extracting or utilizing that knowledge. However combinable the outcome, the skill of discriminating is a radically different skill from extracting the same information from language. The crucial issue for instruction then becomes one of deciding which skill one wishes to cultivate.

What of these skills? As we pointed out earlier, they are frequently rendered invisible by our habitual focus on the knowledge specified through the activity. As we examine a rock by turning it over in our hand, we are aware of the fact that we acquire knowledge about the rock, but the skilled manipulation that gave rise to the knowledge of the rock is transparent to us. Our earlier example was that in carrying out any activity, such as kicking a ball, we are learning not only about the ball, but about the act of kicking. Carrying out that act across widely divergent objects or events would be responsible for the development of a skill of wide applicability. But if we look at the general skills that make up our cognitive or intellectual ability, we see that they are marked by the same property. Verbal, numerical, and spatial abilities reflect skills in such cultural activities as speaking and writing, counting, and manipulating Euclidean space.

Consider these skills in more detail. It is enormously to Piaget's credit to have insisted and demonstrated that the structure of any ability must be conceptualized in some major part in terms of "internalized activity." Activities one carries out in the physical world—rotating an object in space, lining up objects to form a straight line, ordering objects serially—come to be internalized or carried out mentally. There is not only an internalization of operations, but an increasingly economical representation of diverse events operated upon. A face looked at from various angles comes to be represented as a single face. Even more important are the temporal ordering operations that permit an appreciation first of physical order, then of logical relations. Once we can convert back from a changed state to our original one, we come to appreciate that such reversibility is a logical possibility or property of events and not simply an act one performs. In turn, such operations make it possible to transform a novel event into some standard or base event, or convert some base event into a new structure more appropriate to novel contexts.

The operations specified by Piaget were largely those appropriate to the manipulation of real objects in the physical environment. His basic premise is that their internalization not only produces the groundwork for logic, but assures that logic will be appropriate to the state of the world one experiences. Such operations, consequently, have a wide range of applicability and appear to be almost universally relevant to problem solving. But internalized activity related to the physical environment does not begin to describe the range of activities of the human mind.

Specifically, it leaves out of account how we learn to cope with the cultural or symbolic environment. "Learning from the culture," like learning from physical activities, involves the act of picking up information to decide among alternatives; it also involves skills, and results, finally, in a biased knowledge of reality. Sentences, for example, to the extent that they are *about* something, carry information common to the other forms of experiencing and are comprehended and spoken in terms of those general underlying structures of knowledge we discussed earlier. But the skills involved in using sentences are unique to that particular mode of expression and communication. The skillful use of a symbolic system involves the mastery of both its structure and its rules for transformation. Once mastered, these skills may be considered to be "intelligence," primarily because the range of their applicability is virtually open.

This wide and expanding range of applicability is further indicated in the arts, which may be viewed in part as the creative attempt to expand the limitations of a particular medium or symbolic system. These expanded symbolic systems may then be applied to nature, if appropriate, much like the binomial distribution was found to be an appropriate description of the range of human variability. In this way our use of symbolic systems, like our practical activities, results in a version of "reality" appropriate to that activity. There is no objective reality to "copy" or to "imitate," but only a selection from that reality in terms of the kinds of practical and symbolic activities in which we engage. Thus, Nelson Goodman is led to say that "the world is as many ways as there are correct descriptions of it." Similarly, Cezanne pointed out that the artist does not copy the world in his medium but instead recreates it in terms of the structure of that medium. So, too, with the ordinary man operating in the various symbolic systems of his culture. Whorf was among the first to argue that we "dissect nature along lines laid down by our native language." But it probably goes even beyond that to something comparable to Gide's advising young poets to follow the rhymes and not their thoughts (Goodman, 1971). For the child, as for the creative artist, the use of the culture involves the process of expanding and refining the code, of defining "lawful" or "comprehensible" or "possible" options as he goes. This is the heart of skill in the use of symbolic codes. Even our failures in understanding new media, as McLuhan has pointed out, came from a failure to recognize that they require different skills than the medium they replace—as in going from an oral to a written code or in going from print to television.

Man in culture, like the artist, is in continual search for ways of applying symbolic systems to his ordinary experience. One would be in a precarious position to affirm that the translation of experience into any one medium has any more validity than the translation into any other. A historical account of the intellectual roots of the industrial revolution may be of a different medium but not necessarily of greater validity than Yeats' famous epigram:

> Locke sank into a swoon;
> The garden died;
> God took the spinning jenny
> Out of his side.

Yet a scientific and technological culture such as ours has put a premium on translation into a few symbolic systems—ordinary language as in literature and explanations, in logical and mathematical statements, and in spatial systems such as maps, models, graphs, and geometry. We will continue our attention to a few of these.

We would argue that it is not only scholars, poets, and scientists who seek constantly to cast experience into symbolic codes. Our conjecture is that there is a form of metaprocessing that involves the constant reorganization of what we know so that it may be translated into symbolic systems. It is a matter of "going over one's past experiences to see what they yield" (Dewey, 1916), both for the purpose of facilitating the communication that is required for the survival of the culture and for the purposes of rendering one's own personal experience comprehensible. We may label this form of activity as "deutero-praxis" or second-order information processing. Deuteropraxis is elicited not only by failed communication but also by conflict and difficulties in attempting to carry out an action or solve a problem. It occurs whenever there is information processing capacity available not demanded by the task in hand. Deutero-praxis is involved in all translations of specific experience into general accounts. It can occur in any mode but it is best represented by the poet's or essayist's search for the appropriate phrase or the summarizing aphorism and by the scientist's search for the most general mathematical statement.

Merleau-Ponty (1962) has this in mind when he suggests that all intention wants to complete itself in saying. It is deuteropraxis that is responsible for the radical economization of the experience of the tribe or nation in a few great myths, and more generally, for the world view implicit in one's native language. It should not, however, be assumed that this is simply one of translation. It more generally requires that the creator have more information available than was required for the ordinary direct experience of that event. This can easily be seen when one attempts to draw a map of a territory that one knows quite well or when one tries to give a description of a friend's appearance. One looks at or otherwise experiences an event somewhat differently when dealing with it for different purposes; that is no less true when those activities are different cultural activities than when they are different physical activities. It follows that in drawing an object, one requires somewhat different information about that event than one does for manipulating or describing that object. In this sense, media of expression and communication are exploratory devices—a point of immense importance to an understanding of the child's acquisition of knowledge.

Finally, deuteropraxis makes possible the organization of information into a form that is particularly appropriate for cultural transmission to the young. Since it is often difficult for the child (or an adult, for that matter) to put together what he had done with the consequences that followed, he is greatly aided by the deuteropraxic account of the event. Such accounts, whether in the form of an abstract equation, a principle, a noiseless exemplar or an appropriate model have the effect of "time-binding" or virtually simultanizing temporal events and thereby surpassing ordinary experience.

The very accounts that render experience comprehensible render it instructable. But there are limits in the degree to which such representations whether in language or other symbolic systems can substitute for or extend ordinary experience. As a summary of experience, they are indeed powerful; as an alternative to experience, they are sometimes woefully inadequate. One can learn or memorize the summary without having a grasp of the information summarized. Aphorisms, like new vocabulary items, may serve as pointers to which experience is progressively assimilated. When done at an appropriate level, such instruction may be successful in aiding the assimilation of experience. Experience in this case "instantiates" the categories created through the symbolic code.

TECHNOLOGICAL REALIZATIONS

We return again to Figure 2. The column headed Technological Realizations indicates in a rough manner the media appropriate to each of the modes of experiencing the world. Learning through contingent experience may be facilitated through rearranging the environment to render the consequences of activity more obvious. Structured environments (including laboratories), simulations, toys, and automatizing devices of various sorts have the advantages both of extending the range of a child's experience and of making the relations between events observable or otherwise comprehensible.

Observational learning is realized through the provision of a model—"This is how you break out a spinnaker." As we pointed out earlier, carrying out a performance for its own sake and carrying it out so as to instruct another are not identical. A good demonstration makes explicit the decisions made in the course of the activity—thus a good demonstration shows the student what not to do as well as what to do; a skilled performance makes these same decisions invisible. Technological media can greatly facilitate these processes by highlighting in various ways the critical points in the performance; slow motion or stopped action as well as descriptions and drawings, including caricatures, may have this effect. Such instruction, while it may convey some of the same information that would be apprehended through direct contingent experience (by virtue of its shared deep structure) is never complete in itself but instead specifies some of the major features to be looked for when actual performances are attempted. That is, these forms of instruction rely heavily on prior or subsequent experience to "instantiate" that information.

The instructional effects of a model are greatly increased by tying that demonstration to an appropriate symbolic representation as in the provision of a few mnemonic rules. "Keep your weight on the downhill ski," coupled with a demonstration, will render the demonstration more comprehensible—the observer knows what to look for. But even this will not perfect the novice's performance; direct contingent experience is required to "instantiate" that instruction. Indeed, it is probably not until such instantiation occurs that the proposition is fully comprehended. One says (after the first fall), "So that's what he meant." Hence, all modes of instructing are in some sense incomplete or inadequate for achieving full performatory power or efficiency and knowledge in the last analysis is tied to one's own experience.

Learning through the various symbolic systems including language, graphs, mathematics, and the various systems of visual representation is realized through books, graphs, maps, models, and so on. These media make strong assumptions about the literacy of the learner. The properties of a "good" explanation, description, or portrayal are complex subjects worthy of study in their own right. But to untangle the educational effects of these symbolic systems we again have to differentiate the knowledge of the world conveyed through the system from the skills involved in the mastery of the structure of the medium itself. Recall our aphorism; instructional means converge as to the knowledge conveyed, but they diverge as to the skills they assume and develop. As to the knowledge conveyed, these systems are useful for the partitioning of alternatives or the conveying of information in a way that is fundamentally compatible with the information picked up from other types of experiences.

However, in regard to the skills they develop, symbolic media differ radically from contingent experience and modeling as well as from each other. The old experiments contrasting discovery with expository learning or those simply comparing media such as TV with textbook fell wide of the mark in that they assessed only the knowledge conveyed—the level at which all these systems converge—and overlooked the skills developed—the point at which all systems diverge. The choice of an instructional medium, then, must not depend solely on the effectiveness of the system for conveying and developing knowledge; it must depend, as well, on the realization that the medium of instruction, the form of experience from which that knowledge is gleaned, has important consequences on the kinds of intellectual skills children develop.

We return then to our point of departure. Knowledge as the primary goal of education can be seriously questioned. The analysis we have developed points to the joint importance of knowledge and skill in any medium of expression and communication. To neglect skill is to forget about how one acquired and used knowledge. What balance of knowledge and skills and in what media—this is the essence of the debate that is needed.

REFERENCES

Baldwin, A. L. *Theories of Child Development*. New York: Wiley, 1967.

Bandura, A. *Principles of Behavior Modification*. New York: Holt, 1969.

Bruner, J. S. "The Nature and Uses of Immaturity." Invihed address to the American Association for the Advancement of Science. Philadelphia, December 1971.

Bruner, J. S., R. Olver, and P. M. Greenfield. *Studies in Cognitive Growth*. New York: Wiley, 1966.

Bryant, P. E. "Cognitive Development," *British Medical Bulletin 27*, (1971), 200–205.

Bryant, P. E., and T. Trabasso. "Transitive Inferences and Memory in Young Children," *Nature, 232* (1971), 456–458.

Chall, J. S. *Learning to Read: The Great Debate*. Carnegie Corporation of New York, 1965.

Dewey, J. *Democracy and Education*. New York: Macmillan, 1916.

Duncker, K. "On Problem Solving," *Psychological Monographs, 58* (1945), 270.

Frye, N. *The Critical Path*. Bloomington: Indiana University Press, 1971.

Garner, W. R. "To Perceive is to Know," *American Psychologist, 21* (1966), 11–19.

Gelman, R. "Conservation Acquisition: A Problem of Learning to Attend to Relevant Attributes," *Journal of Experimental Child Psychology, 8* (1969), 314–327.

Glaser, R. *The Nature of Reinforcement*. New York: Academic Press, 1971.

Goodman, P. *Speaking and Language: A Defense of Poetry*. New York: Random House, 1971.

Guthrie, E. R. "Conditioning: A Theory of Learning in Terms of Stimulus, Response and Association," in *The Psychology of Learning*, NSSE, 41st Yearbook, 1942, Part 11, pp. 17–60.

Halford, G. S. "A Theory of the Acquisition of Conservation," *Psychological Review, 77* (1970), 302–316.

Hansen, N. R. *Patterns of Discovery*. Cambridge: Cambridge University Press, 1958.

Herbert, J. J., and C. M. Harsh. "Observational Learning by Cat," *Journal of Comparative Psychology, 37* (1944), 81–95.

Jensen, A. R. "How Much Can We Boost IQ and Scholastic Achievement?" *Harvard Educational Review, 39* (1969), 1–123.

Jones, S. "The Effect of a Negative Qualifier in an Instruction," *Journal of Verbal Learning and Verbal Behavior, 5* (1966), 495–501.

Levine, M. "Hypothesis Theory and Nonlearning Despite Ideal S-R Reinforcement Contingencies," *Psychological Review, 78* (1971), 130–140.

"Living and Learning." A report published for the Ontario Provincial Committee on Aims and Objectives of Education in the Schools of Ontario. Toronto: Newton Publishing Company, 1968. Co-chairmen: E. M. Hall and L. A. Dennis.

Maier, N. R. F. "Reasoning and Learning," *Psychological Review, 38* (1931), 332–346.

Maier, N. R. F., J. A. Thurber, and J. C. Janzen. "Studies in Creativity: The Selection Process in Recall and in Problem Solving Situations," *Psychological Reports, 23* (1968), 1003–1022.

Mackintosh, N. J. "Selective Attention in Animal Discrimination Learning," *Psychological Bulletin, 64* (1965), 124–150.

Masters, J. C., and M. N. Branch. "A Comparison of the Relative Effectiveness of Instructions, Modeling, and Reinforcement Procedures for Inducing Behavior Change," *Journal of Experimental Psychology, 80* (1969), 364–368.

Merleau-Ponty, M. *Phenomenology of Perception*. London: Routledge and Kegan Paul, 1962.

Olson, D. R. *Cognitive Development: The Child's Acquisition of Diagonality*. New York: Academic Press, 1970(a).

Olson, D. R. "Language and Thought: Aspects of a Cognitive Theory of Semantics," *Psychological Review, 77,* 4 (1970), 257–273. (b).

Piaget, J. *Biology and Knowledge*. University of Chicago Press, 1971.

Pribram, K. H. *Languages of the Brain: Experimental Paradoxes and Principles in Neuropsychology*. Englewood Cliffs, N. J.: Prentice-Hall, 1971.

Richards, I. A. *Design for Escape: World Education Through Modern Media*. New York: Harcourt, 1968.

Skinner, B. F. *The Behavior of Organisms: An Experimental Analysis*. New York: Appleton-Century-Crofts, 1938.

Sokolov, E. N. "The Modeling Properties of the Nervous System," in *A Handbook of Contemporary Soviet Psychology*, I. Maltzman and M. Cole (eds.). New York: Basic Books, 1969.

Sperry, R. W. "Cerebral Organization and Behavior," *Science, 133* (1961), 1749–1757.

Sutherland, N. S. "Visual Discrimination in Animals," *British Medical Bulletin, 20* (1964), 54–59.

Tolman, E. C. *Purposive Behavior in Animals and Man*. New York: Appleton-Century-Crofts, 1932.

Von Holst, E. "Vom Wesen der Ordnung im Zentralnerven System," *Die Naturwissenschaften, 25* (1937), 625–631, 641–647.

Vygotsky, L. S. *Thought and Language*. Cambridge, Mass.: The M.I.T. Press, 1962.

CHAPTER 15

Formative Research in the Production of Television for Children

EDWARD L. PALMER

There is currently in the United States unparalleled interest in the systematic use of broadcast television to promote the social, emotional, and intellectual growth of young children. Support for this movement lies in the recognition that television is ubiquitous, reaching into 97% of all United States households; that young children are exposed to upwards of 30 hours of television fare each week; that while they learn a great deal from what they watch, there have been far too few significant attempts to plan program content in order to address important areas of learning and development systematically; and that no other approach can promise to deliver so much to so many at so small a unit cost. An important feature of this movement is its emphasis on "formative" planning and research, whereby important objectives are first clearly identified and systematic audience tests are then carried out in order to evaluate progress toward their achievement during the actual course of a program's production.

The Children's Television Workshop (CTW) was created in 1968 to produce a series of 130 hour-long broadcast television programs for preschool children, with special emphasis on the needs of the urban disadvantaged child. The result, the now well-known "Sesame Street" series, is currently completing its fourth production season. During a prebroadcast period of nearly a year and a half, and through the current season, the producers have made extensive use of formative field research to improve both the program's appeal and its educational quality. The formative research methods developed and

229

applied in the production of "Sesame Street" have been reapplied and extended by CTW in planning and producing its second major program series, "The Electric Company." This series, designed to teach selected reading skills to children from 7 to 10 years of age, is currently completing its second broadcast season.

At the beginning of the "Sesame Street" project, the functions formative research could serve and the field methods it could apply were not at all clear. There were no precedents of sufficient scope and generality, either from the field of educational television or from the field of educational planning and research in general, to provide any clear guidelines. What is to be presented here is a case study of the approaches to formative planning and research taken by CTW in the production of "Sesame Street" and "The Electric Company." An effort will be made throughout to discuss these approaches in ways that will suggest their potential usefulness in the development of other new educational products and practices, both of a broadcast and nonbroadcast nature.

Among the unusual circumstances associated with the Workshop's productions, some, no doubt, had quite a direct bearing on the effectiveness of the formative research. For instance, the two Workshop projects were well funded, each budgeted in its first season at upwards of $7 million for production, research, and related activities. This level of support made it possible to utilize high-level production talent and resources and to make extensive use of expert educational advisors and consultants. In addition, both projects enjoyed unusually long periods of time—in each case, approximately 18 months—for prebroadcast planning and research. Time and resources were available to plan their respective curricula very carefully and to state their educational objectives in very explicit terms, so that producers and researchers alike, as well as the independent evaluators who were carrying out pre- and post-season achievement testing projects, could proceed without ambiguity of purpose and in a coordinated fashion. Had there been ambiguity, either in terms of the particular objectives to be addressed or in terms of the commitment of the producers to direct each segment toward the achievement of one or more of those objectives, the formative research could not have been useful, because there would have been no clear criteria against which to evaluate a program segment's effectiveness.

Also unusual in the CTW case were the organizational and interpersonal relationships between the on-board research and production staffs, and the policy followed in production recruiting. All of the key producers came from commercial production backgrounds. None had formal professional training in education or experience in educational television production. Yet they were given the responsibility for final production decisions. They did not work under the researchers, nor did the researchers work under them. The intended function of the formative research was to provide information that the producers would find useful in making program design decisions, relative to both appeal and educational effect.

To the extent that the formative research worked, it worked in large measure because of the attitudes taken toward it by producers and researchers alike. The producers were committed to experimenting with the cyclic process of empirical evaluation and production revision, and tended to have the creative ability not only to see the implications of the research, but to carry these implications through into the form of new and revised production approaches. Accordingly, the usefulness of CTW's formative research has depended not only on the qualities of the research itself, but also on the talents of those who put its results to use. Moreover, the producers never expected the research to yield full-blown decisions; they recognized that its function was to provide one more source of information among many. From the research side, because the responsibility for final production decisions resided with the producers, it was necessary to develop and apply only methods that they themselves found useful. Accordingly, the producers were involved in all research planning, from the earliest possible stage. No observational method was ever persistently applied and no specific study was ever taken into the field without their participation.

At this time, there is no tradition of accumulated knowledge in the area of formative research practice. This is partly because so little research of this type has been done, but it is even more a result of the fact that it has only recently come to be recognized as a distinct field of endeavor. A limited number of articles have contributed to the early general conceptualization of the field (see Cronbach, 1963; Hastings, 1966; and Scriven, 1967). In addition, formative research studies associated with specific product improvements are reported for various media, content areas, and student levels: for example, for televised instruction (Gropper and Lumsdaine, 1961); for programmed instruction (Dick, 1968); for kindergarten instruction in conceptual skills (Scott, 1970a); and art (Scott, 1970b). However, the present scope and depth of the formative research literature is in no way commensurate with its promise for education. The promise of the approach is that it will provide designers of educational products and practices with empirical data far more directly pertinent to their respective media, materials, and conditions than are the results of traditional, more basic research. For the field of educational television in particular, it offers ways to help bring about planned effects. With mass broadcast distribution making it possible to reach hundreds of thousands or even millions of viewers (in the case of "Sesame Street," the daily audience is estimated to be about 8 million children), it behooves the producers to employ every reasonable means for helping to ensure in advance that the programs will achieve their objectives.

As experience with formative research procedure begins to accumulate and the conditions for success or failure become more clear, many of the factors considered significant in the work carried out at CTW must almost certainly receive prominent continuing attention. Those considered most significant here are the subjects of the sections that follow. In brief overview, these include the overall operational framework within which CTW's formative research proceeds, the strategies and rationale for the design of formative field research methods, organizational and interpersonal conditions, and similarities and contrasts between the functions and the methods of formative research on the one hand and those of more traditional research approaches on the other.

THE CTW OPERATIONAL MODEL

The principal activities undertaken in the production of "Sesame Street" have come to be viewed by CTW as a model, and this model was again applied in the production of "The Electric Company." If there is a single, most critical condition for rendering such a model of researcher-producer cooperation effective, it is that the researchers and the producers cannot be marching to different drummers. The model is essentially a model for production planning. More specifically, it is a model for planning the educational (as opposed to the dramatic) aspects of the production, and the formative research is an integral part of that process. In the case of "Sesame Street" and "The Electric Company," at least, it is hard to imagine that the formative research and curriculum planning could have been effective if carried out apart from overall production planning, either as an isolated *a priori* process, or as an independent but simultaneous function.

The activities included in the model are presented in the following section in their approximate chronological order of occurrence.

Curriculum Planning

As the initial step toward establishing its educational goals, CTW, in the summer of 1968, conducted a series of five three-day seminars dealing with the following topics:

1. Social, moral, and affective development.
2. Language and reading.
3. Mathematical and numerical skills.

4. Reasoning and problem solving.
5. Perception.

The seminars, organized and directed by Dr. Gerald S. Lesser, Bigelow Professor of Education and Developmental Psychology at Harvard University, were attended by more than 100 expert advisors, including psychologists, psychiatrists, teachers, sociologists, filmmakers, television producers, writers of children's books, and creative advertising personnel. Each seminar group was asked to suggest educational goals for the prospective series and to discuss ways of realizing the goals on television.

Behavioral Goals

The deliberations of the seminar participants and the recommendations of the CTW Board of Advisors were reviewed in a series of staff meetings from which a list of instructional goals for the program emerged. These goals were grouped under the following major headings:

I. Symbolic Representation
    A. Letters
    B. Numbers
    C. Geometric Forms
II. Cognitive Organization
    A. Perceptual Discrimination and Organization
    B. Relational Concepts
    C. Classification
III. Reasoning and Problem Solving
    A. Problem Sensitivity and Attitudes Toward Inquiry
    B. Inferences and Causality
    C. Generating and Evaluating Explanations and Solutions
IV. The Child and His World
    A. Self
    B. Social Units
    C. Social Interactions
    D. The Man-Made Environment
    E. The Natural Environment

Specific goals under each of these broad headings were stated, insofar as possible, in behavioral terms, so that they might serve as a common reference for the program producers and the designers of the achievement tests. Appropriate coordination of production and evaluation thus was assured.

Existing Competence of Target Audience

While the statement of goals specified the behavioral outcomes the program hoped to achieve, it was necessary to ascertain the existing range of competence in the chosen goal areas among the target audience. The Workshop research staff therefore undertook as its initial formative research effort a compilation of data provided in the literature, as well as some testing of its own, to determine the competence range. The resulting information helped guide the producers in allocating program time and budget among the goal categories and in selecting specific learning instances in each goal area.

Appeal of Existing Materials

To be successful, CTW had to capture its intended audience with an educational show whose highly attractive competition was only a flick of the dial away. Unlike the classroom teacher, the Workshop had to earn the privilege of addressing its audience, and it had to continue to deserve their attention from moment to moment and from day to day. At stake was a variation in daily attendance that could run into millions. Measuring the preferences of the target audience for existing television and film materials was therefore crucial in the design of the new series.

Experimental Production

Seminar participants and CTW advisors had urged using a variety of production styles to achieve the curriculum goals adopted. Research had confirmed the appetite of the target audience for fast pace and variety. Accordingly, the CTW production staff invited a number of live-action and animation film production companies to submit ideas. The first season of "Sesame Street" would eventually include the work of 32 different film companies.

Prototype units of all film series produced by or for the Workshop were subjected to rigorous preliminary scrutiny and empirical field evaluation. Scripts and storyboards were revised by the Workshop producers on the basis of recommendations from the research staff; further revisions were made after review by educational consultants and advisors; and finished films were tested by the research department with sample audiences. Some material never survived the process. Four pilot episodes were produced for a live-action film adventure series entitled "The Man From Alphabet," but when shown to children the films failed to measure up, either in appeal or educational effect, and the series was dropped. Sample videotaped material went through the same process of evaluation, revision, and occasional elimination.

By July 1969 a format for the program had been devised, a title had been selected, a cast had been tentatively assigned, and a week of full-length trial programs had been taped as a dry run.

Completed prototype production elements were tested by the research staff in two ways: (1) appeal for the CTW material was measured against the appeal of previously tested films and television shows, and (2) the CTW material was tested for its educational impact under a number of conditions. For instance, field studies were conducted to determine the effect of various schedules of repetition and spacing, of providing the child with preliminary and follow-up explanations, of presenting different approaches to a given goal separately or in combination, and of the relative effectiveness of adult versus child voice-over narration. Extensive observation of viewing children provided information regarding the child's understanding of various conventions of film and television technique. Upon conclusion of each research study, the results were reported to the producers for their use in modifying the show components and for guidance in the production of subsequent elements.

Production, Airing, and Progress Testing of the Broadcast Series

The evolution of "Sesame Street" did not end with the first national broadcast on November 10, 1969. Formative research studies conducted throughout the six-month broadcast period continued to guide the development of new production techniques, format elements, and teaching strategies. As before, these studies had two foci: (1) the holding power of entertainment techniques, and (2) the effectiveness of educational content.

Earlier and continuing studies of individual program segments, while useful, were necessarily limited in scope. With the onset of the broadcast season, it was possible to examine the impact of continuous viewing of entire shows over a period of time. Accordingly, the research staff instituted a program of progress testing of the show's effectiveness. Using the instruments designed by Educational Testing Service (ETS) of Princeton, New Jersey for a national summative evaluation of the series, a sample of day-care children, predominantly four- and five-year-olds, was pretested prior to the first national telecast. One third were tested again after three weeks of viewing the show; the first third and a second third were tested after six weeks of viewing; and the entire group was tested after three months of viewing. Comparisons between experimental (viewing) groups and control (nonviewing) groups at each stage of the testing gave indications of strengths and weaknesses both in the execution of the curriculum and in the production design. Appeal measurement and informal observation of viewing children also influenced production decisions during this period.

Summative Evaluation

The summative research and evaluation carried out by Educational Testing Service (see Ball and Bogatz, 1970; and Ball and Bogatz, 1971) followed a plan developed in consultation with CTW staff and advisors. Participation of ETS representatives in all main phases of prebroadcast planning helped to ensure coordination between program development and follow-up testing.

ETS developed and administered a special battery of 11 tests covering the major CTW goal areas to a sample of children from Boston, Philadelphia, Durham, and Phoenix. The groups included three-, four-, and five-year-olds in urban and rural settings, from middle- and lower-income families, in both home and day-care situations. A special side study related to children from Spanish-speaking homes. The 11 tests were as follows:

1. Body Parts Test.
2. Letters Test.
3. Numbers Test.
4. Shapes and Forms Test.
5. Relational Terms Test.
6. Sorting Test.
7. Classification Test.
8. Puzzles Test.
9. What Comes First Test.
10. Embedded Figures Test.
11. Sesame Street Test.

Other measures assessed home conditions, parental expectations for the children, and the like. Where the results of the first season's summative research feed into production decisions for the second season, they take on a formative function.

Writer's Manual

As the producers and writers began to develop scripts, animations, and films addressed to particular behaviorally stated goals, it became apparent that the goal statement was not a wholly adequate reference. After having been given several successive assignments in the same goal area, they began to express the need for extended and enriched definitions that would provide creative stimulation. Gradually, through trial and error, a format for a *Writer's Manual* was developed that the producers and writers themselves found useful.

Suggestions were developed for goal areas as requested by the producers and writers. Involving advisors and consultants in the creation of these suggestions afforded one more opportunity for making use of expert input. In addition, the

Manual provided a place and a format for collecting the ideas of the in-house research staff and a channel for helping to ensure that these ideas would be seen and used.

FORMATIVE RESEARCH METHODS

A great deal of information useful to educational television producers can be acquired through the use of a few quite inexpensive and informal methods of field observation. More sophisticated methods can provide considerable additional information in some cases, but often add little to that which can be obtained more simply and economically.

The selection of research methods is particularly critical, because the attributes focused on by these methods tend to become prominent among those focused on by the producers. This is particularly the case when the producers themselves have participated in selecting the methods, and thus in identifying the attributes deserving of their special attention.

CTW's formative research methods presently focus on four principal program attributes, all considered instrumental in producing lasting instructional effects, namely: (1) appeal, (2) comprehensibility, (3) internal compatibility, and (4) activity eliciting potential. These will be discussed in more detail in the following sections. First, however, it is important to note that these attributes are identified and used at the Workshop and are presented here strictly for their heuristic value. It is convenient for the producers and researchers to have a small number of highly significant program attributes with which to associate both the host of related program design features and the many similarly related field research methods. Not only does reference to a limited number of attributes provide a manageable checklist for evaluating materials under production, and a convenient category system, but it also invites researchers and producers alike to identify new attributes, and, for each attribute, additional field research methods and program design features. A more elaborate treatment of these attributes, which couches them in the framework of a general theory of presentational learning, appears elsewhere (see Palmer, in press).

Formative Research on Program Appeal

The appeal of a program has to do with its ability to capture and hold the attention of the intended viewer. In the case of both "Sesame Street" and "The Electric Company," there was no "captive" audience. The programs were designed to attract the largest possible number of at-home viewers. This meant that they needed to be sufficiently high in appeal to draw the children back to the set from day to day and week to week, and to compete with popular entertainment programming available on other channels.

Because the viewer could turn away at any time, and because the two programs were designed according to a magazine format, with successive brief segments addressed to very explicit educational objectives, it was important to maintain high program appeal on a moment-to-moment basis. Accordingly, research methods capable of focusing on appeal from moment to moment throughout the course of a program were developed and used.

Appeal research bears upon a wide range of program design decisions (see Rust, 1972). It reveals the effects of various forms and applications of music, and of music as compared with other types of elements. It indicates the most popular and least popular forms of live-action films, animations, puppets, and live performers. It indicates the attention-holding power of various types of individual or interpersonal activities, such as showing one person guiding another through a difficult task in a supportive versus demeaning manner; presenting conflict resolution through the arbitrary use of power versus cooperation; revealing the simultaneous perspectives of different characters; and portraying the struggle of an individual toward an achievement goal or toward improvement on his own past performance, to mention a few.

Appeal research also helps to indicate for various conditions the amount of time over which attention can be maintained; the optimum amount of variety and the optimum pacing of events; the relative holding power of program elements that are and are not functionally relevant to the action; the ability of a segment to bear up under exact repetition; the most and least salient (memorable) characters; and the effectiveness of special techniques such as pixilation, fast and slow motion, and unusual camera angles. In addition, research on appeal can show growth or decline in the popularity of specific program elements over time; the most and least effective uses of dialogue, monologue, and the voice-over technique; the relative effectiveness of ordinary or caricaturized voices; and the effect of sparse and pointed versus sustained verbalizations. It also can reveal the effects of incongruity, novelty, or fantasy as compared with straightforwardness, predictability, and realism; the effect of different motives or intentions on the part of characters; of episodic versus linear styles of continuity; and also of familiar versus unfamiliar conventions and symbols dealing with time, sequence, interpersonal relationships, and the like. Finally, this type of research can be used to investigate characteristic individual or group preferences *vis-a-vis* such program design features.

Formative Research on Program Comprehensibility

The comprehensibility of a program segment has to do with the manner in which it is interpreted or construed by its viewers during the actual course of its presentation. It is concerned with the program-design features, on the independent variable side, which are responsible for viewer comprehension on the dependent variable side. Simply stated, tests for comprehension are those which reveal the viewer's ability to follow the development of the instructional presentation, and to grasp what the producer intends that he should grasp. The objective of formative research on comprehensibility is to discover the principles which link program-design variables to viewer comprehension. It is instructive to the producers to have an empirical check on their own assumptions about the comprehensibility of program design features they are employing, and even limited amounts of field research can help them to maintain a generalized sensitivity to this important attribute.

Although CTW's research in the area of comprehensibility has just begun, and although the ultimate objective is to identify and set down specific program design principles, it may be useful at this time to mention a few of the program features to which these principles could relate. Among these are the production approaches that can help to clarify the relationship between an event occurring on the screen and the theme, the plot line, or the logical progression of the dramatic component, or between the instances and noninstances of a concept, the referents and nonreferents of a term, or the most effective and least effective of a set of proposed solutions to a problem.

The unique conventions and capabilities of the television medium are frequently used to convey special meanings. The manner in which these conventions are presented will determine their comprehensibility to the viewer, and thus their effectiveness in communicating the meanings intended. These include the use of the flashback technique, of special lighting effects or special combinations of music and lighting, the use of various camera perspectives, of fast or slow motion, of pixilation, and of the matched dissolve between objects. They also include the close juxtaposition of events in order to establish a metaphoric or analogic relationship between them, and the use of conventions having to do with fantasy, such as presenting puppets and cartoon characters who move and talk like humans. Still others include the creation of "magical" effects, such as making an object instantly appear or disappear from a scene, or grow smaller or larger. Other conventions that can be used in more or less comprehensible ways are the speech balloon, the rules of games presented for instruction or entertainment, and rules involved in reading, spelling, mathematical operations, the interpretation of maps, and the like.

Still other facets of comprehensibility relate to timing, sequencing, and the use of redundancy, as in repeating an event exactly or with an illuminating variation, in restating a point from alternative perspectives, and in making use of introductions or reviews. The list could go on indefinitely, a fact that itself suggests the significance of this attribute in educational television research.

A strength of comprehensibility testing relative to traditional forms of summative evaluation is the opportunity it provides for discriminating between the most and least effective of the many individual segments devoted to a particular achievement objective. A potential but largely surmountable limitation is the tendency for these methods to produce biased results. Because comprehensibility testing is performed as the program is being viewed, and because the viewer knows he will be questioned, there is typically an overestimation of a segment's effectiveness. In practice, this bias can be subjectively discounted, at best, and must further be weighed against the possibility that segments that produce no measureable learning when presented in isolation may be effective in combination or when presented along with an appropriate introduction or review. However, these limitations do not detract seriously from the usefulness of such methods. The bias can in fact be turned to an asset, as when it can be shown that a segment of questionable value fails to make its point even when evaluated by means of a liberally biased method.

Internal Compatibility

Internal compatibility is a program attribute that has to do with the relationship of different elements appearing within the same segment. The basic strategy underlying both "Sesame Street" and "The Electric Company" is to attempt to effect instruction through the use of television's most popular entertainment forms. To this end, it is essential that the entertainment and educational elements work well together. Without the entertainment attention strays, and without the education, the whole point of the presentation is lost. In segments where these elements are mutually compatible, the educational point is an inherent part of the dramatic action and often is actually enhanced in its salience as a consequence. In others, the entertaining elements override and thereby actually compete with the educational message. Other cases in which the relationship of elements becomes a concern have to do with auditory-visual, auditory-auditory, and visual-visual compatibilities. The objective of formative research in this area is to shed light on the program design features that make for a high or low degree of compatibility. As in the case of the other major program attributes discussed here, internal compatibility can be evaluated by means of a number of different research methods. In one, a panel of judges is asked to rate each segment of a program according to a predetermined set of categories defining the extent to which a segment's entertainment either facilitates or competes with the instructional content.

Another method involves eye-movement research, which is especially useful in the case of "The Electric Company" because of the extensive presentation of print on the screen and the desire to find ways of motivating the child to read it.

Activity Eliciting Potential

A widely expressed point of view about television as an instructional medium states that because of the passivity of the viewer, the medium is virtually powerless to produce learning. There is no question that the medium has limitations in this regard. However, since it is patently obvious that television can and does teach, a more constructive point of view is to examine conceptually ways in which this capability comes about, and operationally, ways in which it may be exploited. The position taken here is that the activity eliciting potential of the medium, no matter how limited, is nevertheless the chief basis for whatever effectiveness it has.

One significant form of activity television can elicit is intellectual activity. Others include verbal behavior and gross physical acts, ranging from television-modified performance on tests of attitudes and achievements to the imitation of televised models. It is important to note that the concern of the medium can be either to exploit these effects as instruments of instruction or to foster them as instructional objectives.

Intellectual activities include integrating separately presented items of information, anticipating upcoming events, forming new concepts, imputing the motives and intentions of characters, following progressively developed dramatic and instructional presentations, and guessing answers to questions. The viewer also actively evaluates relationships between premises and conclusions, between information given and interpretations made of it, and between codes of behavior and the actual behaviors carried out by the performers. The viewer also frequently relates new information from a televised presentation to his own prior experiences and to his future plans. These are only a few of the many possible instances.

It is understandable that from a superficial look at television's potential as an instructional medium many educators underestimate its ability to recreate the conditions known or presumed to be essential for learning. Tentative indications from formative conceptualization and research on the activity eliciting capabilities of the medium suggest that many of its presumed limitations may be at least partially surmountable. For example, it is often assumed that learning through trial and error or through trial and reinforcement cannot occur through one-way televised presentations, on the basis that there is no opportunity for reinforcement or information feedback to be tied to an action of the learner. This is not a trivial issue, from a practical standpoint, since vast amounts of money may yet be spent studying the use of two-way communication systems in connection with televised instruction. It turns out that *conceptually* it is possible to effect trial-and-error learning through one-way television simply by the use of "if" statements. That is, the viewer may be offered a choice among provided alternatives, given time to make his choice (his point of most active involvement), and then given reinforcement, or an accuracy check, of the form: "If you chose thus and so, you were correct (incorrect)." Empirical studies may or may not support the viability of such an approach, but it certainly deserves further investigation.

Similarly, the notion that certain activities containing a motoric component can be learned only through direct experience is in many instances questionable. For example, direct experience in the construction of alphabetical characters may have its most significant effect on learning by controlling the scan of the eye over the configuration of the letter, by providing extended or repeated exposure to the letter, or by providing an occasion for the most common errors to be made and corrected. But all of these are features one-way television can either duplicate or simulate. We need to know more about the effectiveness of such features when produced by one-way television. We also need to know more about the entry skills required under such conditions in order for learning to occur, and about possibilities for the facilitation of subsequent learning.

All this is not an argument in favor of unduly widespread substitution of television for physical activity among children, by the way, nor is it intended to deny the great importance of extensive direct experience in learning, especially in early learning. It is intended, instead, to urge more open and positive consideration of some of the potential but not yet systematically explored capabilities of the television medium.

ORGANIZATIONAL AND INTERPERSONAL FACTORS

As technologically sophisticated forms of instruction come into increasing prominence, it will be necessary to make increased use of production teams whose members possess a diversity of highly specialized talents. In anticipation of this trend, we need to know more about related organizational and interpersonal conditions. These conditions deserve attention in any attempts to establish a working partnership between television research and production groups, and they play a strikingly more prominent role in the formative research context than in the context of more traditional approaches to educational research. To illustrate briefly the many and different types of factors involved, a major one in CTW's case has been the opportunity afforded by an 18-month prebroadcast period for the members of the research and production groups to learn about each other's areas of specialization. Another has been the attitude that every new formative research approach is an experiment, to be continued or discontinued depending on its merit as evaluated by the producers themselves.

The fact that CTW's researchers and producers possess not the same, but complementary skills is also significant, largely because it provides for clear and distinct functions on the part of each group. Still another factor is that the producers, before joining the project, made the commitment to try to work with formative research. This prior commitment helped to support the cooperative spirit through the early, more tentative period of the effort. Also, research never takes on the role of adversary, to be used against the producers in winning a point or pressing for a particular decision. The producers hold the final power of decision and are free to ignore research suggestions if production constraints require it.

In all, the factors consciously dealt with in the interests of researcher-producer cooperation have ranged from the careful division of labor and responsibility to housing the two staffs in adjacent offices, and from patience and diplomacy to occasional retreat.

THE DISTINCTIVE ROLE AND FUNCTIONS OF FORMATIVE RESEARCH

The most important factor underlying the distinctive form and style of product developmental research is its role as an integral part of the creative production process. It is important to maintain a clear distinction between this type of research, on the one hand, and that undertaken in order to test the validity of a theory or the measureable impact of an educational product or practice, on the other. Research undertaken in the context of scientific validation is concerned with effects that have been hypothesized, *a priori*, within the framework of a broader deductive system; with the use of empirical and statistical procedures well enough defined so as to be strictly replicable (at least in principle); and with the highest possible degree of generalizability across situations. In contrast, while research carried out within the formative context *can* possess all these same characteristics, it need not necessarily, and typically does not. The only pervasive criterion for formative research recommendations is that they appear likely to contribute to the effectiveness of the product or procedure being developed. It is neither expected nor required that they be validated by the research out of which they grew. Establishing their validity is the function of summative research.

As this view implies, to achieve the objectives of formative research it is often necessary to depart from traditional research practices and perspectives. This is not to say that experimental rigor has no place in the formative context. However, for example, even where strict experimental and control conditions have been maintained, there is seldom anything to be gained by using tests of statistical significance. The creative producers often prefer to work directly with information about means, dispersions, and sample size.

In the area of sample selection, it also can be useful to depart from the traditional practice of including all age and socioeconomic groups for which the educational materials are intended. In general, where biased methods of sampling and biased methods of testing are more efficient than unbiased methods, and where the objective is not to make accurate population estimates, it is often useful to exploit the very biases that quite properly would be avoided in other research situations.

Formative studies must first address the information needs of the product designers and not primarily the individualistic or special theoretical interests of the researchers. Covering a wide range of empirical questions may deserve priority over rigorous reporting or establishing careful experimental conditions, where it is economically impossible to achieve both, and where the usefulness of the results is not unduly compromised as a consequence. Broad, speculative interpretations of empirical results are typically more useful than interpretations limited to the most strict implications of a study. However, in these departures from standard research practice there is a risk of producing misleading results. Accordingly, it is essential that the production recommendations be very carefully qualified.

Formative research, in the view taken here, is properly eclectic and pragmatic. In these respects, it is highly compatible with the current trend toward the very explicit definition of instructional objectives, followed by the development through systematic trial and revision of instructional systems for achieving them. This approach holds that a useful step between basic research and educational practice is additional research of a formative sort, which is more directly concerned with specific combinations of educational objectives, instructional media, learners, and learning situations. This is not to say that formative research is exclusively concerned with putting theory into practice. An equally valid function is that of starting with practice and transforming it into improved practice. Still another is that of providing hypotheses for further research and theoretical development.

One long-standing point of view in education is that theories and results growing out of the "mother" disciplines of psychology, sociology, anthropology, and the like, will filter into effective educational practice if enough educators have been trained in these basic disciplines. While this approach has been useful to a degree, it has not produced broadly satisfactory results. Meanwhile, creators of new educational products and practices have proceeded largely without the benefits of measurement and research. This is partly because skill and training in these areas have been linked to the process of theory construction and validation, and partly because of an inappropriately rigid adherence to traditional research practice in connection with product development. Formative research procedure promises to help in creating a mutually constructive relationship between these two overly-isolated realms—the science and the technology of learning.

REFERENCES

Ball, Samuel, and Gerry Ann Bogatz. *The First Year of Sesame Street: An Evaluation*. Princeton: Educational Testing Service, 1970.

————. *The Second Year of Sesame Street: A Continuing Evaluation*. Vols. I and II. Princeton: Educational Testing Service, 1971.

Cronbach, L. J. "Course Improvement Through Education," *Teachers College Record, 64*, (1963), 672-83.

Dick, Walter. "A Methodology for the Formative Evaluation of Instructional Materials," *Journal of Educational Measurement, 5*, 2, (1968).

Grooper, G. L., and Lumsdaine, A. A. *Studies of Televised Instruction*. Metropolitan Pittsburgh Educational TV Stations WQED-WQEX and American Institute for Research, 1961.

Hastings, J. T. "Curriculum Evaluation: The Why of the Outcomes," *Journal of Educational Measurement, 3*, 1, (1966), 27-32.

Palmer, Edward L. "Formative Research in the Production of Television for Children," in D. R. Olson (ed.), *Media and Symbols: The Forms of Experience, Communication, and Education*, 73rd Yearbook of the National Society for the Study of Education, Chicago: University of Chicago Press, in press.

Rust, L. W. *Attributes of "The Electric Company" That Influence Children's Attention to the Television Screen*. In-house research report, Children's Television Workshop, 1972.

Scott, R. O., and Martin, M. F. "The 1969-70 Classroom Tryout of the SWRL Instructional Concepts Program," *SWRL Technical Memorandum*, 1970 (a), No. TM-3-70-4.

Scott, R. O., Castrup, J., and Ain, E. "The SWRL Kindergarten Art Program," *SWRL Technical Memorandum*, 1970(b).

Scriven, Michael. "The Methodology of Education," in R. W. Tyler, R. M. Gagne, and M. Scriven (eds.), *Perspectives of Curriculum Evaluation*, AERA Monograph Series on Curriculum Evaluation, Chicago: Rand McNally, 1967.

CHAPTER 16

The Magic Lantern: Metaphor for Humanistic Education

BRUCE R. JOYCE

The complexity of our society has outdistanced the adaptive capacity of many of our social institutions. However, we also have a rare opportunity to redesign our educational institutions so they become a major force for personal fulfillment, common enterprise, and the humanizing of society.

Education, as an institutionalized social process, is in many ways analogous to a medium of communications. It structures relationships, facilitates certain behaviors and inhibits others, and provides a more or less friendly habitat for the communications modes of the times. This analogy has considerable value when we come to examine communications, media, and technology as they relate to education. Currently, the dominant institutional mode in education is not supportive of several important communication modes, and thus prevents the development of a synchronous relationship between education and society.

Therefore the schools are undergoing changes. Guidance is required to ensure that these changes are not simply superficial reforms that, by perpetuating the present system in a more efficient form, merely exacerbate the problem.

WHY IS THE SCHOOL CHANGING, AND WHY SHOULD IT CHANGE?

The present form of the school was actually created for preparing children for preindustrial village life. Like the village, it repels innovation and must be replaced by a fluid, powerful form of schooling for lifelong education. The recent innovations in the form of the school reflect attempts to evolve a structure that more fully meets the demands of our contemporary society.

The obsolescence of the old forms of education can be seen in terms of purpose, substance, and form. The purpose was to homogenize society and to provide upward mobility through an industrial system. The substance was the substance of a primitive media world in which communication was largely through reading and writing, and through oral communication within the village. The form was mediation largely by textbooks and through the multipurpose functionaries called teachers.

Today the primary setting of education, the classroom, and the chief mediator of instruction, the multipurpose teacher, are obsolete. Five propositions are offered to account for the obsolescence of the present system.

1. *The technetronic revolution has created a new social world that outdistances the response modes taught in most present schools. New models of learning are required to provide strategies for collective action and personal development* (Brzezinski, 1970).

The "global village" has become our normal habitat. The nature of this world has been created by changes in our media technologies and the process of media change is circular. As media transmit messages, they transform social life. In turn, this transformation is content for transmission by yet other transforming media.

The messages of form and those of substance are both information, but the media constitute not only information but make up also the environment in which we relate to all other aspects of our world. Only 200 years ago the events that triggered wars were reported by a few eyewitnesses to heads of state, who in turn reported these and interpreted them to the "people." The medium that was accepted was one of symbolic transmission by a very few people whose verisimilitude could not be judged. At present, television cameras and motion picture cameras move onto the scene of events and the resulting images are transmitted to the bulk of the people in terms of the significant events as these are perceived by cameramen, producers, and editors. The media technology world is different from the more linear media both in the substantive message that is received and in the social situation that has evolved.

When our present educational system was created we were very much separated from one another in that very few of us could contact the individuals who had been the eyewitnesses to events. In addition, we were accustomed to living in a world characterized by fragmented forms of communication. We knew that very few of us could have images of an event except through print or oral communications transmitted through many intermediaries. At present, however, we live with documentation of a different order, and we have grown accustomed to living where we could perceive events from media that minimally distort the reality that is being reported and that gives us a sense of participation.

At the same time, the very verisimilitude of pictorial images and recorded sound has caused us to become acutely aware of the function of the hands that hold the camera and the microphones and that, more important, edit the products thereof.

The problem is *not* that contemporary media require intellectual skills that were not developed formerly; I see little evidence that the "decoding" processes of the newer modes (Gross, this volume) are fundamentally different from those of the oral and print modes. What *has* happened is that participation in social life requires substantially different modes of behavior than were adequate in small isolated communities dominated by print and oral communication and relatively primitive graphics.

Current educational systems as a whole simply are not teaching people how to live in mass, electronically connected societies.

2. *We cannot improve education simply by applying new knowledge and technology to the old educational modes but must instead create new ways of educating and support them with our new technology.*

Why can we not relate our advances in media and other technologies to the future of education simply by thinking of their application to traditional educational purposes and methods? Certainly the application of film, television, computer, and new types of print to the traditional teaching tasks of schooling *is* an important enterprise and one that is sorely underdeveloped. However, with other developments, they have created for us a new world that has to be lived in and comprehended. While creating frightening complexities and prospects, they have given us power that we did not have when the familiar forms of education were conceived. Media technology and other forces have brought about a new world that requires a new education if we are to realize the extraordinary power we have been given.

We can afford to use only those portions of present education that enhance the perception of our technologies and their psychic and social consequences, and that give us necessary control over our personal and social destinies. This is why the importance of media to education is not that it enables us to do the old things more efficiently; instead, the media have created a different world in which we are presently unable to live effectively. We require a new education if we are to survive in this new world. If we do not make fundamental changes in education, we will drift with the winds of technology amid obsolete social organizations, and probably create a world in which the human dysfunctions of the educational system are magnified by technical advances.

3. *The structure of the present educational system was designed to prepare young people for a status-oriented, slowly changing society, in which con-*

sumption of the world's resources was expanding and the needs of industry dictated the content and structure of the educational system.

The basic forms of the present-day educational system and the most practiced educational methods evolved slowly throughout the nineteenth century and the first half of the twentieth century, when the modern industrial state was being developed. Nearly all of the characteristics of education in practice were derived in particular from the need of an industrial society to provide citizens with the opportunity to make their way within the industrial hierarchy and the social status system that reflected it.

For example, the contents of education (such as the basic skills of reading, writing, and arithmetic) were designed to provide entrance to the economic world and to help people establish and maintain families in that world. Nearly all of the educational reform movements that ran counter to that direction have not been successful.

Many of the liberal reform movements of the last 20 years have tried to extend the possibilities for inclusion in the economic society instead of to change the direction of education.

In addition, the graded form of education fits the need for industrial classification nicely. One's economic future is determined by where he gets off the educational ladder. This has perpetuated and reified the inequities within the system.

In this sense the educational system as presently constituted has a direct functional relationship to the status structure of the society, and as long as that status system persists, the aspects of education that bear an organic relationship to it are likely to be perpetuated. Movements that promote individual development encounter considerable opposition, because these represent the greatest threat to the homogenization required by the still-dominant, early-industrial ways of thinking.

The simple fact is that education for mobility through the system has become a false hope for most people. We are presently producing far more educated people than can be absorbed in the industrial commercial system at the kinds of levels that persons of that amount of education have come to expect. Many of our professions have become vastly overcrowded with qualified people. In addition, it is plain that we cannot continue to expand the industrial system indefinitely without destroying our ecology—both social and biological.

But, in fact, technology has greatly reduced many of the traditional types of jobs while creating a vast number of new ones, particularly by expanding the realm of personal services.

Equally, the possibility of a terminal education has disappeared almost entirely. There will no longer exist simple, well-defined vocational lines that do not require a constant reeducation and even periodic reshifting of careers.

To continue to educate children for the past world is a travesty of educational morality.

4. Industrial-age education was created to homogenize men. Our post industrial challenge is, instead, to increase diversity as well as commonality.

The existing educational system was designed to standardize persons by teaching them the same thing. The "educated" man was one who shared a common body of knowledge with other men. Just as standardization in commerce has decreased product differentiation, so has the spread of the educational system increased the threat to human diversity. In the early part of the twentieth century, the educational system deliberately ignored the ethnic differences of the immigrants to the United States in an effort to create an American melting pot.

Standard procedures are necessary for efficient functioning within any technical area, but when standardizing techniques are applied to education, they are a threat to the personal nature of education and to the continuing identity of any group that lays claim to a distinct cultural heritage.

The expectation that in the long run an international American-style culture will ultimately provide a uniform value base is threatening to the existence of other cultures. Current curriculum innovations, such as the development of Black Studies programs and areas of concentration in newer media (such as film making) are all responses to the panoply of social changes that make diversity a desirable thing.

5. The primary setting of education, the classroom, and the chief mediator of instruction, the multipurpose teacher, are obsolete.

First, the "classroom" as the primary "subinstitution" of learning is obsolete. When the school as we know it was created, the primary way of helping someone to learn was to expose him to an older and more knowledgeable person. The things to be taught were largely the familiar symbolic skills of Western civilization and media specialists were content for many years to try to bring a greater diversity of media into this situation. Research into teaching has disclosed a remarkably homogeneous national style of teaching, consisting mostly of exposition and drill (Hoetker and Ahlbrand, 1970) and centered around the most restricted form of print media, the recitation-oriented textbook. Media technologists and other reformers have made the understandable mistake of trying to improve the situation without changing the classroom style of organization. *This style has successfully repelled nearly every form of innovation.*

Alternative forms of staff utilization are gradually coming into existence through efforts to develop settings more conducive to a wider variety of models for teaching and learning. A variety of schemes for team teaching have developed (Joyce, 1968). Systems for teaching built around instructional systems have been developed, tested, and implemented (Smith and Smith, 1966).

Architectural forms have combined learning support centers with a variety of spaces designed for different types of teaching and learning activities (Educational Facilities Laboratories, 1968). In widely scattered settings across North America, England, the Scandinavian countries, and Israel a considerable variety of alternative forms of staff utilization now exist. These represent attempts to evolve from the relatively monolithic, homogenizing atmosphere of the classroom to a more pluralistic environment in which a wide variety of media forms are welcome.

THE FOCUS OF REFORM

The task of reform is the creation of learning environments that permit greater fulfillment of individuals, a fuller actualization of the possibilities of community, and an involvement of citizens in the process of revitalizing and humanizing the society. This is the core of the moral mission of education, which extends beyond the concept of education as a reaction to the other dimensions of social life, to the imperative need for educational processes that have a positive role in the improvement of society.

The concepts we can use to design this education are partly the result of technical advances and partly derived from theories of institutional organization.

The design questions are approached in terms of four concepts that are brought together to generate designs: (1) multimedia systems, (2) learning centers, (3) media functions, and (4) models of learning.

Multimedia Support System

The invention of multimedia instructional and support systems has provided tools for creating centers in which a large number of models of learning can be actualized over a great range of substance with considerable variation in complexity.

What multimedia systems promise are forms that permit the delivery of a range of instructional supports and informational supports. Multimedia systems are not restricted to a single type of learning or instructional model—they can support a range of them.

A comparison of the capability of multimedia systems with ordinary classroom practice is striking in terms of the range of learning and locus of control associated with each. Let us consider the creation of the flight simulator (Parker and Downs, 1961).

This device is striking because it provides the opportunity to learn exceedingly complex skills that are related to sets of diverse and precise theoretical knowledge bases. It uses a variety of media that are brought together with a series of learning tasks. These tasks can be paced by an instructor or by the student, with the aid of tracking systems that provide feedback about learning to either the external training agent or the student acting as the agent.

In a flight simulator the student is presented with a simulation of flight conditions. It can present to the student a sequence of tasks that commence at his entrance into the simulated cockpit, continue through his communication with simulated flight-control agencies, his takeoff and piloting of the vehicle, and conclude with his landing with simulated radar guidance. As he engages in these tasks, his performance is responded to with feedback. If he has difficulty performing some tasks, the relevant phases of the simulation can be repeated until he masters the requisite skills.

On a much simpler scale, Joyce and his collaborators (1970) have developed a learning center based on a set of data banks storing information about a variety of communities representing diverse human societies. This learning center can be used in models of learning that either respond to learner direction or provide structured learning tasks and systematic instruction.

The development of multimedia systems has made the distinctions among various media less striking than the possibilities for the design of complex systems in which an array of media are used in appropriate combinations to support the effort of the learner. The instructional systems developed by Joyce, Weil, and Wald (1972), for example, utilize several media for support of specific learning tasks.

The capability of multimedia systems may be combined with the multimodel concept to provide the base on which we can create a new educational technology aimed not at improving the classroom but instead at replacing it with a flexible array of centers for learning. These centers can serve various educational missions and be arranged so that the education for any given student or group of students can be created by relating him to appropriate combinations of learning centers.

Thus, education need not take place in specific, multipurpose institutions called schools, directly linked to economic advancement, but can be organized in terms of learning centers to which people have a lifelong relationship. These learning centers can be directly related to the needs and purposes of a contemporary education.

Functions of Media Forms in Storage and Instructional Systems

In the design of informational support and instructional support systems, we can distinguish a variety of dimensions that affect the type of support that is made available. They can be employed in both of the two general types of educational support systems—as storehouses of data and artistic products—and as parts of instructional systems that themselves can be stored as an array of instructional possibilities.

There are three major functions of media in storage systems and instructional systems. These functions can be arranged in designs generated by a considerable variety of models for teaching and learning.

> *Task presentation.* Any medium can be used to present learning tasks and a vast variety of learning models can be employed to generate the tasks.
>
> *Feedback.* The communication system is as important as the media that are employed. Many arrangements for feedback of results can be employed and the media can be used for transmission of knowledge about performance.
>
> *Substantive information source.* Again, any medium can be employed to store information units, but the message is affected by the medium that is chosen.

Three styles of media arrangement can be identified; various models of teaching are suitable for different combinations of styles:

1. RANDOM ACCESS. Pure storage, with tasks, feedback messages, and information units being stored in categories from which they can be withdrawn in any order.

The more random the arrangement of tasks, feedback messages, and storage, the greater the active role of the learner in shaping his own educational environment. Control by the learner should not, however, be necessarily thought of as either good or bad.

2. LINEAR. Sequential ordering of media types in terms of various functions. For example, a programmed sequence orders tasks, feedback, and information sources to induce sequential learning. The more linear the arrangement, the greater the control of the system over the behavior of the student.

3. DYNAMIC INTERACTIVE. Arrangement of media functions within a communication system which provides tasks, feedback, and substantive information in a pattern which permits instruction to be regulated according to learner performance and motivation. The pilot simulator and the language laboratory are examples of dynamic systems.

Dynamic arrangements are suitable for complex learning where specific training is necessary, but differences in learner behavior require system adaptability so that each student can increase or decrease his practice of specific skills and integrate several skills into a complex behavioral pattern.

The Array of Media Possibilities

When we consider the media types and the functions to which they can be put, we obtain an array of media possibilities instead of an analytic set of concepts that distinguish media from one another.

Motion pictures, for example, while they generally are used for linear information transmission, can be used for random access information transmission, especially when a series of motion pictures on a particular topic is stored under a category system to which the student has random access. Similarly, feedback can be provided by motion pictures as in the driver simulator, where a learner who turns the wheel to the right or left sees an image that provides him with the information he needs for corrective action. In addition, motion pictures represent an artistic form and the products of the artistry can become the content of study.

The early literature on audiovisual instruction made much of the unique capabilities of various media, but the relevance of most of these distinctions has been greatly reduced by the concept of the multimedia system. This is not to say that distinctive uses of particular media will not emerge as practice and research generate more direct comparisons in the years to come. But it must be noted that the emphasis on uniqueness was necessary when the teacher was seen as the primary mediator of instruction, and it was important to show teachers that they might use graphics and motion pictures (in addition to the chalkboard, textbooks, and oral communication) as primary media of teaching.

Now that we can envision organizing schools and other education opportunities so that personnel are employed in many specific educational roles instead of as generalists, we can introduce the concept of multimedia systems in which a variety of media are combined to perform various functions with respect to any educational objective.

In fact, the greatest change this is going to generate in the future is that only a small proportion of instruction will be primarily teacher mediated, and we can expect a correspondingly greater proportion of the instructional load to be carried by media in various combinations. In the past we have wasted the skills and resources of human agents in many roles for which they are not well suited. In the schools of tomorrow, education will not be limited simply to what agents can or will do at any given point of time, and they will not be wasted on roles better suited to mediated agencies.

MODELS OF LEARNING AND TEACHING

One of the great humanistic issues of our time is how to use technologies, and especially media, to design the entire education milieu so that the learner obtains increasing control over his behavior.

The fundamental proposal we shall make is to provide the learner with a variety of models of learning that he can utilize for his own purposes. The really helpless learner is not one who is controlled from outside, but one who is unable to control his behavior because his own personal repertory is so limited. A student in possession of a variety of learning models is in a position to construct his own education. If he does not have these, he will eventually come under the control of others or will simply fail to learn because he does not have the ability to put together a meaningful education for himself.

Media types and functions are shaped by the kinds of models of teaching and learning that are employed to design the educational environment. The concept of models will be discussed in terms of four types (Weil, 1973; Joyce and Weil, 1972).

1. The Personalist Model

"Personalists" focus primarily on the individual's construction of his own reality. Thus they emphasize the development of the individual and speculate about environments that might affect his personality or his ways of relating to the world. Therapists, especially, tend to be concerned with the distinctive ways in which each person constructs his world.

2. The Synergistic Model

The second type focuses on the processes by which groups and societies negotiate rules and construct social reality, and views education as a process of improving society. Synergistic models suggest ideal models for social structures and interpersonal relations, and procedures for creating educational systems that can help to transform such models into a social reality.

Others who emphasize social behavior have concentrated on interpersonal relations and the dynamics of improving them. The approaches to education in either case have a distinctly social character.

3. The Information-Processing Model

The information-processing model is concerned with developing the information processing system of the student. It includes the procedures that are designed to increase general thinking capacity. It also focuses on ways of teaching students to process information about specific aspects of life. For example, many educational theorists believe that a major objective of education should be to develop approaches to the teaching of the academic disciplines, so that a student may learn to process information like an academic scholar and thereby achieve the intellectual power of scholarship.

4. The Cybernetic and Behavior Modification Models

The fourth group focuses on the processes by which human behavior is externally shaped and reinforced. The major efforts in this area have been devoted to understanding the shaping of human behavior by environmental forces and how education can be built on an understanding of such processes.

These models of teaching and learning represent alternative approaches that can be employed in the design of learning centers and of the informational support and institutional systems that comprise their major elements.

THE DESIGN OF LEARNING CENTERS

Let us visit a possible school of the future. The design of this school is more illustrative of the kind of educational system we wish to create than are specific activities or subject matters taught within it. Such an institutional complex would teach as much by its form as by its substance. By actually managing their own learning, through the use of a complex of supportive systems, students would have an opportunity to develop competence in various communication modes and to acquire the skills required to cope with an advanced technetronic society.

Our school is not housed within a single building. It is organized as a series of learning centers that occupy a variety of physical locations. To visit it we must move from one center to another, although we find that some technical support systems are common to them. In fact, a general storage and retrieval system is designed so that students can retrieve information in several media and also instructional systems from their homes as well as from the learning centers. (Figure 1.)

I will discuss five of the possible learning centers that such an educational system might include.

Idiosyncratic Centers

These serve the students on their own terms. They are staffed with counselors who deal with students as equals, helping them formulate their goals and procedures. These facilitator-teachers help the students relate to a wide variety of part-time teachers—members of the community who serve, largely on a voluntary basis, as tutors, resources, advisors, and teachers of short courses. In addition, they help students relate to the other centers where other teachers and tutors can serve them.

The Idiosyncratic Centers are also supported by a multimedia "library" and data bank, mostly automated and employing microfiche and microfiche copymaking units to bring access to virtually all the material available in the Library of Congress. Many of the automated storage facilities are shared by all the "schools" of the region. The library supports all activities of the other centers.

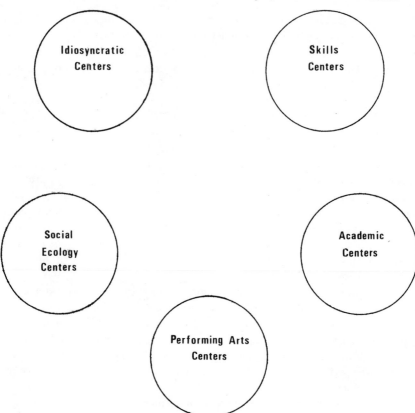

Figure 1   Learning centers by purpose.

The center is also supported by an Instructional Systems Bank, which consists of an array of self-administering multimedia instructional systems in the most common areas of study. A modular plan permits students to select among the offerings and assemble sequences to serve specific purposes.

Thus, the Idiosyncratic Centers consist of counseling areas, where students (of all ages) make contact with counselor-facilitators who help them define their own goals and procedures and relate to the support services they need to actualize their plans. (Figure 2.)

The services of the Idiosyncratic Center are available to students from early childhood until senescence. A student can use the counseling services to obtain personal counseling, diagnosis of needs, facilitation of career education, to pursue hobbies, or to obtain advanced training in academic areas or performing arts. The Idiosyncratic Center provides only 25 to 30% of the schooling experience for children, while it is the focus of most education for individuals in middle age.

The Human Ecology Centers

Whereas the Idiosyncratic Center is designed to facilitate personal growth and to enhance individuality, the Human Ecology Center is devoted to the process of improving society. It is organized to facilitate problem-solving groups who study social issues and problems, examine and improve their own interpersonal behavior, and generate social action to alleviate social problems and initiate improvements in societal relations.

Using the library, data bank, instructional systems, and academic centers for additional support, the Human Ecology Center employs simulators and an information retrieval system to assist students in the study and solution of social and biological problems. An urban simulator supports the study of community problems, an internation simulator provides service to the study of international problems, and an earth resources simulator is used to study biological support systems. (Figure 3.)

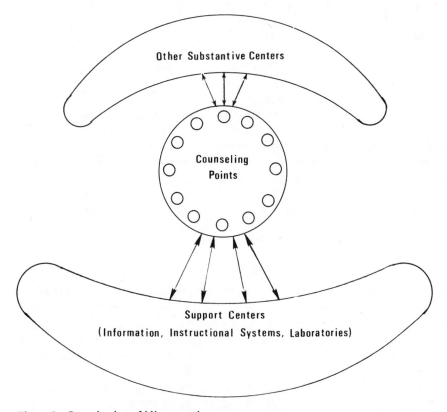

Figure 2   **Organization of idiosyncratic center.**

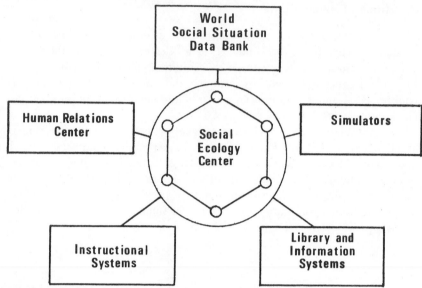

Figure 3   The social ecology center.

The teachers in the Human Ecology Centers are, for the most part, group leaders, skilled in human relations and in the use of teaching models that facilitate dialogue on social problems.

Students are in contact with the Human Ecology Center from the earliest years, but at first they concentrate only on neighborhood problems and face-to-face human relations. Gradually they increase their scope, studying ecology, urbanization, government, and the creation of an international community. The simulators enable them to study social processes and to try alternative modes of social behavior. "Human relations" exercises help them to explore ways of interacting with one another and of organizing themselves to improve social conditions.

The Skills Center

In the Skills Center students find diagnosticians who assess their communications skills and basic areas of knowledge, helping them to relate to instructional systems and to tutors.

While younger children spend the most time in the Skills Center or pursuing related activities, persons of all ages relate to the center, improving their skills and learning new ones.

Communications skills in all media are included in the center. Seminars on form and substance are correlated with the study of encoding and decoding skills so that the structure of media and symbol systems can be understood. At the advanced levels, studies would include training in the comparative analysis of media and symbol systems and the creation of messages in alternative modes. (Figure 4.)

The Skills Center would also include training in the use of the support systems that facilitate each of the learning systems. Training in the use of multi-media instructional systems, information storage and retrieval systems, and diagnostic and management systems are embedded in the Center. The Skills Center includes technical training in self-education, which complements the counseling in the Idiosyncratic Center.

Academic Centers

In the Academic Learning Centers (in the humanities, aesthetics, empirical studies, and mathematics) students join groups of other students for three types of courses. One is survey courses in specific areas, conducted by teachers with support from the Instructional Systems Center. These are followed by

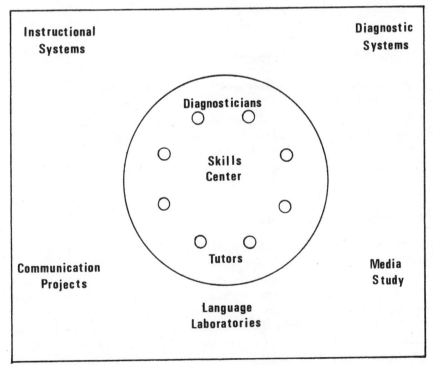

Figure 4   The skills center.

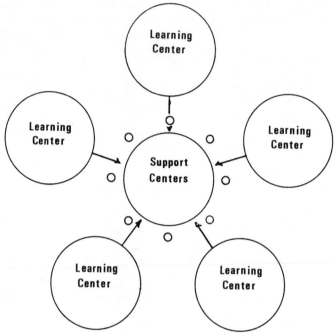

Figure 5   Learning centers and support centers.

inquiry courses in which students work with academic teachers in the modes of inquiry of the disciplines. Advanced students relate to academic tutors who help them to construct plans of personal study and to relate to groups of similarly advanced students. These centers are housed in laboratories that are especially constructed for the discipline (as physics laboratories, art workshops, etc.) and are supported by the Library and Instructional Systems Centers in the same way as the other centers.

The Performing Arts Center

Music, drama, television and film production, dance, athletics, and the other performing arts are housed in a network of laboratories, workshops, and little theaters throughout the community. Students may relate to the Performing Arts Centers in a variety of ways, ranging from recreation to skill development to long-term creative and expressive ventures.

The school would contain other learning centers, but the number described thus far is probably sufficient to provide a concrete idea of the concept on which it is developed. (Figure 5.)

The essence of such a school is the student's participation in a variety of models or strategies of learning to maximize his control over the educational process. "Nurturant" teams of counselor-teachers would help the younger children to do this and would provide a stable environment for them in "home base" locations. The older students would turn to their Idiosyncratic Center counselors for assistance. Adults would operate on their own unless they wished to consult their Idiosyncratic Center counselors. Matching models such as Hunt's (1971) would be used by counselors to locate the learning models that would most facilitate the student's development. Learning centers and support systems would always be changing to meet emerging educational needs, but a warm and facilitative social system would be needed to provide stability for the student.

CONCLUSIONS

The design elements we have been discussing do not mandate that particular learning models or educational missions dominate the school. What we have been discussing are the emergent elements of a design technology that can support a wide variety of educational missions and means.

The politics of educational reform have always been heated. The organic relation of school and society makes it inevitable that any change in schooling will create more dislocation and resistance. These are necessary concomitants of reform in any social institution in a complex society.

However, this chapter is based on the assumption that educational institutions that utilize multimedia systems, learning systems, and instructional and informational support for a variety of learning models would be in line with the demands of postindustrial society and that its ultimate emergence, in some form, is inevitable. In fact, all of the design elements and learning center options described in the last few pages have real-world exemplars that can be interpreted as the beginnings of adjustment of the school to contemporary society.

The proposals in this paper have to be distinguished from proposals that would use education to reform society. We are trying here to create design elements that can reduce the distance between educational forms and an already changed society.

To phase in change, one might well begin with the types of learning centers that reflect familiar educational purposes, follow with those designed to facilitate social adjustment (continuing vocational education), and proceed toward those centers that aim at less familiar educational objectives.

Given a system in which a diversity of instructional models is available to the student for the pursuit of individual and social objectives, the definition of purposes is a particularly serious matter. Communities must elect to emphasize the goals they value most highly (personal development, social purpose, creative expression, etc.) rather than leave such important decisions to the mindless workings of administrative pragmatism.

REFERENCES

Brzezinski, Zbigniew. *Between Two Ages*. New York: Prentice-Hall, 1972.

Educational Facilities Laboratories, "Educational Change and Architectural Consequences," (New York, 1968).

Hoetker, James and William P. Ahlbrand. "The Persistence of the Recitation," *American Educational Research Journal* VI; 2 March 1969.

Hunt, David E. *Matching Models in Education* Toronto: Ontario Institute for Social Studies in Education, 1971.

Joyce, Bruce R. *Men, Media, and Machines*. Washington, D.C.: National Education Association, 1968 for a description of the relation of support centers to facilitative teams.

Joyce, Bruce R. and Marsha Weil, *Models of Teaching*. Englewood Cliffs: Prentice-Hall, 1972.

Joyce, Bruce R. and Marsha Weil, Eds. *Perspectives on Reform in Teacher Education*. Englewood Cliffs: Prentice-Hall, 1972.

Joyce, Bruce R. et al. *Data Banks for Children*. New York: Teachers College, Columbia University, 1970.

Joyce, Bruce R., Marsha Weil, and Rhoada Wald, *Basic Teaching Skills*. Palo Alto, CA.: Science Research Associates., 1972.

Joyce, Bruce R., Marsha Weil, and Rhoada Wald, *Three Teaching Strategies for the Social Studies*. Palo Alto, CA.: Research Associates., 1972.

Parker, Jr., James F. and Judith E. Downs, *Selection of Training Media* Washington, D.C.: Office of Technical Services, U.S. Department of Commerce, 1961.

Smith, Karl and Mary Smith, *Cybernetic Principles of Learning and Educational Design*. New York: Holt, 1966.

Weil, Marsha, *Deriving Teaching Skills from Teaching Strategies: A Paradigm for Competency-based Education*. New York: Teachers College, Columbia University, 1973.

CHAPTER 17

Teacher Image in Mass Culture; Symbolic Functions of the "Hidden Curriculum"

GEORGE GERBNER

"The figure of the schoolteacher," wrote Richard Hofstadter in *Anti-Intellectualism in American Life* (1963, p. 309) "may well be taken as a central symbol in any society." But a symbol of what? Searching for an answer is like opening Pandora's box with its host of evils. My examination of the evidence suggests that teachers, schools, and scholars project a synthetic cultural image that helps to explain—and determine—the ambivalent functions and paradoxical fortunes of the educational enterprise in American society. The clues that point to that disquieting conclusion (which also raises questions about anticipated extensions of the present structure of culture-power) have led me to new reflections. These concern the illusions and reality of schooling, the nature of symbolic functions, the lessons of national and cross-cultural research on the teacher image, and the role of that "hidden curriculum" in social policy.

The basic features of American schools, as of our society, have been fixed for more than a century. Spectacular changes transformed the "quality of life" through unfolding and extending those features into every aspect of existence. Among the most dramatic of the changes has been the rise of institutions of cultural mass production—the mass media—exempt from the laws of public but not of private corporate development and authority. These institutions have taken over many functions performed in the past by the parent, the church, and the school. The media's chief impact stems from their universality as the common bond among *all* groups in our culture. The media manufacture the shared symbolic environment, create and cultivate large heterogeneous publics, define the agenda of public discourse, and represent all other institutions in the vivid imagery of fact and fiction designed for mass publics.

Teachers and schools no longer enjoy much autonomy, let alone their former monopoly, as the public dispensers of knowledge. The formal educational enterprise exists in a cultural climate largely dominated by the informal "curriculum" of the mass media.

New developments in communication technology may both individualize and globalize the penetration of mass-produced messages into the mainstream of collective consciousness. Before we can consider what that new transformation might portend for our schools (and for our culture), we need to take a fresh and sober look at the omens from the past.

Many of those who would correct the evils of society slowly and painlessly have long argued for educational extension, improvement, and reform. And, for more than a century, schooling has been extended, improved, and reformed. Yet it is still compulsory, unequal, class-biased, racist, and sexist. From the Coleman report on *Equality of Educational Opportunity* (1966) through the Jencks report on *Inequality* (1972) study after study demonstrates that the schools, even when "equal," tend to justify instead of rectify the child's fate as defined by the culture of the home, the street, and the television.

Eminent public figures declare schools a "disaster area" and "a pathological sector of the economy," meaning that even money cannot cure what ails them (although that remedy has never really been tried on a large national scale). A few call for their abolition. Instead of becoming the social corrective that idealistic reformers sought and democratic rhetoric promised, the more schools change the more they streamline their induction of young people into their places and roles in the existing social structure. The Mason-Dixon line has been abolished between the states, but its modern equivalent now rings every city and the few bridges busing children across it may be dismantled. Schools still provide custodial drill for the poor, enrichment for the rich, and equal instruction for those with equal economic or political clout.

We are just beginning to understand that these harsh facts result from no accidental aberration or cultural lag. A new generation of "revisionist" historians has exploded "The Great School Legend." Colin Greer's book (1972) by that title shows how the perennial "crisis" of the schools, like the perpetual "problem" of the slums, is in fact more functional to the existing social order than would be its elimination. Michael B. Katz's book, *Class, Bureaucracy, and the Schools; The Illusion of Educational Change in America* (1971), documents the recurring phenomenon of school reform movements that engage the zeal and energies of those who would attack the iniquities of society, only to find that each wave of "reform" harnesses their schools to the dominant interests of the times.

Nearly 20 years after the Supreme Court ordered school desegregation "with all deliberate speed," feeble efforts at enforcement bog down in political controversy, just as before that the parochial school aid controversy was used to defeat proposals for massive federal financing. Busing is claimed to threaten the fiber of society when it brings children of different races and social classes together, but not when used to keep them apart, its traditional use. If "campus unrest" is followed by recession and cutbacks, who is to blame? The schools that should redeem us teeter at the edge of bankruptcy. It seems that when citizens consider what is nearest and dearest to them, such as children, they are most vulnerable to the deceptions of their culture.

SYMBOLIC FUNCTIONS

We are keenly aware that messages intended to persuade usually serve the purposes of those who create and disseminate them. Less obvious but perhaps even more crucial are the purposes served by news, fiction, drama, and other storytelling designed with no other obvious or conscious intent than just to "inform" or to "entertain." The social tasks to which presumably "objective" news, "neutral" fiction, or "nontendentious" entertainment lend themselves are what I call symbolic functions. They are the consequences inherent in the way things "work" in the symbolic world of storytelling.

These functions usually do not stem from individual communications or campaigns but from the general composition and structure of the bulk of the symbolic environment to which an entire community is exposed. The consequences typically are not those of change but of continuity and resistance to change. Given a stable social order, the functions are usually to selectively cultivate existing tendencies, and perhaps to deepen and sharpen them.

Symbolic functions differ from those of nonsymbolic events in the ways in which causal relationships must be traced in the two realms. Physical causation exists outside and independently of consciousness. Trees do not grow and chemicals do not react "on purpose," although human purposes may intervene or cause them to function. When a sequence of physical events is set in motion, we have only partial awareness and little control over the entire chain of its consequences.

The symbolic world, however, is totally invented. Nothing happens in it independently of man's will, although much that happens may again escape individual awareness or scrutiny. The reasons that things exist in the symbolic world, and the ways in which things are related to one another and to their symbolic consequences, as in a play or story, are completely artificial. That does not make their production any more arbitrary or whimsical than the events of the physical world. But it means that the laws of the symbolic world are entirely socially and culturally determined. A character in fiction "dies" not because he has lived but because it serves a purpose to have him die. Intended or not, that purpose is the only reality of the story. The causal link is not between life and death but between a creator's or producing organization's position in life and society and the significance of that death. No TV badman ever dies a natural death, nor can the hero of a western serial be cut down in the prime of life. To be "true to life" in fiction would falsify the deeper truth of cultural and social values served by symbolic functions.

Whatever exists in the symbolic world is there because someone put it there. The reason may be deliberate and planned or circumstantial such as an "unrelated" marketing or programming decision, or a vague feeling that it will "improve the story." Having been put there, things not only "stand for" other things as all symbols do, but also *do* something in their symbolic context. The introduction (or elimination) of a character, a scene, or an event has functional consequences. It changes other things in the story. It makes the whole thing "work" differently. Dynamic symbol systems are not "maps" of some other "real" territory. They are our mythology, our organs of social meaning. They make visible some conceptions of the invisible forces of life and society. We select and shape them to bend otherwise elusive facts to our (not always conscious) purposes. Whether we know it or intend it or not, purposes are inherent in the way things actually work out in the symbolic world. Even when men and institutions lie, they cannot do so without giving off signs of the purposes of their lying, at least in the long run; otherwise why lie? More problems arise from communicating hidden purposes than from "failing" to communicate at all.

How things work out in mass-produced symbolic systems, as in all collective myths, celebrations, and rituals is indicative of institutional interests and pressures. Various power roles within and without the institution enter into the "decision-making process" that prescribes, selects, and shapes the final product. In the creation of news, "facts" impose some constraints on invention; the burden of serving institutional purposes is placed on selection, treatment, context, and display. Fiction and drama carry no presumption of facticity and thus do not inhibit at all the candid expression of social values. On the contrary, they give free reign to adjusting facts to the truth of institutional purpose. Fiction can thus perform social symbolic functions more directly than can other forms of discourse.

That is why in fiction and drama there is no need to moralize. The moral is usually in the "facts" themselves. For example, if a social inferior (lower class, "native," black, etc.) usurps the place of a superior (through marriage, business deal, combat, etc.), he or she can have an "unfortunate accident," thus avoiding overt bias and yet performing the symbolic function of enhancing the superior life chances of "superior" characters. Violence in the mass media— unlike life—is usually among strangers, permitting the lessons of social power (what types stand to win or lose in a conflict) to emerge unhindered by close human ties. Fiction can act out purposes by presenting a world in which things seem to work out as they "ought to," regrettable or even terrible as that might be made to appear.

Characters come to life in the symbolic world of mass culture to perform functions of genuine social import. These functions need not be planned or perceived *as such*. They need not even conform to any overt rationalizations or moralizing. The functions are implicit not in what producers and audiences think they "know" but in what they assimilate of what the characters of the symbolic world in fact *are* and *do*.

THE "HIDDEN CURRICULUM" AND ITS EFFECTS

The "facts of life" in the symbolic world form patterns that I call the "hidden curriculum." It is the framework that makes the notion of "effects" sensible as those changes that can be observed within a stable structure. The *prior* preoccupation with "effects" is misleading. However, it only betrays greater concern with marketing tactics than with the basic allocation of values in our society. The post World War II movement within social science reflected more concern with buying or voting "behavior" than with meanings that govern *all* behavior. Now social scientists are taking another look at the relationships between social structures and those general frameworks of knowledge and values that, in turn, shape the meanings and the efficacy of particular messages. As humanists have always known, no society designs its religions or its customs or its schools on the basis of a comparative assessment of the "effects" of various factual or philosophical statements. That would put the cart of tactics before the horse of basic aims and functions. Any assessment of "effects" must assume the existence of a standard of measurement against which different or changing quantities and qualities can be measured. That standard is implicit in the value structure of a culture. Should that be immune from inquiry? The contention that the existence or meaning of an action or communication should not be assessed until its "effects" are established is tantamount to the assertion that the structure of a culture should not be investigated; only its tactics are to be subjected to "scientific" inquiry. Far from being scientific, this is itself a symbolic tactic attempting to define what is "scientifically" reasonable and respectable in a way that serves only the most dominant, pervasive, and taken-for-granted social interests.

The "hidden curriculum" is a lesson plan that no one teaches but everyone learns. It consists of the symbolic contours of the social order. One cannot sensibly ask what its "effects" are any more than one can ask about the "effects" of being born Chinese instead of American. Culture-power is the ability to define the rules of the game of life that most members of a society will take for granted. That some will reject and others will come to oppose some of the rules or the game itself is obvious and may on occasion be important. But the most important thing to know is the nature and structure of the representations that most people will assume to be normal and inevitable. Having established some features of the "hidden curriculum," one can then ask how its specific lessons are internalized and which of its functions serve what purposes.

Every culture, as any school, will organize knowledge into patterns that cultivate a social order. The fundamental lessons of the curriculum are not just what pupils learn in math, history, physics, and so on, but also the fact that *those* are its commonly required subjects and not basketweaving, harmony, or Marxism-Leninism (except where *that* is required). One cannot ask about the "effects" of that pattern of required learnings except by comparing it with the functional dynamics of other patterns. The structures themselves and most of their symbolic functions are inevitably assimilated if there is to be anything like a relatively stable social order. Culture *is* that system of messages that makes human society possible. After grasping the implicit agenda of discourse, scale of priorities, spectrum of valuations, and clusters of associations that most members of a culture come to assume as the overall framework for most of their thinking and behavior, we can begin to observe the fluctuations and reversals within that structure. Only after that can we ask the "effects" question.

The question of "effects," properly phrased, inquires first into individual and group selectivities by dipping into the currents and crosscurrents of the cultural stream. Second, "effects" research can investigate the contributions that particular types of messages make to the processing of particular conceptions within given frameworks of values and knowledge. We are a long way from being able to answer the second question. The answer will be of strategic significance once we know more of what the game is about.

The prior need is to examine the framework implicit in the "hidden curriculum." We must first go beneath the explicit and fragmented significance of individual images available to casual personal scrutiny and find the symbolic patterns and functions that entertain (in every sense of that word) the collective morality and the dominant sensibilities of the social order.

The image of schools and scholars is that part of the "hidden curriculum" in which all members of society learn about learning itself. Its symbolic functions relate images of learning, and of the formal institutions of learning, to basic human values and to the locus of power in society. I think that the figure of the schoolteacher is a central symbol of the uses and control of popular knowledge. Its most telling features touch upon questions of vitality and self-direction, social relations, morality, and power.

HISTORICAL IMAGES

When he is not the Ichabod Crane of literature (scared out of town by the virile males of the community, with a pumpkin smashed over his head), the typical teacher in American novels is "stooped, gaunt, and gray with weariness. His suit has the shine of shabby gentility and hangs loose from his undernourished frame." (Foff, 1956, p. 21) That is, until class is out and memory rings the school bell when we say a tearful "Good Morning Miss Dove," or bid a nostalgic "Goodbye Mr. Chips."

In his study of the college professor in the novel, Michael Belok (1958) noted that American fiction uses teaching to "unsex a woman." Even being a teacher's wife may be unenviable. Theodore Dreiser characterized Donald Moranville Strunk, A.B., Ph.D., Professor of History, as having had "one of the homeliest women for a wife I ever saw." College students responding to a survey (O'Dowd and Beardsley, 1960) characterized the schoolteacher as a person "who cannot even command an attractive wife." Love eludes even the attractive, eager "Our Miss Brooks" and the owlish but smart "Mr. Peepers"; sex degrades the neurotic fascist Miss Brodie and destroys Professor Rath of "The Blue Angel."

For Americans, the prestigious title "professor" resounds with mock deference. *The Century Dictionary and Cyclopedia* for 1899 gave one definition of professor as " . . . any one who publicly teaches or exercises an art or occupation for pay, as dancing-master, phrenologist, balloonist, juggler, acrobat, boxer, etc." From there it was not too far to the piano player in a brothel, or as Henry L. Mencken euphemistically recorded in *The American Language* (Supplement II), "a house musician." In time the usage mellowed to permit any prominent orchestra leader to be called professor, as those who remember Kay Kyser will recall (cf. Coard, 1959). Recent media fare is replete with such phenomena as the movie "The Nutty Professor," TV's "Professor Backward," the cartoon "Professor Wimple's Crossword Zoo," and Pat Paulsen's Laugh-In "Professor."

Belok could find only about 200 novels since 1900 in which college professors appeared as characters. Major American novelists, wrote Lyons (1962), either have avoided the Academy or wrote novels that are basically anti-intellectual. An English review of the American scene observed, however, that the college novel is now a "cottage industry." "And so it seems," commented Charles Shapiro (1963, p. 37) noting the entry of writers into the universities, "as book after book assaults us with tales of assorted hypocrisies committed under the name of higher education."

Hofstadter has also observed that the American teacher has not become an important national figure, worthy of emulation. Historical reasons may partly account for the fact that the scholar, as Dixon Wecter also noted in his *The Hero in America* (1941), "has never kindled the American imagination."

Until the industrial and national revolutions that changed the map of Europe, teachers were likely to be recruited from among the misfits of society. When the common schools were established in Russia, the theological seminaries dumped their "undesirables" to be the teachers. In the Prussia of Frederick the Great, it was the army that disposed of its invalids by appointing them schoolmasters.

"The low opinion of the rank-and-file schoolmaster in Europe spread to the New World, and a seventeenth-century Rector of Annapolis recorded that on the arrival of every ship containing bondservants or convicts, schoolmasters were offered for sale but that they did not fetch as good prices as weavers, tailors, and other tradesmen" (Wittlin, 1963, p. 750).

The national revolutions of Europe had a popular cultural character. Many of the leaders were writers and poets rising through the ranks of the intellectuals closest to the people, the teachers. W. G. Cove, the British teacher, strike leader, union president, and member of Parliament, once wrote that "At the head of every continental revolutionary movement, or near the head of it, stands an ex-teacher."

Until perhaps the emergence of the black liberation movement, which, for reasons peculiar to American culture seemed to propel clergymen rather than teachers into leadership, there has been no comparable historical force to add a heroic dimension to the traditional image of the American teacher. The forced pace of industrialization in the nineteenth century and the consequent pressure for extending public education created the monitorial schools, according to Wittlin "to fit the early state of industrial civilization."

"Pupils were cheaply mass produced, down to $1 per year. The scholars, who first learned their lessons from the teacher, conveyed exactly the same lesson to other children, ten to a monitor... In 1916 a book appeared in Boston on *Public School Administration*, by E. P. Cubberley, in which it was stated that '... the schools are, in a sense, factories in which the raw materials are to be shaped into products to meet the various demands of life.' According to this philosophy the educator was allotted the modest role of the copyist of patterns" (Wittlin, 1963, p. 751).

During the next 50 years, the cultural forces that shape the common images of society became largely mechanized, centralized, and commercialized. Teacher-power emerged as an organized force, and education became a political battleground. But the social function of the teacher image in the new culture remained the traditional one: to cultivate mistrust of the intellect on the loose.

TEACHER AND SCHOOL IN UNITED STATES MEDIA

There are 2.5 million teachers in the public schools in the United States. They range from 22 to over 65 years of age, and come from all states, classes, religions, and ethnic groups. Of course, they have some characteristics as a group: they average 39 years of age, 12 years of professional experience, and about $9000 a year. Two out of every three are women. Teaching is the largest profession; its members run the gamut of human types.

But not in popular fiction and drama. The raw facts of life are not the truth of social and institutional purpose. Frequency of symbolic representation is not the reflection of census figures. The casting of the symbolic world has a message of its own.

Studies of occupational representation among mass media characters (e.g., DeFleur, 1964; Gerbner, 1969), celebrities (Winick, 1963; Hazard, 1962), and even movie titles (Verb, 1961) agree that teachers, the largest profession in life, rank among the smallest in the media world. About 2 to 3% of all identifiable professional references or characterizations go to media teachers. DeFleur's classification of occupational roles found the same number of educators as taxi, truck, and bus drivers in the televised labor force.

Most of the literary studies delineate a teacher image created for elite audiences. Except when mellowed by misty memories of childhood, it is generally cruel and unsympathetic, as if in revenge for the intellectual and social pretensions of the hired hand. Much of that image found its way into the mass media, somewhat relieved by the populist fantasy of the "good" if not too enviable teacher.

Studies of media images were conducted by a group of researchers at the University of Illinois; the work is continuing at the University of Pennsylvania. For a number of years, our focus was the portrayal of teachers, students, and schools in the mass media. Some studies dealt with one medium, like Schwartz's study (1963) of Hollywood movies and Brown's study (1963) of magazine fiction; others ranged more widely. The U.S. Office of Education supported my analysis of over 1400 feature films, television and radio plays, and popular magazine stories featuring 2800 leading characters in the mass media of 10 countries (Gerbner, 1964, 1966). The National Science Foundation, UNESCO, and the International Sociological Association jointly sponsored a study that I did of the "Film Hero" involving one year's feature film production in six countries (Gerbner, 1969). I will draw on the summaries of these and other studies to piece together some basis for reflecting upon the symbolic functions and social role of the image of the "teacher" in mass culture.

Schwartz's (1963) study of Hollywood movies found that the presence of a teacher tips the odds three to one in favor of the movie being a comedy. Mass media teachers, creatures of private industry depicting public agents, suffer from signs of a cultural power conflict in which the media have the upper hand. The study's comprehensive review of research concludes:

Teachers in books, drama, magazine cartoons, and films were depicted as tyrannical, brutal, pedantic, dull, awkward, queer, and depressed. The few attractive teachers remained in the profession only long enough to find a mate. Teachers had a difficult time getting and staying married. One investigator noted that two-thirds of the teachers were portrayed as emotionally maladjusted. Another writer noted that "to succeed as a teacher one must fail as a man or woman." [Schwartz, 1963, p. 4]

Love and the Teacher

Love and sex are dramatic symbols of vitality and power. How a profession fares in love in the mass media is a good measure of its symbolic stature.

The mass media teacher pays a price for professional success. The price is impotence, and worse. The "schoolmarm" image hits women especially hard. Love and marriage are women's chief media "specialities" and typical reasons for existing in the stories at all.

Female characters in the world of mass fiction and drama are limited to a narrow range of parts. That is why media males not only dominate media females (except in the home where males *prefer* to be incompetent) but also vastly outnumber them. The average ratio is four men for every woman. But the proportion varies by theme. Love, marriage, and bringing up children are themes that utilize women characters in parts that do not require special explanation.

Studies of school-related stories in all media (Gerbner, 1964) found both women teachers and love playing prominent parts—but rarely together. Almost half of all media teachers are women; this is a high female ratio for the media, but still lower than the two thirds of all real-life teachers who are women. The school stories are more likely to feature romance than are stories in general. But the romance rarely involves the teacher, and least of all the woman teacher. Typical is Miss Dove, who is so devoted, so selfless, so excruciatingly *good* that she passes up her opportunity to marry in order to pay back $11,430 her dead father has "borrowed" from the bank where he worked.

Nearly half of all media adults but only 26% of male teachers and no more than 18% of female teachers are married. Despite all the romance and happy endings in the stories, teachers rarely inspire love or fall in love, especially with each other. The most common condition of love is that the teacher finds a partner outside of education. The typical pattern has her quitting a New

England high school and a biology teacher fiance to "find herself" and a *man* in New York. Or it has him leaving a dull musical chair at a Western college, along with a straight-laced professor girlfriend, to be taught something about music and love in Tin-Pan Alley.

Failure in love and defeat in life permit most media teachers to be fully dedicated to the profession. The media teacher leaving for another specific occupation knows the road to success in the media world. Five times out of six the road leads to show business.

Poverty of the Schools

In the film study, 25 of the 470 movies portraying some aspect of education were found to show the financial plight of the schools. The deficiencies are usually in extracurricular activities such as entertainment and sports. There is never a need for more teachers or laboratories or classrooms. Profits from sports events, successful musical shows, and unexpected bequests of the rich are the usual solutions to academic poverty. Only two films show schools to be public responsibilities publicly financed. One deals with support for West Point, and the other depicts the building of a school in a remote New Zealand village.

Only one film shows the financial problem as one of low salaries for teachers. A wealthy Texas rancher is shocked to find his son trying to raise a family on the meager salary of an instructor. He secretly negotiates with a local butcher to sell his son meat at half-price. He also tries to prevent his son's promotion, confident that he would return to the ranch. When he does not, the father solves the problem by donating enough money to the school to provide a pay increase for all teachers (Schwartz, 1963, pp. 49–50).

An analysis of teacher characters in *Saturday Evening Post* fiction found them in more frequent financial pickle. This was usually explained by showing that they strive less than the other characters. About one third of the magazine's teachers solve their financial problems by quitting the profession. No teacher is ever given a salary raise. No student is supported on a public scholarship. No community takes the initiative to raise taxes or to build or improve the schools. When there is a suggestion of improvement in the poverty of the schools it is likely to be a private solution such as finding a rich donor or holding a fantastically successful show or sports event (Brown, 1963).

The School Sports Story

School sports is an arena of "early male socialization" (Booth, 1972). Extensive friendship ties are linked to participation in games. The winning team is also a symbol of an institution's ability to attract talent and display power.

Winning scores have been found to relate to legislative appropriations and certainly to alumni giving. An article in the *Philadelphia Magazine* (May 1972) quotes the head of the Alumni Society as saying that "The Alumnus in Oregon or Texas is going to read about Penn's basketball team in his home town paper, not some professor's finding old ruins in England."

There is no doubt that the most frequent appearance of schools and colleges in the American Press is on the sports page. The magic words of American higher education are Ivy League and Big Ten. *Saturday Evening Post* readers loved the stories of George Fitch. The first of these, published in 1908, began

> Yes, sir, it's been seven years now since old Siwash College has been beaten in football. . . . We've shut out Hopkinsville seven times—pushed them off the field, off the earth, into the hospitals and into the discard. We've beaten six State universities by an average of seven touchdowns, two goal kicks, a rib, three jawbones and four new kinds of yells. We put such a crimp into old Muggledorfer that her Faculty suddenly decided that football developed the toes and teeth at the expense of the intellect and they took up intercollegiate beanbags instead. And in all those seven years we've never really been scared but once. . . .

The school sports story, with its violent terminology, strong group spirit, and concern over the rules of the game, is the most likely vehicle for community enthusiasm, teamwork, and the mixing of different classes and races in a common cause. It generally demonstrates the ethics of skill and power among those who achieve equal status. (This can be contrasted with the symbolic functions of the crime or spy story displaying the game of power among those of unequal status or those who do not play by the same rules.)

A sketch of boys' sports fiction (Evans, 1972) described its symbolic functions as integration into the virtues of unquestioning participation, hero worship, inviolable hierarchy, sorting winners from losers, and a sharp sense of authority, belonging, and superiority. The school becomes society and the game the system at its dramatic best. As the novel of life at Rugby by English author Thomas Hughes (1854), *Tom Brown's School Days*, which introduced the genre to American boys in 1870, pointed out quite explicitly: "Perhaps ours is the only little corner in the British Empire which is thoroughly, wisely, and strongly ruled just now." Several new media and 100 years later, the functions are the same, even if the tactics a bit more sophisticated.

The film study shows sports to be the central theme in twice as many movies as deal with study, science, or research, and to depict virtues never seen in a portrayal of scholarly activity. The school sport story serves its symbolic functions in three ways: (1) as the means by which youths from different walks of life find acceptance in the group; (2) as the chief symbolic unifier of students, faculty, parents, and alumni; and (3) as teaching the importance of passing a realistic test of social and ethical "maturity."

The largest single group of stories concentrates on the third, the socio-ethical lesson. The films warn that romantic illusions lead to cynicism and despair. They counsel realism and vigilance lest "alien" ideologies take advantage of and subvert "our" flexible rules for "their" purposes.

"The most common presentation of sports," reported Schwartz (1963, p. 59), "was that it was a much less glamorous and honest activity than student-players were at first led to believe.... For the sake of victory, schools were shown to sacrifice their honor by depending upon extra-collegiate sources for both personnel and financial support. This dependence upon outside sources was not portrayed as unethical in all films dealing with sports—in fact, several films portrayed this dependence in a vein of lighthearted comedy which, if not condoning the practice, did not take the unethical aspects of the situation seriously. However, in the films to seriously treat the unethical practices of the sport and their demoralizing consequences for students, the portrayal of school sports was likened to a *rites de passage*. Sports were shown as analogous to the battleground upon which a young initiate experienced teamwork and struggle, despair and disillusionment, victory and defeat."

The typical school sports story is a morality play that shows a sort of pragmatic "democracy in action." The rules will bend within reason, and anyone can play, as long as the game is just a game and the prime source of power is clearly understood. Abuse the rules and the tone changes. In a group of films gangsters try to manipulate players and even faculty to reap large gambling profits. In another, radicals "disguised as students" (described in a contemporary *New York Times* review as "namby-pamby, bushy haired, and wearing tortoise-shell glasses") plan to overthrow capitalism, beginning with the college football team. The local hero falls briefly under their spell, but recovers in time to win the "game of the year" and the respect of "normal healthy Americans."

Community and Power

When they cannot relate to "the game," in which students play the lead, teachers usually do not "belong" at all. Typically presented as alien to the community in which they live and work, and often in conflict with its values, teachers may be seen as well meaning and kindly if impotent, or dangerous and evil if powerful, but rarely both good *and* effective.

Studies by Bowman (1938), Boys (1946) and Springer (1951) trace community conflict and antagonism through 50 years of magazine publishing, general fiction, and Broadway drama. Brown's (1963) study of *Saturday Evening Post* fiction found that teachers "act differently" even when trying to conform. The film research concluded that all but 6 of the 28 films touching on relationships between school and community portray a teacher as the target of hostility, ridicule, or ostracism. The offending teachers are usually shown as "outsiders... with their own set of values often aiding in isolating them from the community" (Schwartz, 1963, p. 40).

Nonconformist media teachers usually come to see the error of their ways. One movie depicts a socialistically inclined economics professor striking it rich. He changes his mind about radical causes and returns to his job a millionaire.

Most instances of unreconciled conflict between teachers and community involve the cardinal sins of trying to change society rather than the schools (usually labeled "communism") or of finding a source of wisdom outside the approved community context (usually represented as "atheism"). Sex often appears as a malignant obsession when sought by such unlikely characters as teachers. A cynically explicit portrayal in a 1937 movie shows a southern mob lynch a "yankee" teacher convicted of assaulting an attractive student. The district attorney does not believe the teacher guilty, but prosecutes vigorously because of the political value of the case for his own career.

In casting about for occupations to delineate hero types who are both right and mighty, mass media authors rarely pick teaching. Smythe's (1953) analysis of television drama found teachers outstanding among all TV occupations in being the "cleanest" and the "kindest." But they were also rated the "weakest," the "softest," and the "slowest."

The more potent teacher risks turning into that symbol of evil intellect—the mass media scientist. On television, the scientist was rated as the most "deceitful," "cruel," and "unfair" of all professional types.

Personality ratings used to assess students' images of real-life teachers tapped mass-cultural stereotypes. O'Dowd and Beardsley (1960) found that the student image of the school teacher is that of an unselfish, uninteresting, unsuccessful and effeminate person. The scientist, on the other hand, presents the image of the cool, cruel, hard-driving intellectual and often a loner who cannot be trusted.

Similarly, Gusfield and Schwartz (1963) concluded that the teacher image presents "the sharpest contrast between elements of esteem and status, on the one hand, and those of power and income on the other." The teacher ranks as the most "honest" and second most "useful" of 15 occupations, and also the "weakest" and "lightest." The scientist again appears to be cool, tough, and antisocial as well as "irreligious" and "foreign."

Publish and Perish

It is not surprising that the dramatic uses of scholarship and research contrast sharply with those portraying sports and other entertainment. Academic research leads to murder in nearly half of the 25 films found to portray teachers conducting it. Film teachers invent poisons, revive prehistoric monsters, or train other creatures to do away with suspected enemies. One famous movie of the 1950s shows a psychology professor hypnotizing gorillas to murder the girls who rejected his advances. The typical plot has some obsession drive the demented intellect to invent an instrument that gets out of control and destroys its maker to the relief of all mankind.

In a group of nine films dealing with research, the experimenting teacher or professor falls victim to his own delusions and exposes the stupidity or hypocrisy of scholarship. Typical is the movie in which the professor of Egyptology incorrectly deciphers an ancient tablet and the false message sends him on a series of comic adventures.

Research and experimentation fare better in the hands of amateurs. Student scholarship is usually foolish but never evil or selfish. Incidentally, classroom scenes hardly ever exhibit learning or scholarship. They are used to display problems of authority and discipline.

The teacher struggling for discipline in the school is often brutal and sadistic. In films of more recent vintage, students (as if representing the avenging forces of society) strike back in kind. The "class struggle" is one in which the teacher rarely comes out on top.

IMAGES ACROSS CULTURES

Through a series of cross-cultural comparative studies we tried to understand our own images better by comparing them with those of others (Gerbner, 1964, 1966). Four countries of Western Europe and five countries of Eastern Europe (including the Soviet Union) provided our comparisons. A plot sketch from each country's sample will give something of the flavor of the material.

"Red Castle" is what townspeople call the new headquarters of the Teachers' Recreation Center. It was a baron's palace before the revolution. A priceless collection of jewels is still stored in the Castle. One day a precious stone is missing. The shadow of suspicion falls on Professor Zach, a frequent visitor at "Red Castle." But the clever deductions of his students (turned amateur detectives) vindicate the Professor, and the real culprits are caught. [Czechoslovakia]

Word gets around that the attractive new teacher is carrying on with the well-known high school jock. And in the locker room, too. She is nearly ruined before it develops that the student, himself the victim of a psychopathic, scandal-mongering father, only tried to rape her in an unguarded moment. [United States]

The tactlessness of a dry and dogmatic school director drives one of the students of the elementary school of Borsk into the clutches of a religious sect. The teachers' collective is dismayed. A timid young instructor is drawn into the struggle against the sect. Emboldened through her efforts to demonstrate that religious dogmatism defeats the goals of free education, she realizes the great role of the teacher in public life. [Russia]

The humane methods of the new teacher in an East End slum school lead to disaster. "Spare the rod . . . " gloat the hardened old disciplinarians. The teacher is about to give up and leave when a glimmer of student response at the end of the term gives him second thoughts. [England]

The impoverished peasants of a village refuse to work for starvation wages on the Count's estate. But the gendarmes have a firm grip on this treasonous activity. The peasants are ordered to the railroad station to welcome the arriving Count. They come. But they come to pay respects to the departing teacher who is being run out of town as the chief troublemaker. [Hungary]

A utopian idealist teaching in a lycee becomes so involved in his pacifist schemes that he neglects his family. Reality finally deals him a tragic but sobering blow: his daughter has a lover, has taken part in a robbery, and is about to run away. [France]

Orphaned, hungry, and demoralized, a gang of boys terrorizes the countryside at the end of the war. A former partisan leader, now teacher, turns them into useful citizens. [Poland]

Teacher Goals and Fates

We found that the Russian and other socialist media teachers are depicted as more "learned," "democratic," and "manly" than those of the West. Eastern mass media stories of schools and teachers stress the ideals of service to community and nation more than three times as frequently as United States and other Western media.

United States media portray a higher proportion of women teachers on all levels of education than do the media of other countries. Our media also depict a composite image of the teacher as less professional and less likely either to advance or to slip on the social ladder than the media teacher of other countries. The United States media teacher is more easily frustrated and victimized by the much higher level of violence and illegality prevalent in her world than is the media teacher of the other countries.

Teachers are quitting the profession in about 28% of United States and Western media stories and 14% of Eastern media stories. The main reasons for giving up teaching in Western media are the frustrations and conflicts of the job, and marriage. Eastern media teachers leaving the field of education are most likely to be fired, retired at the end of their service, or advanced to positions of higher leadership.

Teachers stand out everywhere in seeking intellectual values more often than do the other adult characters in the same fictional environment. But Eastern European media characters, and especially teachers, are different from those of other countries in their much more frequent pursuit of goals of social morality (justice, honor, public service, a better world).

We analyzed the barriers that stand in the way of achievement and found that the one major difference between the problems of United States media teachers and those of the other countries lies in the teachers themselves. Only in United States media are teachers more likely than other adults to be depicted as handicapped by their own weaknesses and fears.

The fears may be justified. Over one third of all United States media teachers commit violence and nearly half fall victim to it. This is low by United States media standards, but it is roughly twice the mayhem found in Western media and about six times that found in the media world of education in Eastern Europe.

A happy ending is symbolic insistence that justice triumphs despite all troubles. American media stories are the most insistent. Conditions of success, however, are more indicative of its functions than frequency alone. We compared the goals of unambiguously successful characters with those who clearly fail.

Only in American media are successful teachers depicted as less likely to pursue aims of social morality than are teachers marked for failure. Many United States media teachers who do tackle social goals are naive, comic, and even mad, and most are crushed by some misfortune fictional fate throws in their paths.

The Role of Students

Being a student is a long and varied stage in life. The range of opportunities for portrayal is great. The institutional and social forces that shape the representation of teachers in the mass media also affect the depiction of students. But the potential diversity of the student image leads to extraordinary differences in scope and function.

American mass media are unique in not earmarking significant resources to young people. They treat children as a low-income, high-profit, quick-turnover market where the message of social power (police, violence) can be sold in its cheapest and crudest forms. As if to underline the analogy to the slum, the trade journals call the children's program segment on television the "kidvid ghetto."

Market considerations also account for the fact that children and youth (as well as old people) in leading roles make the product a "specialty story." They presumably fragment audience appeal and need special exploitation. American youths become universally employable for dramatic purposes (as otherwise) when they *leave* school.

An international study of the "film hero" (Gerbner, 1969) classified students as an occupational group. Entertainers head the list of occupations with 18% of all leading characters. Students are next to last with 4%. (The last were laborers.) The Western European pattern is similar, although students are more numerous than in United States films.

The films of Eastern Europe offer striking contrast. Students are in *first* place on the same list of occupations with percentages ranging from 20% in Poland to 24% in Yugoslavia and 28% of all leading characters in Czechoslovakia.

The diversity of the portrayals permits few generalizations. Focus on childhood and adolescence in American media requires specialized story values. They are often found outside the regular social context. Several stories are about mentally ill, retarded, and physically handicapped youngsters. One revolves around a little boy "playing Cupid." Another deals with a sadistic teenage gang leader. A sociology student's research requires her to pose as a prostitute. Youngsters complicate life for attractive widowed fathers or mothers. A hard-boiled manager of a gambling house finds himself the guardian of a six-year-old orphan. A good-hearted mute befriends a homeless prostitute and her little daughter. Six homeless waifs camp out in an unused shack on the Connecticut estate where a glamorous but exhausted star seeks peace and quiet.

Students in the media of Eastern Europe are not only more numerous but also move in the thematic and moral mainstream of their symbolic world. This is a world in which a mountain youth pressed into hard labor by the lord of the manor joins the outlaws to fight injustice—as had his father before him. A crippled and lonely student finds amusement in shooting birds from his wheelchair, until he downs a homing pigeon awaited by a little girl and her fishermen friends, and begins the slow, painful road to recovery for both the pigeon and himself. Three boys on a school outing steal away into the woods and come upon a partisan hideout; their teacher demands an explanation for their absence, but he is the local commander of the native fascist militia! A theft of puppets from the school theater sends a group of youngsters on a wild chase involving an unpopular boy who plays detective, unaware that his schoolmates suspect *him* of the crime. A school girl's vacation love affair, her first, sets her on a course of competition and conflict with her attractive aunt. A school boy longing for a bicycle stumbles upon lost money—and discovers the difficulty of making a moral choice. A group of classmates decide to expose the hypocrisy and stealing going on at their collective farm—but what to do when they find some of their own parents among the culprits? A student poses as a German sympathizer in order to obtain information for the Resistance; the antifascist patriots are out to kill him, but his mission demands that he maintain silence. A young pupil is falsely accused of having stolen his classmate's pencil, and confesses to escape the ridicule of his accusers—only to make matters worse.

In these stories, school is often the center of social and moral struggle. Behind the authority of the teacher stands the power of the state. Analysts rated the media schools of Eastern Europe as "related to real life" and learning as "of immediate benefit" about twice as frequently as in the media schools of the West. Eastern European media students are shown as "interested in knowledge," as "leaders and organizers," and as "participating in community affairs" from two to three times as often as those of the West. Eastern European media students are depicted as taking examinations three times as frequently as United States media students, but the latter were observed "dominating classroom activity" four times as frequently as the former.

KNOWLEDGE AND ITS CONTROL

An episode of the television serial "Wild Wild West" features a geology professor who, imbued with noble if (naturally) impractical ideas, goes West in the employ of a rich prospector in order to alleviate his own genteel poverty. But the prospector lets him down. Feeling betrayed (with some justification), he becomes obsessed with thoughts of revenge. His knowledge, now out of control of an employer, becomes a menace to society. He plots to destroy the state through a series of earthquakes triggered by dynamite blasts at critical points in the fault line he mapped through the area. "I have turned the tide," he cries. "Employed nature for my own use—now I want my reward." Brawny agent West and his brainy sidekick (!) make sure that he gets it. We last see him scrawling equations on a chalkboard as he holds "class" alone in his jail cell.

All societies suspect what they need but cannot fully control. Symbolizing such uneasy symbiotic relationships are ambivalent images of oracles, eccentrics, witches, alchemists, and others "possessed" of independent knowledge, as well as teachers. The teacher image is likely also to fall short of the Mandarin ideal or to suffer from the human tendency to denigrate "outgrown" authority.

Beyond such similarities, however, differences in mass-mediated symbolic functions reveal and cultivate significant social distinctions. As we go from West to East, teachers stand out in their own fictional environments as more distinguished in learning and in qualities of personal and social morality. The terms of this morality are not necessarily comparable across cultures. The ethic of individualistic liberalism is not the same as that of socialist morality or the Soviet concept of "the moral development of the child," even if some of the same terms are used. Nevertheless, the image of the teacher in the socialist media reflects a happier fate and more stable, purposeful, and socially meaningful existence in its own fictional "world" than it does in the West.

Differences in social organization account for some of these distinctions. Mass media are cultural organs of industrial society. Their ownership, management, and clientele—extending the institutional order into the cultural sphere—shape their outlook and functions. The organizational and client relationships of Eastern European media interlock with other public institutions, including the schools, the party, and the state itself. The "hidden curriculum" serves the same institutional interests as the overt one; both are agencies of planned social transformation. This places media images of schools, scholars, and the knowledge they symbolize in the mainstream of the symbolic world undergoing a cultural revolution. In performing their symbolic functions, socialist media can take advantage of their legacy of intellectual leadership in nationalist and proletarian movements in which teachers have had a prominent place for centuries.

Organizational and client relationships of American media also reward development of a particular selection from prevalent cultural patterns. The selection manifests the dual character of private-enterprise views on public enterprise. On one hand, schools are a necessary cost factor whose value is limited to its direct usefulness to the investment in current products, practices, and outlooks. On the other hand, schools represent political capital and popular aspirations for mobility, equality, and social reform. The concept of knowledge and its role in and control by society are caught in the cross fire. The most enduring and pervasive images of teachers in American mass culture are those that humiliate and depress them. Failure in love and impotence in life permit them to be "good." Or they can be vigorous but evil or perhaps only ridiculous.

Poverty is normal and probably desirable for a dependent institution that should not develop a strong power base of its own. When cut loose from corporate, military, law enforcement, or other established power, even the "miracles of science" turn into "mad scientist" horrors.

No school or culture educates children for some other society. Giving teachers a messianic mission and having schools soak up all the dreams and aspirations citizens have for their children doom the enterprise to failure. No social order can "afford" to make good such a promise. The illusion itself contains the seeds of the "noble but impractical" image. It becomes only "reasonable" and "realistic" to show teachers full of goodness but sapped of vitality and power. Turn on the power and the impotent figure becomes a monster, only confirming the doubts and suspicions inherent in the ambivalent image.

Unlike the army and the police, the schools do not appear to be a major public responsibility at all. They are shown as places of controversy and conflict, except when the goal is winning for "the team." The school sports story provides a dramatic framework for learning the rules of order and life in a community dedicated to skills directly applicable to competitive power.

American media scholars symbolize the promise of learning on behalf of noble and idealistic goals, and undercut that promise by being strange, weak, foolish, and generally unworthy of the support of the community. The "hidden curriculum" cultivates the illusion of social reform through education and at the same time helps pave the way for the perennial collapse of its achievement. As things work out in the symbolic realm, the bankruptcy of the schools is their own fault. The invidious distinction between teaching and "doing" is maintained. The promise of a productive society to place the cultivation of a distinctly human self-consciousness highest on its scale of priorities is again betrayed.

American media are cultural arms of private enterprise in the public sphere. The images they project have a dual character. They attempt to be serviceable (or at least not inimical) to the commercial and other interests of private enterprise and, at the same time, represent those public ideals that give them universal attraction, currency, and credibility. That is why the study of capitalist mass media and their symbolic functions presents a particularly complex and challenging task. The task is to discover the actual laws of symbolic behavior in a field of conflicting institutional interests and to assess their real contributions to the cultivation of human conceptions and social policy.

I doubt that the nature of education, the role of knowledge, and the prospect for real changes in school policy can be fully grasped until that assessment is well underway. New developments in communication technology have the potential of altering social patterns of knowledge, as did the "old" developments. The question is whether they will merely extend even further the scope and reach of the existing structure, or begin to change them. That, of course, is not a technological but an institutional question. Institutions use technology in communications and culture for their own purposes. The image of the schoolteacher in the "hidden curriculum" of the mass media may continue to be a useful indicator of those purposes.

REFERENCES

Belok, Michael Victor. "The College Professor in the American Novel, 1940–1957." Doctoral Dissertation, University of Southern California, 1958.

Booth, Alan. "Sex and Social Participation," *American Sociological Review, 37* (April 1972), 183–192.

Bowman, Claude C. "The Professor in the Popular Magazines," *The Journal of Higher Education, 9* (October 1938), 351–56.

Boys, Richard C. "The American College in Fiction," *College English, 1* (April 1946), 379–387.

Brown, Roger L. "The Fictional Presentation of Education in *The Saturday Evening Post* and *Woman*." Master's Thesis, University of Illinois, 1963.

Coard, Robert L. "In Pursuit of the Word 'Professor,'" *The Journal of Higher Education, 3* (May 1959), 237–245.

Coleman, James S., et al. *Equality of Educational Opportunity*, Washington: U.S. Office of Education, 1966.

DeFleur, Melvin L., and Lois B. DeFleur. "The Relative Contribution of Television as a Learning Source for Children's Occupational Knowledge," *American Sociological Review* (October 1967).

Evans, Walter. "The All-American Boys: A Study of Boys' Sports Fiction," *Journal of Popular Culture* (Summer 1972).

Foff, Arthur. "The Teacher as Hero," in *Readings in Education*, Arthur Foff and Jean D. Grambs (eds.). New York: Harper, 1956.

Gerbner, George. "Mass Communications and Popular Conceptions of Education; A Cross-Cultural Study." Cooperative Research Project No. 876, U.S. Office of Education, 1964.

Gerbner, George. "Images Across Cultures: Teachers in Mass Media Fiction and Drama," *The School Review*, 74 (Summer 1966), 212–229.

Gerbner, George. "The Film Hero: A Cross-Cultural Study," *Journalism Monographs*, 13 (November 1969).

Greer, Colin. *The Great School Legend; A Revisionist Interpretation of American Public Education*. New York, London: Basic Books, 1972.

Gusfield, Joseph R., and Michael Schwartz. "The Meanings of Occupational Prestige: Reconsideration of the NORC Scale," *American Sociological Review*, 1963, p. 270.

Hazard, Patrick D. "The Entertainer as Hero: A Problem of the Mass Media," *Journalism Quarterly*, 39 (Autumn 1962), 436–444.

Hofstadter, Richard. *Anti-Intellectualism in American Life*. New York: Knopf, 1963, p. 309.

Hughes, Thomas. *Tom Brown's School Days*. New York: Harper, 1870.

Jencks, Christopher, et al. *Inequality: A Reassessment of the Effort of Family and Schooling in America*. New York: Basic Books, 1972.

Katz, Michael B. *Class, Bureaucracy, and the Schools; The Illusion of Educational Change in America*. New York, Washington, London: Praeger, 1971.

Lyons, John O. *The College Novel in America*. Carbondale: Southern Illinois University Press, 1962.

O'Dowd, Donald D., and David C. Beardslee. *College Student Images of a Selected Group of Professions and Occupations*. Cooperative Research Project No. 562 (8142), U.S. Office of Education, Department of Health, Education, and Welfare, April 1960.

Schwartz, Jack. "The Portrayal of Education in American Motion Pictures 1931–1961." Doctoral Dissertation, University of Illinois, 1963.

Shapiro, Charles. "The Poison Ivy League," *Saturday Review* (October 19, 1963), 37.

Smythe, Dallas W. *Three Years of New York Television, 1951–1953*. Urbana, Ill.: National Association of Educational Broadcasters, University of Illinois Press, 1953.

Springer, Roland A. "Problems of Higher Education in the Broadway Drama: Critical Analysis of Broadway Plays 1920–1950." Doctoral Dissertation, New York University, 1951.

Verb, James. "An Analysis of Movie Titles with the Intention of Finding the Occupations Which Are Listed in Them and the Words Which Are Most Commonly Associated With Them." Unpublished class paper, University of Illinois, 1961.

Wecter, Dixon. *The Hero in America*. New York: Scribner's, 1941. p. 478.

Winick, Charles. "Trends in the Occupation of Celebrities: A Study of Newsmagazine Profiles and Television Interviews," *The Journal of Social Psychology*, 60 (1963), 301–310.

Wittlin, Alma S., "The Teachers," *Deadalus* (Fall 1963).

PART IV

The New Field of
Urban Communications

CONTENTS

INTRODUCTION

BERTRAM M. GROSS

It sometimes seems that every day brings a new crisis to the people of America and the world at large. But even in this world of crises, the current "urban crisis," as it encompasses and epitomizes so many other crises, stands as one of the major challenges to this generation.

In dealing with the diverse problems of the city, it is natural to turn to history for guidance and precedent. From history we gain the perspective for understanding the role of the city in human endeavor. The purpose of the first urban settlement was the same as the purpose of cities today: to bring people together face-to-face for the exchange of goods and information and the integration of large-scale activities extending far beyond the city's boundaries. Although history provides invaluable perspective, it can also be misused. This is seen in the unfortunate tendency for generals to fight old wars—and the tendency for policy makers who hold responsibility for the future of our urban areas to look at current problems with the eyes of the past.

Urban planners seem most comfortable when dealing with urban problems in terms of transportation. Indeed, the most advanced techniques and the most sophisticated, "scientific" body of knowledge readily available to such decision makers are those of transport. Consequently, when the common problems of the city such as congestion and pollution are discussed, the discussion is likely to focus on transportation as the integrating factor in urban life.

However, as early as the end of this decade, new communications technology, particularly in the form of the "wired city" (the use of coaxial cable for television and other communication functions), may change the role of the city by making nonface-to-face relations as practical and efficient in many cases as face-to-face relations. This in itself can operate to reduce congestion and the kind of consumption that leads to pollution. Transportation depends on communication and communication, thanks to newly available technology, will be more and more able to serve both as a supplement to and substitute for transportation in interurban and intraurban contexts, not only by reducing the number of trips, but by facilitating those trips to be made by better coordinating transport systems.

Perhaps of even greater importance when considering the new communications technology are the unprecedented opportunities for interaction between groups who have never faced each other before through media never before used to achieve such ends. Although there is the potential for tremendous good resulting from these opportunities, futures predicted on them can also be quite frightening. In the crisis areas of racial, ethnic, and socioeconomic stratification, cable technology may play its most influential role—but that role may be to lead American society to racial apartheid. New opportunities for interaction may become new chances for political, consumer, and social behavioral control through computer-based, pinpoint advertising and politics as broadcasting is replaced by "narrow" casting to specific, well-defined audiences. The direction in which the cable goes is in the hands of state and local governments in terms of the regulations they develop as franchise conditions. This will be the subject of debate for the next several years and represents urban communications' most salient battleground.

These very real possibilities underscore the need for creative departures from the established routine now found in media production and regulation. The urban planners, the policy makers of our cities, and concerned citizens must become aware of the problems and possibilities of urban communications. This is partially the responsibility of academic institutions. While none exist today, in the next few years, universities will offer courses and doctorates in urban communications. In this regard, the following chapters are an important beginning.

The six chapters of this section represent the work and thinking of a careful blend of established experts in the field and young people, new experts if you will, reporting from the battlefront. Three deal with urban communications in all societies using advanced communication systems and three deal with the particular immediate problems being faced in the cities of the United States.

Among the first to recognize the critical role of communications on the urban scene, Melvin Webber deals further with the inherent relationship between urbanization and communication. Webber, along with contributor Richard L. Meier, has already been a vigorous force in reshaping the academic approach to urban communications.

By fostering increasing cultural pluralism, Webber says in his chapter, communication will continue to erode the localism characterising human societies throughout history and allow communities based on common beliefs, interests, or fashions rather than residence. In this and other ways, Webber suggests that the occurrence of the new communications systems coincides in time with other major historic developments, all working to reduce the functions of the traditional city. He warns, however, of the inequity of communications technology—unless proper measures are taken, it will result in another example of the "rich get richer" syndrome.

Mark L. Hinshaw, an urban planner with a background in architecture, offers a broad perspective on the major alternatives for the urban structure within the context of communication and their consequences for alternative urban life styles. Of the many possible futures, Hinshaw elaborates on two life styles: one assuming a continuation of current trends, and the other postulating a fundamentally different level of societal existence facilitated by new communications technologies.

Noting that ethnic minorities will demand a greater share of the communications industry, Oliver Gray, a black urban planner now with the Urban Coalition, urges that communications must be viewed as a resource in establishing and fulfilling priorities for disadvantaged minorities. Gray also gives us the historical perspective of the role of blacks in the mass media and reports on some of the most creative CATV work in the country now being originated and carried out within black communities in various cities. Like Webber, he warns of the establishment of a new technological elite and its consequences, should it exclude minorities.

Among the opportunities afforded by cable technology is the possible conversion of our currently passive television audience into an active, producing one. This requires more than free air time contends Theodora Sklover, founder and director of New York City's "Open Door," the first public service agency designed to educate private citizens in the use of cable technology. Training in the actual production skills and education in the use of the medium is essential if television, like the written word, is to become a personal communicative tool, she says.

Reporting on the first preliminary studies on the interrelations between communications and transportation, Great Britain's Peter Cowan says that the degree to which communications can act as a substitute and supplement for transport will become an increasingly important factor in terms of city growth, evolution of urban life styles, and our ability to control the future evolution of cities and society. The effects of communication will occur on social, cultural, economic, and political levels in relation to both inter- and intraurban patterns of interaction and settlement.

As strained as human beings are in the urban environment, it is reasonable to ask if the human nervous system can tolerate the additional stresses imposed by the increased communicative capacities of the new technology. Richard L. Meier, another of the pioneers in this field, notes that the stress is already intolerable for certain individuals who occupy specific positions in society, but offers hope in the projection that more information processing, interaction, and decisions that are now considered mental drudgery will be accomplished by machine. Meier also questions the "progress" that leads to increased human stress in the city, but confesses, "there is no way yet for the metropolis to turn back."

Like the metropolis, the media seem unable to turn back, even if established media industries seek to maintain the status quo. Paradoxically, while the media set the agenda of issues to be faced by our society, they have not yet dealt with the imperative issues of quality urban communications, particularly in terms of the opportunities implicit in the new technology.

It is essential that our *media be used to raise the issue of how media should be used*. This should be considered a matter of public policy. Existing governmental regulatory machinery, including the FCC for broadcasting and various state and local agencies for cable communications, and governmental machinery yet to be established must require the media to publicly deal with this issue.

Let us therefore offer a proposal: that as of 1975, there be no license renewal for any current franchise unless publicly advertised prime viewing time has been available on at least three occasions during the three months prior to the expiration date of the franchisee's license for the open and public discussion of how the franchise holder has contributed, or might contribute, to the quality of community communications and bringing the members of the community into contact with the cultural and political issues that affect their lives. Each of the issues raised in this section—the proposals of Gray and Sklover; the perspectives of Webber and Cowan; and the warnings of Hinshaw and Meier—should be among those to which the media address themselves.

This creative departure from the established routine will help us to deal with the ramifications of the technology that we have developed to deal with our problems. While this need not be an inevitable process, it is what we face on the new urban battleground of communication technology.

CHAPTER 18

Urbanization and Communications

MELVIN M. WEBBER

We are living in the midst of a golden age of scientific discovery, technological invention, and creativity in the arts. Never has a society learned so much so quickly. Never has so large a proportion of any population been engaged in the intellectual pursuits. Never have the means for satisfying intellectual curiosity been distributed on so wide a popular base. Already somewhat over 40% of American youth are going to college. (In California it is well over 80%.) With mass distribution of books, magazines, music, painting, radio and television programming, with ready access to schools of extraordinary variety, and with increasing ease and lowering cost of travel, America is becoming a knowledge-hungry society. Learning has become the prime occupation for many young adults, and it is now a major avocation for the millions of older adults who attend night schools or follow personal programs of study. In the middle-range future, learning might become the dominant activity for the mass of Americans. In part, this eventuality would follow as a direct consequence of the revolution now abuilding in the various communications technologies—in two-way cable television, video-casettes, facsimile transmission, electronic access to libraries, and so on.

Knowledge is a rather special type of resource because it has the capacity of effectively infinite expansion, and it is enhanced by being consumed. In future decades when high per capita incomes, high rates of productivity, and high proportions of leisure time combine to permit discretionary use of time and discretionary choice of activities, it seems a safe bet that Americans will devote themselves increasingly to the intellectual endeavors. Knowledge has already become the critical economic resource for industrial and governmental suppliers of goods and services. It is fast becoming the critical resource for consumers as well. If America should become a nation devoted to learning, instead of to the production of goods, the national character and the character of urbanization would both be likely to differ markedly from those we have known during the industrial era of the past century.

There can be little question that a major generator of change in the coming decades will be the continued expansion of the sciences and of the technologies that follow scientific discovery. Already in the years since World War II, the knowledge-based industries have become the most active within the manufacturing sector of the economy. Developments in electronics have created a vast new capability in information handling, in computation, and in communications. The repercussions are spreading to affect every aspect of the society. Whole new industries have developed. Managerial capacities of governmental and private organizations have been greatly expanded. The information content of the goods and services is constantly growing, as information is substituted for labor and materials as primary resources. The distribution of information is being vastly improved as television, telephone, and the other communications media are tying virtually everyone in the nation into a single communications network. It is likely that we are approaching a major transformation in educational methods, resulting from the application of computers to instructional programs. The long-range consequences of the communications revolution are yet to be uncovered, of course. But one thing seems clear: it is fostering increasing cultural pluralism, and it is triggering an unprecedented national integration that in turn will continue to erode the localism that has characterized human societies throughout history.

The continuing rise in real personal incomes over the coming decades will inevitably mean an ever-expanding rise in average living standards and an ever-increasing opportunity for discretionary spending. If recent consumer behavior can be taken as an indication of future households' choices, we can expect continued expansion of home ownership, including ownership of second houses; greater demand for communications gear, personal transportation, vacation and travel, wider interest in literature and art; and, of course, increased consumption of educational and health services of all sorts. In parallel, those who consume these preferred goods and services will enjoy the culturally enriched life that travel, education, health, communications accessibility, and personal mobility afford; and they will have improved access to the employment opportunities that, in circular fashion, carry higher incomes and greater personal satisfactions.

A corollary of expanding knowledge, of more refined occupational specialization, and of experimentation in the arts is the increasing pluralism of the American publics. The more pieces of information available and the more ideas produced, the more combinations and permutations extractable from the enlarging repertoire; hence, a wider range of possible fashions in thought and styles of life.

High-scale society is distinguished by the diversity of social groups that compose it and by the variety of belief systems, behavioral patterns, living preferences, activity choices, styles of music, clothing, housing, and the rest. Groups held together by common beliefs, interests, and fashions are becoming the primary communities to which people belong, displacing the communities of residence as locus of loyalty and fealty. As the numbers and varieties of these communities have been rising, America is becoming a nation of minorities. Each person nowadays finds himself a member of many groups, each with its peculiar set of beliefs, behavioral norms, and obligations. Each of these groups is a minority within the larger society. Indeed, these many minorities are the modern society.

In preindustrial societies, populations tend to be homogeneous—to share beliefs, to follow similar life paths, and to pursue similar occupational and leisure habits. With industrialization the degrees of freedom were multiplied as the cultural diversity that accompanied urbanization opened opportunities to be different from one's community of birth. Now and later, as those opportunities become further expanded, we can expect a dramatically increasing diversity among the increasing number of subpublics that will be living in American metropolitan areas. Each will surely have its own set of environmental preferences, and, with their increasing capacity to command them, most are likely to be supplied with them as well.

I am suggesting that the popular commentary about the coming "mass society" and about the homogenization of suburbia is largely wrong. It appears that just the opposite is happening. And, if that is so, we shall need to prepare ourselves for a highly diverse array of urban settlement patterns, for a diversity of demands for public services, and for a heterogeneous variety of social communities that these would serve.

The growing scale of the society has been marked by a finer division of labor and hence by expanding interdependencies among specialized groups. As improvements in transportation and communication have eroded the frictions of geographic space, specialized persons and organizations have found it feasible to satisfy their interdependencies by dealing with others located considerable distances away. Networks for interchange of goods, ideas, information, and services have by now become spatially extensive so that individual firms count the entire nation or the entire world as their trading district. Some of these firms have grown to huge size, and, with branches located in many countries, have become truly supranational in character.

Large corporate size is mirrored by large size of other groups—labor unions, trade associations, churches, professional societies, civic clubs, and, of course, large governments. In the most developed nations, the countries whose societal scales are greatest, levels of income and accumulated wealth have been growing to extraordinary levels. The size of individual investments is growing through time as both public and private investments call for huge commitments of capital. In turn, large investments rely on large organizations, large markets, and large, intricately complex communications networks that connect each organization to the many others laced into the national and international webs.

Ease of interaction over great distances has made it possible and thus imperative that individuals and organizations find their partners to transaction wherever they might be located. Improvements in transportation and communications are already so weakening the constraints of geographic separation that suppliers can be located thousands of miles from their customers, even in low unit value product lines and when steady-flow delivery schedules are required. Such spatial separation is financially feasible because cost differentials for transportation and communication between any two places in the nation are approaching zero. Where transport costs were once powerful determinants of location, they are probably now a rapidly weakening force in many of the new industries, losing their dominance to considerations of comfort, amenity, and prestige, and to communication access.

So interdependent are business firms located on the east and west coasts, so dispersed are members of families, so interwoven are the affairs of local and federal governments that it is no longer possible to deal with any of these as identifiable, unitary, social organizations. Regional economies are being so intricately interwoven into the national economy as to obliterate their local identities. Local politics are merging into the national politics as the national government assumes greater responsibilities for the affairs of the national society and as the distinctions between the old layers of the federal system get increasingly blurred and marbleized.

The consequence has been to merge the various organizational sectors of the society into an integrated national society. At best, each organization is but a subsystem within an integrated national system whose components are spatially distributed over the continent but intimately involved in the daily affairs of the whole. Similarly, the interaction paths among the many cities are by now so ubiquitous as to have erased many of the lines that once distinguished one city from another—save unique location, of course. The high volumes of communication among the various cities and metropolitan areas suggest that these now compromise a single urban system across the 3000-mile-wide continent. Allowing for the distance costs that remain and for the vestiges of cultural variances among the regions, it is as though the new societal scale has brought the geographic parts of the United States together into a single city.

A telling mark of the postindustrial economy is the proportionate shift in its inputs from natural resources to informational resources. Information and intelligence are becoming the major generators of growth and hence are becoming the most coveted properties. New manufactured products and services devour information the way early factories absorbed trainloads of coal and ore. The congealed information compressed into such contemporary products as electronic equipment and into such activities as those of the Council of Economic Advisers or corporate management must be something like several-thousandfold greater than a generation ago. In turn, expectations for ever-expanding demands for information and knowledge are encouraging huge expansions in product lines and in activities that will further enhance information-handling capabilities—improved computers, higher education, research and development, and improved communications systems of varied types.

Unlike the natural resource inputs to prior generations of industry, information inputs are easily transported over geographic distances. Electronic transmission lines comprise one group of means for carrying information, and the technological capabilities here will undoubtedly be greatly expanded in the near future. Another major carrier is the human mind, packaged in a human body. In a mobile society of large scale, these bodies are very easily moved about, too. Those who engage in the information-rich industries thus have a far wider range of locational options opened to them than do the producers of coal or pig iron, say, whose options are constrained by high transport costs. The old constraints that required locations near deep water, along rivers, or adjacent to mineral deposits are weakening for the modern industries. Increasingly the natural endowments that affect locational decisions are pleasant climate, handsome landscape, and recreational opportunities. A clear derivative of the shift to the information-rich occupations of the postindustrial service economy is vastly increased locational freedom.

Cities develop *only* because proximity means lower transportation and communication costs for those interdependent specialists who must interact with each other frequently or intensively. Throughout human history, people have been moving near each other, trying to increase their opportunities for mutually satisfactory transaction. Of course, shortened distances are equivalent to higher densities and are inevitably associated with some considerable costs of congestion, high rent, loss of privacy, and so on. Nonetheless, over all historic time and over all the world's space, there has been a nearly universal movement to villages, towns, cities, and, now, megalopolitan concentrations.

The trend is apparent in the newly developing nations where preindustrial rural folk are now overflowing the cities' capacities. This has long been the classic mark of the Western industrial nations, where measures of industrialization have been interchangeable with measures of city growth. But must this also be true of postindustrial nations? Is it reasonable to expect that, for the first time in history, high-scale societies may develop that are not city based? It was the demand for ease of communication that first brought men into cities. Could the forthcoming and unprecedented demands for long-distance communication combine with the space-spanning capacities of the new communications technologies to concoct a solvent that could dissolve the city?

The cities we are familiar with in America were largely built in response to the rapid industrialization that followed the Civil War. They were, in effect, adjuncts to the factory, providing houses for workers, railroads for raw materials and finished products, offices for industrial management, and ancillary shops and services to accommodate the city's occupants. The city took the shape dictated by the transportation technologies of the times. Initially they were highly compact, reflecting the limited accessibility afforded by horse-drawn vehicles and street cars and the absence of effective substitutes for face-to-face communications. The overall configuration was concentric, with most factories and commercial establishments near the center and with densities of resident populations declining unimodally to the nearby city edge. That pattern was strikingly similar to the ones that had marked cities of preindustrial times as well, for these, too, had been dependent on pedestrian or slow-vehicle modes for movement, and on direct conversational encounter.

The automobile, the paved road, and the telephone have changed all that. Once auto ownership became widespread, once the road system became an extensive network, and once telephones became commonplace, constraints on location were dramatically reduced. By now, with paved roads leading virtually anywhere that people choose to locate their homes or their workplaces, and with metropolitan populations continuing to rise, cities have been overspilling their traditional boundaries. Cities of the late industrial period, here and throughout the world, are spreading far into the surrounding terrain, as factories, offices, houses, and shops have been located in the more spacious and amenable environments that the automobile has opened in the suburbs and the exurbs. By now the spatial extent of individual settlements can be delineated only by imposing such arbitrary definitions as municipal borders and Census Bureau districts. The real boundaries around the physical cities have already been eroded away. The freestanding settlement of an earlier era is disappearing into spatially extensive settlement held together by a network of roads and communication channels. The settlement pattern of the late industrial, early postindustrial period is clearly the megalopolitan one. This may not persist into the mature postindustrial era, however.

The newfound locational freedom is being taken in two ways. New metropolitan areas are being built at such distant places as Houston, Phoenix, and Denver. In both new and old metropolitan areas, additional developments are being placed at the low-density extensive fringe of prior settlement. The effect is to retain the city as a cluster of populations and buildings in physical space, albeit in new-style patterns, even as their communication habits engage the inhabitants into the lives of their many nationwide and worldwide communities.

Despite the new ease of interaction over geographic space, metropolitan areas remain the destination of migrants from rural areas and small towns. During the past decade metropolitan areas of between 1 and 2 million inhabitants grew fastest, suggesting that here the agglomeration economies may be most nearly optimal. Costs of ready access to the constellations of business services that modern enterprise requires must now be among the important considerations that enter the locational calculus. Small metropolitan areas seem to be too undersupplied with readily accessible suppliers of specialized informational services. The larger ones may be too encumbered by congestion costs—expressed as high rents, competition for road space, pollution effects, crime, and the array of other ills that have come to mark the large contemporary metropolis.

Cities have always been essentially communications switchboards—the locales where the greatest numbers of connections are joined and where the transactional business of the society is most easily accomplished. Communication has always sustained the social community, molded social organization, and maintained the connections therein. Contemporary urban settlements are archaic in these primal respects.

With similar adherence to old ways, the communications lines that comprise cities follow two major patterns. *One* connects the local to other locals through a broad array of channels—friendship links, family ties, business connections, chance encounters, planned meetings, telephone and mail services, face-to-face conversations, and so on. These media serve the day-to-day, high-frequency routinized interchanges that comprise the highest volume of transaction among persons and corporations. These are the interchanges that support daily family life and the daily, repetitive activities of firms and their employees. The *second* pattern ties these establishments to the rest of the nation and the rest of the world. These interactions tend to be more specialized, more infrequent; nonetheless, they may also be more consequential. In selecting from among the available locations, one must inevitably weigh his anticipated demands for communication channels of both types, then select from among the assortment of available metropolitan areas and intrametropolitan places that location most likely to match his communication habits and transactional requirements.

With the present array of organizations and populations, it is clear that the big, old-fashioned central business districts such as Manhattan and Chicago's Loop remain attractive to many headquarters offices. Affording large numbers of face-to-face contacts with related firms and easy access to multiplicities of business-support services (lawyers, banks, accountants, hotels, nightclubs, and so on), these concentrated business districts have been continuing to grow, despite suburbanization of employees' residences and of the firms' manufacturing activities. I can guess only that the exponential increase in volumes of information passing through these offices has made their information-sensitive executives wary of being removed from the communications crossroad. As Richard L. Meier once put it, the validity of intelligence received by word-of-mouth in a face-to-face, handtouching encounter can probably be more sensitively appraised than if it arrives on a piece of paper or through a telephone voice.

The continued expansion of Washington, D.C., is the nation's clearest example of the new importance of information in the management of contemporary affairs. Washington remains among the ranks of the nation's fastest growing metropolitan areas, because the expansion of the roles of federal government has made it important for virtually every trade association and other nationally organized interest group to have its representative inside that communications switchboard. Washington is probably the first purely postindustrial city, being almost totally occupied with handling information and in producing services. Its prime raw materials are information and intelligence (in both meanings, i.e., as cognitive abilities and as "hot-poop"); its major products are restructured information and more inside information. (Over 85% of the employed residents within the metropolitan area are engaged in the service occupations. The others are busy putting up new buildings to house the expanding numbers of service workers and in running the transportation and communications systems that, in turn, serve the services. Only 4% are engaged in manufacturing, the lowest in the nation.)

In its occupational structure, Washington may be our preview of the future city. Its spatial structure is very traditional, however, for it has been built around the same transportation, communication, and building technologies that have shaped cities across the continent. But, because it is so thoroughly involved in postindustrial activities, it may supply us with an observation post from which we can monitor the consequences of the new communications technologies as they become manifest. Establishments like the White House, Pentagon, CIA, NASA, and embassies adopt the fanciest and most effective communications equipment that is available as soon as it becomes available. Other Washington groups having less demanding communications requirements may be slower to adopt new systems, but they are not likely to be far behind. To provide fast feedback, someone should establish a new kind of Washington-watch, aimed at supplying early alert on the use and effects of new communications systems.

Virtually all the technological developments of industrial times (and many, such as the sailing ship, the astrolab, and the compass, before) have had the effect of reducing the constraints of geographic space. During this century, the automobile and the telephone have made possible the suburb and, when combined with the railroad and the airplane, the development of the western reaches of the continent.

Although it would be difficult to demonstrate, it does appear that the patterns of settlement have been more directly influenced by the transportation systems than by communications. If so, it is probably a reflection of the uneven distribution of transport lines and terminals that make some places more accessible than others. International airports exemplify this. In contrast, phone lines have long been virtually ubiquitous. Roadways are only now evening out the accessibilities among all places within any given region of settlement.

The forthcoming developments in communications, however novel and significant, will fit directly onto the historic trend line of the pony express, the telegraph, and the telephone. By supplying better channels for transmitting information and meaning, they will improve the capacities of partners in social intercourse to transact their business at great distances. With the further prospects of holography, a quality of presence may be brought into the transaction of those distant partners that begins to approximate the capabilities of the luncheon table in mid-Manhattan. All impending communications systems will be space eroding in these ways. If combined with the promised transportation system improvements, their effects would be compounded. In either case, the prospectus is clear. Many of the constraints that have limited locators to choices among metropolitan areas and places within them will be greatly relaxed. For the first time in history, it might be possible to locate on a mountain top and to maintain intimate, real-time, and realistic contact with business or other associates. All persons tapped into the global communications net would have ties approximating those used today in a given metropolitan region.

I am guessing that early in the next century, settlement patterns will be spread broadly over the continental surface, localized at those places where the climate and landscape are pleasant. Densities are likely to be at the scale now occurring at the exurban fringes of the eastern metropolitan areas. High incomes will permit most families the household equipment that typifies the exurban upper-income home today, but the household's communications apparatus will surely be more sophisticated and competent than that currently in use at the White House.

It was the desire for improved communications that brought people to the cities at the outset. By getting geographically close to the opportunities concentrated in the city, the migrant was able to better his life chances, to improve his lifetime earnings, to learn a wide variety of skills, and to get connected into the complex social, economic, political, and cognitive networks that comprise urban societies.

It might seem paradoxical or a perversion of these long-term historic trends that the move to high-density settlement should soon be reversed. I think not. It was not a desire for migration per se that induced farmers, peasants, and small towners to move to the big city. The city, with concentrations of people and buildings, held comparatively little attraction to migrants anywhere or at any time. It was the opportunity to get connected up, to become tuned into the culture of urbanized society that was attractive; and the city has always been the place of greatest connectivity, the spot where it was easiest to get hooked up to the affairs of the time.

But if the connections should become nearly as good in more pleasant settings, many will be willing to trade off some of the opportunities offered by high connectivity for the opportunities offered by high amenity. A great many already have, as the moves to the suburbs and exurbs suggest, and as the growth of the Arizona desert demonstrates. It may not take very dramatic improvements at the margin of communication capabilities to trigger a major sector of the commercial, industrial, and residential populations into following them in pursuit of the desert life, too. The postindustrialization of the economy will, after all, make it unnecessary for many to remain at the deep-water ports and railheads that were the initiating causes of their current locations. In the high-scale society that is now emerging, it will increasingly be the ties to the national and global networks that will be determining, and the culture of localism will be likely to decline still further. I am suggesting that the occurrence of the new communications systems coincides in time with some major historic developments that will also work to reduce the functions of the traditional city. In concert, these are likely to be indomitable forces.

As the processes of cultural pluralism work further to differentiate the population into minority groups, there is likely to be another trend counter to the dispersion-inducing ones I have been discussing. We are now struggling with the current manifestations of spatial segregation among groups distinguished by race, ethnicity, national origin, and social class. Most pronounced, of course, is the spatial concentration of lower-class Negroes in the centers of large metropolitan areas. The fast-paced suburbanization of middle-class whites is a direct reaction to the immigration of the blacks, especially to the shifts in school enrollment compositions. At this stage in the nation's development, that trend is still unturned, and virtually everyone now expects an apartheidlike settlement pattern in many metropolitan areas in future years.

It is not only the white-black segregation that occurs, however. Where large numbers of youth associated with the "counterculture" have settled in an area, older residents have found it less comfortable, and some have left. Where ethnic differences have been wide, voluntary segregation has commonly followed. The post-World War II suburbanization has led to the incorporation of thousands of suburban municipalities, each with its own population group, typically comprising a narrow socialclass spectrum, typically marked by its own peculiar stylistic preferences for housing, schooling, public services, and, above all, neighbors. Suburban municipalities exist to help assure each of the plural groups a life setting it prefers and to help assure the exclusion of groups it does not prefer.

Reinforcement of the dispersion-inducing influences will likely also be a reinforcement of the segregation-inducing ones. In this context, the new communications systems may disserve those at the lower reaches of the income distribution and the lower social strata. Middle-class white folks can now get all the way to Westchester to avoid proximity to Harlem. Later they may be able to move all the way to Phoenix.

The city has always been the destination of rural-deprived families who aspired to greater opportunity. Once they arrived, the city became the equivalent of a school—an institution where one learned urban ways, acquired salable skills, and became acculturated into modern urban society. At the turn of the twentieth century, it was pretty easy to make the leap from preindustrial to industrial status, partly because the requisite skills were not difficult to acquire, and partly because the communication channels were relatively open to newcomers. At the turn of the twenty-first century it might be a great deal harder to accomplish that leap. Salable skills take a lot longer to learn nowadays; extensive schooling is necessary if one is to become adept at the information-handling jobs. Compared to the needle-trades, for example, computer programming, machine servicing, or even cashiering are cognitively difficult occupations. Some of these learning difficulties will undoubtedly yield to improved teaching techniques, and later the gap may not be as wide as it is now. But the spatial gap between lower-class districts and middle-class ones may become too wide and thus too difficult for the typical person to bridge.

The new communications systems would seem to be inducing a major social problem, partially of their own doing. In addition to making it easier to transmit credible intelligence among distant places, in addition to providing hard copy by wire, to opening the library to home-video review, to transmitting three-dimensional images, and the many other promised wonders, means must also be found to assure that these channels are opened to all sectors of the pluralistic population. Can we be assured that the communications media of the next decades will accomplish for the underprivileged youth of the year 2000 what the free public library and the free public school did for the immigrant youth of 1900? Is there an inherent trait of the new technology that assures its widespread availability? Or, like the suburb, is it likely that the very considerable benefits will be distributed to those who are already well off, leaving the less advantaged relatively the worse for the improvement?

In our fascination with the improved efficiencies that will accompany the coming revolution in communications, in our enchantment with the marvels of the new hardware, it would be well that some of our attention be directed to the equity consequences, as well. The distribution of benefits is not likely to fall out directly. No self-governing generator of equity is built into the new electronics. The issues of distributive justice will need to be imposed from outside the hardware system. These, it seems, will be by far the most intractable of the research and development problems accompanying the new communications systems and their consequences for urbanization.

CHAPTER 19

Wiring Megalopolis: Two Scenarios

MARK L. HINSHAW

It is becoming increasingly evident that we are in the midst of a tremendous societal transformation. Students of social change have begun in recent years to examine its form and substance and to make predictions as to its consequences for human existence. In an effort to give it an appropriate historical identity, scholars have christened this systemic transformation out of the industrial era variously as the arrival of postindustrialism, the coming of a superindustrial age (Toffler, 1970), the Age of Discontinuity (Drucker, 1968), the dawning of a Universal Civilization (Ribiera, 1968), the evolution of Consciousness III (Reich, 1971), and emergence of the Technetronic Age (Brzezinski, 1970). At least two authors have identified this social phenomenon as revolving primarily around the invention, use, and proliferation of new communications technologies and processes. Robert Theobald (1970) maintains that we are entering into nothing less than a full-blown Communications Era, while L. Clark Stevens (1970) applies the title of Electronic Social Transformation.

In the area of urban affairs and planning, few attempts beyond those of Richard Meier and Melvin Webber have been made to analyze the impact of communications on urban change. Among the myriad of conferences, symposia, books, and journals examining current and future urban development, planners have given virtually no recognition to the consequences of communications for alternative urban life styles. As Jerome Aumente (1971) has noted: "Professional planners who should know better persist today in conventional predictions of future land use and population movement without sufficiently examining the new set of communication variables that turn their predictions topsy-turvy." Indeed many planners may well feel that communications technology will have little or no effect on urban development. Virtually any recognition at all of the relationships between urbanism and communications has come from academicians and professionals outside the fields most directly involved in urban analysis and policy development. Most of the literature coming from such sources, however, treats communication and information-generating hardware seemingly as the means of solving most of the urban problems with which we are presently confronted.

It is imperative that communications resources, goals, and potentials be included in the urban planning process, taking into account local, regional, and national needs. The development of communications technologies and communicative structures is intimately related to housing, transportation, social services, and the political economy. Communications systems must be considered a major component of the urban infrastructure, both as a public resource and as an integral part of urban movement systems involving people, goods, energy, and information. There is a clear need for substantive analysis and synthesis of urban change in terms of concomitant communications developments.

The development of cable communication technology promises to interject a host of services into the urban environment including localized television, facsimile, data transmission, computer-aided instruction, telemedicine, two-way audiovisual communication, special interest programming, and so forth. In addition, cable communications has the potential of destroying most of the structural limitations of present mass media, as seen in the "lowest-common denominator" mind-set of current television fare. The economics of cable simply do not dictate that it be a "mass" medium at all. It is possible for cable linkages to focus on a specific geographical area such as an urban neighborhood or to join spatially dispersed people with common interests. Cable is also unique in that it can afford people direct control over the content and nature of the programming they receive through the cable. So with all this, the current inaccurate, rearview mirror misnomer of "cable television" may well be replaced by a more appropriate designation such as "broadband cable networks," or more simply "the cable."

Cable communications has particular import for urban change in that it has the potential for radically altering the very concept of the urban community. Entirely new perceptions of community life may develop. In addition, it may well be a key to determining the ability of urban inhabitants to understand their individual and collective problems and deal with them effectively. However, it should be pointed out that predictions of the emergence of "the wired city" are clearly shortsighted in that they fail to realize that with so extensive a communicative system, the very term "city," will no longer be a useful term for symbolizing the urban way of life. Indeed, as Melvin Webber (1968) has already pointed out, we are even now in a "post-city age."

Nicholas Johnson (1970) has commented that communications will be the primary technological determinant of urban life in the next several decades.

"Communications will be to the last third of the twentieth century what the automobile has been to the middle third." Such a statement is as foreboding as it is promising. Forecasts of the development of communications media already range from eloquent prose about the tremendous potential of new media (Youngblood, 1970; Shamberg, 1971) to horrifying suggestions of a future society unprecedented in the degree of control and repression (Gross, 1970). Cable communication in particular probably has as many potentially negative consequences as it has positive ones. Cable technology is so imminently powerful that it deserves immediate assessment with respect both to its effect on urban institutions and related technology and the effect of the institutions and technologies on cable itself.

TWO POSSIBLE FUTURES

It is obviously difficult, if not hazardous, to attempt to make forecasts about changes in the nature of urbanism brought on by such a rapidly changing area as cable communications. Peter Drucker (1968) has noted that in the future "the unsuspected and apparently insignificant (will) derail the massive and seemingly invincible trends of today." Nevertheless, it is important to engage in an anticipatory delineation of first, second, and third-order consequences of various alternative developments. Of the many futures that are possible, I will elaborate on two.

The first alternative is essentially an extrapolation into the next few decades of the events, developments, and value systems of the present. This assumes a continuation of current social trends. We will witness a rapid growth of megalopolis possibly developing into Doxiadis' world of ecumenopolis: a continual global city. We will, in addition, continue to see the flight of upper income groups, together with industry and the economic base, to exclusive suburban areas. Older urban centers will then become massive human sinks with palliatives being perenially applied through quasibenevolent, welfare-state policies. Complex bureaucratic institutions will continue to proliferate, becoming diffused and interwoven throughout all areas of society. Finally, with social disorganization increasing, environmental degradation reaching a new high, and clamor for security and control mounting from all sides, government and its corporate cohorts will look to research organizations and academia for solutions in systematic applications of a new and powerful union of the social, behavioral, and technological sciences.

The second alternative assumes that the forecasts of increasing exponential change are wrong; that we are instead entering into an historical era in which exponential curves begin to flatten into logistic or S-shapes—an era of evolutionary change into a fundamentally different level of societal existence. This future assumes an eventual emergence of a corresponding shift in values, with voluntary reductions in overall consumption levels, a redefinition of individual rights and responsibilities, an acceptance of cultural diversity, a recognition of ecological interdependence, and a critical attitude toward the possibilities and the problems of technology. There will be simultaneous undertakings to create a variety of new patterns of urban habitation, with access to life support systems and services being more and more seen as a basic human right. Cable communications and its attendent services will be recognized as a medium for the creation of wholly new communities and as a tool for exchanging socially usable and useful information.

The following scenarios attempt to expand on these two alternatives in terms of an overall societal framework.

Scenario I

Six months after the end of the Vietnam War in mid-1973, it seemed fairly evident that the much hoped for diverting of funds from military expenditures to domestic social problems was not going to materialize in any significant amount. Dissidents began to turn their energies to the inefficiencies and insensibilities of corporate practices and headlines were soon occupied with news of sporadic disruptions in factories and corporate offices around the country. A wide variety of groups from laborers to consumers to racial and ethnic minorities turned to militant tactics to make known their demands.

Within urban areas the crime rate had reached an all time high in June, 1973 with the vast bulk of crime consisting of thefts of personal property and street mugging, much of it violent. There also was an exponential increase in the number of apparently senseless crimes—random shooting and knifing of people in all major American cities.

By 1974 blacks essentially had control of two major cities, and militants in at least one other large city and a half dozen smaller ones were in the process of trying to wrench control from bureaucrats and civil servants who lived outside their communities. Demands for immediate community control came not only from blacks and Spanish-speaking peoples, but from poor and middle-income ethnic white areas as well. Many reacted with violence at attempts by decision makers to change the character of their areas. The chief concern of many politicians was the very real prospect of widespread social disorder occuring before and during the upcoming Bicentennial Celebration. Most people regardless of their race, income, or ethnicity felt such a crisis demanded immediate and drastic action.

So it was that in 1976 a president was elected on a "Security and Stability" platform and together with a sympathetic congress instituted a number of swiftly implemented measures. The National Internal Security Administration was created and under the Urban Communications Act of 1977 the Department of Communications (DOC) was added to the Cabinet. DOC was empowered and given funding to immediately establish a National Communications System, or NATCOM for short. Each megalopolitan complex was to see to the construction, by public or private means, of intraurban cable networks to feed into NATCOM. The scheme developed by national communications planners was multifold.

First NATCOM was devised so as to enable government, military, and police operations to function swiftly and effectively in a widely dispersed pattern. Information about potentially dangerous people or groups was databanked and made instantaneously available. Computers were utilized to collate personal information and activities and to predict by simulation the probability of a particular disruptive action. Thus those potential dissidents who could not be co-opted or otherwise cooled out could be closely monitored. A proposal made back in 1971 for mobile transmitters implanted in the brains of habitual criminals was being implemented experimentally.

The personal crimes in urban areas that were not eliminated by local heroin distribution programs NATCOM sought to minimize by installing miniature video cameras at strategic points on streets. One of the major reasons for the popularity of two-way cable television was its burgular protection service. Privacy from electronic surveillance ceased to become a major concern; after all, it was felt, no decent citizen had anything to hide.

Second, NATCOM could help satisfy public demands for greater localized control through the establishment of intracommunity cable systems within urban areas. By the end of the 1970s almost all urban places over 2500 were fully wired. Planners maintained that by encouraging intense involvement in local cable systems a sense of control over local affairs and participation in local matters could be produced. (Behaviorial research by several prestigious institutions had shown that only a *sense* of participation was necessary to satisfy most people.) With attention so intensely focused on local developments, higher levels of government could thus be freed to pursue their activities unharrassed.

Third, NATCOM facilitated the formation of eight regional superagencies to control urban population distribution, housing, transportation, environmental resources, land use, and internal security. The formerly sticky issue of metropolitan government was skirted by instituting not a new level of government, but instead, technical service agencies empowered to set policy without the chaotic process of public involvement that had bogged down the implementation of so many plans in previous decades. Possible objections to such an arrangement were largely forestalled by the strategy of including potential dissidents in the agencies.

Finally, cable communications was seen by NATCOM planners as a means of eliminating the propensity of mass media for unnecessarily inflaming emotions about particular events and for raising aspirations and expectations of people beyond what corporate enterprises and government could practically provide. This led in the early 1980s to the custom tailoring of packaged information and entertainment to fit the unique characteristics of particular cultural and social groups. Not that this was unwelcome; the previous decade had seen a widespread clamoring for programming more relevant to the experiences of specific racial, ethnic, and economic urban subcultures. NATCOM enlisted the aid of former advertising and public relations specialists, social and behavioral scientists, video artists, and communications experts to research the needs of various publics and to prepare carefully designed pieces of programming for distribution by cable and cassettes. NATCOM operated in close partnership with the three former broadcast networks, which by the mid 1980s had turned their investments entirely from broadcasting to broadband communications. These corporations discovered entirely new areas of profit making by marketing cable hardware and producing programming for video cassettes (particularly with the tremendous demand for violent sports and pornography).

By the mid-1980s the results of the Emergency Housing Act of 1978 were being seen. The Act has provided for the simultaneous construction of 45 new towns and 20 linear megastructures within megalopolitan areas entirely by rapid industrialized methods. Such a massive urban development effort was unprecedented in scale and scope.

At the same time, national obsession with the automobile was being gradually replaced with an equally if not more intense obsession with personal communication systems. Status began to be measured by the number and type of equipment one could wear or affix to home cable terminals: wall-sized plasma screens, quadrasonic sound systems, biofeedback units, cameras and video recorders, colorizers, CAI terminals, facsimile attachments, and other paraphernalia. Waiting on the horizon, holography promised yet another addition to personal "telecoms." Not that the automotive corporations simply disappeared; they, like former broadcast networks, transformed themselves. Megalopolitan living in the 1980s demanded new forms of transportation— personal rapid transit, gravity-vacuum carriers, "people-movers," aerobuses— all of which required both sophisticated transport technology and highly developed and coordinated cybernetic communications systems.

Other corporate institutions were transformed under the impact of universal cable communications. It did not take long for marketing analysts to discover that vastly greater profits could be made by designing information about products and services for particular consumer groups. Even channels devoted entirely to consumer reports, at first resisted by corporate structures, eventually resulted in greater sales because they further encouraged high consumption patterns. Electronic home shopping with instantaneous credit accounting proved to be a particular boon to commerce as impulse purchases soared.

The 1980s also saw the advent of educational cable networks. Experiments conducted by a number of independent academic centers and research sponsored by the Department of Communications had proven conclusively that cable communications learning consoles utilizing stimulus-response and reinforcement patterns could significantly increase certain computational and reading skills. It was found particularly suitable for students who showed, through early testing methods, little capacity for more than basic skills. By putting the earlier theories of B. F. Skinner into practice, educational psychologists found that such learning units could also be structured to produce a certain degree of satisfaction with a particular role in society. Frustrations and anxieties due to unmet expectations could thus be minimized.

Two-way cable was soon recognized by social, behavioral, and demographic scientists to be a blessing. Not only was a continual census possible, but researchers were afforded a means by which to gather wholly new varieties of information about the activities, behavior, and characteristics of people. Never before had such accurate statistical data been available to social scientists and planners. Government and corporate decision makers, seeing the enormous potential of such statistical data gathering, defined this as a major element in public participation in policy making, a method by which government could continually determine the needs of its people. This was deemed much more effective than the mere voting on issues and candidates. Therefore, 1995 was set as the target date by which time all homes would be required to have at least one basic, two-way cable terminal.

In America the beginning years of the last decade of the twentieth century saw an unprecedented era of social stability brought about by strictly imposed government policies. Although conflicts and disturbances periodically arose, they were largely localized, short-lived, and had little effect on society as a whole. The 1990s also saw the gradual formation of a new type of social stratification based on differing degrees of access to certain types and qualities of information. The Kerner Commission and political scientists who in the late 1960s had warned of a racially divided society had not foreseen the impact of localized community communications. This permitted urban communities to defend themselves against intrusion by people they considered undesirable, resulting in a vast array of exclusive subcultural urban enclaves. Many communities formed around economic levels, while others formed around ethnic, racial, or work-role distinctions. Local cable systems facilitated the emergence of rigid in-group/out-group attitudes within communities while helping to legitimize and reinforce their particular beliefs and values. Such community atomization permitted government to identify and isolate potential trouble spots and deal with them without upsetting the larger society. The degree of social stability within America was, however, in sharp contrast to the increasing intensity of social, political, and ecological chaos in many other parts of the world.

Scenario II

Urban America in the last quarter of the twentieth century was the locus of a series of widespread social and institutional changes. The mid-1970s saw the breaking down of restrictive zoning laws in suburban areas while the general movement to outlying urban areas continued. Increasingly entropic conditions in central cities due to an overload of population concentration and diseconomies of overly complex institutions gave rise to desire throughout all economic, social, ethnic, and racial groups for alternative environments and life styles. Even while the popularity of suburban living continued to grow, however, it too was beginning to be seriously questioned as a suitable choice.

Concern for the environment and the quality of goods and services, initiated at the end of the 1960s, had by the middle of the 1970s expanded to a greater concern for the total living environment, including housing, transportation, services, community, and social inequities. Demands for a more humanely organized society were echoed by feelings that the megalopolis had passed the point of diminishing returns and that different choices were sorely needed.

Moreover, people began to realize in the last few years of the decade that full and responsible participation in decisions affecting their lives and their communities demanded access to means of generating, receiving, and exchanging ideas and information. Only in such a way could common areas of concern be discovered and cooperative efforts at problem solving be attempted. Adequate and easily available methods of intercommunity and intracommunity communications were necessary for effectuating mutually beneficial change.

By the end of 1976, cable communications systems had been installed in enough areas that people in many communities began to see their potential for facilitating collective action. Awareness of the potential of community cable resulted not only from the increasing availability of the medium, but from educational campaigns conducted by universities, video groups, and citizens' organizations that explained that the cable was not merely an extension of further refinement of television, but an entirely new means of communication.

With an acceptance of the value of subcultural diversity within the larger society, the abundance of cable channels and internetworking of community systems permitted sharing of experiences, customs, and artistic expression among various urban groups. Local cable systems and portable video recorders helped foster community awareness and self-development. With the steady proliferation of switched two-way systems in the early 1980s, cable communication was gradually seen as an indispensable tool for local planning.

By 1977 the shift from employment in primary and secondary economic activities to employment in services was virtually complete. Fully 75% of the work force was engaged in such tertiary service activities. It was also becoming clear that the single term "services" was inadequate, for cybernation had begun to reduce employment even in many service catagories. At the same time, there was a dramatic increase in the need for people engaged in human care and community development activities such as health services delivery, education, and child care. Simultaneously a desire for performing socially useful roles that permitted more choice and flexibility instead of a single life-long occupation was pervading all sectors of the population. Moreover, the very concept of what activities constituted "work" came under intense criti-cism, with a wide range of people from housewives to students at all levels arguing that they performed functions that made a valuable contribution to the resources and development of society. Finally, the awareness of the fact that American society had decades ago shifted from an economy centered around competition for scarce resources to one of an abundance gave rise to a wide-spread belief that the provision of basic goods and services required for a life of dignity should be a right of citizenship. The collective force of such events and demands resulted in the institution in 1978 of a guaranteed annual income to all persons.

The cumulative effect of such structural changes in society as a more equi-table distribution of goods and services, a reduction in levels of consumption, a more careful use of resources, a blurring of distinctions between leisure, work, and education, and concurrent changes in technologies of information, energy, transportation, and housing was to diminish the necessity for megalopolitan concentrations of people. Two-way cable communication services played a vital role in facilitating the formation during the 1980s of a great variety of urban environments. "New" towns, medium-sized urban areas, community clusters, communal settings, and former small towns and rural areas were receiving emigrants from the denser urban complexes. This expanded range of different environments encouraged more involvement with alternative social relationships such as extended families, family clusters, learning groups, group marriages, and religious groups that had previously enjoyed only limited experimentation. Interactive cable systems with ownership having been sepa-rated from programming in the mid-1970s permitted people to maintain link-ages within and between differing types of communities, some geographically concentrated, some spatially diffuse, others transient and based solely on temporary convergence of interest. For the first time, people were able to enjoy both the benefits of smaller, intimate communities and the access to and par-ticipation in larger, more culturally diverse urban environments—national, transnational, and global. By the mid-1980s the former model of the urban-rural dichotomy had all but disappeared from sociological theory; participa-tion in urban ways of life no longer depended on habitation within an area arbitrarily defined by population, density, or political boundaries but was instead determined by the access to communicative and informational nets.

The maturation of cable communications and its ancillary services aided in the emergence of a full-blown postmassconsumption/production urban economy. Advanced cybernation with computer operations capable of rapid reprogramming was permitting a return to high-quality crafted goods designed and produced to fit unique criteria. Housing, for instance, could be built to meet the specific needs of particular communities or even individual families. Urban planners and designers saw cable as a means of receiving information about the needs and preferences directly from potential user groups. Cable was also seen as a medium of presenting simulated alternative environments and housing configurations and eliciting reactions to them. Outcomes of various policy choices were projected and compared in terms of their possible long-run ecological consequences. Thus it served as a valuable tool for the creation of more responsive and responsible designs.

Interactive cable systems permitted the development of more individualized interpersonal, intracommunity, and transcommunity communicative services as well. People involved in kinetic and visual arts used cable and related technologies of portable video and cassettes to introduce other people to the process of expressing images and ideas. Many people became involved in the production and distribution of entertainment for specialized audiences. Still others engaged in gathering, arranging, and presenting widely varying types of informational materials to meet the demands for more useful and useable knowledge. Multiple-access retrieval systems through cable gave rise to large groups of people engaged in reading, reviewing, cataloging, and abstracting literature and research documents for users who had been suffering from an overload of data and were in need of more manageable forms of information. Completely new forms of exchanging and presenting information were created, centering around methods for understanding interrelationships of societal changes. Still other people became involved in various types of community development, organization, advocacy, individual and group therapy, and the analysis of problems, goals, and potential areas of conflict and cooperation. Finally, others engaged themselves in the communication of customs, beliefs, events, and cultural contributions of the particular communities of which they were a part. Members of communities that were mobile used cable to form ties with those that were geographically stationary. With the realization that urban communities were socially interdependent, cable nets enabled the creation of shared pools of information and ideas and the joining together of disparate groups of people in collective attempts at bringing about desired changes.

During the 1980s an indirect by-product of a universally accessible urban communications medium was the gradual replacement of the former two-party political structure with a political environment containing a multiplicity of active interest groups, each possessing differing value patterns and community myths. In some cases political associations coincided with physically identifiable communities, while others cut across separate communities. Interactive broadband communications networks permitted these groups to coalesce, separate, and recoalesce around particular issues as the need for effective action demanded cooperative group efforts.

One of the many proposals for government reform that had enjoyed public popularity during the Great Debates of 1976 was a voter response feedback system. As in earlier proposals, it had been suggested that the system could be implemented through two-way cable. At that time, however, cable linkages had been made with only a small proportion of the total number of households. An argument at that time against the system was that such a readily available access to a voting mechanism would effectively discriminate against those who did not have cable. By the late 1980s, however, cable penetration had approached 95% and the voter system became politically practical. By that time, since the hardware was essentially in place, all that was necessary for full implementation was a computerized accounting apparatus. However, once the system had been in operation it soon became clear that a simple yes-no response to proposed policies and candidates was entirely inadequate. Such a system of "feedback" had been based on the notion of "feeding" reactions back up to representatives and administrators involved in public policy making. What was needed, it was claimed, was an interactive, truly participatory structure that would give individuals and groups the opportunity to originate and present proposals. This subsequently brought about a movement during the early 1990s to replace the system of representation with more direct and cooperative decision-making mechanisms.

The development and proliferation of interactive cable communications as an urban information utility influenced the development of more fluid, diverse, and participative social environments during the late 1970s and 1980s. The 1990s began to see the impact of ubiquitous information access on the physical environment. Static, fixed, and technologically obsolescent building forms were increasingly replaced by flexible, user-controlled environments. One manifestation of this was the construction of basic life support infrastructures providing water, climate control, waste recycling, and communication services that would be designed to last for a relatively long period of time. Attached to these infrastructures or service grids could be virtually an infinite variety of housing types that would either be designed intentionally with short life-spans or with the capability of being modified when the needs of the inhabitants changed. Many forms of shelter and community facilities even became entirely mobile, some entirely self-sufficient, others requiring links with service networks. Urban architecture like communications had become more process-oriented, individualized, adaptive, and diverse.

The last decade of the twentieth century witnessed a general trend toward more dispersed, polynodal patterns of urban habitation and away from large concentrations of population. Several large urban complexes like New York and San Francisco were maintained because of their unique qualities, but were considerably diminished in population, as they became simply alternatives in a wide range of urban configurations. Locational decisions and choice of life-style became based more on preferences for different environmental or cultural characteristics than on economic determinants. The majority of people were engaged in such activities as interpersonal care and development and coopera-tive crafts and it was discovered that these activities could be performed well in smaller urban units.

By 1995, it was clear that many of the earlier predictions concerning the impact of communications were being proven wrong. Travel had hardly decreased; instead, it saw a net increase as communication about different urban cultures, subcultures, and environments encouraged direct experiential visitation. Predictions 30 years earlier of people communicating instead of commuting to work had also not been borne out, for the very nature of work changed as it fused with localized community service and education. Routin-ized travel did indeed decline; but travel itself was transformed from mere movement from one point to another to an integral part of the total learning process. Finally, electronic communication did not, as had been forecasted, replace such activities as shopping, for people valued the social function of the community marketplace and recognized the importance of tactile, olfactory, kinesthetic, and spatial experiences. Indeed, the proliferation of communica-tion technologies resulted in *more* direct human interaction instead of less interaction; there was a great increase in the demand for places facilitating direct human interchange. The interrelated effects of transportation and communication technologies, economic change, and political decentralization were bringing about the simultaneous phenomena of soicetal dispersion and integration—dispersion into a multiplicity of diverse communities and the integration into a national (and increasingly global) urban culture.

CONCLUSIONS

The scenarios I have presented are only skeletal images, verbal sketches of two possible futures. There are, I am sure, elements in each that could be consid-ered undesirable by some people. Indeed, the scenarios are not necessarily mutually exclusive; a synthesis of conditions from both might well come to pass. Both might be dismissed as mere extreme utopian or dystopian fantasies, although I believe both to be realistically possible. Neither "future history" is entirely probable, although I feel that the first alternative may be more likely. (Another more probable future that was not discussed is one of increasing ecological chaos culminating in global devastation.)

Unforeseen innovations and events during the coming three decades might explode all present projections. Nonetheless, the normative task of attempting to arrive at desirable futures necessitates an ongoing analysis of the multifold potentials, negative as well as positive, of emerging broadband communications. Only in such a manner are we presented with effective charts for helping to guide urban change in the present.

REFERENCES

Aumente, Jerome. "Planning for the Impact of the Communications Revolution," *City* (Fall 1971).

Bagdikian, Ben H. *The Information Machines*. New York: Harper and Row, 1971.

Bailey, James. "Cable Television: Whose Revolution?" *City* (March/April 1971).

Biderman, Albert D. "Kinostatistics for Social Indicators," Bureau of Social Science Research, Inc., Washington D.C., 1971.

Bock, Peter. "Responsive Democracy," Proceedings from the Sixth Annual Symposium of the Association for Computing Machinery, 1971.

Bowdler, Dick and Keith Harrison. "The Future of the City as a Communication Center," in *Citizen and City in the Year 2000*. European Cultural Foundation, 1971.

Brzezinski, Zbigniew. *Between Two Ages: America's Role in the Technetronic Era*. New York: Viking Press, 1970.

Drucker, Peter. *The Age of Discontinuity*. New York: Harper and Row, 1969.

Feldman, N. E. "Cable Television: Opportunities and Problems in Local Program Origination," Rand Corporation paper, September 1970.

Goldhammer, Herbert. "The Social Effects of Communication Technology," Rand Corporation paper, 1970.

Gross, Bertram M. "Friendly Fascism: A Model for America," *Social Policy* (November December 1970).

Johnson, Nicholas. "Urban Man and the Communications Revolution," *Regional Urban Communications*, Metropolitan Fund, Inc., Detroit, 1970.

Meier, Richard. *A Communications Theory of Urban Growth*. Cambridge: MIT Press, 1962.

National Academy of Engineering. *Communications Technology for Urban Improvement*. Washington D.C., June 1971.

Reich, Charles. *The Greening of America*. New York: Random House, 1970.

Ribiera, Darcy.*The Civilizational Process*. Washington D.C.: Smithsonian Institute Press, 1968.

Shamberg, Michael, and Raindance Corporation. *Guerrilla Television*. New York: Holt, Rinehart and Winston, 1971.

The Sloan Commission of Cable Communications. *On the Cable: The Television of Abundance*. New York: McGraw-Hill, 1971.

Stevens, L. Clark. *EST: The Steersman Handbook*. New York: Bantam, 1970.

Theobald, Robert. *An Alternative Future for America*. Chicago: Swallow Press, 1968.

Theobald, Robert. *Teg's 1994*. Chicago: Swallow Press, 1971.

Toffler, Alvin. *Future Shock*. New York: Random House, 1970.

CHAPTER 20

Minorities and the New Media: Exclusion and Access

OLIVER GRAY

The 1960s have been climactic in American race relations and prophetic in the development of communications technology and mass media in the United States. We are told that a greater potential exists through the use of cable television for a more personalized use of the nation's airways resulting in a complete new arrangement as regards the economic and social benefits involved. Network television will decline in importance and new corporations will replace them. The young, the old, the poor, the intellectuals, and the students will compete for a more significant role in the use of the new media. However, it is not these groups who will pose the most difficult challenge to existing over-the-air broadcasters or the new communications-entertainment complex that will replace or compete with them. The ethnic minorities will increasingly demand a greater share in every aspect of the communications industry of the United States.

Questions immediately arise around the issue of priorities confronting the disadvantaged minorities. Is it not more relevant to speak of percentage of the GNP accruing to the ethnic poor, including housing, employment and health care, instead of their relation to the new media? The answer is clearly in the affirmative. But, too often minority people give low priority to the broadcast media as a source of education, a means of projecting culture, and a medium for economic development. As an urban planner and member of a minority group concerned with the allocation of resources, I view each of the aforementioned areas as a resource and feel communications must be viewed in a similar light. If it is not, minorities will find themselves in exactly the same position they find themselves in today 20, 30, or 50 years in the future. The role of minorities in the media is marginal at best and virtually nonexistent in other areas, such as production and ownership.

I believe that the electronic media are a source of education for the ethnic minorities of this country. Those of us concerned with countering negative images or projecting cultural awareness will not find a more powerful medium. Beginning with the early days of radio and moving forward to today I shall attempt to trace the history of minorities in mass media in terms of their ability to gain access in various capacities.

LOOKING BACKWARD

Involvement of blacks in broadcast media extends back to its earliest days and is believed by some to be the sustaining force in those difficult days of commercial radio. The early years of radio broadcasting were characterized by extensive airplay of music, but the music of black America was rejected as having "an immoral influence" (Barnouw, 1966, p. 131). However, jazz and blues attained respectability and suitability for airplay when whites performed watered-down versions of this cultural offering of black America. Paul Whiteman was making a handsome living from jazz before blacks were allowed to perform it on radio. This treatment of black creativity and talent became standard for the relationship of minorities and the communications industry.

Music declined as the major offering of radio and was supplemented by drama during the year 1928 to 1929. While the initial offerings in radio drama were reflective of the expansive nature of the medium, once again blacks dominated thematically.

During 1928 to 1929, a program, the classic "Amos'n'Andy" first appeared on radio. Two white men, Freeman Fisher Gosden and Charles J. Correll, the acknowledged creators of "Amos'n'Andy," made the transition from traveling road shows, which included minstrel show and carnivals, to the nation's airways.

Gosden and Correll first traveled separate routes but eventually worked and roomed together, and began to work up a blackface act. When they were promoted to the Chicago headquarters, it gave them a chance to try out for WEBH in the Edgewater Beach Hotel. Still working for the Joe Bren Company, they began to broadcast weekly blackface routines, receiving free dinners. [Barnouw, 1966, p. 226]

The dignity of blacks in America was dealt away and a childhood relationship exploited for a daily plate of hotel food. Tracing the careers of Gosden and Correll we find an interesting combination of petty entrepreneurship and media interest that propelled them from relative obscurity to the national spotlight.

"Amos'n'Andy" might have produced spin-offs for blacks and other minorities in radio in terms of jobs as writers and actors, or possibly stimulated the creation of other programs of its type. But the spin-offs for blacks were negative.

Johnny Lee, who became a comedy-lawyer on Amos'n'Andy, said: "I had to learn to talk as white people believed Negroes talked." According to actress Maidie Norman: "I have been told repeatedly that I don't sound like a Negro." But when she applied for other roles she was rejected without a reading. [Barnouw, 1968, p. 111]

White men with the capacity to imitate black men forced at least four decades of blacks to play stereotyped roles if they were able to secure a role in the mass media of the United States. Of course, nothing has been said of the almost total exclusion of blacks and other minorities from participation as owners, writers, producers, pages, and technicians who service and control the media as well as derive profits from it.

A confluence of events, the end of war controls on the production of television receivers, FCC relaxation of control over television, and a rising black awareness led to the first recorded black attack on the mass media. In 1951, the NAACP condemned "Amos'n'Andy" at its annual convention. The provocation for the attack was the continuance of a policy of disregard for the collective image of black America in the extension of the program from radio to television. "In 'Amos'n'Andy,' CBS was bringing Negroes to television—but Negroes trained by Freeman Gosden and Charles Correll in the nuances of stereotype," (Barnouw, 1968, p. 247). Black actors were taught how to imitate what the nation thought was true of blacks.

Radio institutionalized racism in the mass communications industry. Black America was involved as participant and/or observer in the beginnings of a wave of struggle and protest against racism and prejudice in other areas of national life beginning in the 1950s.

While black America fought for its freedom, the television networks moved through fantastic growth, consolidating their gains and reaping huge profits. Blacks were rarely if ever seen outside of the news and sports programs, although an occasional black danced or sang his way across the wasteland.

Television was almost a completely white medium until the mid-1960s when Bill Cosby cracked the barrier and became the first nonsinger, nondancer, nonbuffoon black television star. Cosby was followed on television by some blacks playing the role of schoolteachers and nurses in specially oriented programming. In addition, blacks began to appear as newscasters and news anchormen.

But Federal Communications Commission figures for May 1971 indicate that only 8% of the total employees in broadcast jobs were minority group members.

Blacks have not done significantly better as creative, production, or managerial employees during the period of television's greatest growth. The significant difference in black and minority activity in this period as compared to radio is the move by and on behalf of minorities to seek redress from discrimination in the technical, creative, and managerial sides of the media.

Beginning with a successful license challenge to blatantly racist programming and exclusionary hiring practices of a Jackson, Mississippi, television station in 1965, minority efforts have extended to many cities. In addition to license challenges, a few black and Spanish Americans are attempting to buy radio outlets in areas in which they live. It is more difficult to obtain ownership of television stations.

As leaders of the communications revolution, the ethnic minorities must recognize the need to address themselves to the future. Looming on the horizon is the next communications era. Cable will grow rapidly as either a supplement to or replacement for over-the-air broadcast television. Minority Americans must focus their attention and energies here as well.

CABLE: THE NEW TECHNOLOGICAL ELITISM

CATV is now challenging regular broadcast television for a significant share of viewers. This development is important to ethnic, social, and intellectual minorities who have increasingly found themselves unserved by the existing broadcast industry. In fact, new hope has grown among these groups as they attempt to marshal whatever is required in the way of funds, lawsuits, and FCC license challenges to insure that their interests are represented in this new alignment of the broadcasting industry.

The nature of commercial television in the United States is that of a mass medium that attempts to provide programming for the largest audience possible. This arrangement, which requires large amounts of money, influence, and connections to those in power, cannot accommodate groups such as the poor, ethnic minorities, and intellectuals. Another by-product of this arrangement is the astronomical cost of time on network television.

In the face of such an unyielding monolith, blacks and others have begun to see CATV as a viable and necessary alternative. CATV is attractive because there are still franchises available and the increased channel capacity will lower the cost. The danger is that the lure of maximum profits and the actions of public officials, large corporations, and new interest groups may prevent any significant minority inroads into CATV and result in the development of a new technological elite.

With some 6000 franchises already awarded, how can the ethnic minorities gain access to cable television? A number of strategies are being employed in different communities. At hearings held in January 1971, in Washington, nine companies presented proposals for a CATV system. A number of blacks testified among the 50 witnesses. Some proposals emanating from these hearings have significance for minority communities. They include suggestions that

special provisions be made to subsidize CATV service for the poor, and that profits from CATV system operation be used to increase CATV's access to all people and diversity of programming. Black efforts for Soul in Television, BEST, stated that the cable system should be wired free of charge into every home and should be out of the hands of existing media. Howard University presented a proposal for a subscriber owned company to operate what is expected to be the nation's first extensively wired city at a cost of $15 to $20 million. [*Black Communicator*, January 1971, p. 9]

In San Diego, Charles "Chuck" Johnson, a black entrepreneur, has leased a channel from Mission Cable Television, Inc., the local franchisee, and is providing local programming to the black community. In Brooklyn, a state senator is one of the founders of Integrated CATV, one of several Brooklyn groups competing for a cable television franchise. In addition, the Bedford-Stuyvesant Restoration Corporation has completed a feasibility study indicating a high potential for developing and implementing a minority owned and operated system in Brooklyn's largest minority community.

Teleprompter Corporation has entered into an agreement with the Reverend C. T. Vivian, director of A Black Center for Strategy and Community Development for a cooperative effort to utilize cable television for more effective black communications. Since Teleprompter controls the franchise covering Harlem, this agreement is significant.

The Mafundi Institute of Watts and Communications Associates from Santa Barbara, Calif. joined together to establish Watts/Comm, a CATV system to service 5000 subscribers with 30 channels of commercial and local programming.

The television system planned for Watts/Comm is a new concept in "interaction television." It not only permits voice feedback to the originating sources of the television program—i.e., the offices of elected officials, storefront performing arts centers, schools, CAP centers, et cetera—but allows the viewer at home or in the storefront, or wherever the television set is placed, to plug in an inexpensive TV camera to send out his own video message. The system provides the capability of distributing many programs simultaneously with different sub-groups of subscribers having the opportunity to talk with the program source and, through the source to each other. Two way audio and video communications can generate dialogue, and dialogue can lead to community involvement and change. [*Radical Software*, 1971]

A major concern of those favoring as well as opposing minority ownership of CATV outlets is the programming to be presented on such channels, including the issue of whether owners should program their channels at all. Arguments against this are based on the cable system owner's desire to receive a profit from his investment and his possibility of selling programming similar to that offered by existing networks, directed toward a mass audience. This approach is anathema to those who favor community-based program origination. They argue one of the major benefits of the increased channel capacity is the potential for considerable diversity of programming to meet the needs of highly differentiated groups.

The specter of a form of electronic provincialism can be partially refuted by the experience of Canadian CATV. Canada has 300 cable companies providing service to over 4 million persons. A Thunder Bay citizens' production unit is providing one evening's programming a week (programs are made at the request of local groups). Beloeil (Quebec) cablecasts city hall meetings live, Hamilton (Ontario) has conducted intensive Pollution Probes over the community channel, London (Ontario) has special Indian programs, Fredericton (N.B.) citizens prepare their programs with their own cameras (National Film Board of Canada, 1971).

An important feature of locally originated CATV broadcasting is the reduced cost and direct access to the media for minority political aspirants. An undesirable direction of American politics has been the almost total reliance on media campaigning and resulting high costs. Once the scarcity of channels is overcome and the reduced cost of operation is passed on, politics may become more democratic.

Ellis Haizlip, the black producer of "Soul," stated when queried regarding his planned use of the cable that he would like to get stories such as the daily trial action of the New York Panther 21, the New Haven Panther trial, and the Angela Davis trial on tape for rebroadcast in the black community. He feels the type of coverage currently provided for these stories is sensational in nature instead of informative and investigative. Haizlip went on to say he feels cable is an excellent vehicle for "sensitizing" the community.

TOWARD THE FUTURE

The Ford Foundation has announced the funding of a cable information center. The Sloan Commission has issued a major report on cable television. The Federal Communications Commission has decided how it will handle CATV. The FCC regulations are of very real significance to the disadvantaged minorities. They do not halt the growth of cable but may slow its growth until those who stand to be hurt most can reorganize. Blacks and other minorities must realize this and move to prevent or at least impede the third communications giveaway in this century.

REFERENCES

Barnouw, Erik. *The Golden Web*, Vol. 2. New York: Oxford University Press, 1968.

Barnouw, Erik. *A Tower in Babel*, Vol. 1. New York: Oxford University Press, 1966.

Barnouw, Erik. *The Image Empire*, Vol. 3. New York: Oxford University Press, 1970.

Challenge for Change. National Film Board of Canada. Montreal, 1971.

Dedrick, H. S. *Telecommunications in Urban Development.* Santa Monica: The RAND Corporation, 1969.

CHAPTER 21

The Open Door Policy
on Television

THEODORA SKLOVER

If we consider the impact of network television on the majority of American viewers, one observation would be that watching television is the national spectator sport. Conditionally, it is the panacea for the bored housewife, company for the insomniac, and teacher of the preschooler. On any one night, approximately 90 million Americans are united in the mundane ritual of televiewing during prime time. Communications experts claim that television has created a "national village" by programming shows cast with an identifiable population who problem-solve man's universal travails. What these shows tend to accomplish is reinforcement of the "everyman" image not support individual differences.

Is the citizen of this "national village" content with network fare? According to the three major networks, viewers are not at their program directors' mercy. Instead, aired shows are the choice of this theoretical village and not an imposition of network policy on the masses. They would punctuate this position with empirical evidence—the Nielsen ratings. But if we probe deeper than the Nielsen statistics, we find a line of paradoxical information; the viewed program may never truly reflect a person's real taste at all because his choice is so often made by default. NBC's ratings expert, Paul Klein, calls this type of televiewing the "L.O.P." formula; the dial is consistently on the move until the viewer decides what, to him, is the "Least Objectionable Program."

A high percentage of this "national village" apparently indulges in mindless viewing. They absorb what Alvin Toffler, in "Future Shock," calls engineered messages, mass media's presentations of "real life" that is information-rich, edited to the bone, paced for action, and rippling with stylized dialogue. And who promulgates this conception of reality that is supposed to unify and represent 90 million people? Three or four powerful sources that provide America with Hobson's choice—the major networks. They determine program structure with limited content, offer a negligible amount of thought-provoking community issues, and leave open few avenues for feedback. However, collective viewing does not necessarily make us a village. McLuhan and others have described how television has compartmentalized us and destroyed natural communication between people.

PUBLIC ACCESS

The role of television must change by relating to this "national village" as a responding public that can experience the media *actively*, not passively.

Open Channel was created to proselytize and make effective the concept of public access on both a local and national level. It works toward these goals by supplying necessary information and technical expertise to groups who may then undertake their own television production. Production processes are fully explained and volunteers from Open Channel's impressive talent pool serve as instructors in the use of inexpensive, portable video equipment. Open Channel is founded on the premise that the solution for active participation in television by the viewing public is public access TV. This means two-way communication of "narrowcasting" instead than broadcasting, and with one important result—feedback. What public access insures is the right of any individual to present his message on television in his own terms with no threat of the misuse of information, as with broadcast stations.

For public access to exist as more than a theoretical concept in communications, its revolutionary importance must be brought to the public's attention in terms of its informational and creative potentials. Proffering free air time is not enough. That is much like giving a typewriter and ream of paper to someone who has never learned to write and requesting a novel. Training in production skills and education in the use of the medium is intrinsic to the survival and long-term acceptance of public access. This is Open Channel's prime function. Television is a tool, and like the written word, it's expression can turn into a natural extension of the participant. Public access works for person-to-person contact—community action groups, special programs in sign language for the deaf, and performances by small neighborhood dance troupes, to name a few examples.

Ironically, one problem that Open Channel has confronted is convincing people that they can produce a valuable television show. The difficulty lies in the inability of most people to conceive of themselves as television producers. This occurs in large measure because people's experience has been limited to slick commercial broadcasting—the one-way, passive response system. What we need is an educational and promotional effort to alert the masses to the opportunities provided by public access channels and financially stable institutions to train them in the use of this technology.

Public access is made possible technologically by the enormous channel-carrying capacities of CATV systems. The fundamental idea for allowing public access can be achieved by treating the cable as a neutral roadway for messages dispatched by the public for the public. This "common carrier" status has already been applied to the telephone industry. If the cable functions as a tool for message carrying without discrimination (social or political), the cable carrier should be freed from legal liability that may be the result of the users' content or presentation. These conditions are the converse of broadcast policy, where the broadcaster is legally responsible for the content of its programs.

The status of public access is currently ambiguous. While users of public access can exercise nondiscriminatory freedom of speech with no outside control of content, the cable operator is responsible for this content and is liable for suit. Yet, the FCC regulations for this type of programming state: (1) " . . . one dedicated, non-commercial public access channel available without charge at all times on a first-come, first-serve non-discriminatory basis . . . "; and (2) " . . . the cable operator . . . must not censor or exercise program content control of any kind over the material presented on the leased access channels." To date, this has not been the case. Open Channel believes that the cable operator must be released from this awkward and contradictory responsibility if public access is to fulfill its original intent—freedom of expression.

In January 1971, the FCC required cable operators with 3500 or more subscribers to produce shows of local origination as well as carry the standard over-the-air signals. Whereas previously the cable operator was a technological source, the FCC decision forced him into creative production. Shows of local origination were supposed to be the product of the cable operator himself. One of the FCC's objectives in requiring local origination was to foster community dialogue on a microscosmic level—for example, televised local school board meetings. Some cable operators made a mildly courageous effort to provide this kind of local public service. Others aired rented old films, five-year-old "I Love Lucy" reruns, and any other inexpensive shows that meant minimum production difficulties and guaranteed financial successes. Instead of community-oriented programs of local origination, these cable operators reneged by showing stale network fare.

If cable operators adopt the broadcast networks' *modus operandi*—producing shows that will encourage a maximum viewing audience, then the commitment to localized television will become an empty promise. Public access does not seek the mass audience. Instead, it works to permit the community to create its own programs thereby insuring the goal of localized television.

The influence of competitive business forces on the use of the media forces us to consider the financing of public access. There is no tangible evidence that public access will increase the cable operator's profits. This fact, reinforced by his liability for provocative content, leaves no incentive to initiate public access. At least for the present, production, training, and promotion are left to others' design. Unless resources are made available from other resources, those most needing this access will not be able to obtain it. Consequently, a financial base to support these activities must be developed on a nonprofit basis.

GOVERNMENT REGULATION ON PUBLIC ACCESS

New York City has had the chance to try a viable arrangement for genuine public access. This was made possible by the New York City cable franchise, which now covers the two cable companies (Sterling Manhattan and Teleprompter) operating in Manhattan with a total subscription of 100,000, and is, in many ways, a model ordinance. Based on a report issued by the Mayor's Task Force on Telecommunications Policy, headed by Mr. Fred Friendly, the franchise provides for 17 channels, two of them public access. Projected figures show that by August, 1973, 24 channels will be required to fulfill the following specifications: four public access, three governmental, and four channels that can be used for additional commercial services.

The New York City experiment with public access cable television is important—particularly because it is the first of its kind. The distinction between vague policies that would glibly promise local participation and its enforceable franchise stipulation guaranteed public access. Other cities have, in fact, taken New York City's ordinance as a practical model for evolving their own.

Interim rules for day-to-day regulation of public access channels in New York City were drawn up in the Bureau of Franchises office. They were written to cover an experimental period of six months, ending in December 1971, but have been extended to allow for more trial and error before the final revision. Mr. Morris Tarshis, director of the Bureau of Franchises, has been instrumental in framing guidelines for public access that would insure the greatest flexibility. Rules that were advanced for a temporary franchise were: time shall be leased on a first-come, first-serve basis; programming shall be free from any control by the company as to program content except as is required to protect the company from liability under applicable law; the company shall endeavor to lease channel time to as many different users as is practical, it being the intent of the parties that such public channels serve as a significant source of diversified expression.

While these interim rules permit the two cable companies to preview materials before telecasting, the laws applicable to liability are not clearly enough defined. We may confidently presume that the more common threats would be invasion of privacy, defamation, fraud, copyright infringement, obscenity, and, contemporaneously, the possibility of incitement to riot. Although criminal and civil suits can be brought against the operator, the FCC posits the small likelihood of their success. The FCC stated in its Cable Rules and Regulations (1972):

> We recognize that open access carries with it certain risks. But some amount of risk is inherent in a democracy committed to fostering uninhibited, robust, and wide-open debate on public issues (*New York Times Co.* v. *Sullivan*, 376 US.S 254, 270, 1964). In any event, further regulation in this sensitive area should await experience.

To provide diversified channel usage, New York City's two public access stations serve different purposes. Channel "D" is reserved for one-time events, such as a community action group in session, or for last-minute requests. "D" has often been represented as the city's "soapbox" station, but this is an inaccurate evaluation. For example, some programs are on-the-spot coverage, or real-time events. Rent control hearings, which by their nature tend to be longer than the standard half-hour time slots and therefore cannot normally be accommodated on television, fit this category. Also, "D" is ordered so that emergency situations can get immediate coverage.

Channel "C" is more akin to the traditional broadcast station procedures. Shows are aired on a regularly scheduled basis—either daily or weekly. This method of consistent scheduling might well be the ideal transition necessary for establishing a loyal audience as with traditional television.

Whether or not public access will be a success is still a matter for conjecture. Any judgement on this new innovation in televiewing would be premature since this is still a time for experimentation. Public access is a fresh concept that will require an extensive educational process in the matters of both use of video equipment and the development of programming plans. Yet, what can the future of public access bring?

An examination of early radio and television developments would reveal how active public participation in programming was seriously considered as a legislative regulatory goal. However, since industry power was influential and not counterpoised by public interest forces, the dominant factor influencing development was the economic concerns of the industries. There are those public interest advocates who insist that the advent of cable television is the third such similar transition period in communications. The prime concern of Open Channel is that public access provisions will never definitely be grounded on first amendment freedoms, but that they will continue to be dominated by the economic interests of industry or the convenience of legislatures convenience.

A landmark first amendment precedent involving freedom of expression was set in the case of *U.S.* v. *Associated Press*. In this case, Judge Learned Hand noted:

. . . right conclusions are more likely to be gathered out of a multitude of tongues than through any kind of authoritative selection. To many, this is and always will be folly, but we have staked upon it our all. [*U.S.* v. *Associated Press*, 1943]

Open Channel hopes to insure this decision through public access over television.

Regulations offering similar national guidelines for mandating channel usage have been recently announced by the FCC. The requirements carry this contingency for each channel consigned to over-the-air broadcast signal usage: these nonbroadcast uses must, by agreement, include a public access, educational, and governmental channel.

However, continuing public access in the future will require a specific provision in these FCC regulations applying directly to the growth in public access channels—the "N + 1" rule. "N + 1" requires that whenever demand for public access arises, an extra channel must be activated. This is tentatively defined as about 80% usage of the existing systems time allotments. Therefore, when an 80% capacity of one public access channel is reached, another is made available. When the required percentage of that one is filled, another is opened. What this provision does, then, is protect the true intent of public access. An arbitrary limiting of channel space cannot be utilized as justification for disqualifying potential users. More importantly, such limitation cannot be employed as an excuse to enforce a policy into effect that would mean choosing programs on the basis of content.

PIONEERING AT OPEN CHANNEL

Experience at Open Channel in the first few months indicates that "narrowcasting"—programming for the small, self-selected groups—may yet be the most typical of telecasting patterns. Although we are not able to make any claims as to the long-range accomplishments of public access, the extraordinary variety and plans of interested groups are encouraging. Presently Open Channel has over 200 requests for programming. Among the many diverse groups anxiously awaiting our help are the Museum of Modern Art, the Student Struggle for Soviet Jewry, Psycho Education Department of the Coney Island Hospital, Society for Creative Anachronism, Vietnam Veterans Against the War, American Arbitration Association, the Fortune Society, the Association of American Dance Companies, and innumerable church organizations and community action groups. In attempting to identify and encourage creative use of public access channels, Open Channel seeks to identify active groups, to locate community leaders, and to check the availability of television equipment. In each community we look to the neighborhood newspaper, a community center, district leaders, and block associations.

Open Channel is working toward the establishment of neighborhood production centers throughout the two franchise districts. Fortunately, the New York City Franchise stipulates that by the summer of 1974, both Sterling Manhattan and Teleprompter must subdivide each of its existing districts into 10 subdistricts. This will insure their provision for discrete signals in each of these community subdistricts. Open Channel has been active in insuring that these divisions are developed from within the communities instead of from without. What must evolve is community autonomy in production. Cable operators or organizations (including our own) should not be the ones to set up facilities and direct the operation. If that were to occur, the result of public access would be far less valid than if the community leaders cooperating with their constituency determined the location and facilities for production. Open Channel exists to assist in these matters, not manipulate the formation of policy. In its Report on Cable Communications, the Sloan Commission (1972) stated that . . . demands for use of the public channels must be nourished. There will be need for promotional forces within the community and for technical assistance and talent to assist in the production of programs. Open Channel's efforts are directed to nourishing the demands for use of the public channels.

An important function of Open Channel is to encourage formation of local viewing centers available within the neighborhoods. These are for those people who do not have cable connection in their homes. Schools and churches have been especially cooperative in providing space, and hospitals, museums, and libraries look promising. Unusual locations such as laundromats, bars, and police stations are being investigated as possible viewing centers, also. For the larger community, Open Channel offers advice for the establishment of a Public Access Fund, or a fund that would provide financial backing for production, training, equipment, and program promotion. Such a fund can be established with the cooperation of the franchising authority (usually the city), which collects a franchise fee from the cable operators. In New York, over $250,000 will be collected in 1972 and part of it could be allotted to support public access.

Since public access means free expression, and free expression that will motivate a reaction, Open Channel seeks to develop formats that would be natural dialogue between people—not stilted repartee as heard through an editor's perspective on life. Editorializing by paring down the content should probably stay where it is, at the networks and in local news coverage. Public access use of this visual tool should be to communicate directly between people in their own way.

However, there are still legal issues relating to freedom of expression. Originally, the user was required to obtain releases for any copyrighted materials, and assume legal responsibility for minors under 18. Both Sterling Manhattan and Teleprompter insisted on a list of persons appearing on the program, a statement of their purpose, whether or not commercial material was included, and required a prescreening of the programs for libelous statements. Recently, these formalities have loosened up considerably. Groups that appear regularly now are taken at their word that no materials will threaten subsequent legal action.

"Catch 44," a program produced by WGBH in Boston, is, incidentally, the single example of open access we are aware of in *broadcast* television. Groups may appear after promising to follow WGBH's rules, which are:

1. Do not attack identified private persons, unless they have become publicly associated with the issues being discussed.

2. Do not use air time to initiate violence.

3. Do not use language or gestures that people consider indecent or profane.

4. Do not use air time to appeal for money to promote commercial ventures.

This station believes, as does Open Channel, that users who are given the responsibility will not abuse the privilege. Experience to date has not been disappointing. Recent FCC rules require the name of participating groups only and supercede the ones of the New York City Franchise.

Ostensibly, public access differs greatly from network productions. An objective of Open Channel is to convince users that they are *not* in competition with a network market that demands slickness. Initially, most people want to be on television in the same kind of format as Walter Cronkite. Only after repeated use are they able to be more experimental in presenting their case. However, Open Channel does provide expertise so users can utilize the medium to their best advantage.

Examples of the Application of Public Access

"Real-time" programming is a major factor in achieving the goals of public access. A fine example of real-time programming on a local level was accomplished by the Alternate Media Center at New York University in which a 33-hour marathon community school planning weekend was telecast from the University's Greenwich Village location. Teachers, architects, parents, and social scientists worked for three days and nights developing plans for an experimental school. Television monitors were set up outside the buildings to interest passersby in joining the session. These meetings were also cablecast to homes so that if one could not attend for this extended period of time, home viewing and participation, were made possible. When an urgent issue is extended beyond the confines of one room and brought directly into the home, a sense of real community can be gained.

One of the more noteworthy groups which has worked with Open Channel was the *Inwood Advocate*, a volunteer-run community newspaper located in a heavily cable section of upper Manhattan. For the last few months the *Advocate* staff has been presenting a monthly video news show produced and directed by them. Their original intent was to use television to demonstrate their newspaper production in the hopes of enlisting volunteers to join them. Within an eight-week period, a series of half-hour programs documented the steps in newspaper production—the planning, editorial meetings, paste-ups, and following of reporters on the beat. Simple production techniques were used. Two cameras were used for neighborhood meetings and a portable camera for location pieces. To insure an audience, the *Advocate* published the programs' cablecast schedule and arranged for another viewing at an Inwood Community Action meeting. The experience of the *Advocate* staff was summarized in a letter to Open Channel:

Videotape is so much easier to work with than typesetting that we have seriously considered dropping the paper and doing just an open access news show. Video seems especially well suited to transitory and evolving events where it is unimportant that a permanent record be made. Neighborhood response was good. We haven't heard from our enemies yet, but people who like us were very excited about seeing the shows. I am sure that as the Channel "C" audience grows, we will be able to exercise a rather strong influence on public concern over issues and also scare our lazier politicians into some muchneeded action.

Another example involved the Puerto Rican Pentecostal Church. We found their representatives to be quite sophisticated in media concepts—especially distortions of ideas through editing—a phenomenon that typically occurs to many black and Puerto Rican communities. Minority groups, very sensitive to the issue of mispresentation by the media in past experiences, requested that we show only the complete unedited version of the church services and the final review before telecast was retained by the church. (This final edit is guaranteed to every group that Open Channel assists. The client has final control of content.) The result was a well-attended, appreciative, and animated audience. On another occasion, Open Channel taped the entire 2 1/2-hour session of the "Forum on Censorship" conducted by the New York chapter of the National Academy of Television Arts and Sciences. This timely, very important issue that affects everyone was given a few minutes coverage on the commercial station news shows. WNYC, Channel 31, the cityowned station, borrowed the tape to broadcast the meeting for itself.

CONCLUSIONS

Both Sterling Manhattan and Teleprompter, who have been cooperative on a daily basis, have slight variations in their policies for studio space. They must offer free air time to comply with FCC rules, but it is Teleprompter that provides free studio space one day a week. This includes a simple camera, sound setup, and technical assistance. Sterling Manhattan charges about $100/hour for this same equipment. These services exceed the specific minimum requirements of the New York City franchise for a free 5 min including studio space for anyone who asks for it.

During the first four months of public access, 75 individuals and organizations produced public access programs. This accounted for an average of 20 to 30 hours of public programming per week. But, the ratio of inquiries received to the number of actual shows produced has been five to one. Most groups using public access are presently supported by foundations except for those that have taken advantage of Teleprompter's free services or Open Channel's free assistance. Groups that need encouragement and should be seen on public access cannot afford to go on unless there is easy, unhindered use of the channels. Without it, we are forced back to the same economic machinations of over-the-air broadcasting.

The cable companies are not motivated to promote public access channel use by either educating the public or offering a greater amount of services. Open Channel sees the need for an independent nonprofit structure that will support public access programming. Public access use should not compete with cable operators for the time claims on their own studio facilities. Instead, the establishment of independent production facilities specifically reserved for public access use is needed. Foundation support and volunteer help are absolutely indispensible as a beginning. Open Channel could not have survived this far without them.

There are several feasible avenues for positive development of public access that should be kept separate, if possible. They include a franchising authority, the cable operator, the community access committee to promote channel use, the potential community user groups, and the scheduling for time allotments. (This is similar to the Public Broadcasting Service's arrangements, but would be applied on a *local* level.) The community access committee would have the responsibility of supplying information, equipment, and production talent— possibly volunteer-run with a minimum of paid staff.

Yet it must be emphasized that neither the cable companies nor Open Channel controls public access in New York City. People must understand that they are the programmers, not any authority-figure organization in a control position. Open Channel's goal is to motivate more people into employing this technology that would, in turn, foster deeper social interaction.

Now we have an opportunity to bring into existence a unique communications system through cable television. We must not falter and risk its loss. Traditionally, stockholders and advertisers have shaped broadcast television with the profit motive. We must find a way to prevent these same market conditions from decimating public access. If profit making remains the imperious force behind television, then the present commercial broadcast system will smother public access. The only way to capture the diversity and fabric of our society is by public access. We cannot walk the fine line to please big business or compromise ourselves to the legislature.

Power is no longer measured in land, labor, or capital, but by access to information and the means to disseminate it. As long as the most powerful tools (not weapons) are in the hands of those who would hoard them, no alternative cultural vision can succeed. Unless we design and implement alternate information structures which transcend and reconfigure the existing ones, other alternate systems and life styles will be no more than products of the existing process.

Television is not merely a better way to transmit the old culture, but an element in the foundation of a new one.

Our species will survive neither by totally rejecting nor unconditionally embracing technology—but by humanizing it: by allowing people access to the informational tools they need to shape and reassert control over their lives. [*Radical Software*, 1971]

REFERENCES

Federal Communications Commission, Cable Rules and Regulations, February 12, 1972.

Radical Software, November 1, 1970.

Report of the Sloan Commission on Cable Communications, *On the Cable: The Television of Abundance*, 1972.

U.S. v. *Associated Press*, 57 F. Supp. 362 (S.D.N.Y. 1943).

CHAPTER 22

Moving Information Instead of Mass; Transportation versus Communication

PETER COWAN

Some 50 years ago E. M. Forster wrote a story called "The Machine Stops." He described a society of the future in which the movement of people from place to place had all but stopped. Each individual spent his whole life, from cradle to grave, in a small cell, being fed, cared for, and entertained by services piped to him from a central machine, which served the whole world. Such rare trips as could not be avoided were taken by airship from one part of the globe to another and were regarded as at least eccentric and probably highly dangerous. Local journeys never occurred. At last the machine broke down and caused the collapse of society.

Much more recently John R. Pierce (1967) wrote of the future of a communication-dominated society, as follows:

> We no longer need to live in the heart of the city to live comfortably. But we still travel by plane to meet other people, to use rare facilities or to see or participate in rare events. I believe that, in the future, improved communication will enable us to avoid much onerous travel. We can live where we like, travel for pleasure, and communicate to work.

Here we have two statements, separated by many decades, about communications, transport, and the future of society. The contrasts are striking. One statement is pessimistic, the other optimistic. One is written by an Englishman, the other by an American. One is the view of a novelist, the other is the view of a scientist. Which forecast is correct? Or are neither? Or if one is correct, then is this a good thing? How far will the ongoing revolution in communications continue, and where will it take us? Will different cultures react differently to the changes that are going on, or will society become more

339

homogeneous? These questions are among the most important facing society today, for they address a central issue, that of the relationship between technology and human values. Speaking at a conference on the Social Implications of Technology and reported in the *Times* on November 6, 1971, Professor E. Mesthene of Harvard University said:

> The classroom teacher is just as likely to let the new piece of audio-visual equipment gather dust in the corner as to take the trouble to keep it in repair and learn how to use it . . .
>
> For a century we in the West have concentrated on industrial technology for economic growth as a necessary precondition of the good life, only to meet a generation that finds our conception of the good life a bit bizarre, and that believes man can retain his humanity only if he succeeds in reversing recent technological trends.
>
> I would not want to count on the quick advent of wired (cable television supplied) cities—even if the cable technology exists—in the face of youth who display a taste for the simple life and communal living.
>
> Nor would I count on a friendly reception from the business man or public official whom I tell that communication technology has now made travel redundant so that he can henceforth conduct his business and attend his meetings from his console and videophone and ultramicrofiche reader without leaving his home. He may like to travel; and his wife may not want him home all the time.
>
> One should not exclude the possibility that people may decide that there are more valuable things in life than being assaulted by 20 television channels instead of two or five; seeing the face as well as hearing the voice of the unsolicited telephone caller; and having a mahogany console in the living room instead of the piano.

Once again we can see that the relationship between transportation, communications and urbanization, and society is intimate, complicated, and of primary importance for the future. The impact, acceptance, and use of new technologies may be very different from what we expect, for the quirks of human nature are difficult to predict.

THE VIEWS WE HOLD NOW

What are the current views of the future of communications, transportation, and urbanization? How and where did they arise? How do they differ from previous views? What are the promises and pitfalls that they contain? The place to begin any discussion of such questions is with the great series of studies that appeared in the mid- and late 1960s from a number of brilliant scholars and planners in the United States. Webber, Meier, Deutsch, Artle, Meyer, Kain and Wohl, Hoover and Vernon, and others laid the foundations for a seminal view of the problems and possibilities facing our urban areas, which has sustained a body of knowledge and research ever since. These views questioned past orthodoxies and showed urban development in a new light.

Before this anyone attempting to forecast urbanization would have thought mainly in physical terms. If left to itself, most planners thought, the pattern of urbanization would result in large high-density concentrations of population in the larger cities while "ribbon development" would spread around the edges of urban areas. The factory was the main agency of metropolitan growth. In Britain the policies recommended to deal with the coming situation were the same as had been recommended at the turn of the century by Howard, Geddes, Parker, and Unwin—new towns of moderate size located at sufficient distance from major cities to reduce commuting journeys. These new towns would house people from the large cities, whose growth was restricted by green belts. Thus, both the problems of the future and their solution were stated in physical terms.

We can now see that such forecasts were wrong. The automobile has brought about tremendous changes not only in the shape of the city but also in the style of life of the nation. Many of the most pressing problems of our cities today are not physical but social in origin, and social development is at least as important as physical measures for the future. Finally, we now know that one of the major growth forces in the city is not manufacturing industry but service employment; the explosion of office accommodation in great cities indicates a different set of factors at work in urban growth.

The American scholars whom I have mentioned perceived this shift in our urban life and spelled out the consequences. Very broadly, their views ran as follows. During the next few decades we can look forward to a population that is increasingly mobile, both physically and socially. Car ownership will continue to increase. Automation will have a profound effect on labor, and we can expect a great increase in the amount of spare time available to the majority of the population. Educational opportunities will be increased, and mass higher education will spread widely. These events will combine to bring about a pattern of urbanization that is more diffuse and less centered on "place" than in the past. People will travel more, especially for pleasure. The city will be less centrally oriented, and many different nuclei will be linked together in an "urban area." Hierarchical systems of cities will give way to a more complex and dynamic pattern. Finally, urban areas will merge to produce supercities stretching for hundreds of miles across the face of different parts of the nation; in Britain such an area may stretch from London to Manchester—"Megalopolis England." The problems posed by these developments will be both physical and social. Physical problems will relate to achieving effective linkages between the variety of locations that each individual will use in his new way of life. Social problems will arise as the population is confronted with the need to change from a workoriented society to a leisure-oriented society.

The wisdom of these changes of perspective is clear and cannot be denied. What is more, in the United States at least, events have justified the forecasts.

The flight to the suburbs, the decay of downtown areas, the omniscience of the automobile, and the decline of the railroads can all be seen in major metropolitan areas in the United States. And one can begin to detect the same patterns emerging in Britain. But we must be careful to check our facts before assuming some kind of "affluential determinism," which holds that surplus wealth can or will only be used to buy increasing mobility, and that this in turn must lead to a certain kind of settlement pattern. I shall examine this question later.

But let me return to the main theme and the point that I have emphasized earlier. The studies made in the mid-1960s showed quite clearly and unequivocally that communication, transportation, and urbanization are closely linked, and that the future of one factor will affect and be affected by the others. Jean Jacques Servan Schrieber (1967) has summed up this view very well in his book, *Le Defi Americain*. He says:

> urbanization, automation and communication have obviously separate effects, but much of their impact interacts. We may consider them as convergent forces which will forge a new society, which is the accomplishment of the second industrial revolution.

THE PROPENSITY TO CONSUME TRANSPORTATION AND COMMUNICATIONS

There is no doubt that world consumption of communications *and* transportation will continue to rise during the next decades. Even if technological economic and social factors remained constant, the needs of an expanding world population would generate more consumption of transportation and communications. And, of course, technological economic and social factors will *not* remain constant.

I have suggested earlier the notion that there may be some kind of "affluential determinism" at work in society. Much empirical evidence supports the view that more money means greater mobility and a higher use of both transport and communications, especially in the United States. In other words, as the standard of living rises, more money is spent on these items. But several questions arise. How has the utilization of transportation and communication changed in various countries? What factors determine such changes? Can we compare expenditure on transportation and communications with expenditure on housing, which might seem, on the face of things, to be a more serious international problem?

Let us first try to make some comparisons between expenditure on shelter and on transportation and communications for various countries. Using data from both national and international sources, it is possible to show very roughly some of these trends over the past 15 years. Although it is difficult to make exact comparisons among different nations because of the disparity between modes of collection and organization of data, it is just possible to discern some general patterns (Table 1). First, we see that the ratio of expenditure on housing compared to transportation and communications varies widely, according to the state of development of a country, and especially according to the level of personal income. Although these data are too diffuse for hard conclusions to be drawn, we can see that the ratio of family expenditure on housing to expenditure on transportation and communications is con-

Table 1

Town	Census Data	Population (Millions)	Percent of Consumption Expenditure Devoted to		Ratio of Transportation/Communication and Housing
			Housing	Transportation/Communication	
Sweden (large cities)	1958		21.8	15.5	0.7
United States	1960				
Other cities over 5000 population			24.3	16.1	0.68
Central cities			25.6	15.3	0.6
Nairobi		0.197	20.7	11.6	0.58
Toronto	1959	7.63	26.8	14.6	0.54
United Kingdom	1964				
London		11+	27.6	19.6	0.53
Tel Aviv	1963	0.57	20.9	9.6	0.46
Sao Paulo	1961–62	13.4	23.7[a]	9.5[b]	0.4
Rio de Janeiro	1961–62	3.6	21.6[a]	8.2[b]	0.38
Paris	1956	7.0	21.6	7.8	0.36
United Kingdom urban areas	1964		25.6	9.2	0.36
Karachi	1955	1.3	15.6	4.5	0.29
Athens	1951–58	1.5	25.8	6.2	0.24
Buenos Aires	1960	5.75	10.5	2.5	0.24
Abidjan	1956	0.127	24.3	5.7	0.23
Calcutta	1958	7.5	23.6	5.0	0.21

Source: Various Official Household Income and Expenditure Statistics 1950–1964.
[a] Including postal and telephone service.
[b] Transportation only.

Table 2

City Population	Transportation as Percentage of Consumption Expenditure (United States)		
	Automobile	Other Travel	Total
Over 1 million	10.4	2.2	12.6
¼–1 million	12.2	1.8	14.0
30,000–¼ million	13.3	1.4	14.7
All	12.3	1.8	14.1

Source: Clark (1967).

nected both with the size of a city and with the stage of development it has reached. Other factors enter into the equation, such as householder's income level, which we shall deal with later; the low ratio found in Japanese cities also needs further explanation. But in general, cities with ratios of expenditure on transportation and communications below 0.3 compared with housing tend to be in less well-developed nations, the exceptions being Athens and Nairobi. These exceptions tend to be accounted for mainly by differences in percentage expenditure on transportation and communications instead of housing. Also note that within a certain culture, size of city also affects the percentage of consumption expenditure devoted to transportation, as shown in Table 2.

How do people's expenditure patterns change as their income increases? It has been possible to assemble some crude data from various household expenditure studies (see Figure 1). Although direct comparisons are difficult, some interesting points emerge, but before discussing them, remember that these figures refer only to *urban* household expenditures.

Since the definition of "expenditures on housing" here includes all related expenditure such as fuel and light, household equipment, and domestic help, it is natural that these expenditures should rise with income. In the case of the United States, the slope of the curve may be influenced by the greater mobility of the population. As their income rises, Americans may move about more, and they may not place so much value on the home. Turning to expenditures on transportation and communications, we can see quite clearly the differences that occur among various parts of the world, and perhaps in particular between India and Japan and elsewhere. Essentially we can see that in most of these countries expenditure on transportation and communications as a percentage of total expenditure rises steeply with increasing income, with some subsequent leveling off. The proportion of expenditure devoted to housing behaves in a more erratic fashion. Sometimes it rises steeply and then levels off

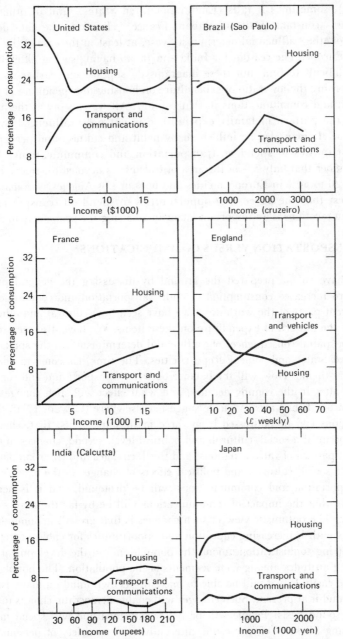

Figure 1   Source: Llewelyn-Davies and Cowan "Science and the City," *Science Journal*
(October 1969).

(Rio), sometimes it falls (England, United States), and in other cases it remains constant (Calcutta, Japan, France). However, the data do tend to support the "affluential determinist" case, at least in the case of the Americas and Europe. The reasons for India and Japan having such a different pattern are difficult to find, but these diagrams do support our earlier observations concerning the low ratio of expenditure on housing to expenditure of transportation and communications in that country. All we can say at the moment is that the pattern of family expenditure is not only influenced by level of income. It may also be influenced by habit and culture, and perhaps by the opportunities to spend on transportation and communications. After all, remember that only 5% of the area of Calcutta is devoted to streets, compared with 20 to 30% in European cities and 66% in Los Angeles! We may perhaps suggest that the *absence of* opportunities to spend on transportation and communications is an important factor in assessing the propensity to consume.

TRANSPORTATION VERSUS COMMUNICATIONS?

We have so far prepared the ground by discussing the general propensity toward increased consumption of both transportation and communications in different parts of the world, and we have seen how cultural and other factors affect the pattern of expenditure on these items. We have also seen something of the spatial consequences of "affluential determinism" in the spread of most United States and some European cities. The time has come to address our central theme. How will the balance between transportation and communications alter in the coming decades, and what effect will such changes have on life in our urban areas? I have suggested elsewhere (Cowan, 1970) that these questions can be addressed from three main standpoints: first, changes that will occur at a social, cultural, and political level; second, changes in the interurban pattern of various nations; and third, changes in intraurban patterns.

The social, cultural, and political effects of changes in the balance between transportation and communications will be profound, and it is here that I believe that the impact of communications will be by far the most significant factor. The optimistic view of such changes is that growth in communications brings with it increasing education and opportunities for choice, together with a growing sophistication among the population and the development of questioning attitudes among wide segments of the population. The population will not be easily led; it will be able to make up its own mind on many issues that confound it at present. Such changes will enable democratic choices to be made on a sound basis. People will be better informed and will spend more time choosing all sorts of interesting alternatives on a variety of personal, social, economic, and political issues.

Pessimistic forecasts take a different view. They suggest that changes in information handling mean invasion of privacy, thought control, and other evils. They point out that technical improvements call for social inventions that are usually directed toward the centralization of information, which makes the individual vulnerable to spying.

We must recognize the overriding importance of these social, cultural, and political effects of the rapid development of communications. The effect will be to spread ideas even more rapidly and to contribute to the blurring of distinctions between "urban" and "nonurban" ways of life. Already most people in Britain are able to know, hear, and see the latest news and current events very soon after they happen. Radio and television networks and the rapid distribution of national newspapers mean that most people in Britain share a common background of information, and this assists in the formation of a common culture.

M ay people regret the passing of urban and rural cultural differences while others say that the advantages in terms of equalizing opportunities and perhaps in reducing the dominance of particular groups through wider diffusion of information far outweigh the disadvantages.

On the whole I believe it sensible to suggest that improved or more communication opportunities provide means for both greater collaboration and greater conflict. The more we get to know about other people, other regions, or other institutions, the more we may grow to like and understand them. On the other hand, understanding someone else does not *always* make it easier to like them. "Communication or conflict" may be the major social issue raised by the ever-increasing flow of information in our culture.

Again, bear in mind the differences in the cultural and political impact of the development of mass communications between developed and developing nations, and between cultures that are politically stable and those that are unstable. For example, the communications industry as such is not labor intensive, and therefore in a developing country cannot be used to create wealth through work. In fact, the general effect of the development of communications and automation is to reduce the number of jobs and/or working hours. On the other hand, the educational consequences of mass communications, together with their cultural effects, are to create aspirations for a different and better life-style. If this occurs too rapidly in developing countries the political consequences can be severe, and the problems of control over communications media in these countries are difficult and delicate. We must, I am afraid, return an open verdict about the effect of mass communications in developing nations during the short term, although the long-term global effects are most likely to be of value in stabilizing world development.

In a strict sense the development of mass communications means that traditional distinctions between "urban" and "nonurban" places and patterns no longer exist. In Webber's (1964) sense, all Britain is urbanized. As I noted in the preceding section, most people in the developed part of the Western world share a common culture, which is highly urbanized. Nevertheless, even in

Britain, some people live in cities and some outside, despite the fact that more and more of the population is migrating toward major urban areas. This situation, in which it is possible to distinguish physically between urban and nonurban spatial patterns, will be with us for several decades to come. It is therefore legitimate to ask what the effect of communications on this pattern will be.

Many major inventions and innovations will have little effect on the overall pattern of settlements in Britain. For example, telephone, television, radio, videophones, and many other technical advances are all essentially "blanket-type" inventions. There is no evidence to show that the telephone, by itself, has had any major influence upon the overall settlement pattern of Britain, and yet it has been in use since the late nineteenth century. On the other hand, the telephone and telegraph assisted greatly in the exploitation of other aspects of the late industrial revolution. Seen in this light, the development of new means of communications may have a marginal but significant effect on the growth rates of settlements in different locations. But such innovations will have a much greater effect on the intraurban pattern, which I shall discuss later. For the present we can simply suggest that the effect of these innovations on the overall pattern of urbanization in Britain will be small.

Innovations in transportation may have considerable effect on settlement patterns in Britain and other countries. This is not the place for a detailed list of all the possible developments in transportation that are just around the corner, so I will merely try to sketch in some of their possible effects on the interurban pattern.

High-speed, rapid-rail transport, automated motorway guidance systems, and VTOL and STOL aircraft will all affect the interurban pattern. Most of these developments will speed up movement *between* places, and it is unlikely to be worth stopping on route, except at a major center. On a wider scale, changes in transport technology may have considerable international effects, especially in the tertiary sectors. In those levels of industry where face-to-face meetings remain important, changes in transportation technologies may have a striking effect on the location of meeting places, and this in turn will be affected by political changes in the years to come.

As the affairs of the world become even more interlinked, the location of these meeting places may change. The importance of some cities will decline while others will move up the scale. With the development of South America, Japan, Africa, and Southeast Asia, their capital cities may become primarily nodes in the world network of transactions. Such global changes may be complemented by the growth of new city centers at airports, hoverports, or other interchanges. It is at this level that the major effects of transportation *and* communications on patterns of urbanization will be felt. And it is here that we need once again to consider the substitutability question. So far all the evidence points one way—as people use more and more electronic communication, they use more transportation. As Hall (1966) says:

Indeed if a researcher were to make an international historic comparison extending over the last hundred years, he would probably find that the two curves (of transportation and communication) march very closely together, and appear to be related to the general level of economic development. The psychology of person-to-person communications is very complex, and it is probably safest to conclude that as advanced western economies become more sophisticated, they seemingly multiply the opportunities for rapid and economic inter communication of all sorts.

From our own work we know that telecommunication is cheaper than rail travel in Britain for distances of over 150 miles. Of course, such estimates are very sensitive to the value of time, and to some extent ignore the *effectiveness* of telecommunications as compared with face-to-face contact. However, they do indicate that, despite the current evidence, if telecommunications can be made cheaper there may be some quite positive economic advantages over travel and transportation at the intermetropolitan scale.

The effects of future changes in communication patterns at the intraurban level will be considerable. Nearly all the forecast changes in communications have to do with patterns of work, and since cities are closely bound up with work patterns, their form must be affected by these new innovations. It seems to me that the changes and innovations that are most affected by intraurban patterns are those that do not affect the scale of interurban distribution. The telephone and its associated inventions and other inventions such as automation and computers will have a major impact on urban form because of their association with particular kinds of jobs that will become more and more important in the urban society of the future.

We can already see the particular impact of automation and computers on manufacturing industries. These inventions are now affecting service industries. The impact of computers on office work will continue to grow. This is most important, because it is the office sector that seems likely to be the major focus of urban work in the future.

As with transportation at the interurban scale, these inventions will speed up messages *between* activities. *Within* activities, messages are, at present, slowed down and processed by hand. The telephone, telegraph, and videophone enable contacts to be established quickly, and orders and messages can be flashed from one office to another with great speed. But once they arrive in a particular organization, messages and orders are dealt with one at a time by manual labor. The object of the office automation movement is to speed up some of these hand-processing phases. The movement seems likely to gain ground, although there may be a time when the retention of manual work in offices may be necessary to occupy a certain amount of surplus labor. But, of course, the final decisions, which are the key part of most offices, must be taken by hand, and it is this particular decision function that may govern the future pattern of urbanization under the impact of communication innovations.

These changes are affecting two distinct sections of the office functions. First, the decision makers, the leaders of industry, finance, communications, and so on—what has been called the "elite group"—are finding it more and more necessary to locate in the center of the city. These people's businesses and, indeed, survival, depend on rapid communications; they flourish on the newest ideas and must react instantly in order to maintain their place in the system. These people, the elite, also must be near, or have immediate access to their own information, which is contained in the more routine parts of the office organization. The second trend is for those more routine parts of the office function to become more footloose: able to locate in the suburbs, near to the female labor force, and away from the high rents and access difficulties of the center.

The office function splits, and the two elements have different location and space needs. Once again the importance of communications becomes clear. Elite groups locate in the center where they can have immediate face-to-face contact with a wide variety of people. Routine functions may be suburbanized at the end of an airline, a telephone wire, or a television circuit; the finer nuances of communication are not necessary for the kind of messages that flow between executives and the routine office plant.

For the next few decades this pattern is by far the most probable but are there any longer-term prophecies we might make? Or are there details of this general picture that we might refine?

The answers to such questions will depend once again on how far we think electronic communications can become an effective substitute for face-to-face meetings in the future metropolis. There is a considerable divergence of views on this issue. At one end of the scale are those who believe that a total substitution is possible, and that in the future we shall all live entirely separate lives physically, although bound together by electronic links of all kinds. On the other hand, there are those who think that very little advance can be made on our present systems of communication, and that the investment in existing cities is so great that they cannot be changed much, even over a very long period. Most observers fall somewhere between these two positions. Developments in communications technology are selective in their effects. Although some more routine types of transaction such as the instructions necessary to pass orders from an executive to his staff or from an office to the factory floor may be carried out by remote control, others, especially those requiring a sensitive awareness of a multitude of surrounding circumstances, will still require direct contact among individuals. Certainly there seem to be no economic grounds for choosing telecommunications as a substitute for face-to-face contact at the intrametropolitan scale. The personal contact is often cheaper for the extended meeting. This is true of conventional telephone calls. For more expensive and elaborate systems such as videophone, the cost factor could be a major stumbling block for many firms.

CONCLUSIONS

The general conclusion must be that communications and transportation are mutually reinforcing. Although it is true that there has been a decline in mass transit within Western cities, the actual number of movements of the population seems to be on the increase. As cities decentralize at a metropolitan scale, they reconcentrate on a regional and subregional level. So far as one can judge, the effect of communications may reinforce or result in an increase in the number of journeys that people take. What is more, telecommunications in part are competitive with transport for distances over about 150 miles, but within this distance the difference in cost, taking time into account, is not great, and for long meetings involving a number of people, telecommunications is more expensive than face-to-face contact. The advantage of communications over travel is, of course, further reduced if more elaborate and expensive systems such as the use of videophones is adopted. Thus, in the West, at least, we must consider very carefully the validity of the assessment that communications will be capable of substituting for travel on any large scale. The essential point to bear in mind is that communications is in fact not a substitute for travel but simply a different medium of interaction. Clearly, if communications is regarded as a substitute for face-to-face contact, it will always be inferior by definition. If we are to move ahead in assessing the impact of communications on personal travel in our metropolitan areas we must know a great deal more about its specific advantages as a medium in its own right. This could include, among other things, a privacy that is not possible in face-to-face contact. But as Reid (1971) has pointed out, telecommunication combines in one medium the advantages of mail and face-to-face communication. Like mail, it is indifferent to distance, thus providing an enormous geographical range of potential contact. But many activities that require the rapid exchange of small packets of information cannot be conducted through a system that inserts such long delays between transmission and reception. Face-to-face contact, on the other hand, requires all the participants to be in the same place at the same time, a constraint that restricts the variety and flexibility of an individual's communication network. The individual may make some limited extensions to this network by traveling, but only at the cost of considerable time and effort. The ability of telecommunications, therefore, to transmit small packets of information rapidly to a variety of destinations represents a unique capability, which must not be regarded simply as some kind of inferior substitute for face-to-face contact.

The research being carried out by our Communications Studies Group is striving to learn more about these qualities of telecommunications, because we believe that it is here that we must bring a major part of our efforts to bear if we are to understand fully how people will use urban areas in the postindustrial society.

REFERENCES

Clark, C. *Population Growth and Land Use*. London: Macmillan, 1967.

Cowan, P. (ed.) *Developing Patterns of Urbanization*. Glasgow: Oliver and Boyd, 1970.

Hall, P. *The World Cities*. London: Weidenfield and Nicholson, 1966.

Pierce, John R. "Communications," *Science Journal*, October 1967.

Reid, A. "Needs Technology Effectiveness and Impact," (mimeo) Communications Studies Group, Joint Unit for Planning Research, London, 1971.

Servan-Schreiber, J. J. "*Le Defi Americain*." Paris: Editions Denoel, 1967.

CHAPTER 23

Urban Ecostructures in a Cybernetic Age: Responses to Communications Stress

RICHARD L. MEIER

What follows is a progress report on an environmental process that is ordinarily regarded as healthy—the evolution of more stable communities and social systems, better able to adapt to changing circumstances, through increases in the interaction and integration of living populations. However, this discussion will concentrate on the excess of interaction and the responses people and their organizations make to it to reduce the resulting stress. Expansion of the communications flow required for such interaction and integration constitutes an ill-defined threat to important roles and substructures in urban society. Recent experience contributes to the improvement of the diagnosis as well as to the treatment.

The metropolis must be regarded as a giant feedback-mixer for society. It produces amplification and reverberation for certain attention-grabbing signals and messages; it supports a myriad of homeostasis-maintaining resistances for others that might otherwise damage the social fabric. Messages and signals so exchanged are interpreted politically in some specific locations, financially in some others, and culturally in quite a few places, but in most parts of the city the meaning transmitted has all of these components combined with others. The overall level of interaction seems to be advancing several percent per year per capita, so that the average doubles approximately with each generation growing up in the city. The speedup is primarily found in machine-interposed communication instead of in face-to-face interaction (Meier, 1956).

Buildup in communications flow is certainly far from uniform. There are hot spots in the city where the bombardment with messages can sometimes reach a volume that is intolerable. Inevitably these stressful occasions produce "burned out cases," which may be individuals, groups, teams, agencies, or organizations. For example, Toffler's *Future Shock* deals almost entirely with features of communications stress. In this chapter concerning the threat of communications overload, I wish to concentrate on the who, where, why, and when of the phenomenon, along with the best proposals developed to date for coping with the pressure.

WHO ARE THE HURT?

Reflect on the school board member's predicament when controversy flares. He puts in a standard day of work, often doing more work than a business or professional man, but must deal with a dozen or more emotional, conspiratorial, or strategic telephone calls per day in addition. These begin to arrive when he is still at work and accelerate as soon as he is due to be at home. This is an addition to the normal 10 to 50 routine calls he expects to handle when at work. Most people are unthinking in their harassment of the pivotal figures and are strongly interested in seeing that their own interests prevail.

Messages arriving through other media are even more numerous: the daily mail, the fliers, which may be delivered by hand or stuffed into the mail box, and brown envelopes containing mimeographed proposals appearing just prior to meetings. At meetings there will be printed giveaways, usually serving as background material. Certain parts of the newspaper must be watched, especially the local editorials, letters to the editor, and the women's page. More and more programs on local television and FM radio must be monitored. Even lapel buttons, bumper stickers, and posters in storefronts should be included on this list.

In addition, huge blocks of time must be allocated to meetings—closed, open, formal, informal, ad hoc, rump, caucus, confrontation, and others where one must be continuously alert to one's appearance as well as one's words because signs of weakness may be exploited.

A school board member is an unpaid volunteer, but the prominence of his position in the community operates to keep him from resigning office. The involvement is too great. For all but a few, however, one term on the board is enough. Public life nowadays often becomes too demanding. One commonly sees this kind of burned-out case in the weekend lodges and other retreats from suburbia.

Consider also the editor who, each year, sees his in-basket piled higher and the number of apparently significant community organizations increase in number, while minutes of reading time and columns available for transmission of information remain approximately constant. This editor is forced to make more decisions per unit time in a communications-saturated context than any of his predecessors, and (what is more important to him) many more judgments per hour than he did before. Presently he must read 5 to 12 times as much as he is allowed to set into type, while the telephone tends to ring with increasing frequency. Almost anyone filling such a focal position becomes irritable, "high strung" from the point of view of people whose work he directs, and eventually develops peptic ulcers or high blood pressure with indications of atherosclerosis. Editors must also find socially accepted means of escape from communications stress. Many trade in the job for a public relations post (an extreme specialism within journalism) after they have been pressed to the limit of endurance.

Every scientist is working in a profession that is pure communications. He usually becomes part of a sociocultural enterprise called a research team that is engaged in decoding the signals presented by nature to the human race. His fellow scientists are publishing 3 to 5% more each year concerning new methods, equipment, and concepts, not to mention the findings, all of which seem to be relevant in some way. At some point in the race to absorb and produce information, the scientist must give up. As he cannot compete with the newest generation of investigators, by concentrating on a very limited problem he develops blinders that keep the world relatively simple, or he takes advantage of his hard-won information about organizational contexts and becomes an administrator.

A favorite escape for scientists is the wilderness, preferably mountainous. A remarkably high proportion of the visitors encountered in the wild high country are scientists getting away from the tensions associated with fast-paced knowledge acquisition and production.

WHERE IS THE ACTION?

The people filling such roles are concentrated in the most highly organized portions of the metropolis for a good share of their active life. The stage has been set in these locales by the instrumentation of the capacity to communicate by careful programming of interrelated activities, by competitive pressure exerted by one group on another, and by an elaborate reward-punishment system for achievements and errors.

The most well-known locale is the central business district (CBD) with its business headquarters, government offices, department stores, commodity and security markets, banks and theaters. However, most CBD's in North America are losing their momentum to other foci in the city as many activities move out of downtown locations and are replaced by slower-moving functions (e.g., parking, warehousing, auto service, etc.). Within the CBD the current hot spots of communications stress are sites of change, such as newly occupied office buildings, suites with a newly obtained time-shared computer terminal, expanding radio-television studios, and sometimes the convention center.

Airports have recently become the sites of the most frenetic activity. Growth in transaction rates (purchases, telephone calls, meetings, exposures to advertisements, etc.) regularly exceeded 10%/year in terminal areas and its feeders until some time in 1969. For two years airports experienced a traffic plateau, but by Christmas 1971, the original trend seems to have been resumed. Metropolitan airports have gained hotels, small convention centers, recreation areas, technical education, and much branch office space. They are now adding department stores, comprehensive health services, and ultramodern light industry based on air-freighted components; they are becoming multishift, round-the-clock organizations capable of smoothing out the peaks in both output and input. Nevertheless, a combination of weather and a high demand for travel can result in delays in the flight schedule that can overload the control tower team and challenge the capacity of a number of other services.

The airport has become more a true metropolitan center, promoting tourism along with intercity collaboration for building and maintaining modern organizations, than the regional shopping centers that specialize in retail distribution. Although airports have been engaged in continuous reconstruction and expansion on site to meet the concentrated demand, regional centers have colonized thinly occupied space. The growing transactional load at the metropolitan periphery has been spread almost evenly among them with the aid of freeways.

In all but a few North American metropolises, cultural centers have moved to the vicinity of the major universities. Downtown drama, symphony, and opera tend to be imitative and ritualistic, while innovations, experiments, and revivals on and near the campuses have vitality and tremendous variety. Thus, many recognized but not yet famous artists and critics are under heavy pressure in this environment. The rise in cultural activity is spurred on by the growth of science in the university milieu. In a number of instances, campus researchers have spun off private enterprise that settled down in the neighborhood. As a result, a large, appreciative audience has been created. Many students, particularly the graduate and professional categories, are given assignments that press them to the limits of their information-processing capabilities; most faculty members and administrators are competing for prestige and status, and the games they play require sophistication in the communications techniques. Benchmark data of a decade and two decades ago show a very significant speedup in books read, images that can be identified, visual impressions absorbed, computations completed, concepts mastered, and new knowledge produced by the individual participants in the university society. Many work huge amounts of overtime because of the challenge presented by their situations.

Similar in many ways to the university community and sometimes adjunct to it is the medical center. Most such centers have multiplied their floor space many times and have been forced to take the initiative in planning their neighborhoods. Doctors also work overtime, as do many nurses, technicians, and other participants in the medical community. Medical centers have had to cope with electronic instrumentation, information-generating machines, and the knowledge explosion. Their reactions to harassment by relevant data are far more inhibited than the reactions around the periphery of a university. Plans for medicine suggest some decentralization of personal services for patients, but further centralization of records and crucial operations, so that the operation of a medical center will increasingly resemble that of a private corporation headquarters or the offices of government agencies.

In general, communications stress appears where the intellectual challenges are raised, big problems need to be solved, and new enterprises start from small beginnings to acquire substantial proportions in a short period of time.

WHY IS STRESS INCREASING?

The sources of the pressure to interact can be described from many overlapping viewpoints, each of them able to capture an added part of the truth. It is possible to offer at this time a half dozen explanations, but all of these together do not explain the phenomenon.

Economic growth constitutes an inexorable force for growth in communications. The categories of inputs into the production process—labor, capital, natural resources, knowledge, and organization—have exhibited either constant or rising marginal costs over time except for use of one major natural resource and the factor of organization. Exploitation of the electromagnetic spectrum for telecommunications has yielded rapidly declining marginal costs for transmitting information for the last half century. Moreover, the new technology still in the laboratories, but on the way to wide-scale application, promises to extend this trend for another 10 to 20 years. Therefore, every firm and well-managed government agency finds it profitable to innovate by discovering methods of getting the job done with less labor, capital, energy, materials, water, and land, but more knowledge, communications, data processing, and planning. When the initiators of an innovation succeed, their competitors must imitate or improve on them and their suppliers must adapt to their needs. The laggards are steadily squeezed out of the scene. Thus it is not surprising that the growth of machine-interposed communication must exceed that of the true rate of growth of GNP. Empirical studies all over the world show more correlation between economic growth and communications than for any other single factor.

Students of social organization note that many large human organizations are growing still larger, and at the same time many more organizations are being spawned. The largest complex organizations are taking on new activities yet do not seem threatened with dissolution. A relatively satisfying explanation, as elaborated by Ben Bagdikian in his book *The Information Machines* (1971), is that the response time separating a central event and appraisal of it by participants in the organization has been continuously reduced over the past generation. At present more than 90% of the members living in urban environments anywhere in the world can be apprised of such an event within an hour. With this immediacy to the important news, together with the opportunity to respond by telephone, mail, or casting a vote, people can develop a sense of belonging to a large organization. This attachment is far more adaptive than the earlier use of banners, images, uniforms, rituals, slogans, and authoritative pronouncements for bringing about concerted action. Moreover, it is possible for people living in a modern city to maintain multiple memberships in a variety of independent social organizations. Analysis of this behavior shows that it is accomplished primarily through the communications media. Thus, increased participation means receiving more messages from a greater variety of sources.

Another strong trend is the drive for self-improvement by acquiring formal education. In cities, the largest increases are in higher education and constitute a series of exercises aimed at expanding the repertory of concepts of the student while enhancing his ability to generate precise reports. The accumulating investment in education increases the range of the receivers of communications and expands the number of senders practiced in the arts of transmission. It is believed that well over half of the urban population of the United States now less than 21 years of age will have done at least some college level study before the year 2000, as compared to about 20 to 30% among those presently over 21.

There is a natural drive to make friends, which means finding people with compatible personalities and mutual interests. Cities provide a rich environment for choosing friends; social networks are built up to much larger dimensions when it is so easy to arrange face-to-face meetings. As one gets older and the effort expended in making new contacts diminishes, existing networks are maintained by means of telephone and correspondence. Thus, part of the growth in communications is explained by the past successes in building up circles of friends.

Underlying the need for friends we may note the present mechanisms for forming bonds among humans. They are almost always initiated by face-to-face contact, allowing both parties to judge the extent to which they can trust the other. The second contact usually needs to be arranged, so that several messages must be sent. On that occasion, mutual interests are revealed and generate opportunities that one calls to the attention of the other—which means more messages transmitted in the interval between face-to-face contacts. The messages serve as substitutes for trips undertaken to "visit" or "meet someone." Each trip *not* taken requires about 10 communications to maintain the strength of the bond of friendship.

Finally, there remains a technological explanation for the increase in communication. Many organizations undertake a systematic search for ways of gaining the attention of individuals as a means of selling products and services. Carefully constructed computer programs exist for comparing marginal costs with returns so that messages continue to be launched until it no longer pays to do so. Equipment for engaging attention is improving its capability (already the urban adult is bombarded with more than 300 exposures/day on television, mostly in color, perhaps 100 more of similar type in the newspaper, and receives several pieces of mail advertising) and computer software is fulfilling its objective with even greater efficacy.

WHEN WILL THE LIMIT BE REACHED?

The answer to this question is clear—the end is not in sight. I concluded in my study, *A Communications Theory for Urban Growth* (1962), that when humans became tired of receiving routine messages they would have machines do it. Ways will be devised for making automatic decisions. Meters and instruments originally designed to be read by people, who would then adjust equipment so as to approach the optimum, will report to automata that make simple decisions and keep records. Exchanges among automata already use a larger share of the intermetropolitan telecommunications capacity than humans do for long-distance calls. Perhaps some time in the 1980s a number of metropolitan areas will have advanced to the point where the amount of interaction among the local automata will surpass that of human residents. Growth in the cybernated sector can continue for many decades without encountering diminishing returns, as no limiting resource constraints exist at this scale of information exchange.

People will be relieved of the information processing tasks they most dislike. A very large share of accounting and bookkeeping would be reduced to electronic data processing. The credit card phenomenon makes this development easily understandable. The chore of "drilling" students in the elementary mental manipulations of numbers, characters, and symbols will be eliminated. More difficult, but more and more common, will be the translation of symbols from one medium to another (e.g., speech or music converted into print without the aid of a stenographer-typist or a key puncher). Many of the easy optimization tasks normally undertaken by professionals and middle managers (such as inventory control, programming of organizational activities, and personality assessment) are also likely to be taken over by automata.

When drudgery in information processing and communications is relegated to the machines and the networks of channels, stress on urban residents is expected to increase. At present, boredom ("stimulus deprivation" is the term used by behavioral scientists) serves as a kind of dilutent; routine tasks act as a breather that separates the difficult choices. In the future the challenge will be more continuous as an increasing proportion of messages raises complicated issues. Responsible members of corporate organizations, the professions, and the community will have to remain always vigilant, as mistakes will have greater consequence. Worry about such mistakes, compounded over time, seems to set the conditions for stress-induced diseases (*WHO*, 1971).

STRESS RELEASES FOR CITIES OF THE FUTURE

The novel feature in the picture of the metropolis of the future, a result of the continuing growth in communications, is the elaborate interlocking web of fatigue-resisting sensors and automata. The growing community of automata will serve and be served by the vast majority of urban residents of all ages. People at the interface will be vulnerable to sudden spurts in the flow of significant communications initiated by other people needing help.

The first response to a flurry of transactions is to become more efficient, organizing effort to minimize the costs to others. Queues can be set up, their priority ascertained so that the most important messages are processed first, the lowest priorities can be scrapped, and the standardized operations can be contracted out. Time and attention previously assigned to long-range issues and to planning may be drawn on to get over immediate crises. Branches or out-stations may be set up closer to the origins of the incoming messages so that the bulk of the transactions can be intercepted and shortcuts applied. Only more complicated ones gain serious attention centrally. Equipment may be sought to increase capacity for response to requests for service. Waste may be reduced by carefully defining standards of service to be performed. This challenge is one that virtually all educated persons would meet with some exhilaration, since it constitutes a test of the capabilities of one's group. People like to be noticed, and an organization rises to the occasions when their services come into demand.

The shock comes when all this effort and good will seems fruitless and standards of performance need to be reduced. When such situations are imminent, people try to escape. A vacation is one solution, delisting the telephone is another, and getting sick is quite common. The privacy shield is then raised much higher, so public life is prevented from spilling over into private life and completely usurping the time allocated to it.

Planners and designers must become much more sophisticated about the semipermeable filters that make up the concept of privacy. Some of this cleverness should show up at the interface with the telephone, and more of it at the interface with the console. Organizations aiming to penetrate the privacy barrier should be taxed, preferably according to a schedule that is roughly proportional to the amount of attention they waste. Repetitious messages are one important source of overload. Redundant messages probably cannot be stopped by taxation, so decision centers in organizations need to screen them out by various devices, some of them mechanical. In the positive sum game played between sender and receiver, we must restore balance by providing the receiver with a few more defenses at the risk of losing the obtained joint benefits.

Individuals must be allowed to escape by disconnecting from the wired-up city that contains most of the records and expedites the huge volume of transactions. As much as a quarter of the population may reject life styles based on an ever-accelerating volume of communications. These people, who strongly prefer a slower pace, will be willing to accept a much lower real income. They will be concentrated in such activities as the fine arts, personal services, growing perishable foods, part-time construction, and salvage. The "burned-out" cases will also have this supportive milieu to drop into, should they be forced to retire from most of their ties to the system.

Those who insist on living under low communications pressure will most often choose the aging inner city communities for residence and employment. They will also take over the dying resorts, villages, and farms, once out of daily contact with the metropolis but now being linked to the freeway network and serviced by feeder airlines. Most of the time these people will live simply, consuming much less than suburbanites. In emergencies, however, they will rejoin the system long enough to use it to advantage. In outlook, they will remain urbanites and could resume an active role at any time their desires change.

Planners and designers can lessen the damage done by communications growth by thinking more carefully about the connections of any role or group function with the rest of the world. Average rates of interaction are not enough for designing niches in the future urban system, but the economic responses to peaks must be thought through very carefully and a number of defenses provided in advance. Repeatedly, the best solution will expedite the reduction of parts of the task to programs so that they can be carried out by automata connected to the telecommunications networks. Stricter standards of privacy will become one of the best sources of protection at the interface. In this manner people can be saved from overcommitting themselves.

At the same time, refuges of the society must be preserved so the overstressed will have places of escape. Although consuming much less than the average, many people living in these protected environments will still need to be supported by the larger system, so these plans should be considered administrative arrangements instead of designs for places and new roles. A variety of therapeutic communities and open spaces can provide safety valves for a cybernetic age.

At this stage, some may ask "Why continue to research, develop, organize, urbanize and threaten with overload? Why not stop now?" Such questions forget that the increasing scarcity of natural resources must be counterpoised by new knowledge and its large-scale application. Without it, given man's present trajectory, the Malthusian catastrophe becomes inevitable. Those casualties would be real, painful, and large scale, while communications stress wounds only the will among those who are normally born affluent. There is no way yet for the metropolis to turn back.

REFERENCES

Meier, Richard L. "Communications and Social Change," *Behavioral Science 1* (1956), 43–58.

——— *A Communications Theory for Urban Growth*. Cambridge: MIT Press, 1962.

"Society, Stress and Disease," *WHO Chronicle, 25*, 4 (April 1971) 168–177.

PART V

Global Communications: Cultural Explosion or Invasion?

CONTENTS

INTRODUCTION

HERBERT I. SCHILLER

The developmental process has been a central international concern since the end of World War II. Yet in both industrialized and nonindustrialized countries, development has been measured by what happened on the productive side of society's activity. Commodity and manufacturing indices, workers' productivity, and exports have been scrutinized closely to determine national progress or the lack thereof. This has been the justification, too, for foreign assistance. Significant amounts of resources have been transferred to nations with the ostensible aim of stimulating or assisting economic development. Besides, many states have formulated developmental plans intended to generate a steady economic growth rate.

Nevertheless, there is a sense of disappointment and frustration with the entire developmental record. The continuing and even widening gaps between the privileged few and the many poor states, the dismal character of so much of the development that has occurred in both the advanced and less-advanced nations, and the enormous unmet tasks that still lie ahead have forced some basic reconsiderations.

One of the few clear results of this continuing reevaluation is a revised ranking of developmental factors. If the economic component remains preeminent, it is no longer exclusive. The cultural-communications side of the developmental process has emerged as a vital element. The question *Development for What?* has assumed a new urgency. With it, not surprisingly, comes the recognition that peoples' values and beliefs, their attitudes and opinions, and the social imagery in general that circulate in a community are significant determinants of that society's present and future way of life.

In brief, men and women's general performance are affected by their informational-cultural climate. And, more than the work effort, important as this is, is involved. Nothing less than the realization of community goals is dependent ultimately on the mind-set of the people. It is to the forces that influence this mind-set that attention has begun to move.

In nations whose political processes have evolved over a considerable period of time, social norms have been internalized. This means that the cultural forms that are operative do not have to be explicit in their support of the ongoing features of the society. By merely *not* emphasizing contradictions, irrationalities, inequities, and fundamental disturbances (assuming they exist to some degree), the communications-informational apparatus provides reinforcement for the existing order.

The situation is quite different in many of the newly independent, ex-colonial nations. In these states many of the social forms are not so long lived; some in fact have been changed or at least disturbed in the recent independence struggles. They may be open, therefore, to a variety of influences while the state's institutional character is not yet entirely hardened.

It is in this international environment of uneven economic development among countries underscored with the relatively recent emergence of scores of new nations that the impact of instantaneous global communication must be analyzed. For in this era the role of communications may be central to the character of the social change that does occur or to the condition of social inertia that persists. It is certainly not without significance that there are already 250 million television sets in use around the world and that global audiences in the hundreds of millions view (at least some) common programs. It would seem essential that the sources and inspiration of the messages already circulating globally be explored and evaluated.

Accordingly, Part V considers some of the problems and issues created for the developmental process in various parts of the world by the *existing* flow of international communications, and reviews some alternate models for relating communications to national needs.

In his chapter, Schiller emphasizes the imbalance of the present worldwide communications flow, citing the preponderant role a few industrialized nations (especially the United States) play in the generation and transmission of information and messages. These flows penetrate developed and undeveloped societies alike, although, of course, with differing impacts. What happens, Schiller asks, when the communications technologies and messages of the United States and a few other industrialized nations dominate the cultural space of much of the world? How appropriate to the needs of the new, or small, or poor nations are these messages and the communications technology that accompany them?

Frutkin, writing about the development of communications satellites, details the possibilities of intercontinental message delivery and suggests many benefits that less-developed nations may derive from this new technology including greatly reduced investment costs for domestic communications systems.

Katz believes that the gravitational pull of a powerful society's communications techniques and materials is irresistible to a small country. He sees Israel, for example, pushed relentlessly toward the Western model of nonstop broadcasting, the quest for the mass audience, and "up-to-the-minute" news—all the special characteristics of the American broadcasting system.

Not everyone is so concerned with the source and intensity of the present international communication flow. Some feel that the structure of the national state (that is, how it is organized and who exercises control), ultimately is the determining factor in the quality and content of a nation's cultural-communications condition.

Nordenstreng and Varis state flatly that "the crucial boundaries in the world do not occur between nations but within them," and they emphasize the class role of the mass media. They suggest that attention be focused on the international flow of communication "in terms of the social classes that on the one hand control its production and on the other constitute its destination ('consciousness consumption')."

Nedzynski, too, emphasizes the nonhomogeneous national state in which irreconcilable class interests exist and "media are an expression of the system of dominance and a means of reinforcing it." He notes the systematic exclusion of the working man and his interests in the Western mass communication system.

If the organization of power *within* the nation is the real determinant of that state's communications climate, how fares a country that is in the midst of an internal struggle for social transformation? Mattelart considers the situation in Chile. He claims that the cultural forms of a capitalist society cannot be used as the vehicles for realizing a social transformation no matter how cleverly they are adapted. He sees objectivity, pluralism, and freedom of expression as the instrumentalities for preserving the ideological (and material) status quo. The entire system of capitalistic message making and transmission, in Mattelart's opinion, must be uprooted. Otherwise, it will continue to disseminate, explicitly or implicitly, the messages of bourgeois culture. These are encapsulated in genres, styles, and even in the organization of the media system. Needless to add, imported media material from the world centers of commercialism can only strengthen the internal resistance to domestic transformation.

Dallas Smythe, reporting on his research in the People's Republic of China, found that his experiences in that country suggested a broader definition of communications. It is helpful, he believes "to regard a whole social system as a complex of message systems each of which plays a significant role in perpetuating or restoring an old social system, or innovating a new social system." Smythe describes the efforts of the Chinese to influence people according to this comprehensive view of communications. He notes that the cultural revolution served "to enforce a process of critical scrutiny of past and imported art forms" and he concludes that China has actually benefitted from "its isolation from Western culture and technology since 1949."

In sum, if a nation and its people are not entirely free to choose their developmental course and their cultural condition, neither are they totally defenseless. The technicoeconomic power and penetrative force of the already developed states are great and the international extension of their influence is considerable. The extent to which cultural sovereignty may be safeguarded at this time is still an open question. Still, there seems to be evidence that the goals people choose *consciously* and are willing to defend are not entirely at the mercy of external pressure, no matter how aggressive and forceful it is. This, in any event, is the critical issue of contemporary international communications.

CHAPTER 24

Space Communications and the Developing Countries

ARNOLD W. FRUTKIN

Space communications may well rival the green revolution in impact on the developing world. A decade ago, Latin American and African states were entirely dependent on the European and American metropoles for their external communications, and enjoyed few regional links. Limited and uncertain high-frequency circuits were the rule. A tedious and enormously expensive quarter century for the development of conventional communications systems stretched ahead of them before they could expect to achieve modern communications links.

Today, regional and intercontinental communications ties are ready at hand, traffic volume is soaring, direct regional interconnections have been established, and modest revenues are coming in.

The central factor in the communications revolution is, of course, the space satellite. Its distinguishing characteristics are its multipoint connections, enormous circuit capacity, and low cost per circuit when compared with the cable, and its quality and reliability when compared with high-frequency radio, its cost insensitivity to distance, and its unique capability for international television service.

Two developments spell out the present achievement and the future promise of space communications. The first is INTELSAT, the international commercial communications satellite consortium. The second, looking to the future, is ATS–F, a NASA experimental satellite scheduled to support a community TV broadcast experiment in India in 1974 to 1975. Together, INTELSAT and ATS–F constitute a revolution in communications for the developing world. Besides vastly improved telecommunications services, they offer new opportunities and stimulation economically, politically, socially, and technically.

At the same time, both developments are certain to swell the tide of rising expectations, sharpen popular perceptions of disparities in the quality of life,

369

and subject the governments of less developed countries (LDCs) to increasing pressure for accelerated measures to redress deficiencies.

INTELSAT

The United States pioneered in communications satellite experimentation in the early 1960s and, by the middle of the decade, established a national mechanism, the ComSat Corporation, to organize and exploit the new capability on an international level. As a global enterprise in commercial communications, INTELSAT was brought into being in 1964 on the basis of (a) an intergovernmental agreement open to all nations that are members of the International Telecommunications Union, and (b) a companion agreement between operating communications agencies of the signatory governments. Today, INTELSAT has reconstituted itself in accordance with a definitive agreement concluded in 1971.

Organizational Considerations

The first point of interest for the developing world in INTELSAT is the simple fact that some 40 developing countries joined with the advanced nations as partners in an unprecedented international commercial enterprise. Nowhere else can we find LDCs introduced as members of a multinational activity having a substantial capital subscription, advanced technical operations, immediate social, political and economic impact, and growing revenues.

The relative role of the LDCs in INTELSAT can be quickly sketched. They represent somewhat more than half the 80 member states, they subscribe roughly a tenth of the capital, they hold a similar proportion of the voting strength, but they account for nearly double that percentage of the total use of the system (which simply reflects the growth of LDC traffic beyond the levels that were estimated to obtain when the organization began). The LDCs have no role in developing or producing the hardware end of the space system, but a number operate ground stations.

Politically, a committee of the larger investors directed INTELSAT affairs under the interim agreement. This Interim Committee had 18 members, representing operating agencies or groups of national operating agencies with 1 1/2% or more of the total investment quota. Four countries (the United States, the United Kingdom, Germany, and France) controlled nearly 70% of the voting power. Among the developing countries, Brazil, for example, was a member in her own right, but India was represented along with a number of other countries through a pooling of quotas to qualify a single spokesman.

Together, the 18 members represented some 48 states, obviously including a fair number of LDCs.

Under the definitive arrangement, INTELSAT's political representation will be broadened in a number of ways. The governing board, which succeeds the Interim Committee, is expected to have about 25 members who will continue to cast weighted votes. There will be a general "meeting of signatories," which will essentially represent the operating agencies. A new body, the Assembly of Parties, will represent all member states on the basis of one vote each. The Assembly will concern itself with matters of governmental interest, general policy on the long-term objectives of INTELSAT, measures to avoid conflict between INTELSAT and other multilateral conventions, amendments to the intergovernmental agreement, the expression of *views* on amendments to the operating agreement, the authorization of the use of INTELSAT space facilities for specialized telecommunications services, the making of *recommendations* concerning the establishment of space facilities separate from INTELSAT, and so forth. The Assembly will meet biennially. Questions of substance will require a two-thirds vote. Thus, the definitive INTELSAT agreement affords some new scope to the generality of membership, and therefore to the LDCs. Control, however, rests solidly with the advanced nations.

The participation of LDCs in an organization of such substantive and precedential importance should contribute in some degree to their experience in the world of affairs, to the development of common interests with the advanced North, and to the interest and participation of LDCs in other international activities of substance.

Communications Considerations

For the LDCs, INTELSAT has opened up a large new window on the world. Its key contributions have been direct and independent linkage and reduced costs. Space communications systems have already relieved the LDCs of the need to go through French, British, and United States metropoles to communicate with neighboring LDC regions.

Participation and Growth

A substantial base for LDC use of the INTELSAT system has already been established. Twenty-six of the LDCs have built ground stations for communication through INTELSAT spacecraft, and 19 more are expected to have ground stations in operation by 1974. Other LDCs are now and will continue to be linked to such ground stations in neighboring countries.

On this base, a dramatic growth of telecommunications usage by the LDCs has occurred. ComSat's report for 1970 highlighted an "especially notable" increase in the previous 3 years of earth stations in Latin America, Africa, the Far East, and the Middle East. In the first year of operation of a ground station in Argentina, the volume of telephone and message traffic to the United States doubled (from September 1969 to September 1970, from 200 a day to 400 a day). In Brazil's first year, communications with the United States also doubled, then increased 50% (an equivalent amount again) the second year. Kenya was using a total of 24 full-time voice circuits; Kuwait was using 18 circuits in INTELSAT's Indian Ocean satellite, to the Far East and Western Europe. The late Dr. Sarabhai estimated that India's international traffic more than doubled in 3 months following the coming on line of the country's first ground station. The Indian earth station at Arvi began in April 1971 with 27 half-circuits and increased its circuit usage monthly until October, the latest date for which I have information, when 42 half-circuits were in use, close to a doubling of the original quantum jump in circuit availability.

Much of this growth occurred prior to the doubling of world capacity for satellite telecommunications that came with the launching of INTELSAT IV —which should prompt a further marked increase in usage by developing countries.

As significant as the dramatic increase in international traffic was the gain in regional traffic. For example, ComSat's report to the President and the Congress at the end of May 1971 highlighted a marked increase in Latin American regional traffic. The new opportunities for internal and external communications are illustrated also by the action of European and Latin American broadcasting agencies in establishing, in March 1971, the Ibero-American Television Organization to provide for the exchange of transmissions of sport and artistic programs and particularly news exchanges. The interchange was established as an experiment that was to continue to the end of 1971. The political, cultural, and economic implications are apparent.

Costs of Participation in INTELSAT

The costs of participation for the LDCs are not exorbitant, and in fact the money spent brings early and rapid return. For example, India's equity as of September 30, 1971, was slightly in excess of $1 million, based on a quota of 0.466 of the total. Brazil's equity in the same period was slightly over $3 million, for a quota of 1.4063. It was expected that both countries would realize a return of about 14% (cumulative) by the end of 1971.

Apart from this capital subscription, the Indian earth station at Arvi, built by RCA-Canada, cost in excess of $6 million. Brazil's earth station at Tangua was built by Hughes Communications International for $4.3 million. A significant return is expected from such stations within 2 to 3 years.

On this basis, numerous agencies have been prepared to provide economic assistance to those LDCs wishing to enter actively into INTELSAT operations. AID has guaranteed loans, for example, to construct a ground station in Indonesia. The Export-Import Bank has given assistance to Chile, Cameroon, Pakistan, Thailand, and others. The UN Special Fund assisted in building an experimental dish for training purposes in Ahmedabad in India.

ComSatCorp assists through studies of the potential and economic feasibility of the use of ground stations in less developed countries. Technical assistance is provided for the initiation and operation of ground stations. So far as the United States is concerned, this is consistent with the Communications Satellite Act of 1962, which provided that care and attention would be directed toward providing services to economically less developed countries.

The cost-benefit equation, in the eyes of the LDCs, is clearly regarded as a favorable one. The developing countries have given no indication that they will not renew their memberships under the INTELSAT definitive agreement; the majority of them have been using the system considerably beyond their original quota estimates. A rough calculation based on 1970 data suggests that LDC use of the system was exceeding quota estimates by nearly 50%.

The cost of using INTELSAT satellites is itself steadily going down. The tariff on North Atlantic overseas telephone calls has been cut in half since satellites first came on service. ComSat has reduced its own rates for trans-Atlantic circuits by 25%. Moreover, it is clear that the basis for a continued reduction in rates exists: The first INTELSAT satellite represented an investment of about $25,000/circuit year. The second INTELSAT series cut the cost to about $17,600/circuit year, and the third series cut it further to $2900/circuit year. The current series of INTELSAT IV satellites, with an average capacity of 5000 circuits and design life of 7 years, reduces the investment per circuit year to $1000. Thus, there is likely to be a continuing downward pressure on tariffs.

Types of Use

For the most part, satellite links are used for conventional communications, especially voice and record traffic. Television is only a very minor element in the international satellite traffic to date. Less than 2% of INTELSAT unit hours went for television in August 1971. In absolute terms, television usage had fluctuated at comparably low levels over the previous couple of years. In relative terms, TV shows a declining percentage. Where television was used, it was very largely concerned with sports programs and the Apollo space program. The most extensive single coverage was probably devoted to the World

Cup Soccer Games in Mexico City, representing about half the total TV transmission time that year over INTELSAT circuits.

As Hartford Gunn has said, "Unfortunately the worldwide exchange of television programs has not been expedited by the establishment of the international satellite system. The bulk of satellite television transmissions has served the three commercial networks, their news services from Europe and Viet Nam, and their sports programs to Hawaii and Puerto Rico. Few programs have been exchanged outside the Atlantic basin. Nevertheless, this spring [1971] broadcast organizations from Latin America, Spain, and Portugal have banded together to negotiate special low satellite rates and are operating a daily, live news coexchange between the two continents."

The limitations on the exchange of television in the developing world are, of course, dual: (a) the high cost of television transmissions, and (b) the limited distribution of home receivers. Feldman and Kelley (1971) pointed out that even if the cost of the satellite segment could be reduced to zero, the high hourly cost of TV transmissions from the originating studio to the broadcast transmitter on another continent would decrease by less than 15%. The bulk of the cost is in the terrestrial facilities.

In consideration of the high cost of ground facilities, the ComSat Corporation has been experimenting with small ground receivers that could substantially reduce cost. Further economies will require large increases in power in the satellite itself to permit operation with ground stations very much smaller than the 97-ft standard station now recommended. Besides ComSat's experimentation in this area, Canada and the United States plan jointly to experiment with 200-W traveling wave tubes and to operate at the 12 GHz frequency range. Together, these factors should permit large cost reductions, which in turn increase the potential for developing countries in particular.

THE INDIA-UNITED STATES SATELLITE BROADCASTING EXPERIMENT

The India-United States Instructional TV Satellite Experiment project is one of those widely sought but rarely grasped opportunities to use modern technology in a developing country so as to bypass historical transition requirements. The late Dr. Vikram Sarabhai, who was the guiding spirit in India for the experiment, stated that the use of satellites for direct broadcasting would make it possible to bring televised instruction to illiterate millions in India's 550,000 villages and in 10 instead of the 30 years it would take to construct and extend a conventional TV distribution system on the ground—given the same annual rate of investment.

The United States-India experiment is made possible by the development of a satellite by the National Aeronautics and Space Administration (NASA) that will radiate considerably more effective power than heretofore. The power level is achieved by a high-gain 30-ft antenna to be deployed in space by the satellite and by a high pointing accuracy (one tenth of a degree of arc). This highly focused power will permit small TV receivers on the ground to pick up the satellite's TV signal directly. There will be no need for the large and expensive ground receivers now required to collect weaker signals or for conventional high-powered TV transmitters to rebroadcast the signals locally.

The satellite is designated ATS–F, one of a series of Advanced Technology Satellites in the NASA program. It is scheduled for launch sometime in 1974. Note that the ATS–F satellite is not capable of broadcasting directly into today's home receivers. The conventional home set is indeed the core of the village receiver required for the India experiment, but such sets would have to be "augmented" to receive TV programs beamed by satellite at today's power level. Augmentation includes the addition of a preamplifier, a modulation converter, and a rooftop antenna 8 to 10 ft in diameter. The augmentation cost has been variously estimated at 150 to $300 or more, depending importantly on the production rate.

Genesis of the Experiment

The concept of a full-blown experiment to test the direct broadcasting capabilities of ATS–F was formulated in NASA as early as 1964. India became interested early in 1965 and requested detailed information on ATS–F characteristics (at the Tokyo meeting of a NASA Telecommunications Ground Station Advisory Committee in November 1965). In March 1967, Dr. Sarabhai (Director both of India's Department of Atomic Energy (DAE) and the India Space Research Organization) submitted a proposal to NASA for a year-long joint study of the long-range implications for India of an experiment using ATS–F. The study would consider whether India should start down the road toward a national TV distribution system based on satellites or should instead build a system based on conventional ground microwave links. Thus, the study would develop knowledge of the prospective tradeoffs between satellite and conventional TV broadcasting.

At this point, two additional matters should be clarified. First, India had to assess the widely assumed value of TV as an educational medium. Did the effectiveness of TV justify a costly experiment and an even more costly follow-on national development program? This question India sought to resolve by means of a preliminary test project. Conventional TV receivers were set up in some 80 villages near New Delhi where the only TV transmitter in India is located. TV instruction in agricultural matters was beamed twice weekly to these villages. The effectiveness of this technique was measured against experience in a number of control villages. The results, which have been documented, were considered to confirm the effectiveness of village TV for rapid dissemination of agricultural information, for stimulating early and effective implementation of new agricultural techniques, and for putting more income in the pockets of village farmers.

Second, it should be made clear that India, while not unique, may be especially suited for a satellite experiment in the direct broadcasting of TV. With the exception of the Delhi transmitter noted above, India lacks any TV facilities, although a number are now planned. There is, therefore, no question of conflict with existing systems or interests. The Indian subcontinent is ideal in size and population distribution for satellite broadcasting purposes and for the effective antenna pattern of the ATS–F satellite. And the population is distributed fairly homogeneously in over half a million villages throughout the region.

This contrasts, for example, with the United States, where an established conventional TV system already covers most of the country. It contrasts also with other developing countries. For instance, in Brazil a substantial portion of the population is concentrated in coastal cities, all of which already possess TV networks, while only the scattered inland population lacks TV. Brazil's problems with respect to TV distribution, while real, are therefore quite different from India's.

Joint Study

A Memorandum of Understanding between DAE and NASA was signed on October 6, 1967; it called for a joint study of tradeoffs in providing for village TV. The study group considered (1) a conventional microwave network, (2) a conventional system using satellites for interconnections, (3) a system based entirely on direct broadcasting by satellite to augmented village receivers, and (4) a hybrid system providing for direct broadcasting but, where village proximity and density permitted, using small ground relay stations to receive programs from the satellite and retransmit them to conventional (cheaper) receivers.

The joint study concluded that the direct and hybrid systems would cost between one third and one half as much as the conventional system and that the hybrid system offered some modest savings over the direct broadcasting system. On June 8, 1968, the Joint Study Group recommended proceeding with the ATS–F experiment.

By August 1968, Indian representatives were able to provide extensive documentation on the Delhi tests and the projected satellite experiment at the

United Nations Conference on the Peaceful Uses of Outer Space in Vienna. Approximately 65 nations attending that conference, obviously including many LDCs, were given a preview of the intended test of satellite broadcasting.

The NASA-DAE Experiment

India and the United States mean to demonstrate the potential value of satellite technology for the rapid development of effective mass communications in developing countries, to demonstrate the value of satellite-broadcast TV for the practical instruction of illiterate villagers, and to stimulate national cohesion in India with important managerial, economic, technological, and social implications.

NASA's contribution to the experiment is to make the ATS-F satellite available for Indian use and to provide technical advice and guidance to India in the discharge of her responsibilities. The satellite is to be made available in the following way. After its launching, it will be positioned in synchronous orbit above a meridian intersecting the United States. There NASA and associated United States and other users will, for up to a year, implement numerous experiments made possible by the satellite. Then, the satellite will be moved slowly eastward along the equator until it is stabilized some 15 to 20° east of the Greenwich meridian in line-of-sight of India.

India will then employ an existing ground transmitter at Ahmedabad, together with other future transmitters as desired, to beam TV programs to the satellite on a frequency in the range of 2000 to 4000 MHz. The satellite will receive these programs and rebroadcast them, using a frequency in the 860 MHz range. Augmented receivers in some 2000 remote Indian villages will receive the broadcasts directly. Some 3000 conventional receivers, placed in village areas of greater population concentration, will receive the program through about five relatively small and inexpensive ground relay stations.

India is responsible for the development of instructional TV programs, for the necessary ground-to-satellite transmitters, for the augmented and conventional village receivers, for maintenance and logistics, and, of course, for the organization of village audiences. Perhaps most difficult of all, India will be responsible for the coordinated provision of such seeds, fertilizers, tools, contraceptives, and so forth, whose use will be implied by instructional programs.

The TV ground transmitter at Ahmedabad is important not only as India's prime means of access to the ATS-F satellite but also as a possible door to observation of the experiment by other developing countries. The transmitter was constructed some years ago with UN financial aid as an international training facility for experimental work in the field of satellite communications.

Programming

Even at this early date it seems clear that new techniques in community TV programming may be expected to issue from the Indian experiment. For example, considerable emphasis is being placed on programming procedures that would maximize local village participation in order to enhance identification, credibility, comprehension, and acceptability in the provincial audiences. Mobile studios may be employed to stage and record programs in neighborhood areas, using local personnel and background.

The programs themselves are expected to run from 4 to 6 hours/day (as limited by the satellite's power budget). There will be a single video channel and two audio channels, permitting simultaneous broadcasting in two languages. Through time sharing, this could, if desired, be increased to four and even six languages.

The content of the programs, according to the terms of the agreement, will be directed primarily to increasing farm productivity, supporting family planning objectives, and contributing to national cohesion. Secondary objectives include community hygiene, handicraft instruction, teacher and student training, and related subjects.

It cannot be stressed too much that India and not the United States will be responsible for the TV programming. Thus, India will talk to India in the framework of this experiment.

Responsibility and Stimulus

The provision of village receivers is another Indian responsibility with vast implications. A nascent TV receiver industry already exists in India. (Production levels were recently reported in the press to have reached 1000 per year.) Production rates could continue to go up. Design of an augmentation kit will have to be completed and the kit put into production. If, as presently hoped, all this is done in India and the necessary production and distribution rates achieved, a powerful stimulus will have been applied to native electronics with concomitant development of new skills, component sources, services, and demands.

The distribution of receivers to scattered villages and their local power supply and maintenance will further challenge Indian managerial and logistics capacities. Current thinking is understood to favor the establishment of a number of central depots from which jeeps will operate to recover, repair, and replace receivers. Thus, key technical skills will be conserved. This is important since the training of a half million expert repair men, to place even one to a village, would otherwise represent an enormous task for an ultimate all-India system.

A final responsibility of India will be to evaluate the results of the experiment and to make the results generally available. The evaluation is to be in quantitative terms to the maximum extent possible. Thus, it is expected that the impact of programs on population planning would be tested by comparing subsequent birth rates in TV-equipped and non-TV (control) villages. Farm productivity and income enhancement will be similarly evaluated. In this way, the value of the experiment for the United States and India, as well as for other interested nations, can be assessed.

Costs, Benefits, and Follow-On

It is difficult to cost out NASA's share of the cooperative effort since the satellite's capabilities for the Indian experiment are largely integral to its broader mission. The tangible incremental costs, however, will amount to 1 or 2% of the total ATS-F and -G program budgets.

The costs to India of the year-long experiment may be estimated as roughly $15 million or more. This cost, however, will be offset by the acquisition of thousands of TV receivers suitable for adaptation to follow-on operational programs, by the creation of a new industrial capability, by the enlargement of software skills, by the growth of a systems development and management capability, by advances in village organization and cohesion, and most broadly by the achievement of an ambitious national goal.

India's purpose is to proceed beyond this first step toward an operational (nonexperimental) national system. In its first stages, this will require procurement of the necessary satellite circuits, either from INTELSAT, if direct broadcasting power levels become available, or through purchase of an ATS–F type satellite and its launching. India would hope, through training in this activity, to gain the ability to develop later satellites domestically.

In Sum

The India-United States TV satellite broadcast experiment will be the first to provide direct broadcasting of television programs from a satellite into small village receivers without the need for relay stations on the ground. It is of great significance on a number of grounds. It represents an important experimental step for India in the development of a national communications system and the underlying technological, managerial, and social supporting elements. It is a constructive step forward in cooperation between one of the world's superpowers and a developing nation. For other developing countries, it should serve on a no-cost basis to test the values, the feasibility, and the requirements of a multipurpose tool that could be critical to accelerating their progress in an increasingly technological world.

In studying the Indian experience as it unfolds, observers, especially in the developing regions, will want to keep its distinctive features carefully before them. First and foremost, the NASA–DAE project is an experiment and no more. Second, the experiment is applicable only to augmented receivers and not to conventional TV receivers such as are now in home use. Third, the best technical judgment is that direct TV broadcasting to nonaugmented (conventional) receivers is well beyond the present state of the art and should not be expected on an operational basis much before the mid-1980s, if then. Fourth, the frequencies to be used in the agreed experiment are unlikely to be those ultimately agreed on as best suited for operational TV broadcasting services. Fifth, the experiment raises no questions of broadcasting by one nation into another nation's TV receivers, since reception is not possible without specialized augmentation to accommodate it. Sixth, the experiment may well assist in suggesting pragmatic solutions to such issues. Finally, there is likely to be no better opportunity to understand the benefits, costs, and other requirements of direct broadcasting than that presented by the United States-India instructional TV project.

Public interest in the direct broadcasting of television is premised on (1) the unlimited civilizing, educational, and political values imputed to visual communications and (2) the economic savings that satellite techniques promise in areas lacking established television distribution systems.

On all these counts, the India experiment is enormously exciting and important. It is a natural complement to the dramatic achievement of INTELSAT in expanding telecommunications services in the immediate period for the developing world. It suggests an important direction for further international communications development.

REFERENCES

COMSAT. Annual Reports to the President and the Congress.

———. Annual Reports to Shareholders.

———. "Managers Contributions": "consideration of Non-Standard Earth Stations." August 19, 1971; "INTELSAT IV Follow-On System Studies." April 22, 1971; "INTELSAT Tariff Manual." December 27, 1971; and "Traffic Data Base." July 2, 1971.

Feldman and Kelley. In *Aeronautics and Astronautics* (September 1971).

Fernandez-Shaw, Felix. "Television Relations between Europe and Latin America, I and II." *EBU Review*, Numbers 128 A&B.

Memorandum of Understanding between the Department of Atomic Energy of the Government of India and the National Aeronautics and Space Administration of the United States on the India-US ITV Satellite Experiment Project, September 18, 1969.

CHAPTER 25

Television as a Horseless Carriage

ELIHU KATZ

Some of the problems of television, I believe, are the result of the uncritical transplantation of certain of the norms of radio* It is something like the early automobile, which was designed with features of its predecessor, the horse-drawn carriage. While I do not know whether this process did any real damage to the development of the automobile, I strongly believe that television has been adversely affected. My concern is not so much with the countries that pioneered in the establishment of television and its institutions—although this is where the trouble began—but in the fact that small and developing nations have to bear the burden.

Having introduced the metaphor of the horseless carriage, let me push it a little further along. I want to argue—as a second point in this chapter—that the carriage is not only horseless but driverless as well. Partly as a result of imitating radio and partly because of other things, it is my impression that television (once again, especially in new and small nations) goes off on its own way, despite the best efforts of often capable and willful people. Television in new nation X, with all that nation's great aspirations for indigenous cultural creativity, seems to want to be like television everywhere else, and nobody seems able to stop it. My debate with those who hold that television is over-controlled and overmanipulated by national and international interests is that they overlook this runaway character of the medium. It is possible, of course, that this is what makes it so vulnerable.

* I am grateful to M. Pierre Schaeffer, director of the Service de la recherche of the ORTF, for inviting me to present an earlier version of some of these ideas at a seminar held at UNESCO in April 1971. The comments of the discussants and participants at that seminar and at a conference on "New Frontiers of Television" at Ljubljana in June 1971 were helpful to me in this extension and reformulation.

381

This chapter is speculative and, in effect, calls for collating and comparing the experience of many nations with the introduction of television. In the absence of such systematic data, it is based on observation and reflection, my own and other people's. There are two things that need study: (1) whether I am correct in my description of the character of television in new and small nations; and (2) whether I am correct in designating the horseless carriage complex as one of the causes for what I think I see. For the moment, let us call these things hypotheses.

THE HERITAGE OF RADIO

Television as an institution is modeled after radio. The fact is very easy to understand—so much so, that it has a seeming inevitability about it—and therefore it is all the more important to consider whether there was, or is, an alternative. Might not TV have been more like movies, for example, or more like theater, or even more like television—than like radio?

When I say that television is like radio, I am referring to the professional *norms* governing television programming. Specifically, I will refer to three of these: (1) the goal of nonstop broadcasting; (2) the orientation toward an everybody audience; and (3) the striving for up-to-the-minute news. There are probably many other norms—including very functional ones—that have been transferred from radio to television, but these are three of the basic ones, and I shall refer to them to illustrate my arguments.

But before going into these three norms in detail, another word about *why* television inherited these norms so uncritically. The most obvious reason, of course, is that television was developed by the radio companies. Both in Britain and in the United States, television was perfected within the radio broadcasting organizations; indeed, in the United States, the profits from radio broadcasting were poured into the development of television (Barnouw, 1968, p. 244). It seemed obvious that television was expected to add the visual dimension that was missing in radio.

Then, when early television broadcasting began, executives, technicians, and stars made their way from radio to television. Not all of them lasted, of course, but the administrative and creative structures, the technical infrastructure, even the governmental superstructure (the FCC, etc.), all acted as agents for the transfer of a conception of broadcasting, its content and its norms, from radio to television.

The process by which television simply inherited the norms of radio seems even more inevitable at the level of the audience. There was a time when the radio, like today's television set, was at the family hearth. There was a big radio console in the middle of the living room, and family members gathered round it to hear their favorite comedian, drama series, newscaster, or politician. As early as 1933, morning exercises began at 5:45 A.M. and the broadcast day was 18 hours long. Not long after, the broadcasting day was 24 hours long. During the late 1930s, as World War II built up, news and news analysis were added as basic ingredients of American radio. These were hard-won victories, and particularly in a nation where commercial advertisers reigned supreme, the stories of Kaltenborn, Murrow, Shirer are well worth the telling. People came to depend increasingly on radio news, just as they did on "Amos 'n' Andy," "Jack Benny," and "Big Sister." So it was easy enough for people to adjust to the idea of placing a television set where the living-room radio had been, expecting to be entertained and informed in much the same ways. Just as radio had inherited vaudeville, as Barnouw (1968) demonstrates, television had inherited radio. The nonstop broadcasting, the family entertainment, and the up-to-the-minute news and analysis were all there. People were being asked, in effect, to switch their allegiance from one box to the other. The audience was ready-made, and was being invited to accept more for their money.

Of course, we all know that television soon began doing things that radio could not do. And we all know how radio staged its comeback, by becoming something else that may be very promising. But for the present purpose, it is fair—if oversimplified, to say that television became radio, in the several senses already outlined: in striving for around-the-clock broadcasting, in aiming for maximum audience, and in the competition for up-to-the-minute news and analysis. Each of these norms deserves a few words.

Nonstop Broadcasting

The image of a nonstop service seems unquestioned in professional circles. It is seen as part of the public service, on the one hand, and is stimulated and encouraged by commercial sponsorship, on the other. In a large and industrial country, there is always an audience for a late-late show or for a midmorning show, not to speak of the peak hours of evening entertainment. And there is always a sponsor asking for more.

Even if the norm of continuous service is still largely unquestioned, the arguments against it seem clearer today. For one thing, there is not enough talent—even in a large country—to sustain it. The talent is quickly burnt out.

Another reason is that continuous broadcasting leads inevitably to the never-ending series and serial and to the hackneyed and violent formulas without which this rate of productivity could hardly be sustained. As it was in radio, so it is in television. Original drama has been driven out of television just as it was from radio.

In other words, I am arguing that it is continuous broadcasting that has made television trivial. We used to believe that television demanded the viewer's attention, because unlike radio, it asks for both eyes and ears. We know now that television is used almost as often as radio as a backdrop for other activities—"moving wallpaper," to use Stuart Hood's (1967) phrase.

With hindsight, it is not hard to see that this might have been otherwise—although not everybody need agree that it should have been so. Suppose, for example, that television had opted for discontinuous broadcasting. Suppose that television were built around specials instead of serials and series, and on festive instead of routine entertainment. Suppose television were on the air only for these special events; suppose that television were not on the air every night; suppose that television were on the air only when it had something to say. Or, suppose that television had been modeled after the theater—just as the "Play of the Week" tried, some years ago, to present the *same* play every night all week long. Had television not followed hard on the heels of radio in all the senses noted previously, both television and radio would have been very different media.

The Everybody Audience

By the same token, the family audience of all ages and classes seems the natural target for television, just as it was for radio. Indeed, it seems *more* natural for television, given the expense of producing programs, with or without commercial sponsorship. Even the BBC or the ORTF become nervous when their ratings are low; and, alas, this is true even when a second channel is explicitly designed for minority audiences.

There is no need to belabor this argument; it is familiar enough. But so is the other side. Nobody needs convincing that television has become highly homogeneous both within countries and among them. The choice among channels is often exactly the same type of program. When BBC has its current affairs program, ITV has its current affairs program, and so on. This may be the way for each channel to maximize its audience, but it is no answer for society.

Nor does it take much to remind people that they have special interests that might be better served by television. The problem is that since the days of radio, people have been socialized to accept that programs must sacrifice minorities to majorities. The networks have opted for the kind of competition that leads to sameness, although competition for differentiated audiences, which leads to differences, may be equally viable, even commercially (Rothenberg, 1962). Presentday radio has learned this lesson. But network radio and its heir, television, do not accept it. One wonders whether even public television will be satisfied with pluralistic programming.

Up-to-the-Minute News

Finally, the news. It took many years to overcome the 10 P.M. news on television: the fashionable announcer who did most of the work; the rear projection of a wirephoto; a film of a local fire. Why didn't anybody realize that television could not (and still cannot) keep up with radio coverage of "bulletin" news? Why do we assume that the announcer adds instead of detracts from the news he is reading? Why is it not clear that television cannot get quickly to where things are happening, except if they are scheduled in advance? Why is more thought not given to the difficulties inherent in personalizing the abstractions that are often the events of major importance in the news?

The alternative surely is a different kind of television news—based on visualization of events that cannot be visualized in a few hours. Only in the last few years has the newsmagazine come to prominence on United States television with half an hour or even an hour of visualized news. This is only the beginning, but it cannot be done without work in the film archives, serious thought about the personification of abstractions, and some good documentary-type research. This takes not hours, but days, and sometimes more; television is better at presenting yesterday's news or the news of last week than today's news. It is better at presenting a review of an extended and coherent period of happenings in Vietnam than another 30 seconds of tonight's jungle.

This realization is gradually taking root in television. But for two decades, the pattern for the news was the pattern set by radio.

CONSEQUENCES FOR BROADCASTING IN NEW NATIONS

The point of this chapter, however, is not to analyze the ways in which the norms of radio were transferred to television broadcasting in the United States and Britain. All that has been said so far is a prelude to the statement that this set of norms, which guides professional work in television, has been adopted by broadcasters everywhere, but nowhere more than in new and small nations building both on their own experience with radio and on the model of their more experienced and presumably successful colleagues in the large and developed nations. And nowhere are these borrowed goals less appropriate, I contend, than in these new and small nations.

This is more than a value judgment; I am not simply asserting my own point of view. When I say that these norms are inappropriate to new and small nations, I believe I am saying so from *their* point of view—or at least from the point of view of those leaders who have made public statements about their expectations of the new medium.

In country after country, one hears the Prime Minister or the Minister of Culture and Information rise to proclaim the three expectations that his government has of television. Despite the high cost of the investment, says the Minister, television is being introduced (1) to integrate the diversity of tribes, ethnic groups, or linguistic groups in the society; (2) to contribute to the renaissance of indigeneous culture; and (3) to promote economic development. Having said this, the professionals—or aspiring professionals—take over, together with their technical assistance experts. In a year or so, they will be ready to ask the Minister to inaugurate the new station that will answer his lofty list of Expectation with "I Love Lucy," "Peyton Place," "Mission Impossible," plus 30 second film clips of the world as seen by the editors of the television news agencies. In the short time between the proclamation of goals for the station and the beginning of broadcasting, the goals have been completely overcome by the homogenized formula for television programming that has conquered the world (Varis, 1971; Ainslee, 1967; Hachten, 1971).

How does this transformation take place? It is related, in no small measure, to the striving for achievement of these inherited and transplanted professional norms. The results of this striving have much more far-reaching consequences in the small and developing countries than in the United States or Britain. They are enumerated in the following discussion.

Too Much: The Key Problem of Television in Small Nations

The unchallenged norm of continuous broadcasting is transformed in new and small nations into the striving for more hours. Indeed, there is a tendency to measure the maturity of a nation's television broadcasting in terms of the number of broadcast hours.

Most nations, of course, are very far from the American standard of nonstop television. But even when they have only a few hours of broadcasting—even as few as 4 or 5 hours/evening—it is almost always too much.

Again, I am not simply pronouncing my personal judgment; I think the decision to broadcast a certain number of hours in a given country has immediate and measurable consequences for the programming schedule and is a major key to the process whereby television becomes sidetracked from its proclaimed nation-building goals.

Judging from a few cases I have examined at close range, I think there is a certain regularity in this process that deserves to be elaborated. It appears, in fact, that one of the first public announcements that are made about the introduction of television broadcasting is the number of broadcast hours being planned. This number, it seems, regularly underestimates several things: the amount of money it costs to produce a locally made program; the talent it takes to do so; and the infrastructure—in the visual arts of theater, graphics, film, and so on—that is required by the producers both for supporting facilities and for subcontracts.

The number of broadcast hours proclaimed, then, is beyond the capability of the new station, but corresponds roughly to prime time radio broadcasting, which is then gradually devaluated and phased out. The latent message in this proclamation is that television will provide the viewer with all of the services heretofore provided by radio. The radio professionals naturally rush to television for jobs—and in many cases these jobs are automatically assured because of seniority rights and because it is assumed that people trained in radio represent the best potential for work in television. And so, the viewer in a very far-off country is taught to expect in 1960 or 1970 what Americans expected in 1950, namely, that television be the heir to radio. He is not taught to stay tuned to radio for some things and to television for others. Thus, the norms of radio are transferred to television producers and consumers.

Soon enough, the new station finds that it cannot live up to its proclamation without importing 50 or 60% or more of its programs. Some stations find that the best way is to turn over their entire program schedule to an outside contractor who may, perhaps, have served as technical adviser. Other stations do not buy so high a proportion of their programs outright but, instead, buy the standard program formats wherein one of the large distributing companies contracts to teach the local station to produce "Quiz Program Number 26A" or "Variety Show 14B."

The decision to buy more and more foreign programming is easily rationalized; it is called "opening a window to the world." This sounds like a good thing—and often connotes the sincere intention of bringing the best of television to our new nation. But when the Director of Programs returns from abroad he brings with him windows to the world named "Lucy," "Peyton Place," "Mission Impossible," and "Bonanza." It is not because he did not look for anything else. It is just that there *is not* anything else that will satisfy his need to fill 26 Monday nights from 7 to 8 P.M., or 52 Wednesday nights from 9:30 to 10 P.M. Specials and one-offs will not solve his problem; old documentaries are out of date or irrelevant. For the same reasons, he may well bring back the "Ed Sullivan Show," and when he learns that television tapes exist outside of New York and London, he may return with "Festival San Remo."

All this, of course, is very far from the dream that television will find a way to express the nation-building hopes of the country or become an important center for indigenous cultural creativity. The programming schedules of television the world over have become homogenized, and the more hours, the more homogenization—not just in small poor countries, but in small rich ones as well. There is a telling example, I think, in the fact that children in Israel watch Jordan television to find out what is going to happen a few weeks hence in the same American series being run on Israel television and vice versa.

If I am correct, this situation can be traced to the professional commitment to continuous broadcasting—if not nonstop, then for as many hours as possible. In the context of a new nation, this commitment may now be seen to have a double-barreled effect. First of all, it has an effect on the new station; it sends the program director scurrying abroad for whatever material he can find. And second, it leads him directly into the waiting arms of the merchandisers of all the series and serials that are themselves the by-product of the American commitment to nonstop broadcasting. Thus, the vicious circle is closed.

The Everybody Audience in New Nations

If the norm of nonstop broadcasting has had a uniformly disturbing effect on the institutionalization of television in new and small nations, the norm of maximizing audiences has had a more mixed effect, when judged from the point of view of nation building.

On the one hand, there is something functional in having an entire nation exposed to a shared cultural experience. If groups of diverse background can respond to the same messages and as a result develop a shared set of symbols in terms of which they can communicate with each other, there is obviously much to recommend in it. The case seems strangely different, however, if the program in question is "Bonanza" or "Hawaii Five-0." Are these the shared cultural experiences that justify the huge national investment in television broadcasting? Are these truly the programs that audiences everywhere like best? Certain evidence from Ireland and Israel (Dowling et al, 1969; Katz and Gurevitch, in press) suggests that homemade programming is more popular. Altogether, we know far too little about the presumed universal popularity of American series and serials, and even less about the transferability of their meanings into other cultural contexts.

Apart from the merit of shared experience, which is considerable, the norm of audience maximization must be evaluated against the alternative idea that people might be offered what they are interested in, and that viewing should be more *selective*. It occurs to TV broadcasters only very rarely that it might be legitimate to talk to different people in different ways. In Israel, for example, distinct groups exist that would be interested in such things as language learning, occupational training, and traditional themes (even in prime time); a great deal could be accomplished through television for these audiences. But, apart from the Arabic program (which is an extremely interesting exception), nobody seems able to justify the investment in minority audiences, and the audience itself—as has already been noted—has been taught to expect that all programs are for everybody. This is where the clamor for a second channel begins, but if it is fair to learn from the experience of the more developed nations, a second channel tends to wind up competing through sameness instead of difference. And it is ironic to talk of a second channel when half of what is broadcast on the first channel consists of imported material, which is very largely American series.

There is another paradoxical point worth noting about the problem of the "everybody" audience in new nations. "Everybody" for the broadcaster is often the equivalent not of the society as a whole but of the subgroups within the society with which the broadcaster has contact. These reference groups may be of a certain ethnicity or a certain language, or may be urban instead of rural. Again, this is a kind of double-barreled problem; the norm of the "everybody" audience is adopted from the presumed code of television professionals abroad, but when transferred to the new or developing nation, the image of "everybody" may exclude very large proportions of the population.

The Problem of Homogenized News

The third of the norms governing television broadcasting in general, but having *stronger* effects in developing countries, is the nature of news broadcasting. Inherited from radio, as has been noted, the television news bulletin—complete with announcer and whatever pictures he can get hold of—adds very little to anybody's understanding of the news. New and developing nations suffer all the more from this kind of imitation because the news is so much less *theirs* than that of the capitals of the West.

The news service of a small non-European country is supplied by daily deliveries, usually by airfreight, of snippets of film, much of which is hard to digest in other cultural contexts. The news is necessarily a day or two or three late—and will continue to be late even when delivered by satellite—because television news cannot help being late. But because the professional standard and style for the news bulletin has been set in New York and London, nobody is prepared to face the limitations of the present system and to capitalize on the virtues of a division of labor among the media. In this division of labor, the job of television is to make the news visual. This means research in pictorial archives, cabling around the world for film, and putting together a story that gives insight into yesterday's or last week's news. The minidocumentary newsmagazine looking at a relatively small number of items, would seem to be a better answer than the traditional television news bulletin.

Part of this, obviously, is a generic problem, and part of it is specific to new and small nations. The more specific part consists in how to make television news useful and relevant to people who want to see the world from their point of view, and rightly so. The simultaneous presentation of pictures provided by

UPI, Visnews, or CBS, read by announcers throughout the world dressed in ties of equal width, does not automatically make for one world. Far from it. The problem is how to get a news team from a given country to a place in which that country is interested to see the story from its point of view. The problem is how to get a hearing for the idea that radio can do today's news bulletins better than television, and that the resources of television news might be better directed elsewhere. The problem is how to get a world archive of news and documentary film that can be drawn upon, when the need arises, to illuminate a contemporary news event for people who have not the slightest conception of the background to the event.

SUMMARY AND CONCLUSIONS

In conclusion, I wish again to emphasize that I do not deny the seriousness of the problems relating to control over the media. Political, commercial, ideological, and other interest groups are constantly at work, trying to influence the media, its producers, and its content. What I am trying to say is that these manipulative attempts have an important competitor—the almost gravitational force that pulls television stations into an orbit where program schedules look curiously alike, in the most diverse kinds of broadcasting systems.

In trying to analyze this gravitational pull, I think I have been able to locate some part of it, at least, in those norms of professionalism in television that have been mistakenly transferred from radio. I referred to three such norms: the idea that television ought to broadcast as much as possible; the idea that all programs should be right for everybody; and the idea that the news on television ought to try to scoop the other media.

I might have examined some other norms, such as the heavy investment in studios. Many observers believe that the future of television is outside the studio—where people meet, formally and informally, and where life goes on— and the new, increasingly mobile technology promises to make this possible. Yet, the policy makers in broadcasting, much influenced by the engineers, continue to invest (the last I heard) in safe and well-padded studios. These investments can be justified only if the studios are used. And if the studios are used, programming will be more of the same.

However problematic these norms may be for highly developed societies, I have tried to show that they are all the more problematic in small and new states. I think they are responsible for the incredibly high ratio of imported to domestic programs, for the sameness of programs everywhere, for the triviality or irrelevance of so many programs and so much news, and for the overlooking of indigenous needs and particularistic interests because such things—

educational programming, for example—do not much occupy the professionals in New York and London.

All new and small nations are not trapped in this way. Some have escaped, only to pay a higher price—that of totalitarian control. Others have succeeded, at least to some extent—in some of the countries of Asia, I am told—in keeping their own great traditions predominant.

I am not speaking against the overall importance of professional norms in broadcasting. Quite the contrary. Sometimes freedom's only protection is in the dogged determination of broadcast journalists and program makers to stick by the norms of making public what the public has a right to know, of giving access to minority voices, and of resisting the dictates of arbitrary power in politics and in art. We have witnessed several heroic examples of this in recent years—in the United States, Britain, and Czechoslovakia—so that is not what I am talking about. What I am saying is that certain professional views—certain conventions—about the nature of television seem to me dysfunctional from the point of view of the inherent nature of the medium, and its essential, or at least potential, difference from radio. They are certainly dysfunctional from the point of view of the potential role of television in nation building and in the stimulation of indigenous cultural expression and creativity. I admit to being biased in favor of a more festive use of television, but I have tried to say something more than that.

REFERENCES

Ainslie, Roselynde. *The Press in Africa: Communications Past and Present.* New York: Walker, 1967.

Barnouw, Erik. *The Golden Web: A History of Broadcasting in the United States,* Vol. II, 1933–1953. New York: Oxford University Press, 1968.

Dowling, Jack, et al. *Sit Down and Be Counted: The Cultural Evolution of a Television Station.* Dublin: Wellington, 1969.

Emery, Walter B. *National and International Systems of Broadcasting: Their History, Operation and Control.* East Lansing: Michigan State University Press, 1969.

Hachten, William A. *Muffled Drums: The News Media in Africa.* Ames: Iowa State University Press, 1971.

Hood, Stuart. *A Survey of Television.* London: Heinemann, 1967.

Katz, Elihu. "Television Comes to the People of the Book," in *The Use and Abuse of Social Science,* Irving Louis Horowitz (ed.). New Brunswick: Transaction Books, 1971; abridged in *Trans-action* (June 1971), 42–48.

Katz, Elihu, and Michael Gurevitch, *The Secularization of Leisure: Culture and Communication in Israel.* London Faber and Faber (in press).

Rothenberg, Jerome. "Consumer Sovereignty and the Economics of Television Programming," *Studies in Public Communication,* 4 (Autumn 1962), 45–54.

Schiller, Herbert I. *Mass Communications and American Empire*. New York: Augustus M. Kelley, 1969.

Varis, Tapio. "Approaching International Inventory of Television Program Structure: Preliminary Report of a Pilot Study," Paper read at International Symposium, "New Frontiers of Television." Bled, Yugoslavia, June, 2–4, 1971.

CHAPTER 26

The Nonhomogeneity of the National State and the International Flow of Communication

KAARLE NORDENSTRENG AND TAPIO VARIS

Often the mere title of a scientific chapter contains the essence of the new contribution that the author is using to promote—or confuse—the development of
scientific thinking. Thus the main idea of this chapter is contained in the concept of "the nonhomogeneity of the national state," that is, in the statement
that the crucial boundaries in the world do not occur between nations but
within them.

It is true of course that nations constitute natural and useful units for everyday thinking and scientific analysis, as is pointed out by C. W. Mills (1959).
But on the other hand there is the danger that we will overlook what happens
within nations and merely treat the concept of the nation as a "black box"
without specifying its internal structure and conflicting forces. The problems
related to the international flow of information are often analyzed without
considering the inner structure of societies. This is also true of the most recent
efforts to oppose the traditional principle of "free flow of information": in
practice it is seen to work in favor of the economically strong nations and
against the weak and rising countries, and consequently, the "underdog
nations" are recommended a strategy of "cultural sovereignty," with "cultural
screens" to protect them from expansive foreign domination, "cultural imperialism" (Schiller, 1969; Smythe, 1971).

The ongoing process of international economic integration has, at least in
Europe, drawn more and more attention to the artificial and misleading character of the idea of a sovereign and homogeneous national state. On the one

hand, trade and production is planned and practiced internationally within politically, economically, and militarily integrated blocks, while on the other hand, labor has faced the realities of an international market. In this context it is realized more widely—and not least by the working class itself—that in all countries based on a free capitalist economy there is the same kind of permanent antagonism between the system of production and those who earn their livelihood by running the system, and that the working class is international not only in slogans but in practice, too (Mandel, 1971). Such current developments have caused a renaissance and an increased appreciation of 100-year-old documents such as the Manifesto of the Communist Party:

The bourgeoisie has through its exploitation of the world market given a cosmopolitan character to production and consumption in every country. To the great chagrin of Reactionists, it has drawn from under the feet of industry the national ground on which it stood. All old-established national industries have been destroyed or are daily being destroyed. They are dislodged by new industries, whose introduction becomes a life and death question for all civilized nations, by industries that no longer work up indigenous raw material, but raw material drawn from the remotest zones; industries whose products are consumed, not only at home, but in every quarter of the globe. . m. [Marx and Engels, 1967, p. 46]

In addition to the idea of the nonhomogeneity of the national state, this chapter proposes another basic principle: a distinction between physical and psychological levels and the priority of materialistic and economic phenomena over the phenomena of social consciousness. Thus mass communication should be seen not as an isolated institution for the distribution of information, culture, and entertainment, but as an integral organ of the social body, in which the deepest blood vessels and nervous pathways traverse the politicoeconomic tissue. In order to be able to analyze objectively the present state of world communication flows, we need a historical perspective on the ways in which communication has been integrated into the production system of society and the international structure of nations.

HISTORICAL DEVELOPMENT OF COMMUNICATION IN SOCIETY

In the history of communication, four major turning points can be distinguished. The first of these was the acquisition of language, which meant at the same time the birth of man as a human being. The second was the development of the art of writing alongside of communication based on speech. The third fundamental change took place when reproduction of the written word by means of the printing press—mass communication proper—became possible. And finally, the fourth breakthrough—taking place during our lifetime

—has been the appearance of electronic communication, beginning with the telegraph, telephone, radio, and television, and continuing now with the development of communication satellites.

The field of communication and the turning points in its history cannot, however, be fruitfully approached in isolation from the rest of society. Communication between human beings is only part of that social activity that in the final analysis has been directed toward gaining a livelihood. The two main differences between man and other creatures, after all, are that he is capable of communicating with his fellowmen through symbols and that he uses tools in coping with his environment. Thus, the history of mankind is at the same time the history of communication, and the latter is at the same time the story of man's economic activity.

The history of communication must therefore be examined in terms of the social structure at different historical stages, as has been done by Smythe (1969, pp. 53–55). The history of mass communication in the strictest sense— spanning the last five centuries—covers a very brief period from the point of communication as a whole; furthermore, the basic principles of communication have not changed during this period to any noteworthy extent. The following discussion sketches an outline of history in order to recall the functioning of a national state from the viewpoint of communications.

The most primitive way of gaining a livelihood was a clan, consisting of few families, moving around after animal and plant food. The tribal society was an oral community in which the cultural tradition (beliefs, skills, etc.) was transmitted from one generation to another by means of immediate experience.

The patriarchal organization apparent in the clans of the hunting and gathering era was slightly altered with the shift to the village of the period of settled agriculture. This gave rise to a new group of religious leaders, the priesthood. The great importance of the priests in the village society was because of their replacement of the ancestors of the family or clan as an integrating force in society; their control over the leadership of the group was even stronger than that of the chiefs had been, since the priests also controlled the land of the village. The land was not owned by any individual privately; it belonged to the gods, whose authorized representatives on earth were the priests. The birth of the priesthood meant at the same time the birth of a sacred language, not understood by the other members of society who were not initiated into sacred affairs. Like other communication at this stage of history, this secret communication among priests was still exclusively oral.

The next turning point in human history after the village was the birth of the (national) state some 5000 to 7000 years ago, with the Sumerian conquest of Mesopotamia. Compared to the village, the early state was a relatively large society, with a population of several (tens of) thousands and with a military organization for the defense of the society.

The use of slaves as a work force in such an organized state freed the other members of the society from the need to perform tasks directly related to gaining a livelihood. Thus new occupations constantly arose alongside agriculture, the division of labor increased, and the society became more diversified in its functioning. A new factor in this society is that of commerce, which meant a crucial increase in communication among societies. The merchants carried with them information about other places and transmitted their experiences orally to other men, therefore functioning as a channel of communication from other parts of the world.

The point of view of the merchant was naturally above all that of a seller and purchaser of goods; his observations were grounded in commercial advantage (instead of, for instance, in religious belief), and he also had to bear in mind what he could relate to other merchants and to the other members of his community without endangering his own commercial interests. Along with expanding communication, there thus developed the supression of important information, which had of course been practiced already a great deal earlier; all magic, after all, is based on secret knowledge.

The development of writing is tied to the early process of urbanization; society began to be so complex that its administration without such aids to memory became impossible.

In addition to the need to write down laws, writing was also needed for bookkeeping in the state administration; the priests were the stewards of the gods here on earth and they needed exact records concerning the rental of land and the payment of taxes. Writing was naturally also needed in commerce; measures and values of goods had to be written down and jointly approved by the parties to the exchange.

Writing not only functioned as an aid to memory, enabling men to transcend time; it also improved the possibilities of communication over space. Furthermore, it was now possible to send secret messages, inaccessible to outsiders. The supression of information began to be of great importance in matters of state, and it was easy in practice, since only the priesthood and the scribes in their service knew how to write. Thus secret information was known only to a small group of men. The art of writing was from the very beginning an exclusive privilege of the power holders in society, and served the state.

The people at large, meanwhile, remained isolated from such communication. Among a majority of the population, traditional knowledge was transmitted from one generation to the next orally. This situation remained stable for thousands of years. World view was composed of material derived from the oral tradition and information received from political rulers, priests, and merchants. The latter three classes were well informed, both about the state of their own society and about the rest of the world. They had control over knowledge and over channels of communication, and they exploited these for their own good and for that of the state, thus dominating the psychic reality (consciousness) of the masses.

As technology and the means of communication developed from the time of tribal societies onward, they were systematically used to suppress one sector of society and to help others to concentrate more and more power in their own hands. The creation of the state can, in fact, be seen as an organizational solution to the problems of securing and legitimizing this development toward inequality and nondemocracy.

The increasing diversification of society and the expansion of communication that took place at the end of the Middle Ages required an entirely new technique for the transmission of messages; as the number of readers began to grow, letters and manuscripts were no longer enough. Before knowledge and the power that accompanies it could spread among the masses of the people, an efficient technique for the duplication of messages was needed. The invention of the printing press marked, in fact, the beginning of the history of mass communication; only now did it become technically possible to transmit long and complex messages simultaneously to a large and heterogenous audience.

Another innovation made possible by the printing press, besides the mass-produced book, was that of the newspaper. Newssheets telling about current events were needed above all by merchants for whom such information as the political situation, war developments, the sinking of ships, and so on, was very important. The newspaper played a part in accelerating the stabilization of the capitalist economy. The first regular (weekly) papers were established in Germany, England, Holland, and France—the centers of European commerce and colonization.

The tremendous expansion of industry and commerce during the nineteenth century would before long have come to a dead end without the invention of electronic communication in the middle of the century. Electricity made it possible to conquer the physical obstacles posed by distance; the transmission of a message from one place to another no longer required transportation. Messages could now be transmitted over any distance instantaneously.

The first forms of electronic communication required wires; these were the telephone and the telegraph. By the beginning of the twentieth century, all the more important cities and nations of the world were connected by the network of wire communication. Distance had ceased to be an obstacle to communication; the world had in principle become synchronous. Let us note, however, that the telegraph and telephones were used exclusively for private communication—from one individual to another—and not for public mass communication involving a large and unspecified audience, as was the case with books and newspapers. The wire media served primarily the needs of administrative and commercial communication, both of which involved the suppression of information; for business in particular it was indispensably important to be able to transmit information from one place to another both rapidly and secretly.

In the beginning of the twentieth century, the printed word—books and newspapers—continued to represent the only forms of *mass* communication. By that time the press had developed into a significant institution along with the process of industrialization; the concentration of population in industrial cities created a potential audience for mass circulation and the rise of the labor movement mobilized the printed medium in the service of political action. The press also began to be a noteworthy capital-creating industry itself.

The decisive breakthrough in the field of communication took place only with the beginning of wireless communication, which made it possible to receive the same message simultaneously in an unlimited number of locations. Public and synchronous communication were now technically possible; in fact, radio communication could not be anything but public, since it was impossible to prevent others from hearing what was broadcast. The only requirement for the reception of electromagnetic messages was the possession of a receiver.

In the early public broadcasting of the 1910s and 1920s, the general tone of the programs was usually dignified, instructive, and avoided all extremes. Public broadcasting was used mostly to guide the people and to educate them to a conforming way of thinking instead of providing them with the information that they have lacked throughout history and without which they have humbly and unquestioningly submitted to serve the purposes of those with knowledge and thus with power. Thus from the very beginning, radio programs have contained more material directed toward the people and considered suitable by those in power instead of speeches from the people to the power wielders; citizens were instructed as to what they should think instead of being provided with diversified information enabling them to form their own opinions. This general perspective can be shown to be true of so-called "state-controlled" broadcasting as well as of commercially organized broadcasting.

The introduction of television has not changed the broad general framework to any significant degree; it simply entered the operational setup of radio and became perhaps more commercial than the latter. By and large we may conclude that from the Middle Ages to the present, the print media and later the electronic forms of communication (including film) have in general served the *expansion* of communication from a small group with control over power to practically all members of the society. This however has not necessarily meant a *democratization* of communication in the sense of the group in power giving up using the media as a means of advancing its own interests.

Thus mass communication based on present-day technology continues to follow principles adopted in the ancient temple states. The only difference is that the modern mass media are even more effective—and subtle—in manipulating the consciousness of the people. But there is also another aspect of the situation: groups outside the sphere of power have begun to realize the importance of mass communication and to demand the right to determine what the media are to communicate.

Today the global interests of the United States and other capitalist nations stand in opposition to socialist interest. Several studies have indicated the way in which communication flows in the world from the power centers to the periphery. This ideological dependence of the developing countries on the imperialist countries is best revealed in the case of radical changes. One example is Cuba after revolution; in the beginning the people were only emotionally revolutionary, but highly confused regarding political and social problems. The consciousness of the people had been thoroughly penetrated by imperialist newspapers, magazines, films, books, and other media (Edigiones Gor, July 26, 1970; Declaration on Cultural Activity, 1971). Another example is that of Algeria. Fanon (1967) describes how radio in Algeria before the liberation of the country functioned exclusively in the interest of the colonizers. However, during the liberation radio could also be used to arouse the consciousness of the masses.

The consciousness of the general public—mass consciousness—turns out to be a strategically central phenomenon in the course of human history. On the one hand, mass consciousness has operated in favor of regressive social structures; on the other hand, it may go together with a progressive awareness and practice. There is therefore reason to deal with the theoretical aspects of consciousness before going on to the problems of international communication.

THE ROLE OF CONSCIOUSNESS

In the Anglo-Saxon tradition, the prevalent way of looking at psychological phenomena involving human consciousness is one of isolationism; little reference is made to the physical environment and material conditions that in the final analysis determine the realities of an individual. Culture is understood as a largely independent realm of human society, and (mass) communication is similarly perceived as a kind of stream of isolated elements of consciousness. In fact, the main trend of communication research has been idealistic in its philosophical orientation, thus following the general pattern of Western social sciences.

The philosophical orientation of the dialectic materialism in the Marxist social science, on the other hand, clearly views the psychological phenomena as an integral part of the total material structure. There is a dialectic interaction between the material conditions and an active human consciousness, and furthermore, the consciousness functions above all as a reflector of the material reality. This implies a distinction between subjective reality, that is, that which exists in the consciousness of individuals, and objective reality, which is outside human consciousness and independent of it.

Subjective reality is reality as it exists in the minds of human beings at any given moment; as such, it can be illuminated, for example, by opinion surveys. Objective reality, on the other hand, is that which exists independently of human beings, such as, for example, the statement, "The earth is round." With regard to social phenomena, objective factors include the economic structure of production in society and the laws of its development.

Another point is the way in which reality is changed and the part played in such a change by communication and by awareness. According to Marxist theories of society, people's consciousness is determined by concrete social structure, by a given individual's position within the structure of production. However, although material factors of social structure are always primary, this does not mean that consciousness cannot in turn affect social existence. The relation is not a deterministic but a dialectical one; Lenin, for example, emphasized in his *Philosophical Notebooks* that man's consciousness not only reflects the objective world but also creates it.

In this view the function of communication is to arouse in men a need for social consciousness, to induce them to use their consciousness as a force that can bring about social change. In this sense, Marxism emphasizes the fact that social consciousness should be seen not only as a fulfiller of passive functions, as the defenders, for example, of commercial broadcasting wish to see it, but above all as an active force directed toward the future of man and society (*Philosophisches Worterbuch*, 1971, p. 425). This is reflected also in the Gramscian ideal of man, to which idealist social science has not been able to postulate its own countervailing conception (Gramsci 1970, p. 58):

> ... the question: is it preferable to "think" without having critical awareness, in a disjointed and irregular way, in other words to "participate" in a conception of the world "imposed" mechanically by external environment, that is, by one of the many social groups in which everyone is automatically involved from the time he enters the conscious world ... or is it preferable to work out one's own conception of the world consciously and critically, and so out of this work of one's own brain to choose one's own sphere of activity, to participate actively in making the history of the world, and not simply to accept passively and without care the imprint of one's own personality from outside?

The same idea of a critical and autonomous public as contrasted with the traditional manipulated (and perhaps quasicritical) public has these days been accepted by most intellectuals and even by many political leaders of capitalist countries. This is what the President of Finland said in 1971:

> Democracy cannot function under conditions when independent critical thinking does not prevail among the citizens of a nation, when accepted customs and the pres-

sure of public opinion form the content of people's view of the world. Under such conditions, we cannot speak of the will of the people; this is merely a reflection, an echo of the message which has originated in a small group of privileged individuals who exercise control over the channels of power and of influence. When this is the case, the so-called free market economy, which calls itself the society of free choice, is not entitled to cast stones at so-called totalitarian societies. [Kekkonen, 1971]

In reality, Finland is far from this ideal, as indicated by Littunen and Nordenstreng; the last few years of Finnish broadcasting have clearly revealed that "pluralism in a society with the hegemony of a single social class is a propagandistic fiction rather than an everyday practice" (Littunen and Nordenstreng, 1971, p. 23). The example of Finland shows that even in a modern industrial capitalistic state, certain institutionalized norms and values occupy a dominant position. These dominant values determine the general tendencies and character of most of the institutions of the superstructure (legislation, morality, political institutions, and other ideological activities such as education, religion, and other socialization functions). In a capitalist country like Finland, these institutionalized values are essentially bourgeois and thoroughly opposed to socialist values. Not only is the system of school education built on the bourgeois morality and learning conditioned by the rewards of bourgeois virtues, but the machinery of daily agitation and propagation reinforces this *Weltanschauung*.

We define an existing hegemony as a mechanism that "fills in" people's thinking and *Weltanschauung*. The relation between consciousness and the hegemony is a dialectical one. Consciousness may promote or prevent historical development, depending on the kind of ideological influence that dominates the masses. The term "false consciousness" is used to refer to the historical lag that exists when mass consciousness reflects the reality of a past society in the interest of ruling classes. Incidentally, the concept of hegemony sets the question of supply and demand in mass communication in an essentially different light compared to the prevailing Western commercial principle, which views the recipient's prevailing taste as the ultimate guideline for designing the supply of messages.

Communication as a part of ideological preparation is necessary in order to arouse the oppressed classes. Power is needed to control the reactionary counterforces and to minimize their influence. When a capitalist society is transformed into a socialist one, the freedoms of the old press are usually restricted or abolished in order to break the bourgeois hegemony. A current example of another kind is that of Chile, where the old freedom of the press has been preserved, although we must bear in mind that Chile is not yet a socialist country but only moving toward socialism. Another interesting example in the modern world is the German Democratic Republic. This country is undergoing a fundamental change from a fascist society into a socialist one.

The function of the mass media must be seen against this background, remembering at the same time the influence of the old institutionalized values and of ideological propaganda outside the country; the aim of the mass media is to develop the socialist consciousness. This explicitly ideological basis of communication is criticized in the West as an example of ideological distortion. However, we may ask what kind of ideological distortion is practiced by the Western mass media, what kind of a society they maintain, and what kind of a man they promote—systematically, even if not always consciously.

Against this background, we may ask what the modern mass media—especially television—offer men, what kind of information they provide them with for the development of their social consciousness or to prevent that development.

WORLD IN TRANSITION

We can see that in the historical development the suppressed classes of society first had to be able to analyze carefully (with the help of a vanguard) the conditions of their society in order to be able to break it down and build a new and more progressive system. Human beings were able to fly only after they became aware of the forces that chained them to the earth and of the ways in which they could overcome the force of gravity. Thus, for example, the revolutionary bourgeoisie of France created in the eighteenth century an ideology that in many ways reflected the reality correctly; this ideology could therefore be used as the theoretical basis of the bourgeois revolution. The interests of the rising bourgeoisie were represented as the interests of the society as a whole. Similarly, at a later stage of development, socialism can be seen to break through as the most sophisticated and progressive analysis of the state of affairs and their future course.

The historical development of mankind has lead to the present-day global situation, marked by two divergent theories of human life and social organization: capitalism and socialism. In addition to this ideological dimension there is the problem of the developing countries. The historical evidence tends to show, however, that the relative deprivation of colonialized societies (underdevelopment) is closely linked with the rise of capitalism; neocolonialism is accordingly not a separate problem (Frank 1969 and Smythe 1969). Thus the crucial factor in today's world is that of the roles of the two dominating ideologies with their variants.

In terms of the previous passages, we may state that both of these world systems have mobilized the channels of communication in society in order to gain or maintain a hegemony of consciousness among the intellectuals and the general public. Representing a historically earlier stage of development, capitalism projects itself in a defensive relationship to consciousness: it aims at maintaining the status quo (cf. Magdoff 1969). Socialism, on the other hand, claims an active and regenerating relationship to the humanistic concept of man as well as to his material environment. The consequent ideological antagonism has become a key factor in structuring the information systems of the world. Both systems exercise an effect on people through the media of (mass) communication, although the forms of propagation are quite different (less easily noticeable in capitalist societies), and the ultimate theories of life and ideas of man are very different.

Problems connected with the application and propagation of socialism remain outside the scope of this paper. Instead, we may have a closer look at the role of mass consciousness control in capitalist countries and in their spheres of influence.

With regard to the function and role of the mass media in capitalist societies, three basic observations on bourgeois journalism can be made:

1. The use of the mass media to conceal class antagonism inside society and to compensate for the symptoms of alienation.

2. The use of the mass media to illegitimize the concrete social alternatives to the existing order of society.

3. The profit-making use of the media as a branch of commercial industry.

The first characteristic of bourgeois journalism is best revealed by the everyday practice of the mass media. The obvious tendency to present issues in terms of individuals instead of social groups is an efficient method of reducing social problems to the level of individual ones. Such a presentation affords no basis for group solidarity among the underprivileged; this inclines them to endeavor to improve their fortunes and to seek upward social mobility by means of personal effort and their own industry instead of by collective action (e.g., Ossowski, 1963, p. 154). The emphasized use of "human interest" material and "harmless recreational entertainment" forms a substitute to conquering the alienation which is created by setting man and technology against each other (e.g., Vieweg-Walther, 1970).

The second point is illustrated by the extensive and ongoing news and other campaigns against the socialist countries. This is usually done by taking some aspect of socialism as "good," other aspects as "bad," and the whole system as utopian. The concrete applications of socialism are then compared to the utopian and idealistic concept of socialism. Notice that socialism receives most of its applause in the capitalist press when a conflict can be observed within or between socialist countries (e.g., Czechoslovakia; Solzhenitsyn).

The third central function of mass communication in capitalist societies is its industrial use as a means of making money, analogous to other sectors of the economy. That this is not an insignificant aspect is illustrated by the fact that in Finland, for example, 2 to 3% of the GNP is used to produce mass communication, that is, the same share as that of any significant branch of production (e.g., the textile industry). Of the total money involved in running the mass media institutions, as much as 70% is controlled by private capital and only 30% is under democratic control (e.g., the budget of the Finnish Broadcasting Company). Thus capital is interested not only in using the channels of communication to manipulate the mass consciousness according to its interests, but also in using them directly to produce a profit. Advertising plays a central role in this "alliance between profit making and consciousness control"; it gives the media almost one half of their income and it is an essential component in the stream of messages functioning as a cultural barrier to protect the economic system.

In the present era of economic integration, capital has become more and more internationalized and it is becoming impossible in practice and artificial in theory to distinguish between domestic and foreign elements in mass media: both serve the same functions in profit making as well as consciousness control. As stated in the beginning of this chapter, national boundaries are not as crucial as boundaries within nations, and consequently, discussion of international phenomena—as "the international flow of communication" in our title —should be understood at least partly in terms of intranational phenomena (e.g., class interests). Keeping these considerations in mind, we now turn to some facts about the composition of the flow of television program material.

A study still in progress at the University of Tampere has attempted to establish what may be called an inventory of television program output throughout the world (for a general description of the project, see Varis, 1971). A tentative picture of the global situation is presented in Tables 1 to 3. An overall observation on the program structure in various countries is that the commercial pattern has been penetrating widely even to the parliamentary or governmentally controlled noncommercial TV stations. The general dependence on foreign production in the middle-sized countries seems to account for around 30% of total program time (Table 1). The influence of imported material is more evident in the category of entertainment (Table 2). The crucial point in this international exchange is that the sources of import seem to be highly concentrated in the same centers of origin. The quality and tendencies of these productions are determined by commercial "taste."

The dependence of foreign production with regard to information programs is more difficult to estimate. It seems, however, that similar relations exist here as in the case of entertainment (Table 3). Unfortunately our study has not yet reached the stage of tracing the international "pipelines" of film and video material, that is, the TV "software market." It will be interesting to supplement with TV material the global view presented in the field of feature films by Guback (1969).

Table 1 Percentage of Imported (Exchange and Import) TV Programs in Various Countries (Around 1970)[a]

Country/Institution (if More Than One)	Imported Programs (Percent, Hours)	
Europe		
Finland	40	(1300)
Hungary	26	(447)
Ireland	54	(1202)
Netherlands	22	(782)
Norway	20	(342)
Portugal	24	(856)
Romania	27	(—)
United Kingdom/ITV	13	(—)
Yugoslavia/Beograd	18	(330)
Middle East		
Iraq	52	(1658)
Saudi Arabia/Rivadh TV	31	(832)
Saudi Arabia/Aramco TV	100	(—)
United Arab Republic	41	(2636)
Africa		
Ghana	34	(152)
Asia		
Japan/NHK, General TV	1	(94)
Malaysia	71	(2574)
Philippines/ABS-CBN	29	(3756)
Republic of Korea/TBC	31	(1569)
America		
Chile/Canal 13	46	(—)
Costa Rica	85	(—)
Ecuador	73	(—)
United States	Average 0	

[a] The table is not systematic because at the present stage all relevant data are not available from each country.

The present state of affairs has inspired some observers to look for a "democratic and humane use" of mass communication at national as well as international level (cf. Schiller 1969). Besides looking for another use of the media, we may ask what kinds of organizational solutions are desirable and achievable.

We are only beginning to find some tentative answers. Some main principles, however, are beginning to emerge; there is, for example, no warrant at the present stage for aiming at the kind of decentralized and multilateral

Table 2   Proportion of Series and Films of the Total Programming Time and the Amount and Main Sources of Foreign Production in Various TV Stations[a]

Country/Institution	Percentage of the Total Programming (h)		Percentage of Imported Material		
	Series (Percent)	Films (Percent)	Series (Percent)	Films (Percent)	Main Sources of Import
Europe					
Finland	13	—	87	—	USA, UK, France
German Federal Republic/ZDF	8	9	—	76	USA, UK
Hungary	4	9	93	94	UK, France, GDR, DRB, USA, USSR, Poland, Italy
Ireland	11	(Including series)	100	(Including series)	USA, UK
Italy	4	4			
Netherlands	21	2	72	75	USA, UK, Belgium
Norway	8	6			USA, UK
Portugal	8	4	100	100	USA, France
Romania	(Entertainment 14%)	(Films 12%)			
Soviet Union/Leningrad					
United Kingdom/BBC-1	8	8			—
Yugoslavia/Beograd	2	7	100	87	USA, UK, France

Middle East					
Israel	5	8	—	—	USA, UK, France
Iraq[b]	6	6	100	100	Arab C., UK, France
Lebanon/Tele-Orient	25	24	100	100	USA, UK, Arab C.
Saudi Arabia/Riyadh TV	27	2	100	—	USA, UK, Lebanon
Saudi Arabia/Aramco TV	100	(Films including)	100	(Films including)	USA
United Arab Republic	16	7	90	33	USA, France, USSR
Yemen	61	10	100	100	USA, UK, Arab C., USSR
Asia					
Japan/NHK, General TV	7	1	3	96	USA, UK, Italy
Malaysia	46	10	100	79	USA, UK, Japan, France
Philippines/ABS-CBN	32	9	54	22	USA, UK
Republic of Korea/TBC	13	3	100	25	USA
Taiwan/TV Enterprise	47	(Including drama)	39	100	USA
AMERICA					
Dominican Republic/Corp.	41	8	100	100	USA
Mexico/Telesistema	38	17	100	—	USA

[a] The table is not systematic because at the present stage all relevant data are not available from each country.
[b] The category of "shows" includes 53% total programming time though "series" and "films" form a minor part.

Table 3 Proportion of Current Affairs, News, and Documentaries of Total Programming Time and the Amount and Main Sources of Foreign Production in Various TV Stations[a]

Country/Institution	Percentage of Current Affairs, News, and Documentaries of the Total Programming (h) (Percent)	Percentage of Imported Material (Percent)	Main Sources of Import
Europe			
Finland	47	—	—
German Federal Republic/ZDF	36	14[d]	
Hungary	25	2	USSR, Poland, Italy, Belgium
Ireland	21	37	USA, UK
Italy	25	—	—
Netherlands	25	15	USA, UK, BRD.
Norway	36	20	UK, Scandinavia, USA
Portugal	44[b]	16	USA
Romania	25[b]	—	—
Soviet Union/ Leningrad	16	—	—
United Kingdom/ BBC	22	—	—
United Kingdom/ ITV	16	—	—
Yugoslavia/Beograd	38	—	USA
Middle East			
Israel	22	50[d]	USA, UK, France, Italy
Iraq	6		
Lebanon/Tele-Orient	22	—	—
Saudi Arabia/ Riyadh TV			
Saudi Arabia/ Aramco TV	—	—	—
United Arab Republic	14	13	USA, USSR, UK, France
Yemen	30	100	USSR, GDR, France, UK, Italy

Asia			
Japan/NHK, General TV	39	12	USA, UK, Canada, USSR, France, BRD
Malaysia	24	29	USA, UK, France
Philippines/ABS-CBN	26	29	USA, UK, Japan, France, Australia
Republic of Korea/TBC	26	33[d]	USA, UK
Taiwan/TV Enterprise	15	—	—
America			
Dominical Republic/Corp.	24	100[d]	USA
Mexico/Telesistema	16	—	USA

[a] The table is not systematic because at the present stage all relevant data are not available from each country.

[b] Mainly documentaries (31%).

[c] Different system of categorization. If current cultural programs are included the percentage is 38%.

[d] Percentage of documentaries only.

communication system that, for example, Enzensberger (1970) seems to be recommending and that also is the credo among many cable TV enthusiasts. It is in itself a romantic idea of people communicating in their own terms, but it is hardly realistic in the concentrated and internationalized media field of today; it is historically naive to think that international capital, within the present politicoeconomical framework, would suddenly cede the strategically important mass media to the people.

To our minds, concentration must be accepted at the present stage, and it is not necessarily an evil. This is not the view of the prevalent American thinking, even in the case of progressive scientists such as Schiller. The important questions are *who controls* the concentrated production, *whose goals* does it serve, and *on whose terms* does it function. Capitalist trusts and monopolies— in the field of mass communication as elsewhere—have been outside the reach of control by the mass audience and have been controlled exclusively by capital. At present, however, they are beginning to merge to a greater and greater extent with the state and with the political decision-making process (e.g., the military-industrial complex), in which the people have at least a formal voice. Parliamentary control over production—that is, concentrated control over a concentrated system—is thus at the present stage in history the most sensible strategy, even though such control is not likely to affect the basic direction of production (either in the field of communication or in any other area) to any sizable extent; the bourgeois state has throughout history been on the side of capital and power instead of on that of the masses and labor.

Thus, while aiming at increasingly democratic control of communication, we must follow social developments closely; it is also not necessary to lose faith even if the results of our efforts seem scanty or if the progressive line undergoes momentary reverses (as has been the case in Finnish broadcasting; cf. Littunen and Nordenstreng, 1971). The present stage in the history of capitalist and industrial society is characterized by the increasing and irrevocable exposure of social conflicts, in spite of the way in which the marketing and manipulation machinery unrelentingly fills the consciousness of the population with messages consistent with the interests of the state and of the production system.

The Scandinavian countries are a good example; although they have often been adduced as exemplary democracies, cracks are now beginning to appear in the elegant facade. This is the case not least in Sweden, the furthest advanced of the Scandinavian countries. As though in a last effort, an unprecendently large campaign has been launched there that may be interpreted as a typical case of propaganda of a state-monopoly capitalism; in the name of "social information," the people are taught, with the use of their own taxes, to think in such a way that the power structure is not endangered. But even here, communication cannot remedy the basic problem; the Scandinavian countries will remain manipulative democracies in which social activity is based not on the initiative of the people but on the relentless pressure of the production and marketing machinery. No such society, however, can function for long; sooner or later it is forced to adopt an active and fundamentally renewing approach to its problems.

It has become the fashion to frighten one's audience with Orwellian and Huxleyan images of the future, with visions of a society based on total manipulation (e.g., Nordenstreng, 1970, p. 5). We no longer do so, first because we have lost faith in the omnipotence of communication, and second because we have begun to feel that the Orwellian utopia is a historical impossibility, a horror story for adults. If world history really depended on the gullibility of the people, our development would have come to an end at the level of the pagan rites of tribal society or at the latest with the class divisions of feudalism. The historical fact seems to be that a renewing power always lifts society onward to a higher level, regardless of the resistance of contemporaries.

The world is in transition that in the last analysis is determined by the strains and conflicts within nations—and not primarily by international conflicts or technological innovations as it first appears to be. We view the international streams of communication as a manifestation of the ruling interests of the societies from which they originate and not as a unanimous output of the nations involved. In the course of history, the ruling class in each society has maneuvered the main media of communication (as well as other means of exercising power) to contribute to its interests in various ways; for example, by producing material profit (accumulation of capital) and by controlling the consciousness of the oppressed classes. Accordingly, the main point of this paper is to suggest an examination of the international flow of communication in terms of the social classes that on the one hand control its production and on the other hand constitute its destination ("consciousness consumption").

As nations are understood to be nonhomogenous in their character, the developing countries can no longer be seen as simple and poor national units at the mercy of the abundant supply of the manufactured products (TV programs) of the developed countries. In fact, the national oligarchy of a developing country has very close interests with those who sell the products of the industrialised countries, whereas the material interests of the poor masses are often almost in opposition to those of the ruling class. That the situation in a developing country seldom looks like this is not necessarily an indication of a national harmony but often just a demonstration of an underdeveloped or carefully controlled consciousness of the masses. In this view one should not wonder why Western TV material enjoys such a popular market in the Third World; it is not only because imported material is so cheap but also because it well serves the maintaining of the socioeconomic status quo.

Our approach, then, does not view the media and their content in isolation from the socio-politico-economical structures; we are careful not to overdifferentiate between the establishment (including the government and the private sector) and the media system. It is not sufficient to look at what happens within the media and try to change their policies, since most of the determining forces operate from outside, through institutional structures. However, the mass media should not be viewed as 100% "deterministic," either; there is always a marginal scope of freedom (perhaps some 20 to 30%) to change the practices within the media. Finnish broadcasting may also offer an example of such a more optimistic perspective: a systematic reappraisal of news values and a formulation of policy for news transmission (Nordenstreng, 1971).

We do not deny that something may be done to change the present media situation (and we are constantly working on that). We claim, however, that those who want to limit their devotion in changing the world only within the media are biased in their analysis and policy. Mass media and the international flow of communication should be seen as an integral part of wider national structures; the latter, in turn, should not be seen as a homogenous framework but as a historically determined system of potential conflicts.

REFERENCES

Declaration on Cultural Activity, First National Congress on Education and Culture, Cuba 1971. *Ediciones Gor*, July 26, 1970.

Enzensberger, Hans Magnus. "Baukasten zur einer Theorie der Medien," *Kursbuch, 20* (1970), 159–186; in English: "Constituents of Theory of the Media," *New Left Review 64* (1970), 13–36.

Fanon, Frantz. *A Dying Colonialism*, New York: Grove Press, 1967.

Frank, Andre Gunder. *The Sociology of Development and the Underdevelopment of Sociology*, Monthly Review Press Zenit reprint 1/1969.

Gouldner, Alvin W. *The Coming Crisis of Western Sociology*, New York: Basic Books, 1970.

Gramsci, Antonio. *Modern Prince & Other Writings*, New York: International Publishers Co., 1970.

Guback, Thomas H. *The International Film Industry*. Bloomington: Indiana University Press, 1969.

Kekkonen, Urho. Statement on the Day of Independence, December 6, 1971.

Littunen, Yrjo, and Kaarle Nordenstreng. Informational Broadcasting Policy: The Finnish Experiment, International Symposium "New Frontiers of Television," Bled, Yugoslavia, 1971 (to be published in the proceedings of the symposium by the School of Social Sciences, University of Ljubljana).

Magdoff, Harry. *The Age of Imperialism: The Economics of U.S. Foreign Policy*, New York: Monthly Review Press, 1969.

Mandel, Ernest. *Europe vs. America: Contradictions of Imperialism*, New York: Monthly Review Press, 1970.

Marx, Karl, and Friedrich Engels. *The Communist Manifesto*, New York: Pantheon Books, 1967

Mills, C. Wright. *The Sociological Imagination*, New York: Oxford University Press, 1959.

Nordenstreng, Kaarle. A Policy Approach to Communications Futurology, in *Report on a Round Table "Communication 1980" on Mass Communication Research and Policy*, the Finnish National Commission of Unesco and the Finnish Broadcasting Company 1970.

———. "Policy of News Transmission," *Educational Broadcasting Review 5* (1971), 20–30.

Ossowski, Stanislaw. *Class Structure in the Social Consciousness*, London: Routledge and Kegan Paul, LTD., 1963.

Philosophisches Worterbuch, 1, Leipzig, 1971.

Schiller, Herbert I. *Mass Communications and American Empire*, New York: Augustus M. Kelley, 1969.

Smythe, Dallas W. "Conflict, Cooperation and Communication Satellites," in *International Symposium Mass Media and International Understanding*, Ljubljana: School of Sociology, Political Science and Journalism, 1969.

———. Cultural Realism and Cultural Screens, International Symposium "New Frontiers of Television," Bled, Yugoslavia 1971 (to be published in the proceedings of the symposium by the School of Social Sciences, University of Ljubljana).

Varis, Tapio. Approaching International Inventory of Television Program Structure: Preliminary Report of a Pilot Study, International Symposium "New Frontiers of Television," Bled, Yugoslavia 1971 (proceedings of the symposium by the School of Social Sciences, University of Ljubljana).

Vieweg, Klaus and Willy Walther. Veranderungen in der Informations-struktur des imperialistischen Journalismus, in *Probleme aus Theorie und Praxis des Journalismus der DDR*, 1/1970.

CHAPTER 27

Inequalities in Access to Communication Facilities for Working-Class Organizations

STEFAN NEDZYNSKI

Workers feel a sense of frustration and injustice on those occasions when they are involved in a dispute with management and when their opponents have the mass media on their side.

To see the problem solely as one of direct access to the media for trade unions is to misunderstand its nature, since it speaks to the much broader question of the role of unions in the larger society. This being so, one must begin a study of the media problem by looking at the role of working-class organizations in the society, and by using a dynamic analysis, see that unions arise out of the economic relationship between labor and capital.

After this, one can consider the most important of the mass media: newspapers, radio and television, and the principal ways in which poor access or bias reveals itself. Three reasons for the existence of bias are then considered: (1) the private ownership of the media; (2) the pressure of advertising on the media; and (3) government pressures.

The special problems of trade unions in developing nations are briefly considered before an attempt is made to synthesize an analysis of the measures that are or should be taken by trade unions facing mass media that are inclined to be hostile.

UNIONS AND SOCIETY

The view one takes of the role of labor unions in society must inevitably influence an analysis of the various factors, including mass media, which have a bearing on the ability of the unions to play what one regards to be their primary role in the society. It also appears necessary to state, at this stage, one's views on the nature of the society in which labor unions operate to achieve their essential aim, which is to better the economic position of the workers by constantly trying to correct the distribution of income in their favor. This very broad subject has to be circumscribed to be brought to manageable proportions. Hence we shall deal primarily with the free enterprise society with a parliamentary form of government.

An analysis of working-class organizations and society can be based on one of two paradigms: static or dynamic. With a static view of reality, the principal assumption about the nature of industrialized society is that a consensus exists that can only be temporarily broken and, in consequence, this constitutes a rationalization of the existing industrial relations structure. When the consensus is broken by some event, the event is labeled as disruptive. The analysis carries over to the everyday perception of social reality, as described by Vic Allen (1971, p. 107–109).

There are few things so sacrosanct as a conventionally accepted social theory. There is good reason for this fact. A social theory achieves acceptability and hence dominance not through its capacity to explain reality alone but through its capacity to explain reality in the context of a given power structure. Any theory which leads to questioning of basic power relationships is unsatisfactory to those who dominate the power structure. This is the case at all levels, in all types of power situations.

Having suggested that a static analysis is defective, one must look to a dynamic analysis of society. If one views social behavior as a dynamic force, nothing can be viewed as fixed. Social phenomena are not fixed but open to change, and behavior is environmentally determined. The next step is to isolate the factors in the environment that have the greatest causal significance, and it would appear that economic circumstance is the most important of these. Of the economic factors pertaining to unions, the prime relationship in society is the buying and selling of labor.

Because of this prime economic relationship, there is constant interaction between buyers and sellers of labor because of the constant pressure for them to use the power of their respective positions. Workers depend on the sale of labor for their existence, and thus tend to maximize the price, while employers, to whom it is a cost, have to minimize it in the interest of high profits. (Other factors and considerations also come into the relationship between labor and capital, but the buyers and sellers of labor relationships remain the crucial factor in most situations.) The result is a dynamic conflict situation because of the nature of the two interests, which are largely irreconcilable.

It is sometimes true that both parties to industrial relations have common interests: in higher revenue of the industry, and more generally, in a higher national income. But in these situations the inequality of access to mass media is of no practical consequence. It does not matter then which of the two sides presents the common view and the common interest to the public at large through the use of mass media.

If it is accepted that there is a constant possibility of conflict between management and labor because of their divergent interests, the crucial question is that of the power relationship between the two parties. In and by itself the structure of the capitalist society gives more power to capital than to labor. Labor unions exist to redress this imbalance. They can achieve this objective better in the spheres of activity where organization of the labor force can effectively oppose the power deriving from property and wealth. While this permits the unions to face the employers around the bargaining table or in a strike, it does not necessarily, and certainly not by itself, change the power structure in the society as such, where wealth is a predominant factor. The control of mass media is an important aspect of this situation.

THE MEDIA PROBLEM

The press and broadcasting authorities of the free world claim independance from control and impartiality of coverage as basic to their operation. As Ralph Miliband (1969) has put it:

> In no field do the claims of democratic diversity and free political competition which are made on behalf of the "open societies" of advanced capitalism appear to be more valid than in the field of communications. . . . For in contrast to communist and other "monolithic" regimes, the means of expression in capitalist countries is not normally monopolized or subservient to the ruling political power.

However, as he goes on to point out, the apparent existence of freedom and opportunity, pluralistic diversity, and competitive equilibrium is both superficial and misleading, for the agencies of communications and notably the mass media are in reality essential to the maintenance of the status quo. The absolute right to freedom of expression is unchallenged, but it is mitigated by the real economic and political pressures of these societies; in this context, the free expression of ideas and attitudes means primarily the free expression of ideas and attitudes that are helpful to the prevailing system of power and privilege.

This function of the media is obscured by many features of cultural life in these systems, such as the absence of state dictation, the existence of public debate, and the nature of conservatism as a sufficiently loose body of thought to permit minor deviations and variations without substantial loss of the core philosophy. However, the fact remains that the media fulfill a "functional" role; they are an expression of the system of dominance and a means of reinforcing it.

The Press

The press is potentially the most partial of all the mass media, since it is among newspapers that there is the greatest variation in quality, content, and bias. But, whatever the differences, most mass circulation newspapers in the United States and United Kingdom do have certain qualities in common, namely a prevalence of hostility to changes in the fundamental structure of the society and a tendency to view social conflicts as disturbances caused by labor in an irresponsible pursuit of egoistic group interests. Consequently, the press often tends to be antiunion, although for obvious reasons it frequently strives to create the impression of not being opposed to unionism as such, only when, in disregard for the country's welfare and its own members' interests, a union greedily and irresponsibly seeks to achieve short-term gains that are blindly self-defeating. In other words, newspapers frequently dislike unions when they effectively do the job for which they exist. Strikes, in particular, tend to receive hostile coverage from the press; in fact, the longer the strike lasts, the greater the hostility. Union leaders who encourage these aberrations of behavior are vilified and pilloried, the rights and wrongs of a dispute having been obscured or sacrificed to the community, the public and the consumer, all of whom must be protected at all costs from the actions of men who blindly obey the summons of misguided and, most likely, ill-intentioned leaders.

Radio and Television

Radio and television, whether publicly or commercially owned, theoretically have a statutory charge to maintain "impartiality." This guarantee of objectivity is misleading since it most often applies to subjects within the existing area of political consensus, but does not prevent a stream of propoganda against all views that fall outside it. As Christopher Chataway (*Observer*, 1971) has stated:

No obligation of impartiality could absolve the broadcasting authorities from exercising editorial judgement within the context of the values and objectives of the society they are here to serve.

In other words, the media do and must perform an editorial function to tailor the events that are outside the political consensus in such a way that they support the existing order.

In countries where the large antigovernment unions and political parties form the main political opposition, the notion of impartiality is harder to sustain. Radio and television are much more directly involved in the struggle and have tended to become the instruments of the government, to be used against the opposition line, with little notion that access would be equitable. In

France, the Gaullists have turned radio and television into progovernment institutions, and the same has happened in Italy.

It is important to remember that radio and television are not run solely as agencies of political communication, but are predominantly concerned with "entertainment" and, in the case of commercial radio and television, with the generation of profits. However, even where profits are the main motivation, the content of output may not be free from political and ideological undertones. The media are accused of many defects: cultural poverty, outright commercialism, tendencies toward triviality, excessive brutality and violence, and a deliberate exploitation of sex; even such severe comments fail to capture the fact that programs can be and are used as vehicles for conveying a certain view of the world to audiences. Professor Lowenthal, in his article "Historical Perspectives in Popular Culture," observed that:

A superficial inventory of the contents and motivation in the products of the entertainment and publishing worlds in our Western civilization will include such themes as the nation, the family, religion, free enterprise, and individual initiative.

All of these are basic reference points hostile to the collectivist ideals of trade unionism.

FACTORS THAT RESULT IN POOR ACCESS TO THE MEDIA BY UNIONS AND OTHER SUBGROUPS

The reasons for the media's bias are a number of forces, all of which work toward conformity and conservatism. These forces include

a. Ownership and control of the media.
b. Pressure from advertisers.
c. Governmental pressures.

Ownership and Control of the Media

Much of the mass media is owned privately and is in the part of the private sector that is dominated by large-scale enterprise. Thus, the media are not only business, but big business. The process of communicating information from place to place, given the level of twentieth century technology, means that the communications process is virtually instantaneous but highly expensive. Large amounts of investment are needed, both in terms of physical assets and human capital, and the use of these assets is governed by the fact that they are both costly and scarce. One can see how a large-scale operation is needed to justify and meet the expenses involved.

The pattern of concentration of units familiar in other business sectors is evident in the media sector too; the press, magazines, theatres, cinemas, book publishing, radio, and television have all steadily come under the influence of a declining number of large enterprises, with combined media interests and, often, interests outside the media field altogether.

It is evident that those who own and control the mass media are most likely to be the men whose own ideological viewpoints are soundly conservative. In the case of newspapers, the impact of their views is likely to be immediate and direct, particularly with respect to unions or any organization of workers. Newspaper proprietors have often used the right of ownership to control editorial and political lines and have thus transformed the papers, by constant intervention, into vehicles of their personal views. As Lord Beaverbrook told the Royal Commission on the Press, "I run the newspaper purely for the purpose of making propoganda" (Hutchins, 1956, p. 62). The right of ownership confers the right to make propoganda, and where this right is exercised, it is most likely to be exercised in the service of strongly conservative prejudices.

Because of the amorphous nature of conservative philosophy, it is not necessary that those who own the media should dictate policy in a hard-line way—individual criticisms (or deviations) can be tolerated so long as the whole framework is broadly conservative. A "proper" attitude is, however, usually looked for with respect to free enterprise, conflicts between labor and capital, and the activities of trade unions, since these are the basic tenets of the conservative attitude.

Pressure from Advertisers

Pressures can be placed on the editorial policies of the commercial media by business interests using the manipulation or threat of manipulation of advertising contracts. Advertising revenue is often vital to the survival of newspapers, magazines, radio and television stations, this fact enhancing a general disposition on the part of these media to show exceptional care in dealing with such powerful and vested interests. Also, this source of revenue guarantees that the need to accommodate the "business community" will exceed that of accommodating interests of labor and the trade unions, because unions do not have lucrative advertising contracts to offer to "understanding" media bosses.

Of course, the emphasis is bound to vary from country to country, but it is apparent that it would not be profitable for sectors of the mass media to be violently antibusiness—to do so would simply be bad for business.

Governmental Pressures

A third pressure stems from government itself and, although it does not take the form of imperative dictation in most Western nations, it is nevertheless a real force in a number of ways. The state indulges in "news management" more than ever before and the organs of the government increasingly use the media, supplying the media with explanations of official policy in an endeavor to put its actions in the most favorable light.

In the case of newspapers, the state can use a variety of pressures: bribes or threats based on the promise of information and news, or, as in the "Der Speigel" case, direct repression, designed to present its views most favorably. However, in most cases relating to organized labor, these sanctions do not need to be used since a majority of newspapers tend to share the prevailing view of the national interest.

Publicly owned radio and television are "official" institutions and as such are susceptible to a variety of governmental pressures. They are public bureaucracies and are permeated by an official climate and acceptance of the "public good." This is not a new phenomenon and is accentuated in times of strife. For example, at the time of the general strike in the United Kingdom in 1926, Lord Reith, as General Manager of the BBC, wrote to the Prime Minister, asserting that "assuming the BBC is for the people and the government is for the people, it follows the BBC must be for the government in this crisis, too" (Reith, 1959, p. 108).

It would seem that given the economic and political spheres within which the mass media operate, the media cannot choose but to be agencies for the dissemination of ideas and beliefs that affirm instead of challenge existing patterns of power and privilege. The media contribute to the fostering of a climate of conformity, not by the total suppression of dissent, but by the presentation of views that fall outside the consensus as curious heresies, or, even more effectively, by treating them as irrelevant deviations that serious and reasonable people may dismiss as of no consequence.

Wherever the media exist within a "pluralist" society, these pressures are felt, but the nature of them does tend to differ between the developed and the developing nations. All of the pressures operate within the industrialized, developed world, but in the developing world, the problem is complicated by two additional factors. They are

a. A very low income level that may preclude the mass of people having regular access to the mass media.

b. The absence of political pluralism of parties or the incumbence of a single party in power for many years, which may lead to a situation where government dominance and physical control over the media is exercised.

Where income levels are low, the technologically more sophisticated media are often out of reach of most of the people; radio and television are often expensive in relation to most people's annual income. The $25 it would cost to buy a radio may represent a month's earnings and the $175 for a television receiver may be as much as a year's salary to an East African or South American worker. This means that these media will be principally used by the more wealthy middle classes. Even where this is not entirely true, the government of a developing country is usually the only corporate body big enough to finance the building of a radio and television network. In this case, there may be no attempt to create an impartial broadcasting network along the lines of those in the developed countries. The national media have a "duty" to the state, but this duty may have dubious results where the state is synonymous with the ruling political party.

In many cases, the printed word is the most widely available form of mass communication in developing countries, but where literacy levels are low, access is again denied to the bulk of the population. Also, government controls on newspapers and their freedom to print may result in the effective curbing of much comment not in line with government opinion.

These problems create great difficulties for organized labor in the Third World, where the very premise of free trade unions may be questioned and where a conflict between organized labor and business will frequently evoke a strong response from governments that may be ambitious to produce a semblance of consensus to attract development capital.

BALANCING FORCES AND POSSIBILITIES OPEN TO WORKERS' ORGANIZATIONS

The situation would be extremely difficult for the working-class organizations except for the existence of certain factors that can, to some extent, counteract the inherent bias of mass media. These factors can be, and in varying degrees have been, turned into "balancing forces."

In industrial countries where labor unions have acquired appreciable economic resources, they can themselves enter into the field of mass communications. In some countries they have decided that the principal way in which unions can expect to get their views fairly and accurately printed is to print the news themselves. Some labor organizations print and publish journals and news-sheets, and even national dailies. Most unions disseminate their news, comments, and arguments to at least their own members, not only to keep them informed but also to provide them with the ammunition with which to counteract the hostile or potentially hostile views disseminated by mass media.

Labor unions also publish their views by placing "advertisements" in the press or buying television time, but these means are open to only a relatively small number of unions that can afford the expense.

More effective use of mass media consists of the advantage taken of the fact that mass media cannot, and frequently do not wish to suppress news. In a dispute, the eloquence of a hard-hitting, phrase-coining labor leader may well get the union's point of view to the public through mass media. This will be particularly effective if the effort to develop a "public image" of the leader and his organization has been sustained and consistent over a period of time.

These examples show that labor organizations can make use of mass media at least occasionally to practically compel them to present labor's case to some extent. It may be profitable to reflect on the reasons that make it possible and on the limitations of this possibility.

In a free enterprise society where mass media are privately owned, newspapers will be printed only if they are sold and read. Radio and television must have an audience to stay in business. The largest group of "consumers" of mass media in highly industrialized countries are the workers. It is this fact that in conflict situations may force otherwise hostile media to measure their hostility and to present at least to some extent the views of labor.

But the force of this factor is usually circumscribed rather severely. Not many people will give up reading the newspaper to which they are accustomed and they find amusing or generally well informed or containing a good column of special interest to them. Not many people will give up watching a television channel or listening to the radio on account of a subtle bias. In other words, provided the mass media satisfy the public's desire for entertainment in the broadest sense of the term, they can permit themselves to have a general bias against even a large section of their public provided only that it does not come out into the open too clearly or too frequently.

This consideration is of serious consequence to workers' organizations. Even if in an immediate conflict over conditions of employment of their members they may not always be faced with open hostility of mass media, there is little they can do against the general bias toward conservatism, the prejudice against fundamental social changes, and the tendency to defend the status quo.

The unions must also strive more directly to minimize the adverse media publicity they receive in terms of disputes. Such publicity can do great damage to the strength of a union's case, especially where the unions are public sector unions and rely a great deal on public support for the success or failure of a dispute. The media are powerful mobilizers of public opinion, and when this opinion moves away from the workers, it places the unions in a more vulnerable situation.

This situation has at least two extremely important consequences. First, the general public is to some extent conditioned against workers' organizations. One has to observe the public reaction to most major strikes in most West European countries to see the conditioned reflex of this kind when a strike is announced.

Second, the economic and cultural interests of the workers are not, and indeed cannot be, promoted exclusively through changes in wages and other conditions of employment, which can be gained through collective bargaining. Workers' organizations have to strive for social security measures, for policies of full employment, for structural changes in the society that would tilt the balance of power in favor of the workers, and similar broad objectives. It is in these matters that the almost inherent conservatism of mass media is most difficult to combat or attenuate, unless there is an effective measure of control over mass media.

In industrialized countries where radio and television are publicly owned and managed by government appointed boards, the labor movement insists that workers' organizations be represented, in one way or another, on these boards or on other organs of supervision in order to have a chance to correct the conservatism of these mass media.

Where the media are privately owned some kind of supervision by a public organ, with labor representation on it, appears imperative—not to make them a channel of expression for workers' organizations but to ensure a minimum degree of objectivity, and more particularly to make the media more open to ideas involving challenges to the established structure of the society.

If the situation in the economically developed countries is far from satisfactory, it is often very considerably worse in most of the developing countries, particularly the smaller of them. Where the totality of mass media are government controlled, and, still worse, when there is only one radio station and one television channel, they convey only the views of the government and the ruling party and the economic interests that they represent. The measures previously suggested for industrialized countries are in these conditions inappropriate and their application is in any case next to impossible. Labor organizations representing a minority of the population in these countries have no strength to impose their will on the governments. They have at best two alternative courses open to them. One is the formation of a revolutionary movement in alliance with such other groups as the peasants. Some oppressive governments have indeed been overthrown by such movements, but examples of this kind are few and the end result of such upheavals open to doubt.

The other alternative is cooperation with the government and the ruling party in the running of the country, and, in particular, in programs of economic development. In the countries in which governments have a progressive economic and social program, which is, for instance, the case of Singapore (to mention but one example) such an alliance is advantageous to workers' organizations. The effects of this alliance are manifold and the impact on mass media is but one of them.

In other countries government control over mass media is only one aspect of the authoritarian rule and not necessarily the most important one. Suppression of trade union freedom is more crucial.

In the developing countries, much more than in industrialized countries with a democratic type of government, the problem of social inequalities in access to mass media is intricately bound up with the broad and essential question of the nature and structure of the society in the economically underdeveloped countries. The analysis of this kind should go far beyond the admittedly superficial observations made in this chapter. To be really meaningful, it would have to take cognizance of the fact that economically underdeveloped countries present great dissimilarities between one another. For this reason, if for no other, this chapter does not attempt to treat this question.

REFERENCES

Allen, Vic. *The Sociology of Industrial Relations.* New York: Humanities Press, 1971, pp. 107–109.

Blackburn, Robin, and Alexander Cockburn. *The Incompatibles: Trade Union Militancy and the Consensus.* London: New Left Review, 1967, p. 7.

Lowenthal, L. "Historical Perspectives in Popular Culture" in Bernard Rosenberg and David M. White, (eds.), *Mass Culture: The Popular Arts in America.* New York: The Free Press of Glencoe, 1957.

Miliband, Ralph. *The State in Capitalist Society.* New York: Basic Books, 1969, p. 218.

Observer, November 21, 1971, quotation of Christopher Chataway, British Minister for Posts and Telecommunications. Although he was talking within the context of the current conflict in Northern Ireland, the statement does have direct relevance to the whole question of minority and subgroup representation in the media.

Reith, J. C. W. *Into the Wind.* London: Hodder and Stoughton, 1949, p. 108.

CHAPTER 28

Mass Media and the Socialist Revolution: The Experience of Chile

ARMAND MATTELART

Several years have passed since the forces of the "Unidad Popular" assumed the government in Chile by way of election (with only 36% of the votes; the rest were distributed among the candidates of the Center and the Right). This time lapse allows us to draw some conclusions and trace some perspectives that provide new elements for judgment concerning the transformation and management of the mass media in the transition to socialism.

To understand the apparently enigmatic Chilean case, it is obviously necessary to separate oneself from the terrifying vision that the conspiratorial North American news agencies—in collusion with the national bourgeoisie—fabricate concerning the actions of the Popular Government. The living and each day revitalized presence of the bourgeoisie and imperialism (profuse transmitters opposed to the revolutionary effort), is a fact that in some way tends to assimilate the everyday reality of Chile, in spite of the existence of this revolutionary effort, with the reality of other countries of the Third World. It offers simultaneous examples of resistance and offenses by the class enemy; it offers a trajectory of reversion of content and form, perpetrated against and within a system that is still controlled by a dominant culture with bourgeois and imperialist characteristics.

The Chilean experience has become a subject for observation and study in other Latin American countries where the Left can intercede through the mass media to which it has limited access in an ideological struggle against the dominance of reactionary forces. This field of observation is of interest for two fundamental reasons: the discovery of the tactics of the adversary and of his capacity to activate the driving forces of his already dominant culture to recuperate his losses and discredit the efforts of his opponent, who is now exercising political power; and the examination of the tactics and strategy by which the forces of the Left confront the ideological offensive of the enemy and make the total socialist construction effort flow toward and through the media of mass communication.

LEGACY OF A POWER STRUCTURE: PANORAMA OF CONCENTRATION OF OWNERSHIP OF MASS MEDIA AND EXTERNAL DEPENDENCY

In 1969, 10 financial groups controlled 34.3% of all Chilean stock companies and 78.4% of their social capital. The monopolistic network of the mass media copies the monopolistic network of banking, commerce, agriculture, industry, and mines. Press freedom was converted into a freedom of property. Let us look at some illustrations (Mattelart, Piccini, and Mattelart, 1970; Mattelart and Castillo, 1970).

1. The Edwards family clan (in 1969, Edwards was president of the Inter-American Press Society), which controlled more than 60 stock companies and another economic group, Zig-Zag, virtually monopolized the mass production of magazines. Each week they delivered to the national public a total of about 2.7 million units. Confronting this empire were two informative magazines published by leftist groups with a bimonthly circulation of about 15 thousand copies.

2. In the same year the total number of daily papers published in Santiago and in the provinces by the three strongest economic groups (Edwards: *El Mercurio*—Sopesur—Copesa) reached 425 thousand copies per day. On the other hand there were only two leftist daily papers with combined daily runs of around 30 thousand copies. Reinforcing them somewhat was a morning paper that at the time tended toward the political center (circulation: 100,000). It is worth noting that all of the provincial papers (25 dailies) were associated with the dominant economic powers.

3. In the multitude of radio stations that exist in Chile (about 25 can be received in the capital by any type of receiver) these same economic groups controlled the five strongest broadcasting stations, which had the majority of listeners.

4. As for television there existed four channels and about 300 thousand receivers. By legal disposition TV had remained outside the mercantile system from the point of view of property. Three belong to the universities and the fourth—the only one with a network reaching the whole country—is the national channel in the hands of the state. But if television escaped the eco-

nomic powers in the property sense, it was nevertheless highly dependent on them for serials and films imported from the United States. On channel 13, which belongs to the Catholic University of Santiago, 46% of program time (according to minutes of transmission) came from abroad. The cost of renting a foreign program of 60 min fluctuated between 130 and $180, while the cost of national programming often reached more than $2000/hour.

5. Of course, the above-mentioned economic groups reinforced the chain of foreign dependency. For example, the president of the Edwards group was also the president of the Chilean IBEC (International Basic Economic Corporation). Through this stock company numerous national firms were controlled by North American investors, the majority of whom beonged to the Rockefeller group. The same Edwards clan, owner of the paper with the greatest circulation in the country, has exclusive rights to the big news agencies of the capitalist world: AP, Reuter, France-Press, and New York Times. Before reaching the national public, the information originating from these agencies passed through a double filter: in the United States or Europe and in Santiago. Of 120,000 words that the teletypes receive daily, only about 9000 reach the public. All important foreign news—especially foreign political news—was handled by the central management of the paper instead of by the professional journalists of the appropriate sections. Finally, Edwards had exclusive rights of publication for magazines made in the United States for Latin America (*Vanidades, Continental, Selecciones de Reader's Digest*, etc.) and of almost all of the comic strips published in the daily press.

6. Where publicity is concerned, the national bourgeoisie and the imperialists controlled the entire advertising apparatus. Among the 20 agencies there were five Chilean ones affiliated with North American firms, two of which were the most important in the country. In 1968, for example, McCann Erickson Chilena handled the advertising for 51 commercial firms, of which 18 were foreign. The year before the firm handled $2.6 million worth of advertising (as a comparison, in Brazil this amount reached $6.8 million).

The number of advertising firms is only one indicator of the volume of advertising business and many times not the most decisive factor. When one examines the kind of products and services advertised by North American affiliated agencies, one realizes that they handle most of the publicity for the strategic sectors of the economy and, especially, that of their own foreign affiliates, which are also North American.

In this context, the pressure that the foreign firm could bring to bear through its publicity apparatus was evident. Thus, 45% of the paid advertisements on channel 13 belonged to foreign firms, and the great majority of radio news programs were sponsored by foreign firms (Anaconda, Esso, Braden, Life, General Tire and Rubber Co., etc.)

The Functionality of the Dependent Culture

The narrow focus of this chapter does not allow us to go into a deep analysis of the ideological models underlying the imported material, nor to show how they are reflected in the national mass products. In a recent study (Dorfman and Mattelart, 1971) we have brought to light the key to the ideological discourse of the Walt Disney comics that circulate on a mass level in Latin America. We have shown that the imaginary childishness is the political utopia of a class. In all his comics Disney uses animality, infantilism, and innocence to mask the web of interests that form a socially and historically determined and concretely situated system: North American imperialism.

We have also shown that when there is a place in the world where the code of domination and the fantastic creations of Disney are infringed and contradict the exemplary, submissive behavior of the good, underdeveloped savage, the comic cannot silence the fact. It must make flower arrangements and interpret the fact for its reader, especially if he is a child. This strategy is called recuperation, a strategy that openly and dynamically attacks a system and an explicit political conflagration that serves to nourish and justify aggressive repression. The kingdom of Disney is not one of fantasy because it reacts to the world of events. Alongside his strategy of dilution applied to the hippie and feminist movements of the United States, one finds recuperation of the big current conflicts of imperialism. The Caribbean was the sea of pirates 15 years ago. Now the comic has had to adapt itself to the fact of Cuba and the invasion of the Dominican Republic. The buccaneer now shouts vivas to the revolution and is conquered. Let us take a comic designed to blur the reality of the Cuban revolution. Place of action: the republic of San Bananador in the Caribbean. Donald Duck ridicules the children for playing kidnap; these are things that do not happen any more: "Nobody gets kidnapped on ships any more and sailors don't suffer from scurvy. Walking the plank is also strictly prohibited. We face only the inoffensive sea." Still there are places where remnants of these savage customs survive.

A man tries to escape from a boat that he says is terrifying. "It has a dangerous cargo and its captain is a living menace. Help!" Brought back by force he invokes his freedom ("I am a free man! Let me go!") while the kidnappers treat him as a *slave*. Donald, typically, sees the incident as "actors making a film," but then he and his nephews are kidnapped, too. On the boat they live a nightmare; the food is rationed, even the rats are prevented from leaving the ship, there is only the unjust, arbitrary and insane law of "Captain Torment" and his bearded followers; there is forced work—they are slaves, slaves, slaves.

But, isn't all this about some old pirates? Absolutely not. These are revolutionaries fighting against law and order, pursued by the Navy of their country because they intend to bring a cargo of arms to the rebels of the Republic of

San Bananador. "They will try to locate us with airplanes. Switch off the lights. We will escape in the darkness." And with his fist in the air the operator shouts: "Long live the revolution!" The only hope, according to Donald, is "the good, old Navy, the symbol of law and order." By definition, the rebels act in the name of tyranny, dictatorship, and totalitarianism. The slave society that reigns on board ship is meant to be a replica of the society that they propose to install in place of the legitimately established regime. Thus Disney shows that in modern times, the only way slavery can make a comeback is through societies that support insurrectional movements.

COEXISTENCE

An Untouched Power

It is December 1971, and the preceding panorama has not changed substantially. Donald Duck still has more than a million Chilean readers. North American television serials continue to break audience records (in June of this year, for example, "FBI in Action" had 77.8% of the audience; "High Chaparral," 78.7%; "Bonanza," 53%; "Disneyland," 83%). If the forces of the Left have strengthened their mass arsenal, so have the forces of opposition maintained and even increased their powers of mass manipulation. The bourgeoisie has added two morning papers and three magazines to the four they already had in the capital. The economic group that holds the exclusive rights to the North American comics now publishes six magazines instead of four with editions of approximately 700,000 copies/month. There is no protectionist policy. On the contrary, the borders are open to the free import of North American comics printed in Mexico and these importations have almost doubled within the past few months. For its part, the popular government has bought out part of the former Zig-Zag enterprise and currently publishes a popular information magazine, a children's magazine, a youth magazine, a popular political education magazine, and a series of national comics. Recently this government enterprise initiated a policy of mass publication of books. Where radio is concerned the imbalance continues, notwithstanding the acquisition by the "Central Unica de Trabajadores" (CUT) of a broadcasting station in the capital; the broadcasting stations of the opposition add up to a total of 400,120 kW while the stations that favor the government add up to 222,270 kW, 33% of the total. In television, the popular forces have the support of the state channel (the national network) and of the University of Chile (provinces of Santiago and Valparaiso). Channel 13 of the Catholic University of Santiago and channel 4 of the Catholic University of Valparaiso stay, on the whole, outside of this orbit. Nevertheless, ideological homogeneity does not exist in any of these channels. As a matter of fact, the directorate of the national channel includes representatives of opposition forces, a representation that is also noted among the directors of the programs. Finally, it is worthy of note that the five most important advertising agencies are still affiliated with North American firms. So far, only the J. Walter Thompson agency has packed and left.

The Dialectic Response of the Bourgeois and Imperialist Media

It would be a mistake to think only in terms of amplification or quantitative conservation of the bourgeois media. Not only do the bourgeoisie and imperialists try to consolidate their control of the media, but they also work to modify and adapt their day-to-day strategy vis-à-vis the socialist project. Perhaps it is the field of communication that best illustrates what Fidel Castro said during his visit to Chile: "During this first year, the reaction has learned more than the masses." In the context of social communication, Castro's phrase refers to the fact that the mass media in the hands of the bourgeoisie do not act according to an abstract model, set up beforehand once and for all. Instead, they adapt to concrete conditions and respond to new practices of the subordinated classes. In another article I pointed out the first two phases of the bourgeoisie's superstructural recuperation during the process of change. During the first five or six months of the new government, the united right, with one voice, had a single line of frontal attack on the socialist project. The right was disconcerted and threw itself desperately into press efforts that had as their object to continue anticommunist terror, create an image of chaos and illegality and of democracy in danger, threatened by totalitarianism. But from April 1970 onward one begins to notice a division of the conspiratorial work. Two areas appear: the substantially seditious media and other much more tacit and elliptical media working to neutralize the project of the popular government.

What they intend is to subvert the socialist project, turning it into a petit bourgeois movement and reducing it to a purely cultural phenomenon so that the basic contradiction between capitalism and socialism will not be noted. They try to cloud the drastic gap that is the basis of the class struggle. They try to activate within each stratum, from the working classes to the middle and high bourgeoisie, all the driving forces of their culture, working at an everyday level and reinforcing the reactionary concepts of family organization, sex, women, youth, and science. Lately one can notice that this second phase has as its only objective to prepare, to arm ideologically considerable sectors of the population in order to consolidate the bourgeois life-style as a definitive obstruction to socialism. The superstructural recuperation effort of the bourgeois media has shown its capability to fulfill its function of political agitation and to prepare the terrain and win sectors that are available and apt to support seditious plans. The ideological preparation and the psychological pressure resulted in the organization and mobilization of these groups in street manifestations. The tactic of taking refuge in the superstructure to nullify the projects of change (i.e., in a university that has no committment to partisan political struggles; in the woman and her dignity as a housewife) is not just a rhetorical trumpet. It ties into an infrastructure of day-to-day resistance to the process of change, and that is where one notes the coherence of a total and single project of the right. They send their women into the streets to protest the shortages and escort them with armed shock troops that, in the division of the conspiratorial labor, are explicitly seditious. Thus the extraordinary, voluntaristic projects of the organizations of the extreme Right meet in the street with the daily effort to absorb and neutralize the socialist project. Women with pots and pans and young people from "Patria y Libertad" (an organization of the extreme Right) are heads and tails of the same coin, unconscious and conscious facets of the same project of reaction.

CONTRADICTIONS BETWEEN CLASS STRUGGLE AND MERCANTILE COMPETITION: THE IMPASSES

We now propose to explicate the objective contradictions implied by the existence of two criteria within the revolutionary project: (1) the necessity to defeat the bourgeoisie and imperialism in the marketplace and to penetrate a public that had not yet been dominated; (2) the necessity to achieve some level of effectiveness in the class struggle in ideological and cultural terms, and more generally, to advance the construction of socialism. On entering into a market controlled by the norms of the class enemy, the forces of change tried to adopt the bourgeois form in order to subvert it from within. At least this defines its intention. The first supposition that derived from the acceptance of "the laws of the market" is that it was possible at that precise moment to separate form from substance. In some domains more than others this separation was acutely felt.

The Genres

The first issue that arises when one wants to utilize the existing media to accomplish purposes contrary to those of the bourgeoisie is that of genre or format.

In capitalist society, every format divides programs and issues into restricted universes such as feminine, sports, political, and comical, and parcels them into subdivisions like the woman's magazine, the culinary magazine, the romantic magazine, the joke magazine, the comics, and the pornographic magazine. This system separates and creates closed, uncontaminated worlds that follow the lines of fragmentation of reality and of the world that one class offers to make its order seem indispensable. All these forms are totally allergic to any theme that does not correspond to them; for the reader of a bourgeois sports magazine, a political or even a cultural idea seems exotic in this reserved space. These unidimensional genres shelter themselves in the great dichotomies that underlie the mass culture of the bourgeoisie; the divorce of work and leisure, production and diversion, the daily and the extraordinary. Each one, moreover, subdivides each dichotomy matrix, for example, politics versus sports, manual versus intellectual, professor versus pupil. These correspond to a social division of labor and to the formation of a hierarchy of social status that these dichotomies hide, revitalize, and legitimize.

Perhaps one must respect some separation and segmentation of messages. But to start off without questioning the bourgeois heritage in matters of categories and formats, and of their meaning as forms, has a high political cost.

In several areas the State Editorial Enterprises have formulated various magazines taking as accepted, for the moment, the bourgeois genres. This means starting with the disadvantage of a system imposed by the class enemy and his culture. It means taking as a point of reference themes invoked and produced by the established system. Up to now many of the reformulated or newly created magazines have, in the end, responded to the dynamics of the dominant culture. It seems that most of the time one is faced with a concrete image of the adversary in the market, and he imposes his form. (Even if one does not begin with the intention to compete, a magazine reformulated according to existing subject matter themes is inserted into a market that necessarily assimilates it with all the magazines belonging to the same genre.) It is definitely around the genre that definitions and variations are articulated. The idol, for example, whether he appears profusely or not in a magazine, characterizes the publication and places it so firmly in the domain of fan magazines that he establishes antithetic poles: fans and antifans. The adoption of the genre or the format generally turns out at the level of decodification to be a mere inversion of the bourgeois points of view, given the fact that those who continue to sponsor this struggle are the dominant cultural powers. This fragmentation is coherent and functional to the order of the bourgeoisie and when revolutionary forces suddenly vouch for it, they introduce incoherency and anarchy to the class struggle.

The first question that follows from the preceding discussion concerns overcoming the risks of ineffectiveness in the ideological battle: when and under what conditions will an attempt be made to give progressive autonomy to certain domains of mass communication? That is, when will an attempt be made to conceive organisms of diffusion that do not necessarily respond to the cultural themes of the enemy and that, instead, anticipate the new culture, understood as the emergence of the social practice of the masses.

A second question derives directy from the first. What sectorization or what fragmentation of the audience will be kept or introduced? The multiple parceling of the bourgeoisie is, as a matter of fact, only a mask, given that it does not attend to different and opposed class interests; the profile of the receiver who unifies the multitude of messages and who assumes a universal and naturalistic normative character gives way to a model of the aspirations and euphoric possibilities of the petit bourgeois man. Behind the magazines designed for the clientele of sports spectacles and behind the magazines that gravitate to the high and medium sectors of the female population there is an implicit, unifying image that corresponds to the parameters of the dominant culture. It is precisely here that we find the organizing criteria of the separation of clientele groups; our judgement of the bourgeois genres leads us to state the necessity of attending first and foremost to *class interests* and to get away from laws that may seem to apply only to ecological and biological domains. Although one of the fundamental political tasks of the moment is to penetrate and win the middle classes (the nature of which must be determined), one cannot take this objective as the only one and, more important, one cannot view this necessity as an exclusive and exhaustive distributor of norms. On the other hand, it is not a matter of establishing a closed universe of rurality, for example, but one of analyzing the norms of efficient media programming (semantic fields, adequate diagrammation, and previous studies of decodification) to pick up the rural reality and permeate it with the universality of class meanings, of class existence, and of class struggle. This means that a medium of diffusion directed to a rural audience is not allergic to the news and the problems of the urban proletariat (and on the contrary, consolidates the peasant-worker union).

Another question that arises from the adoption of the principle of genres is the lack of flexibility that they represent. Once a closed and rigid relation between a kind of stimulus and a sector of the public has been established, it is very difficult for the magazine or the program conceived according to this or that genre to experience substantial modifications. In other words, there is a latent conflict between the rigidity of the law of the market and the flexibility with which one must approach the class struggle, from the point of view of themes as well as sectors to reach and penetrate. The class struggle prevents a message from being fixed, delimited, or controlled without considering the proper dynamics of a process, its alternatives, and its evolution.

Implicit Contents

Having adopted the bourgeois genres, the government forces try to perpetrate, through the assumed form, a process of reorientation of media contents.

Although the bourgeois genre represents the first form of conciliation with the laws of the market and also—it must not be forgotten—the only form that is known and recognizable by the possible transmitters and the possible receivers, the election of so-called cunning or hidden content is justified through the same arguments. It is redefined as a tactic of penetration in a context of receptivity defined by an adverse, dominant cultural power. It is inserted into the laws of democratic pluralism, which prevent or limit the emergence of another cultural universe seen as antagonistic to the experienced, dominant culture. In sum, the mechanism of penetration that is identified as implicit corresponds to an intention to win over uncommitted sectors; it is the art of conquering the middle levels of the population. The genres that are retained and that continue to appeal to the same basic incentives, the forms that are adopted, and the themes that are treated are all means to a single end: that of not frightening away a penetrable public. This strategy supposes that a public still exists that can be won for the socialist cause, that sectors of opinion are not deeply frozen, and that the sectors of the class struggle still have a certain elasticity. The new youth magazine *ONDA* was adjusted to this perception and this analysis of Chilean reality.

What, then, are the implications of this way of approaching the class struggle? One first source of confusion, the doctrine of hidden contents, contains an inherent fault in an erroneous approach: the approach that adopts the same unilateral notion of politics as the bourgeoisie, a notion that confuses propaganda, party, and contingent politics so that the password of hidden contents many times hides a fear and an incapacity to escape bourgeois cultural forms. The hidden meanings in the bourgeois system of transmission of messages correspond to a style of living and organizing social relations that we experience daily even if we do not let it show. On the other hand, implied meanings employed by the forces of change are part of a superstructural effort, that is, an effort not rooted in generalized ways of daily life. Take the example of a fashion chronicle in a magazine designed to straighten out the expectations created by living in a society that continues to be bourgeois. What happens? We want to reduce the importance of fashion as a collective determinant. We want to show its colonizing character, reducing decisions, tastes, and personal determinations. We want to tell of its direct relations with economic and cultural imperialism. Finally, we want to unmask the nature of alienated social motivation. But the photo that accompanies the chronicle and that shows an expensive, copied, universalized, "alienated" style, is well suited to make one find the girl or woman who wears it sweet, beautiful, elegant, and "in." The demystifying power of the chronicle is nullified in that the daily revitalized bourgeois project continues to affect tastes, appreciations, desires, and dreams in a generalized way. The chronicle is no more powerful than the plots of two Lillputians against Gulliver.

A second criticism that can be raised against the concept of implicit content as an efficient weapon in the superstructural struggle is the following. By concentrating on the objective of ideological penetration of a determined sector of the population, one forgets other sectors, which, through their votes, are supporting the forces of change, but which need a tireless impulse to create conscience and nourish and deepen the process of class struggle. This has nothing to do with supplying these friendly sectors with watchword messages. It has nothing to do with promoting an improvement in their adhesion and their participation in a program of change, but instead with taking into account the formation of the new man and considering all aspects of change and their approximation to the new culture. At this point the concept of implicit meaning must take another definition. It is easier to perceive it through an example. Take the case of a youth magazine that tries to penetrate the preceding levels that we have mentioned. Implicit in this case becomes the inclusion of every fact that logically appears to be an element of a larger structure: for example, the necessity to place the seizure of the university by a rightist group within an analysis of the strategy of these political sectors and within an understanding that goes beyond the trivial and ends with a challenge to the concepts of justice, equality, legality, and violence that these sectors use to justify their rejection of the revolutionary project. In this same example, the explicit would limit itself to the expression of the first level of reality, that is, a recounting of the facts and a partisan, even sectarian, evaluation of the event. In this last example, we see the necessity of always complementing an explicit level with an implicit one in order to create conscience. The implicit contents are, in our concept, those that reveal and disembowel the mechanisms of unconsciously experienced ideological domination, including even the most trivial facts.

The problem of the explicit and the implicit meanings within a revolutionary mass communication policy is intimately associated with the survival of news criteria inherited from the bourgeoisie. Essential elements of the mercantile idiosyncrasy in which communication activities are circumscribed (also) termed criteria or rules of selection) constitute the bronze law of mass culture in "modern society": sensationalism. This law is no longer concentrated in trivial and vulgar form in the yellow press, but is integrated subtly into all products of the mass culture imposed daily by the bourgeoisie and imperialism. Focused through the lens of sensationalism, the fact, the event that makes news, sells products, and competes is kept isolated from other facts that prepare it and allow it to exist and is, at the same time, kept separate from the multitude of actors who made it, leaving in evidence a mere result—an event, the birth of which, like a mushroom, is without roots. An event that makes news unusual, against the nature of things, out of all normality, out of time and space, and separated from the future and the past, becomes an ephemeral and anecdotal fact. It has the transitory character of every object of consumption. The magazine that one buys every week is a closed world, a whole of news and reportages, a whole of "news-making news" that next week already is obsolete. Next week other events will occur, other news that will give one the impression of "another new thing" that polarizes one's attention and so on successively through all the weeks of the year. Once consumed, one can throw away the magazine; the issue that follows does not need the preceding one to be understood and does not participate in the accumulation of knowledge, of conscience. Reality is an immense redundancy of *weekly gossip*.

In this sensationalistic universe, various ingredients—which help this operation of parting from "normal" reality, through strong sensations—are preferred: sex, crime, magic, and sports. These ingredients originate even more closed worlds; sex news, for example, creates a world governed by its own rules and has nothing to do with politics or the daily life of the individual.

The law of sensationalism is no more and no less than compliance with and obedience to the immediate; it is the fact that creates sensation at the moment. To follow this law means to select the relevant facts in an anarchic way. We know that news anarchy is used by the bourgeoisie as a way of subjugating news products to their order. "Objective" information given by the bourgeoisie is in reality always prepared and enters into a plan of implicit organization for the reader. The decodification of the news appeals to social meanings or to connotations that are managed and universalized by the bourgeoisie. It is precisely to prevent the class enemy from supplying his readers with a total and coherent vision of the system that the bourgeoisie has imposed this way of transmitting reality. The daily paper and the magazines permit consumption of false "news" (le faux nouveau), false movements, and false change. This law of sensationalism is what gives the mass communication of the Left an epiphenomenical character, separated from daily life. It also generally subtracts efficiency from the Left in the ideological struggle. This sensationalism is not only seen in its vulgar form in the populistic press, but is generally mixed in the norms that precede the selection of material treated in the products of the state editorial enterprises. This is seen in the absence of pedagogical pages and programs in magazines and television. They choose to deliver news, events and spectacles to fill free time, taking over the leisure function of the bourgeois medium. They do not include formats that favor didactic and demystifying themes on the informative reality manipulated by the media that are against change. Instead, they deliver materials for direct use in the specific practice of certain social groups, for example, chronicles about photosynthesis and the use of fertilizer, included in so-called popular informative magazines. Moreover, there are no magazines or programs directed to specific sectors conceived of from this point of view.

This lack, for the rest, leads to a broad questioning of the forms that mass communication must assume. In the first place, one defines the publics one wants to reach. As we have seen above, there must be criteria capable of substituting for existing genres. In the second place, this new possible definition of mass communication implies asking oneself about the monopoly that the professional group of journalists exercises over the elaboration of communicative material. A sports magazine, conceived in a revolutionary way, that is, a magazine that is not exclusively for consumption of sports events, but that takes into account physical education, health, popular sports, games, and soon, has no reason to be exclusively in the hands of "sports" journalists. A much more adequate place for its elaboration and distribution would seem to be the sports organizations of the state. The same thing, and even more, could be said about a magazine of basic education. This, of course, means that the state editorial effort would converge with the established infrastructure breaking its monopoly hold on the elaboration of messages. This convergence tends to occur in various domains, but suffers serious difficulties in the traditional journalistic domain. The necessity of incorporating didactic materials and deepening the information to escape the superficiality of sensationalism is directly connected with the urgency of providing incentives for the participation of mass organizations as well as state institutions and university research centers in the massive elaboration and transmission of practical and theoretical knowledge. This point of view would have the university redefining its research themes and establishing corresponding organisms to make proper use of intellectual work. With systematic planning the interest of the mass public could shift to high-level research, although one must recognize the possible need to reformulate the materials elaborated by these centers in more easily understandable language. As can be seen, the ideology of journalism is not the only one affected by this reformulation of mass communication. At the same time, the sciences become objects of inquiry. Perhaps an organic collaboration between the elaboraters and disseminators of knowledge would be one field to study in aiding the emergence of a new journalism; a second field might be the association of both research and dissemination with the reality that is experienced by the masses. Thus one could prevent mass communication from becoming an adulteration and a vulgarization of products of an elitist vision that conforms to the practice of leisure.

The Democratization of Distribution

One of the preponderant obstacles to the democratization of mass communication and its utilization as an instrument for cultural agitation is, without doubt, the kind of relationship with the public imposed by the traditional distributional system. This distribution works with a profile of individualistic receivers that goes hand in hand with the proposition of atomizing the mass of receivers and, in the end, demobilizing them. As in all domains, the guilds that hold the monopoly set by the traditional mercantile system oppose any transformation. In the specific case that refers to magazines, for example, the distribution agents and the newsstand owners can boycott new products, yielding to the organized pressure of rightist competitors. It has been proven more than once that there have been organized campaigns to prevent the display of magazines produced by state organizations in Chile. Among the parallel tactics that the Right has used to torpedo the forces of change, one finds that the newsstand owner represents the last link of a chain of domination and manipulation. His vulnerability is graphically expressed by the tininess of his place of business (the popular newsstand), by the ease with which it can be tipped over, and also by the scarcity of capital that he has available. By elaborating such examples one can mark all the obstacles that the leftist media must confront that place them on an unequal footing. First of all, there is an overproduction of magazines. Added to the bourgeois magazines that have circulated for years are the magazines and other editorial products inaugurated this year by the state editorial enterprises.

The Right has also substantially increased the number of its productions, so that the newsstand owner, instead of distributing 10 magazines, as he did a year ago, now has to display and sell 30 magazines. Between August and December 1971, the number of magazines that were available in the newsstand rose from 47 to 81. The Right is adopting a policy of overpopulation of editorial products. Thus, it has been proven that a private person during the last months has thrown six different youth magazines, supporting the status quo, into the market. Four of these show a deficit. The logical question is, "Where do the funds come from?" The magazines accumulate inside his stand as well as on the walls. They neutralize each other and this very sharp competition serves as a point of reference to devise a cover that decisively calls attention to the magazine. Behind this mountain of formats and colors the newsstand owner prefers to buy and sell a reduced number of expensive magazines that occupy less of his physical space and give him a certain profit. A great number of less expensive magazines must be sold to make the same profit. That is to say he, indirectly, introduces more pressure for setting and maintaining a high price for a product intended for popular diffusion. The magazine *La Firme,* which strives for popular political education, has had to raise its price, change its cover, and add colors to its interior since it began competing in the newsstand market. To deepen this example we can perceive a contradiction between the contents and the implicit profile of the reader for whom this political publication is intended and the real buying public, composed of the middle bourgeoisie. These readers come to the newsstand to buy and consume the themes of mobilization and swallow both the greatness and the misery of the proletariat. Fortunately, this obstacle was overcome through the direct association with popular organizations that can use the magazine as a source of information, knowledge, and conscience. Naturally, mass communication will cease to be an object of consumption available in the market place only if the way of access to the public is changed. There must be a conscious reception that does not gravitate to the values of individual entertainment but that assures the incorporation of the magazine as one more element for discussion and ideological growth of popular organizations. We must dwell somewhat on this point in the next section, because it refers to the idea of the creation of popular cultural power.

The Participation of the Masses

We all know the unilateralism of the transmitter-receiver model of the communicative circuit that the bourgeois concept of communication has imposed. From the top to the bottom, that is, from the transmitter that spreads the superstructure of capitalistic production to a receiver who constitutes a base, neither the concerns of the majority nor their own ways of life are reflected, but rather the aspirations, values, and norms that the bourgeois and imperialists deem most convenient. They impose a message prepared by a group of specialists on receivers whose only participation in the program they will consume is to periodically lend themselves to listener polls about the commercial feasibility of an already created product program. These mercantile inquiries are, as a matter of fact, plebiscites about themselves. They are, finally, the network of the numerous sophistics that form the bases of bourgeois domination. A revolutionary process has to do with the demystification of this concept of the colonization of one class by another, inverting the authoritarian terms that generally disguise a paternalistic face, and reestablishing the relationship between base and superstructure. That is to say, it has to do with making the mass communication medium into an instrument through which the social practice of dominated groups reaches its culmination. The message is no longer imposed from above, but the people themselves are the generators and actors. The mass communication medium loses its epiphenomenological and transcendental character by removing the national bourgeoisie and the imperialists from their roles as curators and judges of the culture. Consequently, the notions of freedom of expression and press freedom lose their abstractness and acquire substance. This same process of concretization rescues the privilege of expression from the monopolistic minority. The notion of freedom of expression turns out to be a class utopia.

It is in order to break out of the mercantile system and make the mass communication medium *a medium of communication of the masses* that the "talleres populares" (popular workshops) have been proposed. These would be interwoven with the mass organizations in which one discusses and criticizes and from which rises the cultural creations of the organized masses. In these "talleres" the magazine is only one of many elements to mobilize organized groups.

It is here that the policy of communication finds its real objective, that of mass mobilization. Its rhythm and intensity depend on the revolutionary transformation of the communication apparatus. This is a fundamental problem that communication specialists generally pass over. In a society in transition toward socialism the analysis of the forms and practices of communication must also deal with the relationship that unites the masses with their vanguard lthe political party. But that is another question (Mattelart, Biedma, and Funes,1971).

REFERENCES

Dorfman, A., and Armand Mattelart. *Para Leer al Pato Donald.* Valparaiso, Chile: Universidad de Valparaiso, 1971.

Mattelart, A., and CyL. Castille, *La ideologia de la Dominacion en una Sociedad Dependiente.* Buenos Aires: Signos, 1970.

Mattelart, A., M. Piccini, and M. Mattelart. *Los Medios de Communicacion de Masses: La Ideologia de Prensa Liberal en Chile.* Santiago, Chile.

Mattelart, A., P. Biedma, and S. Funes. *Communicacion Masiva y Revolucion Socialista.* Santiago, Chile: Ediciones Prensa Latinoamericana, 1971.

CHAPTER 29

Mass Communications and
Cultural Revolution:
The Experience of China

DALLAS W. SMYTHE

A nonprofessional "China watcher" approaches the on-the-spot study of the Great Proletarian Cultural Revolution (hereafter abbreviated to Cultural Revolution) modestly, aware of possible gaps in his knowledge of Chinese history and culture.* This feeling of humility is somewhat tempered by realization of how often the "experts" on China have been arrogantly wrong. If, as in my case, he is concerned about studying mass communications in the Cultural Revolution, the problems become manageable. This is because the Cultural Revolution, in its origins, substance, and consequences, has been a communications revolution in a profound and historically new sense.

In its planning and analytical phases this study presented unique methodological and theoretical problems. Could I take the essentially Lasswellian formula of Western communications theory (Who says what to whom with what effect?) as my frame of reference? To do so would mean I was assuming that some "who" people were administering messages through a given heirarchically organized structure of mass communications institutions (TV-radio, newspapers, cinema, books) to some 800 million Chinese (the "whom") and they would experience these messages with "desired" and "undesired" effects. The typical Western view has been that the "who" in this model for China was a tightly knit web of authority known as the Chinese Communist Party and that the remaining 97% of the population were a "target audience" to be manipulated into obeying the "who."

This model has suited the Western capitalist nations in managing their cultures and training the minds of their populations. Paul Lazarsfeld long ago

* The research on which this is a partial report is supported by a grant from the Canada Council.

criticized the inadequacies of such "administrative" communications theory and advocated critical theory to little avail. I anticipated that the conventional "administrative" theory would not fit the Chinese situation, and as appears below, it would not. What was needed was a model that would take account of a process by which some 99.9% of the Chinese population were using all modalities of message systems to transform quickly their culture (or if you prefer, consciousness) into one based on unselfishness and service to mankind.

Was I to study the uses made of mass communications media in the narrow strategic and tactical sense during the Cultural Revolution? Or the transformation of institutions (including, among other things, mass media) during the Cultural Revolution? I chose the latter. The appropriate conceptual tools seemed to be those I have written about earlier (Smythe 1960, 1962). They regard the content and process of communications in an institutional and policy context. So I inquired, "What is the nature of the policy that determines the identity (or nature) of the mass media products and services to be produced? What is the source of control for the communications institutions (i.e., how are they constituted, supported, and managed)? What is the process for innovating new kinds of such products and services? What is the policy on availability and allocation of such products and services? Finally, looking at the fundamental character of the social system in which the nation exists, how do the communications policies and institutions relate to the creation of a new social order (socialism/communism) or the maintenance/restoration of an old social order (capitalism/feudalism)?" (Smythe, 1971).

THE CONTEXT OF THE CULTURAL REVOLUTION

A social system is based not only on a legal system that defines rights to property, but also on its unique and more or less coherent pattern of institutions, traditions, values, and attitudes. Popular culture, fine arts, mass communications agencies, educational, religious institutions, and so on, are specialized social habits that are "carriers" from generation to generation of the values, attitudes, and traditions that give scope and meaning to the lives of the people living in a social system. Even the forms of the family, the methods or organizing business enterprises and recruiting people to work in them, and the methods of doing research and development and of innovating new technology grow out of and condition the future consciousness of a people as to what they should expect from themselves and the social system. In this sense a social system is a mass communications system comprised of innumerable message subsystems, of which we characteristically in the West think only of publishing, TV-radio, and cinema as mass communications institutions.

We should recall that in the West the medieval social system had a certain coherence and ideology that lasted many centuries. And that when the modern nation-state and modern capitalism emerged there was a drastic revolutionary period prolonged over three centuries that changed the institutional patterns, attitudes, and values from those of feudalism to those of our time. We must come to recognize that a transition began in 1917 to a world in which a quite new sort of social system—socialism/communism—embraces a growing part of the world's population. When the Communist Party of China completed its seizure of power on the mainland a short 24 years ago, the only major available model from which it might learn how to build a socialist system was the Soviet Union. Both of those socialist systems faced the necessity of accepting or rejecting in whole or in part the pattern of institutions, values, and attitudes that they inherited from their predecessor system. The Cultural Revolution in China is a rapid, drastic process, initiated by Chairman Mao Tse Tung to conduct this process of screening and reforming the Chinese pattern of institutions, values, and attitudes—or ideology or political consciousness, as they prefer to call it.

To understand the substance of the Cultural Revolution it is necessary to remember a few facts about Chinese culture in 1949. Life for the vast majority of the Chinese population was miserable; death from malnutrition, disease, or exposure was a constant possibility. Consequently, progress for them has been measureable in terms of solutions to problems that long since ceased to be problems for the bulk of North Americans. As many as four out of five Chinese lived in rural areas and about 75% of the population were poor and middle peasants. All of the heads of families in 1949 had spent their formative years at least partly in the generally corrupt system of the Kuomintang and most of the grandparents in 1949 had been born and reared under a Chinese emperor in a feudal system. The very restricted Chinese contacts with Western science, education, culture, and technology had only fanned out from the imperial enclaves that Western nations had carved out of China, mostly along the seacoast, after the English first forced the opium trade on China in 1841. Because intellectuals were a particular target of the Cultural Revolution, it is necessary to look a bit closer at them.

In 1957 there were about 5 million intellectuals of all kinds, or less than 1% of the population (Mao, 1957). Of this total, Mao estimated that from 1 to 3% (50,000 to 150,000) were hostile to socialism. Another 10% (about 500,000) were Communists and sympathizers; the remainder, nearly 90%, ". . . have the desire to study Marxism and have already learned a little, but they are not yet familiar with it. Some of them still have doubts, their stand is not yet firm and they vacillate in moments of stress." According to Chiang Ching, the small number of literary and artistic intellectuals were still influenced by the literature and art of the 1930s in the Kuomintang areas of China where the left wing movement was in the hands of bourgeois nationalist-democrats, many of whom had become antiscocialist.

Apart from their relationship to Marxism, of course, the Chinese intellectuals were influenced by the Confucian tradition by which a very conservative and elitist educational system instilled elite values and a contempt for practical work. It seems obvious that Chairman Mao feared the continuation of this tradition even in diluted form in the bureaucracy that the Party had developed since liberation. In fact, "the old scholar class still existed, controlling educational institutions and administering some of the various organs of state" (Wheelwright and McFarlane, 1970, p. 105).

The middle class and remnants of the upper class also represented a threat to the development of the socialist ethic and program. As Joan Robinson (1971) points out, whereas the old middle class in the Soviet Union was practically wiped out by 1920, the greater part of the Chinese middle class welcomed the victory of the Communists over the corrupt Kuomintang and provided necessary expertise to the new government in running the schools, industries, and government offices that the largely illiterate peasantry was incapable of doing in 1949. As she also remarks, the Rightists in the Party were often of middle-class origin themselves (Robinson, 1971, p. 14). Land had not been nationalized in the cities before the Cultural Revolution and landlords still received rent from it. Of the former capitalists, some were employed to manage industries while many lived on the 5% return on their former investments in business enterprises that had been nationalized.

Finally, to finish this summary of the position of the intellectuals and bourgeoisee, the attitudes in China toward Western technology were ambivalent. There was hatred of the imperialists whose enclaves in Shanghai, Canton, Wuhan, and soon, bore signs: "No dogs or Chinese allowed." There was envy of the elegance, luxury, and comforts that they had seen the Westerners enjoy. There was a feeling of Chinese inferiority and lack of confidence in technological matters where the West had so long a lead. Consequently, technical experts tended to become arcane and arrogant in relation to the workers and peasants.

Against this backdrop, the foreground of the context of the Cultural Revolution is the running policy debate over the mode of national development in the economic sphere between 1952 and 1962. On the one hand were advocates of a gradualistic model of development, resting on massive investment in heavy industry, electrification, and transport. The logic of this position required emphasis on experts in technique to accomplish the desired economic growth. Inevitably this position also implied dependence on administered plans, material incentives, and market control mechanisms. The model for this line of policy advocates was based on Soviet experience.

On the other hand there were those who, with Chairman Mao, insisted on raising the question, Development for what? If the object of the exercise was indeed socialist man, was this the best way to do it? Arguing that the first model rested on the assumption (of Stalin) that when public ownership of the means of production had been established there was an end to class struggle, the advocates of the second course insisted that the Soviet Union had merely substituted bureaucratic managers for the capitalist managers and that socialism was incompatible with an unrecognized subordination of workers and peasants to a class of managers. Alternatively, the second group advocated policies that relied on the "mass line" by which administration would be decentralized and integrated with the activities of peasants and workers. Instead of heavy centralized investment in huge enterprises, there would be decentralized, diversified investment under direction of workers and peasants. Instead of gradualistic growth there would be leaps in development in which qualitative change in the forms and methods of production would accompany the development of new institutional modes based on the ethic: "serve the people." Instead of depending on experts imbedded in massive, heirarchically organized chains of command, socialist culture would emerge from reliance on the transformation that class struggle between workers and peasants on the one hand and bureaucrats and other bourgeois elements on the other hand would provide.

In essence, the second group said, socialist man is a "widget." No one has seen him yet. We must trust the integrity of the process of class struggle rather than assume he will result from the ministrations of a bureaucracy. The problem is to motivate people to devise institutional processes that will lead to the mobilization of collective energies and dedication to service to the whole people. The problem of socialist development is to be solved by finding a procedure for turning the problem over to everyone. Instead of motivating people by material incentives, we must release the creative potential in human beings to serve the common welfare. To do this we must recognize that the key problem is to change people's consciousness. Hence the economic problem of development can best be tackled under socialism by putting politics in command of culture and production. Otherwise, they said, the policy of putting profits (material incentives) in command will surely lead to the restoration of capitalism—by which they meant the abandonment of the socialist road.

Even a brief sketch of the context of the Cultural Revolution must take into account the Chinese fear that they might take the capitalist road, as they considered the Soviet Union to have done. American aggression, by its long-standing intervention in the Chinese Civil War (its military shield around Taiwan), its military adventures in Indochina, and its implacable economic boycott was an everpresent menace. And Chairman Mao and his supporters were critical of the policy of the Party itself. They felt that it was more concerned with its own self-preservation and its members' status and political power than with pursuing the revolution. Elitism and concern with bureaucratic procedures and factional power cliques were leading to revisionist policies. It relied too much on internal disciplinary measures such as mild rectification campaigns and self-criticism and avoided mass criticism and open reliance on the masses of poor and middle peasants, workers, and soldiers. And its recruitment policies favored technical and managerial skills over ideological qualifications. Of compelling significance was the fact that in the 1960s the first generation of postliberation Chinese youth would come of age. In the mid 1960s, 40% of the population were less than 18 years old (Nagel, 1968). They had not been tempered in the hardships and idealism of the anti-Japanese and Civil Wars. Would "new elements" emerge in China to promote liberalization of the regime, as American officials openly anticipated?

Chairman Mao's own writings were an important part of the context of the Cultural Revolution. In 1942 in "Talks at the Yenan Forum on Literature and Art," he had spoken directly to the problem that became the center of the Cultural Revolution. "To defeat the enemy we must rely primarily on the army with guns. But this army alone is not enough; we must also have a cultural army, which is absolutely indispensable for uniting our own ranks and defeating the enemy" (Mao, 1942, p. 1). He saw literature and art as powerful weapons for helping the people fight the enemy "with one heart and one mind." Attacking elitism among writers he urged that they learn from the peasants and recounted how in his youth he had held petty bourgeois attitudes toward manual labor and the dirt with which peasants worked. After living with workers, peasants, and soldiers, "I came to feel that compared with the workers and peasants the unremolded intellectuals were not clean and that, in the last analysis, the workers and peasants were the cleanest people and even though their hands were soiled and their feet smeared with cow-dung, they were really cleaner than the bourgeois and petty-bourgeois intellectuals."

THE CULTURAL REVOLUTION

What was the Cultural Revolution? I begin by speaking of what it was not. To the confusion of those China watchers who think in terms of coups d'etat, personal power factional struggles, feuds over foreign policy and internal economic policy, regional versus centrist power struggles, and soon, it was none of these. A recent book by a group of RAND people says

> While considerations of power thus came to play a role, the conflict was still far from being, at its heart, a classic power struggle. It was not fundamentally an unprincipled strife arising out of conflicting personal ambitions and loyalties. Even if we hypothesize that internal tensions and disagreements within the leadership led to the emergence

of informal alignments and interest groups, there is still no evidence that these coalesced into power factions operating with a distinct and relatively consistent membership and dedicated primarily to the seizure of power. [Robinson, 1971, pp. 86–87]

One of the authors of that book sees it as not a master plan of Chairman Mao but as "a logical series of spontaneous eruptions." In importance they place it alongside the fall of the Ch'ing Dynasty in 1911 and Liberation in 1949.

The Cultural Revolution was a nationwide debate to clarify goals and policies: to define the "socialist road" and the "capitalist road" and to organize themselves to pursue the former. It began as a debate, sparked by Chairman Mao, over specific plays, books, and operas and their ideological effects and implications. It spread to universities and schools, as students criticized the elitism and careerism of their intellectual mentors. It fanned out into criticism of Party functionaries at all levels, culminating in a focus on Liu Shao-ch'i, President of the Republic, who had been appointed as the successor to Mao Tse-tung. It involved the entire government apparatus as well as all mass organizations at all levels. And when Chairman Mao went "to the masses" by putting up his own big character poster, "Bombard Headquarters" on August 5, 1966, it encouraged the nascent Red Guards, workers, peasants, and soldiers to make the debate truly nationwide. Following these years of escalation of the issues in debate, the climactic phase of the Cultural Revolution was from mid-1966 to the end of 1967; this phase involved sporadic outbreaks of violence in some cities. After 1967, with the aid of the unarmed PLA men, the debate was channeled into the creations at all levels of new administrative organs, the Revolutionary Committees, and the process of criticism, struggle, and transformation within and through those committees.

Expressive of the centrality of the arts and all culture in the debate is this statement by Chou En-lai on May 1, 1966:

A socialist cultural revolution of great historic significance is being launched in our country. This is a fierce and protracted struggle of "who will win," the proletariat or the bourgeoisie, in the ideological field. We must vigorously promote proletarian ideology and eradicate bourgeois ideology in the academic, educational and journalistic fields, in art, literature, and in all other fields of culture. This is a key question in the development in depth for socialist revolution at the present stage, a question concerning the situation as a whole and a matter of the first magnitude affecting the destiny and future of our Party and country. [Quoted in T. W. Robinson, p. 93]

While analysis of mass media perfomance *during* the Cultural Revolution is not the purpose of this chapter, several comments on this are necessary.

While there were spectacular instances of denial of access to the press to critics of the party heirarchy, in general Chairman Mao's views found adequate national press coverage, particularly through the organs of the PLA. In the heat of debate and controversy at the level of cities and provinces, control over access to the mass media was tactically crucial. For example, an account of Shanghai experience reported that

> The first move in seizing power was to capture the organs influencing public opinion. The chief daily newspapers of Shanghai were taken over by the rebels on 4 and 5 January. They could begin to make their own propaganda, in the same way as the reactionaries had begun their campaign in 1962 by working on public opinion. [J. Robinson, p. 61]

The long-term perspective that permeates the literature on the Cultural Revolution is one of continuing struggle on the ideological cultural issue.

I quote one typical statement:

> . . . during the whole historical stage of the dictatorship of the proletariat, there exists the struggle between the bourgeoisie's attempts at a comeback and the proletariat's efforts to oppose it. It is either the dictatorship of the proletariat over the bourgeoisie or vice versa. . . . So long as classes and class contradictions exist, class struggle is inevitable, as is the struggle for the seizure of political power. Who will win in the revolution is a question which can be solved only over a very long historical period. Therefore, we must develop the thorough-going revolutionary spirit of the proletariat, and be prepared to wage arduous, long term struggles, to struggle for decades, for a full hundred years, or even several hundred years, until the socialist revolution is carried through to the end and the transition from socialist to communist society is completed. If we do not understand the problem in this way, if we believe that after one or several cultural revolutions there will be peace and tranquillity in the country and thus slacken our militancy and lose our vigilance, we might get on to the path of capitalist restoration, just as the Soviet Union has done. We must not in the least lose our vigilance on this question. [Editorial of *Hongqi*, 1967 in Chiang Ching]

"MASS COMMUNICATIONS" BEFORE AND AFTER

Policy on Service or Product

Before the Cultural Revolution the policy in all the "mass media" had a somewhat bourgeois, liberal character. (All comment on precultural revolution policy comes from present workers in the institutions.) Mass media content consisted largely of "general interest" material conducive to individualism, feudal ideas, and foreign values.

> The radio programs were feudal and bourgeois in tendency. So we broadcast a story of elephants making love in the zoo; about beautiful islands with birds and snakes; instruction in arranging flowers in vases; about cries of animals; instruction in cooking; about the color of clothes and dancing in holidays in parks; on how to play chess. [Interview, Kwangchow]

The editorial policy of the provincial party paper in Wuhan was to deemphasize politics and print something to interest everyone. The Pearl River Film Studios said that in most of their drama films, emperors and beauties dominated the stage. One film, *Five Golden Flowers*, praised love to the skies. After seeing it boys went seeking golden flowers—pretty girls. Films produced before the Cultural Revolution failed to reflect and explain the heroic image of soldiers, workers, and peasants; some distorted those images and made them ugly. For example, *Adverse Wind Blows a Thousand Li* "seemed to mirror heroic deeds of the PLA, but in effect it uglified them and praised and glorified Kuomintang reactionaries." Some newsreels were bad, for example *Silver River and Golden Net* endorsed reliance on the state for capital investment instead of self-reliant hard work. Another propagandized for the free market in the countryside. Some science films praised expert intellectuals instead of the reality of intellectuals being educated by peasants, workers, and soldiers. Film distribution and exhibition in Kwantung province had the profit-making test applied under Liu Shaoh-Ch'i. Rural cinema teams went only where the crowds were and avoided remote rural areas. Wide-screen movies that served few and neglected the masses were introduced in the cities. At that time generators for mobile units weighed 135 kg.

In the Jan Li Primary School, "East is Red" District, Nanking, before the Cultural Revolution, the educational policy considered that intelligence came first. The textbooks were full of feudalist and capitalist material. The teachers believed students should study the books and pay no attention to practice, politics, or current affairs. The tradition was that *(a)* to be a teacher is unfortunate (echoes of "Our Miss Brooks") and *(b)* teachers deserve respect simply because they are teachers. At Wuhan University, the teaching of literature was that it was a minor part of life, concerned with private individual pleasure; foreign languages were taught as dead languages. Our traveling interpreter reported that her English vocabulary had been learned from studying Jane Austen at the Foreign Language Institute in Peking; she had had to reeducate herself after graduation. At Hsinghua University, Peking, students were led to adopt the aim of becoming officials. Some from homes of worker or peasant origin concealed their parents' status when the parents came to visit them. Students "took knowledge as their personal, private property and the professors were responsible for this." In 1958 the majority of students had followed Chairman Mao's line and started to run factories at the University. In 1960, Liu Shao-Ch'i and his agents counterattacked against the 1958 reforms. They said the educational revolution in 1958 was too disorderly and there was too much politics in the University.

In the postal service (in all China) I was told that Liu Shao-Ch'i's policy had been to put "profits in command." In 1959 after the Great Leap, 90% of the communes had post offices and 98% of the production brigades had regular postal service; by 1962, 50% of the former and 23% of the latter had been cut off as unprofitable.

According to Chiang Ching, the policy on literature and publishing before the Cultural Revolution was heavily influenced by bourgeois critical realism. Advocates of "truthful writing (e.g., Hu Feng) sought out the seamy side of life in socialist society and the rotten things inherited from pre-Liberation." There were followers of the "broad path of realism" theory, which held that each author should write whatever he pleased according to his "different personal experience of life, education and temperament and artistic individuality." They wanted writers to abandon the worker-peasant-soldier orientation and explore "new fields which would give unlimited scope to their creativeness." Shao Chuan-lin, formerly Vice-Chairman of the Chinese Writers' Union, held the theory of the "deepening of realism" and advocated writing about "middle characters"—riddled with inner contradictions, summarizing "the spiritual burdens of individual peasants through the centuries," and presenting the "painful stages" of the peasants' transition from an individual to a collective economy. He held that writers should write about "everyday" events to "reveal the greatness in trivial things" and attempt to show "the rich diversity of the world in a grain of rice." Other writers (Tien Han, Hsia Yen, etc.) opposed writing on proletarian themes. They proposed writing on "human interest," "love of mankind," and so on. Chou Ku-cheng argued that "the merging of various trends as the spirit of the age" should guide writers and that the trends merging included "pseudo-revolutionary, non-revolutionary and even counter-revolutionary ideas" (Chiang Ching, pp. 23-25). The performing arts before the Cultural Revolution presented imports from the West such as *Swan Lake*, Western music, films from other socialist countries, and dramas such as Wu Han's on the Ming dynasty officials Hai Jui and Yu Chien, which were understood to be semiovert political propaganda against Chairman Mao.

As a consequence of the Cultural Revolution, the content of these institutions is strictly governed by the "Thought of Mao Tse-tung." The operative formulation of the policy is "either revolutionaries or reactionaries will make propaganda for their own purposes. There is either freedom for the people or for the bourgeoisie. We don't allow bourgeoisie freedom. We put our program openly—to serve the proletariat. This is very different than in western countries where the sham is freedom and the reality is selfishness. The sole task of our propaganda is to consolidate the dictatorship of the proletariat and to prevent China from having a capitalist restoration" (Li Shu-Fu, Responsible Member, Department of Political Work, Kwangtung Provincial Revolutionary Committee).

Apart from foreign news and newsworthy public events (e.g., visits by foreign diplomats or officials, an Afro-Asian Invitational Ping Pong Tournament, etc.), domestic news in newspapers, radio, and TV generally is chosen and edited to pass on to the masses of peasants, workers, and soldiers the fruit of the application of theory for them to practice. That is to say, "typical" examples (Learn from Tachai, Learn from Taching) of socialist construction. "Typical" in this context means a model to be emulated. The process by which such news is obtained and the process for educating the professional journalists will be described below. *Peoples Daily* in Peking prints six pages daily of which the first four are domestic news and the last two international news. Provincial and municipal dailies usually run four pages of which the first three contain domestic news and the fourth international news.

Radio programming—and radio is much more important than TV in China —is about 50% occupied with news in Wuhan of which two fifths is relayed from Peking. There is a total of 13 hours, 30 minutes of programming in three segments per day. Weather and river-level reports constitute 7% of the time. The remainder, 43%, is classed as art. It consists of Modern Revolutionary Peking Opera relayed from Peking, local productions of the same repertoire of operas, revolutionary music (songs from the anti-Japanese and Civil War period, modern patriotic songs), and music from Chinese ballet. This distribution of programs is typical. Total hours per day of broadcasting differ from city to city. In Kwangchow the total is 18 hours/day.

Shanghai may be taken as typical of the TV program fare. It totals about 2 1/2 hours/day, six days a week in evenings. News takes 30 minutes, literature and art take 60 minutes, documentaries take 50 minutes, and children's programs take 15 minutes. Some evenings are devoted, apart from news, to full-length presentations of revolutionary opera, ballet, or concerts. Because facilities for remote pickups are limited, extensive use is made of filmed material in Chinese TV. Hours of service vary between cities. Of those visited, Peking and Kwangchow had the most—about 180 minutes/day, six times a week. Wuhan broadcast four times a week for 3 hours each time. Nanking broadcast on Wednesday and Saturday for a total of 2 to 3 hours.

After watching many hours of Modern Revolutionary Peking opera—in theatres and on film and TV, I can report that it is very colorful, dramatic, tautly staged, well choreographed, and extremely popular with Chinese audiences. The themes are consistent with the Thought of Mao Tse-tung; they deal with the heroic struggles of the masses of workers, soldiers, and peasants against oppression by foreign imperialists or domestic class enemies.

Cinema production after the Cultural Revolution follows the following general principles: consolidate the dictatorship of the proletariat; serve socialist construction; show the great achievements of workers, peasants, and soldiers, and intellectuals and students in the class struggle, and the struggle for production in scientific experiment and in remolding world outlook by study of labor. I saw impressive documentaries from the Pearl River Studios (Kwangchow) and the Shanghai Film Studio. And I heard descriptions of feature length films, for example, one now in production in Shanghai, the *Barefoot Doctor*. My impression is that the cinema studios are just emerging from the phase of criticism, struggle, and transformation. Among the documentaries observed was one on a new commune being built on Hainan Island by volunteers from overcrowded Kwangchow, one on a newly developed technique for making cement ships, one on the use of acupuncture anesthesia, and one on the technique of rejoining severed fingers. I also saw one example of the Chinese alternative to Walt Disney, a feature length film made before the Cultural Revolution, which, in beautiful color cartoon form, told the tale of the two little girls tending their sheep in the high grasslands. The Kwangchow Film Studios now specializes in 16 mm and super 8 for rural audiences.

The Jan Li Primary School teachers are now teaching for the revolution, practicing the Thought of Mao Tse-tung, and there has been improvement in teaching method and textbook contents. There is now a small workshop in the school and we saw children practicing their woodworking by making serviceable furniture for kindergarten use. Now, if the teachers teach something wrong, the students can criticize them. Before, teachers did not pay attention to students after school hours. Now there is a Mao Tse-tung propaganda team for out-of-school activity. This team meets on Sunday and reviews its work after studying Mao Tse-tung's writings and Marx and Engels. The team does some good deeds, helping neighbors to clean rooms, cook, and take care of small children. They try to learn from Dr. Norman Bethune: care for others first and pay no attention to self. The teachers work with this propaganda team. At Wuhan University the study of literature is now concerned with critical analysis of literature in relation to society. Foreign languages are taught as live languages involved in daily life—and they are trying new teaching methods in this field.

The Chinese postal service has restored rural service so that half of China's communes now have post offices, while 97% of the 700,000 production brigades have regular daily pickup and delivery service. As early as 1951 the Chinese post office began to perform the distribution function for newspapers (soliciting and collecting subscriptions as well as physically distributing them); since the Cultural Revolution they have enlarged this educational task (as they view it) to also organize farm reading groups of workers, peasants, and soldiers and nominate a "reading person." Among many other service innovations particularly interesting is the initiation of rural pickup (as well as delivery) service for parcels and remittances. In nine months in Shantung province in 1971 this service saved 1,730 million work days for peasants who otherwise would have had to travel to post offices to mail or pickup their parcels and remittances.

The effect of the Cultural Revolution on literature and performing arts has been to enforce a process of critical scrutiny of past and imported art forms. As the Chinese are quick to admit, it is easy to agree that the policy should be to "make the past serve the present," but it is difficult to carry out in practice. And one finds such policy statements as:

In the cultural revolution, there must be both destruction and construction. Leaders must take personal charge and see to it that good models are created. The bourgeoisie has its reactionary "monologue on creating the new." We, too, should create what is new and original, new in the sense that it is socialist and original in the sense that it is proletarian. The basic task of socialist literature and art is to work hard and create heroic models of workers, peasants and soldiers. Only when we have such models and successful experience in creating them will we be able to convince people, to consolidate the positions we hold, and to knock the reactionaries' stick out of their hands. On this question, we should have a sense of pride and not of inferiority. [Chiang Ching, 1968, p. 12]

As the protracted experience with the Renaissance should have taught bourgeois artists and intellectuals, it takes a long time to develop art forms and spirit hat optimally express the perspectives of a new social system. There will be experiment, debate, and controversy over the appropriate art style for socialism for centuries to come. In China it is recognized that "Literature and art can only spring from the life of the people which is their sole source" (Chiang Ching, 1968, p. 14). Accordingly, their policy is to rely on the masses, "follow the line of 'from the masses, to the masses', and repeatedly undergo the test of practice over a long period, so that a work may become better and better and achieve the unity of revolutionary political content and the best possible artistic form" (Chiang Ching, 1968, p. 15). It is therefore necessary to encourage literary and art criticism in the hands of the masses of workers, peasants, and soldiers and integrate professional critics with critics from among the masses. The present formulation of method is "combine revolutionary realism with revolutionary romanticism."

EXAMPLE: Before writing the script of *Barefoot Doctor*, the creative team (writers, director, cameramen) went and lived and worked with barefoot doctors in communes for many months. From this they learned the perspectives on life of peasants and barefoot doctors. This is practicing revolutionary realism. Then they considered the experiences of barefoot doctors in all of their diversity of personal and group experiences, and from this distilled the heroic model of *the* barefoot doctor that best exemplified his best characteristics. This is practicing revolutionary romanticism. Moreover, there is much emphasis on encouraging workers, peasants, and soldiers to produce original works of art and literature based directly on their own life experience. Such works, in the Chinese view, may have a validity, vitality, and directness superior to those of more sophisticated "professional" efforts.

By January 1972 two related tendencies were obvious in literature and art. On the one hand a diversified genre of publications for different levels of age and education and in minority languages as well as "common speech" is being published to serve directly the soldiers, workers, and peasants. This includes children's and adults' cartoon books, "reportage" telling stories of socialist construction, "morality tales" of selfless heroes among the masses, illustrated scripts of Modern Revolutionary Peking Opera, and scientific and technical books to serve production in agriculture and industry (including *Ten Thousand Whys*—a scientific-technical encyclopedia for young people). On the other hand, intellectuals are being encouraged to edit traditional literature and to write new books. Thus, Kou Ji Kan is editing 24 volumes of Chinese history written in the Ching dynasty and covering 2000 years. Famous Chinese classical works (*Water Magi, Dream of Three Kingdoms,* and *Dream of the Red Chamber*) and literature of foreign countries will be published soon. Indeed, this has already begun to happen since I left China. Hou Zhan, author of a long book before the Cultural Revolution, is now writing *Brilliant Path*. Meanwhile, the publishing of Marxist-Leninist classics and of the works of Chairman Mao continues. From 1966 to 1968, 150 million sets of the four-volume collected works of Chairman Mao were published—14 times the total previously published. The famous "quotations" is available in large type for the poor vision of the elderly and in miniature pocket size for the convenience of the field or factory worker.

Finally, new institutional forms of cultural information, education, and entertainment have sprung from the Cultural Revolution, and traditional forms have been revived, remodeled, or expanded. A junior version of the Red Guards, the Little Red Guards, age 7 to 16, study the Thought of Mao Tse-tung, do "good deeds," and perform patriotic songs, dances, and scenes from Peking Revolutionary Opera. In Hupeh province, with a population of 38 million, there are 80,000 Little Red Guard groups with from 10 to 50 members. In the cities, "children's palaces" are teeming with activity—in Shanghai there are 11. These offer education (moral and political) and activities in the arts (including Peking opera and patriotic songs and dances) and sciences for children age 7 to 16 after school hours for short terms. Admission is by recommendation from the schools. Apart from all this there are 105 art troupes who are professional singers, dancers and acrobats constantly performing, as well as an almost infinite number of such "amateur" groups in factories and schools and communes. And no description of contemporary China should omit reference to the colorful display posters and billboards on socialist construction themes, the "big character" slogans, and the wall newspapers which are very much in use.

Policy on Organization and Control

The Pearl River (Kwangchow) Film Studio exemplifies the effects of the Cultural Revolution on the organization and control of communications and many other institutions. Before the Cultural Revolution, the leadership of the Party had failed to take root in the work of the studio. Experts dominated the studio and technique ruled everything. Motivation•was by material incentives— awards, salaries, and royalties (the same as in capitalist motion picture industries). Isolated from the workers, peasants, and soldiers, the film workers followed their own personal interests and served themselves. Their feelings were more with the emperors and beauties they portrayed than with the workers, peasants, and soldiers. Their outlook was bourgeois and so was their product. After the struggle, criticism, and transformation, the studio repudiated the autonomous experts and now an "open door" policy is followed. Now the masses are the beginning and end of the film-making process. The artists go to work and live with the masses to learn from them. They then make proposals for films to be produced arising from this experience. These proposals are submitted to the Provincial Revolutionary Committee's Department of Literature and Art. This body may also invite proposals on topics appropriate to its propaganda plan. It will also assist in the selection of the "advanced unit" that is to be the topic of the film.

A similar basic procedure is followed in the newspapers and broadcast stations, except that here the relationship has been massively institutionalized. *People's Daily* refers to its network of nonprofessional correspondents in every commune and factory in China as a mass movement itself. They had no statistics on the total but said that there were 300 correspondents in one county alone; as there are over 2000 counties in China this suggests as many as a quarter to a half million correspondents. From them a typical daily flow of articles and reports to the *People's Daily* is from 500 to 600— around 200,000 a year. The initiative arises at either end, for while correspondents send in articles voluntarily, they are also requested to write articles on events of particular interest in Peking. The tactical policy is everywhere to seek out outstanding examples of socialist construction and get amateur correspondents (who are workers or peasants) to write them. The editorial function is to select from this vast flood of reports those appropriate to the short-term strategic line of the propaganda plan (e.g., last year Chairman Mao asked people to read Marxism and apply it, so in that period articles were selected with respect to this line). Finally, individual articles are selected by the editors on the basis of their profundity, vividness, and inspirational and educational quality. An example, given by the *Hupeh Daily* (Wuhan), was a 600-character article, written by a poor peasant who was on a propaganda team at a school. Entitled "Enter the School Three Times," it said that the first time he came to the school, he was a cowboy. It was before liberation and he merely came to look as he did not have the money or qualifications to enter. He was driven away by an attendant. The second time he came to the school it was to enroll as a student after liberation. He was rejected by the bourgeois authorities running the school because of inadequate preparation to enter—and was sent away again. The third time he came to the school was after the Cultural Revolution when he had been invited to come as a member of the peasants' propaganda team. He then told the students how he had suffered under the old regime, using this experience to illustrate it.

A crucially important part of the new policy for operating the communications media is the rotation of their professional staff. While the precise practice differs from place to place, the *South Daily* (Kwangchow) is an example. At any given time one third of the professional staff is on the job at the newspaper; one third is working at the May 7 Cadre Training School for upwards of six months; and one third is working in communes and factories (for from three to six months). All members of the staff study Marxism-Leninism and Mao Tse-tung Thought and run a small farm as well.

All newspapers and broadcast stations have a correspondent system analogous to that of *Peoples Daily*. Broadcast stations also invite peasants, workers, and soldiers to come to the studio to make broadcasts. For example, the Wuhan Peoples Broadcasting radio station presented talks by workers, peasants, and soldiers 530 times in 1971, on their experience and achievements in studying Marxism and applying it to production. Peasants, workers, and soldiers are also invited to the broadcasting stations to make talks to the professional staff for their education. In fact, as the Wuhan Peoples broadcasting station responsible person told me, the masses now run the station, whereas before the Cultural Revolution professionals did.

Similarly, the masses run the schools, the communes, the factories, and the "living quarters" (housing developments, we would say). Hsinghua University's program is now tightly linked with the problems and practice of workers and peasants. It is now administered by a Revolutionary Committee of 31 members of whom seven come from the teaching staff, seven from the students, three from leading cadres, five from laboratory workers and other full-time assistants, one from families living on the campus, three from the Workers' Propaganda Team, and five from the PLA Propaganda Team. Each department has its own Revolutionary Committee, smaller in size but similarly constituted. The offices of president and deans have been abolished; teaching reform groups replace the deans and the responsible person for the University Revolutionary Committee is a young man who was a steel worker until he went to the campus as a member of the Workers' Propaganda Team. Budgets are discussed (and investigated) by the Revolutionary Committee and the whole academic community and finally determined by the Municipal Revolutionary Committee.

Innovation Policy

Possibly nowhere in China is the ingenuity better displayed in making new institutional arrangements during the Cultural Revolution than in regard to technological innovation. At the Institute for Telecommunications Engineering, Peking, before the Cultural Revolution the emphasis was on intellectual knowledge and technique above everything. Students were taught to aspire to be members of an elite class of intellectuals who would make famous careers for themselves in positions in the hierarchy out of private exploitation of knowledge as if it were private property. In the process they were divorced from practice, from the socialist reality, and cloistered with books. Children of soldiers, workers, and peasants were elbowed away as unqualified or dropped as stupid if they failed their examinations. Now, they strengthen political and ideological work and criticize bourgeois ideas. Teaching materials are being reformed to integrate theory and practice. Students are sent to the Institute on recommendation from their work places; workers with five years experience continue to receive their wages; with less than five years experience, workers get free board and lodging as students. The educational and research policy is based on Mao Tse-tung's proposition that practice leads to cognition, which leads to improved practice. So the process is for the educators first to go to the factories and operating systems of electronic equipment to find out what kinds of skilled workers they need and what technical problems need to be solved. They also interview former students to find out what they had not learned in the Institute that they should have. The teaching plan has four stages: (1) teach elementary theory; (2) practice in a simple workshop: learn to handle the productive process and quality controls and actually make integrated circuits; (3) higher theory; (4) further practice: students are required to design and trial produce some process or equipment, that is, to solve a practical problem. As with all other institutions there is a rigorous program of exchanging work experiences between factories and from the factories to the Institute.

Peking University economists told me that before the Cultural Revolution the workers were not allowed to make innovations; experts had to be convinced first, so innovation came only from experts who revered foreign machines and methods. Now the full mobilization of the masses bears on management and innovation. Problems are attacked by "Three-in-One" teams. To solve difficult technical problems the "great battle" is used—bringing in workers and experts from other places to help solve the problem. Emphasis is on finding their own solutions to the problems, which are often simpler and equally as efficient as more complex and costly foreign techniques. There is no patent system. Innovations are spread as quickly as possible by means of ad hoc meetings, the press, and documentary films.

As they are the first to point out, China is somewhat undeveloped technologically. One would expect then to find that they are most concerned to solve problems arising from that fact in the most economical and self-sufficient manner possible. This they do in, for example, cinema. Serving a huge and mostly now inaccessible (by rail, air, and highway transport) rural population, they rely on mobile cinema exhibition teams to bring films to the peasants. Before the Cultural Revolution, a mobile 16 mm projector weighed 74 kg and a mobile generator between 100 and 200 kg. As a result of research and development work during and since the Cultural Revolution, the Nanking Film Machinery Institute has redesigned both of these instruments, using lighter weight material. As a result they reduced the weight of the projector to 32 kg and the generator to 28 kg, and these models are now in quantity production. A still newer model projector weighing 28 kg will go into production in 1972. At the same time electricity consumption of the projector has been reduced by 45% and light intensity on the screen increased. The research and development process involved an overall system analysis (although these workers had never heard of Western systems analysis theory) by a steering committee composed on the three-in-one principle (experienced workers, students from Nanking Science and Technology Institute, and plant technicians with some book knowledge). The steering committee broke the problem down into parts and charged other three-in-one teams with solving them. Some teams worked with film projection teams in communes and mountains. By theory and practice, they solved the problem. How do they learn from Western technology? Mostly they do not. They say they solve problems according to their specific needs. These people had not heard of video cassettes.

TV receivers in China—all monochrome—are almost all owned by factories and communes and are used by groups of 25 to 100 viewers. I was told in Kwangchow that they plan to "popularize" black and white TV throughout the province within five years. They are also considering innovating color TV and have research and development underway in it. They have in trial production, surprisingly, a large screen TV receiver (2 by 3 m), and are currently debugging it. If indeed the Chinese innovate a large screen TV receiver soon they will have a spectacular example of leapfrogging ahead of Western technology instead of "snailing after it" as the "capitalist readers" preferred to do. They are also trying to lower the cost of conventional TV sets that currently sell for 300 to 400 Yuan (roughly 150 to $200).

If this analysis of research and development in elemental terms of the "three-in-one" teams fluidly moving from theory to practice and back to theory and more practice sounds simplistic, it is realistic. In a country as huge as China the genius of the Cultural Revolution seems to be that overall policy is deliberately kept as simple and general as possible so that it invites a maximum of dispersed and diversified initiative. It really is as simple as it seems.

Policy on Availability

If space permitted I would expand the analysis of this topic. For the present, a few highlights suffice. In general, the most significant aspect of the policy change because of the Cultural Revolution is the abandonment of the profit standard that had been imposed by the regime identified with Liu Shao Ch'i, and the pursuit of a general policy of serve the people without regard to microcost considerations. In practice this means extended and innovated services and price reductions.

The wired-radio system has been greatly expanded since the Cultural Revolution in terms of the number of communes and counties with their own program-originating equipment, and in terms of the proportion of production brigades and teams equipped with loudspeakers. The province of which Nanking is the capital now has 68 county-level wire-broadcast stations—10 times the number in 1965; the province also has 8 million loudspeakers as compared with 400,000 in 1958. A parallel expansion of the over-the-air radio broadcast receivers has taken place. The saturation with radio broadcast receivers is impressively high in urban areas (between 80 and 90% of households in Kwangchow and Wuhan, 50% in Nanking) while in rural areas it is lower and statistics are in general unavailable. In Shanghai some 24 different brands of radio receivers are on sale generally at prices beginning about 35 Yuan ($17), and many of these contain short-wave bands. Moreover, radio parts and supplies are commonly available in retail stores and are bought in large quantities. With the general level of technical knowledge what it is, virtually everyone in China with a desire to have access to a radio receiver can have one. No license fees are charged for the use of radio (or TV) receivers, nor is a record kept of who buys radios or the parts with which to assemble them. Noticing this and remembering the Western statements to the effect that the Chinese Party prevents the Chinese from listening to foreign broadcasts, I raised the question in various places. The answers added up to this:

In China quite a number of sets have the capability to receive foreign programs (e.g. medium wave in South and South East China from Hong Kong and Taiwan; medium and short wave from Japan, Okinawa, Guam, the USSR and Korea). No one could stop the people from listening to them if they want to. The people have liberty and can discriminate between what is true and what is not. When the Soviets attack us, we publish the attacks for people to know. In the U.S., the people are denied the truth about China. In China, the people can decide for themselves. We want as many people as possible to have radios. We trust and learn from the people. You have seen how freely they are sold. (Wuhan). The government and the Party do not forbid people to listen to foreign broadcasts. Radio sets are sold without a record of sale. Prices of sets were reduced by 30 percent last year. We did some audience research in our communes. We found some listening to foreign broadcasts, especially by young people. They are curious. The foreign programs said that the Chinese do not have enough to eat or to wear. They knew it was a lie. They asked the older people about it. The older people recalled the past bitterness when it had been true, before liberation. But in general, the Chinese people are not willing to listen to foreign broadcasts; they prefer to listen to their own stations. But if they listen to the foreign stations they have the ability to evaluate for themselves what they hear. If we wanted to prevent listening to foreign stations why would we manufacture complete radio receivers, including allwave receivers? We could manufacture single-channel receivers as the Japanese did during their occupation of China (Kwangchow).

The broadened availability and reduced price of cinema also must be mentioned. In Kwangtung province in 1966 there were 800 cinema projection teams serving the rural areas; by 1971 there were 1600—between 15 and 20 teams per county. All cinema admission prices have been reduced by 40% since the Cultural Revolution. There is still much progress to be made; before the Cultural Revolution in that province each commune could see a cinema every three months; now it is reduced to every two months. And the demand is avid. In the 13 months between August 1970 and October 1971, in Kwantung province (which has a population of about 40 million), there were 140 million admissions to see modern revolutionary Peking Opera films.

Among the temporary casualties of the Cultural Revolution is the ambitious over-the-air TV educational program, which in 1960 had some 35,000 students in 1000 centers in Peking alone. This program is still closed, pending completion of the teaching reform process.

CONCLUSIONS

In light of the fundamental character of the social system in which the nation exists, how do the communications policies and institutions relate to the creation of a new social order (socialism/communism) or the maintenance/restoration of an old social order (capitalism/feudalism)? To raise this question is to take Western communications theorists into generally unknown territory, for their concern with studying the relation of communications to social change is conspicuously undeveloped except in relation to such behavioral changes as advertising might cause in brushing teeth. (Smythe, 1954). To deal with it requires the capacity to handle some theory of social change, for example, Marxism. A fruitful approach would be to regard a whole social system as a complex of message systems each of which plays a significant role in perpetuating (or restoring) an old social system or innovating a new social system.

Looking at the Cultural Revolution in this context I would agree with Professor John G. Gurley.

Communist China is certainly not a paradise, but it is now engaged in perhaps the most interesting economic and social experiment ever attempted, in which tremendous efforts are being made to achieve an egalitarian development, an industrial development without dehumanization, one that involves everyone and affects everyone. [Gurley, 1971, p. 29]

Four salient features of the real situation stand out in considering this question regarding China since liberation in 1949.

In the first place it was a huge, poor, and predominantly precapitalist country in terms of the bulk of its technology and ethos. It is huge both in material resources and, more importantly, population. One of the implications of its hugeness is that it has the potential for self-sufficiency, not only nationally but regionally. And a second, commented on earlier, is that it is almost ungovernable from the center in a direct, authoritarian style; decentralization of power and dispersion of responsibility (with appropriate motivation) is invited by its enormous size. In respect to its absymal poverty at the time of liberation and the changes since, someone has referred to its progress "from misery to poverty." Its misery in 1949 is understated when the Chinese refer to the "past bitterness." And its "poverty" today is only so by Western standards. Driving through Wuhan to the airport I noted the children playing in the streets outside their meager houses, the adults washing shirts in dishpans on the sidewalk (no running water in the houses), and the people carrying celery cabbage home for dinner. I reflected that it is the quality of this material level now as compared to preliberation that proves to every Chinese that socialism is a success. For now everyone has, in effect, an insurance policy against having *no* housing, *no* shirt, *no* food, and preventible diseases. That huge Chinese population was so miserable in 1949 that the progress wrought under the leadership of the Communist Party is to them unsurpassably great. By speaking of China in 1949 as being predominantly precapitalist in technology and ethos I refer to the fact that the great bulk of the population—the peasants and villagers—were not impregnated with the ethic and the discipline of capitalist industrialism. In this respect the Russians in 1917 were much further down the capitalist road of development than the Chinese in 1949. So, in effect, in China after 1949 there was not the predisposition toward the capitalist road in practice to which broader prior experience might have inclined them.

The second salient feature is precisely this "Chineseness" culturally. For 3000 years of unbroken practice, the role of morality, the collectivity, and the law in China has been very different than in the West. As the Nagel *Encyclopedia-Guide* says:

Western traditions place much importance on institutions, rights and the contract which limits individual liberty in the common interest. These ideas have never been prevalent in China; laws were considered as a last resort to be appealed to when morality failed to find an answer to all difficulties. [Nagel, 1968, pp. 135–136]

In short, possessive individualism in the Lockean tradition was not indigenous in China; it was an import from the West in the coastal cities where bourgeois structures and attitudes developed in the century after 1841.

The third salient feature of China's situation since 1949 is its isolation from Western culture and technology. In unintended and unconscious cooperation with the thrust of Chairman Mao's policy, the United States' total embargo on trade and cultural contacts with China, imposed during the Korean War, was probably one of the most valuable assets China had in gestating the Cultural Revolution. Not too long afterward, the Soviet's withdrawal of technical assistance left the Chinese entirely on their own. So in January 1972, Li Shu-fu, the responsible member for political work of the Kwangtung Province Revolutionary Committee said to me:

Our country is still somewhat backward in science and technology. We need a large amount of investment to develop agriculture and industry. We must rely on our own efforts to accumulate capital rather than buying from abroad or exploiting other countries. We can't depend on foreign loans or on exploitation of people. What we can rely on is hard struggle and economy. We can't afford extravagance and waste. The broad masses have developed this good idea: oppose corruption and waste . . . There are many western correspondents who spread the idea of three super powers in the world. From our side we have no such idea. One of Chairman Mao's instructions in the New Year's Day commentary says that the issues in the world should be handled by discussion and consultation. Internal affairs should be handled by people in that locality. It cannot be solved by super-powers. We will never stand beside the USA and the USSR as a super power. We will give firm support to peoples wanting to be independent. We participate in the UN to expose the two super-powers all over the world. And to work with developing nations to resist exploitation by super-powers.

This necessary drive for self-reliance is dynamically related to the first two features that I have discussed; it strengthens them and they it.

The fourth feature necessary to evaluate the Cultural Revolution is, of course, Marxism. And the Chinese seem to be making themselves distinctive among Marxists by their total commitment to its humanitarian, social ethic. "Serve the people" is the slogan that expresses it everywhere in every context in China. To place the interest of others ahead of one's own is the highest virtue in China today. We were reminded countless times of the immeasurable gratitude of the Chinese people to the Canadian doctor Norman Bethune who gave his life, serving with the Red Army in Yenan, and of the fact that Chairman Mao said of him, "We must all learn the spirit of absolute selflessness from him. With this spirit everyone can be very useful to the people." The Chinese are proceeding on the assumption that the cynics who think that most people most of the time must always act selfishly are wrong. Their Cultural Revolution operates on the assumption that self-interest and selfishness are distinguishable, and that if self-interest is harnessed to social ends, not only will social goals be approached faster, but self-interest itself can be transformed into "other interest." So today the purpose of art and literature in China is not to cultivate individual experience, sensitivity, adornment, or idle exploration of human values. It is to guide and coordinate joint human effort to inspire, implant, and cultivate values compatible with collective welfare. By and large, Westerners will be disappointed in the next generation at least if they expect otherwise from the Chinese. Jumping from feudalism to socialism without going through the inescapable individualism of capitalist-style technology puts the Chinese in the position of having to create *sui generis* art forms and styles that are appropriate to unselfish, socialist men and women.

Linked with the emphasis on serving the people is the companion ethic of egalitarianism in the sense that the Chinese insist that the development of socialist man means that all men and women must develop together. Here they depart sharply in both theory and practice from both capitalism and Soviet-style socialism. The Chinese believe that public ownership of the means of production is a necessary but insufficient condition for socialism. One of the prime objectives of the Cultural Revolution was to eliminate the class conflict between the bureaucracy-technocracy and the masses by starting a process by which the masses must become self-reliant masters of science and culture. The economic policy of decentralization of industry and self-sufficiency has the humane purpose of avoiding the alienation of individual men from the bureaucratic, industrial processes of overblown megalopolises. Instead of building on the best (of individual intelligences and abilities, or of natural locations), the Chinese have deliberately opted to build on the least able. As Gurley (1971, p. 23) says:

Experts are pushed aside in favor of decision making by "the masses"; new industries are established in rural areas; the educational system favors the disadvantaged; expertise (and hence work proficiency in a narrow sense) is discouraged; new products are domestically produced rather than imported "more efficiently"; the growth of cities as centers of industrial and cultural life is discouraged; steel, for a time, is made by "everyone" instead of by only the much more efficient steel industry.

They prefer to accept the "opportunity cost" of temporarily lost absolute efficiency because they believe that in the long run the material productivity of the country will rise faster, but more importantly because, by leaving no one behind, they are creating a society of people who will be happier and better able to respond to the world than if they took "the capitalist-revisionist road." For those who worry about the ideological purity of such a policy, we are reminded that Marx anticipated that in communist society there would be no specialization of work as we now know it, and that individuals would practice being masters of many fields of learning and art.

It would be quite misleading to leave the impression that the Chinese have reached Utopia or even are positively bound to do so. There will be many zigs and zags in the development of Chinese policy and institutions. As Chairman Mao has repeatedly reminded people:

> The class struggle is by no means over. The class struggle between the proletariat and the bourgeoisie, the class struggle between the different political forces, and the class struggle in the ideological field between the proletariat and the bourgeoisie will continue to be long and tortuous and at times will even become very acute. The proletariat seeks to transform the world according to its own world outlook, and so does the bourgeoisie. In this respect, the question of which will win out, socialism or capitalism, is still not really settled.

And, as this chapter should by now have made clear beyond the need of amplification, the Chinese consider that the arts, literature, and the manifold media of communication are everybody's business and are particularly an area where as first among equals, politics must be in command.

REFERENCES

Chiang Ching. *Summary of the Forum on the Work in Literature and Art in the Armed Forces with which Comrade Lin Piao Entrusted Comrade Chiang Ching*. Peking: Foreign Languages Press, 1968.

Gurley, John G. "Capitalist and Maoist Economic Development," *Monthly Review Press. 22*, 9 (February 1971), pp.15–35

Marx, Karl, and Friedrich Engels. *German Ideology*. New York: International Publishers, 1939.

Mao Tse-tung. *On Literature and Art*. Peking: Foreign Languages Press, 1967.

———. "On the Correct Handling of Contradictions Among the People." February 27, 1957. Quoted in *Quotations from Chairman Mao Tse-tung*. Peking: Foreign Languages Press, 1957.

China—Encyclopedia-Guide. Nagel Publishers. Geneva, Paris, 1968.

Robinson, Joan. *The Cultural Revolution in China*. London: Penguin Books, 1969.

Robinson, T.W. (ed.) *The Cultural Revolution in China*. Berkeley: University of California Press, 1971.

Smythe, Dallas W. "Some Observations on Communications Theory," *Audio-Visual Communications Review* (Winter 1954), pp. 24–37.

———. "On the Political Economy of Communications," *Journalism Quarterly 37*, 4 (Autumn 1960), pp. 563–572.

———. "Time, Market and Space Factors in Communications Economics," *Journalism Quarterly, 39*, 1 (Winter 1962), pp. 3–14.

―――. "Cultural Realism and Cultural Screens." Paper for International Symposium, "New Frontiers of Television," Lake Bled, Yugoslavia, June, 1971.

Wheelwright, E.L., and Bruce McFarlane. *The Chinese Road to Socialism.* New York: Monthly Review Press, 1970.

CHAPTER 30

Authentic National Development versus the Free Flow of Information and the New Communications Technology

HERBERT I. SCHILLER

Not long ago, Leonard Marks, a former director of the United States Information Agency, observed:

More and more, as a nation of fact-gatherers and distributors, the United States spills out this enthusiasm over its borders. The American share in the world's knowledge industry assures it a special role which is too big to ignore. Sixty-five per cent of all world communications originate in this country. This is matched by a *long* lead in the production of information. [Marks, 1968, p. 9]

There is a need to review the implications of the global hegemony of American communications flows and media. In large part it reflects the shift of the United States business system in the last 50 years from an almost exclusive concern with domestic markets to far-ranging international operations. The bulk of international messages today are business-originated communications, most of which are transmitted by the few hundred United States multinational corporations whose branch plants and subsidiaries dot national landscapes across the world. These facilities represented, in 1971, an estimated $85 billion of direct capital investment (Wilcke, 1971, p. 1). Moreover, the powerful domestic United States knowledge industry, which attends to the production and distribution of information and which shapes and forms its consumers' consciousness, has become internationalized. Marks refers to this:

The output of our national knowledge industry is, of course, a tremendous resource. A problem occurs as this resource produces at a rate that is disparate with that of the rest of the world. If anything, the gap can be expected to widen in the coming years. [Marks, 1968, p. 10.]

In fact, image and information creation, managed by giant conglomerates in the United States, now has a global reach. What follows are only some of its visible dimensions: magazine publishing and distribution (there are overseas editions, for example, of *Time, Newsweek, Readers' Digest, Playboy* and similar items); films, made in the United States and rented and sold overseas, as well as films made outside the United States with American capital and distributed by American companies (revenues from film sales abroad now exceed domestic box office receipts in the United States); television films and programs, some of which are sold and distributed to as many as 100 nations; book exports, as well as joint publishing ventures and acquisitions of foreign book publishers; record sales and ownership of record and/or distributing rights of record companies; English-language newspapers with significant circulations throughout Asia, Africa, South America, and Western Europe; foreign-based United States advertising agencies and public relations firms that handle the media accounts of the foreign subsidiaries of major United States corporations; United States-owned market research and public opinion polling firms located abroad that provide data to whoever will pay for it, including a variety of United States private and governmental organizations.

This says nothing of the on-the-job image creation efforts *within* the overseas plants and business installations of United States companies.

An additional component of the international flow of communications is the still considerable military presence of the United States around the world. There is, of course, the exclusive and enormously developed worldwide military communications system of the Pentagon. On a more general level, and accompanying the global deployment of United States military installations, there are the scores of Armed Forces Radio and Television stations whose transmitters reach local audiences numbering in the millions. There are also the overseas facilities of the U.S. Information Services (Voice of America) whose transmitters broadcast to the world at large.

This by no means exhaustive catalogue of private and governmental international communications activity reflects the current preponderance of American communications media. This preponderance should be viewed in the light of the current global maldistribution of wealth and opportunity. Most recently, the President of the World Bank inquired "how great is child mortality in the developing world?" and supplied these statistics: "In India, there are large areas where deaths in the first year of life number as many as 150 to 200 per 1000 live births. In the United Arab Republic, the proportion of children between the ages of one and two who die is more than 100 times higher than in Sweden. In Cameroon, children under five, although one-sixth of the population, account for one-half of the deaths. In Pakistan, the percentage of children between ages of one and four who die is 40 times higher than in Japan." On the equally grave matter of employment, he reported: "On reasonable definitions—including allowances for underemployment—unemployment approximates 20-25% in most (developing) countries." And, "If past patterns continue, unemployment is bound to become worse" (McNamara, 1971, pp. 7, 11).

Clearly, the conditions of existence that these figures describe are intolerable. They are intolerable in a moral sense, of course, but at the present level of global interdependence they are a threat to the very survival of *all* human existence. If they are allowed to persist vast territories and populations will be convulsed and the conflicts will never be able to be contained. This is the import of The Declaration of Lima, the result of a meeting of 80 developing countries in Peru in November 1971. It warned that "indefinite co-existence between poverty and affluence is no longer possible" (*New York Times*, November 9, 1971).

The international economy must *quickly* offer its expertise and its resources to at least partially mitigate these tragic trends. In such an overall effort, communications systems and technology obviously occupy a high order of priority. But the concomitants of unrestricted technological transfer and of indiscriminate information flow are by no means an unmixed blessing for the developing countries. The prevailing view maintains that technical and scientific gains can be secured from the experience and the expertise of those who have already crossed the threshold of economic development. It holds that a society can in fact, select its advantages and sift out the negative elements in the cultural-informational "package" it accepts from without. Is there justification for this optimism?

It is frequently assumed that the scientific-technical component can be isolated successfully from the sheath of social institutions that encase it. An American-designed steel mill can be built in a socialist country without necessarily importing a structure of private ownership. And the mill's output presumably might be used for social consumption instead of private individual goods production. Similarly, atomic energy processes and biological discoveries would appear to be transferrable from one society to another without accompanying cultural conditions.

But the process is neither simple nor automatic. Theory and abstract research cannot be confused with application and practice. Most of the "borrowing" between societies involves the latter, and it is in the applied technical areas that the social forms of the politico-economic system are most likely to be embodied. It would be most unusual for a technology or a technical process not to bear the marks of the social system that produced it. In the words of Ivan Illich:

We have embodied our world view into our institutions and are now their prisoners. Factories, news media, hospitals, governments, and schools produce goods and services packaged to contain our view of the world. We—the rich—conceive of progress as the expansion of these establishments. We conceive of heightened mobility as luxury and safety packaged by General Motors or Boeing . . .

Rich Nations now benevolently impose a straitjacket of traffic jams, hospital confinements, and classrooms on the poor nations, and by international agreement call this development. [Illich, 1969, p. 20]

The argument can be extended. The automobile, for example, is the supreme artifact of a wasteful and privatistic culture. In importing an automobile-dependent transport system, more than mere mechanical methods of travel are being introduced. "Each car which Brazil puts on the road," observes Illich, "denies fifty people good transportation by bus. Each merchandised refrigerator reduces the chance of building a community freezer" (Illich, 1969).

It may be argued that these are extreme cases and that many other scientific-technical discoveries are less value laden. This may be so, but *key* technological developments are almost certain to reflect the social origins of their place of creation. Besides, it is also possible that the character of the work process, the managerial function, and the scale of the enterprise are socially influenced. Borrowing the technology may introduce social forms in the production process that carry with them their own imperatives.

But for the moment, let us accept with reservation the claim that scientific developments are in the main transferrable without the accompaniment of severely limiting social features. The same can hardly be said of nonmaterial, cultural products. By definition, cultural forms are the ultimate expression of the social character of a society. The image of neutral, value-free cultural forms is absurd. It is therefore, a cause for wonderment that information-communications flows, so inseparably linked to the sociocultural structure, can be promoted enthusiastically and uncritically. Endorsement of the knowledge industry corporations and by the leaders of decision-making bureaucracies are more understandable.

The torrential flow of information and communications, moving domestically and internationally, often passes unwary eyes and ears by claiming to be objective and value free. Concern with the content of the messages is largely eliminated by either becoming a vulgarized McLuhanite or by regarding a large part of the communications stream as "entertainment," a category that seemingly excludes for many even the possibility of influencing human behavior.

To repeat, the crisis in international communications derives from an unwillingness to handle the desperate informational (and other) needs of the global poor, the majority of mankind. This failure occurs at the same time as incredibly powerful new means of mass communications are developed for all sorts of other purposes, few of them essential and many of them undesirable. As noted already, a considerable portion of the communications facilities and technology are in the service of commercial or military aggrandizement. This has a double cost. Vital human needs are neglected. And, their incalculable power is made available for either facilitating systems of instantaneous repression, or for focusing on life-styles, social arrangements, and production objectives that may be disastrous to hundreds of millions of people.

In the handful of affluent nations mass communications and broadcasting in particular were never significantly involved in either the developmental or the formal educational process. Industrialization and mass literacy were generally achieved before the advent of radio. In the West, in varying degree, radio and later television turned their considerable capabilities to servicing commercial enterprise. Now, as an instantly accessible world emerges and scores of nations and billions of people are being made aware of possibilities for new living arrangements, the choice of meaningful communications-developmental models is critical.

The simple notion that the methods and the social arrangements of the few technically advanced societies are applicable to and desirable for the still developing states deserves the closest scrutiny. So, too, does the uncritical acceptance of the belief that the role communications has played in the West is relevant to the informational/developmental needs of the excolonial world. A rethinking of objectives is essential.

Today, communications *without selectivity* may be dangerously inappropriate for most of the international community. By virtue of their unique developmental history, the communications institutions of the Western industrialized nations may be worse than irrelevant for newly emergent nations.

THE NEW AWARENESS

For almost a quarter of a century a host of international agencies, largely dependent after World War II on the goodwill and support of the United States, created the *appearance* of a genuinely workable international community. These bodies functioned in the political (United Nations), in the cultural (UNESCO), and in the economic sectors (The International Monetary Fund and the World Bank), and in a variety of other specialized areas. For the most part, their common characteristic was an espousal of multilateral internationalist behavior. They encouraged free trade, free movement of capital and the free flow of information.

Undergirding all these international associations was the sometimes unacknowledged mid-twentieth century fact of politico-economic life. One country, by virtue of its economic and military power, was in the ascendancy. Most of the participating national states were either supplicants, clients, or at best, wholly absorbed with their own developmental problems.

The objectives of these very unevenly weighted international organizations reflected, not surprisingly, the aspirations of its most influential participant. Well-schooled in matters of public presentation, an outgrowth of domestic marketing experience, the goals of the United States policy managers were consistently couched in terms of universal advantage.

Most effective was the appeal to freedom, denied so thoroughly almost everywhere in the preceding war period, and so continuously absent from the centuries-old empires recently dismantled. Freedom was a theme that drew its supporters from disparate camps and at the same time nicely satisfied the needs of a newly emerged expansionist system that could scarcely rely on conventional forms for its outward-thrusting inclinations.

Accordingly, the free flow of information became the dominant cultural-communications theme in the postwar period. Attractive to intellectuals everywhere, it provided justification for the globe-encircling wave of commercial-recreational communications that swelled up out of the United States mainland after 1945.

An early indication that United States influence was no longer as omnipotent as it had been for the preceding 25 years was manifested in the first public challenge to the concept of the free flow of information. This came, remarkably enough, at the site of its origin—UNESCO.

In June 1969 at a meeting in Montreal, a report of a group of experts on mass communications and society noted that among many other matters:

One of the priority areas for research is the study of the role of mass media in conveying information and in helping to form attitudes about other people and other countries. While the media have the potential for improving and extending international understanding, intercultural communications does not necessarily or automatically lead to better international understanding. On the contrary, the opinion was expressed that what has come to be known as the "free flow of information" at the present time is often in fact a "one-way" flow rather than a true exchange of information. In these circumstances the need for "cultural privacy" tends to be asserted, and it is considered necessary to protect the cultural integrity of a nation against erosive influences from outside. However, the meeting recognized the dangers inherent in blocking any free flow of communications, and felt the whole subject was worthy of deeper inquiry.

The fact that the production of mass communication materials is largely concentrated in the hands of the major developed countries also affects the role of the media in promoting international understanding. Communication at the moment is a "one-way street" and the problems of the developing nations are seen with the eyes of journalists and producers from the developed regions; moreover, the materials they produce are aimed primarily at audiences of those regions. As a result, not only is the image of the developing nations often a false and distorted one, but that very image is reflected back to the developing countries themselves. [United Nations Educational, Scientific and Cultural Organization, 1969, pp. 3, 5]

A report to the conference by James Halloran, director of the Centre for Mass Communications Research in Leicester, England, made the same point even more strongly than did the assembled experts' report. Halloran declared:

In fact, the whole question of barriers to all forms of intercultural communication merits study. What are the factors—social, economic, political, legal, etc.—which determine the nature and degree of the import and distribution of media material? Moreover, it is usually taken for granted that inter-cultural exchange and improved international understanding go hand in hand. What justification is there for this association? We are not entitled to assume that the latter will automatically stem from the former. [UNESCO, 1970, p. 16]

This cautious calling to account of the free flow of information idea has been followed by other studies that look at the concept with at least equal reserve (*The Uncertain Mirror*, 1970, for example). Does this growing criticism of the mechanics of the free flow of information signal an eventual breakdown or at least a crippling of the worldwide flow of socially important information? Or does it instead represent the rejection of an altogether bogus internationalism and a modest step toward a new beginning?

Something of both interpretations may be sustainable—rejection *and* affirmation, temporary withdrawal *and* eventual reintegration. Which tendency will be the strongest in the time immediately ahead will depend on other developments still unfolding. These developments are visible both in the technically advanced and in the newly developing societies. Despite the obvious strength of America's global communications system—its far-ranging facilities and infrastructure, and its connecting links with basic power wielders such as the Department of Defense, the governmental bureaucracy, and the corporate industrial oligarchy—there are challenges to its authority. Technologically, the incredible developments in electronics have reduced the justification for concentrated control, economic or managerial. A "gale of innovation" has created a new communications technology that is capable of individual operation, local functioning, and low-cost acquisition. True, most of these possibilities are just emerging and the long tradition of private monopolization for enrichment will not easily be broken. But the main features of the new electronics—miniaturization and individuation—facilitate individual acquisition and use. The full dimensions of the technological leap are yet to be revealed. We cannot foresee the character or the magnitude of the social impact. Certainly the days of the communications monopolies are not yet numbered. Their adaptability cannot be underestimated. But now new options are at least imaginable.

Preceding the possible future breakup of monopoly communications control is an already observable and growing uneasiness of the population with its current informational fare. The staggering fact that in 1970 more than three fifths (62%) of the United States labor force was engaged in the production of services instead of goods suggests the structural changes that underly the rapidly shifting cultural currents and social attitudes in the country. Evidently, knowledge workers have different informational needs (Sorrentino, 1971, pp. 3–11, and Drucker, 1968). Although this is by no means a simple, one dimensional development (there are some very unpleasant and regressive features about it of which Agnewism is but one manifestation), it signifies to me, at least, an emerging receptivity to new communications initiatives. Linking such efforts to the technical opportunities now appearing permits a reasonable hope for substantial change in the years ahead.

But these events are still in the future, even if closer than we presently imagine. For the time being, the flow of communications and the structure of information making remain in the hands of the long-standing managers of power. The domestic population and the international public continue to receive the messages, recreational or informational, that sustain the American market economy and its built-in priorities. Change may be over the horizon but current reality prevails. While hoping for improvement, it is appropriate to deal with the world as it is. What does this suggest to those industrialized and politically autonomous states that find their destinies affected by their association with and their penetration by the powerful United States economy and culture?

Consider the cultural communications plight of Canada. This northern neighbor, to be sure, is somewhat atypical, with her 3000-mile boundary with the United States. Yet her condition of information/cultural saturation does represent one extreme on the yardstick of cultural penetration. For example, an American columnist reported on a recent trip to Ottawa:

On a given day, the newstand at the Skyline Hotel has on sale 66 American magazines, 11 French or French-Canadian, five British and four English-Canadian. Last Friday (Nov. 19th, 1971) residents of Ottawa area could choose their television viewing from eight stations, from programs including Romper Room, Captain Kangaroo, Dinah's Place, The Lucy Show, Concentration, The Beverly Hillbillies, Green Acres, Family Affair, Batman, I Dream of Jeannie, Peyton Place, As the World Turns, Guiding Light, Gomer Pyle, Dick Van Dyke, The Flintstones, Lassie, Andy of Mayberry, NBC News, CBS News, Carol Burnett, The FBI, Dragnet, Merv Griffin, and seven movies, six of them American. [Furgurson, 1971]

A similar picture is depicted in an official Canadian study:

The major obstacle to the well-being of the Canadian publishing industry, its authors, and to a Canadian literary culture is the existence of a powerful economy and a dramatic culture alongside us which operates and communicates in the English language. Its books, publications about books, its book clubs, literary reviews, radio, film, TV, its "hard-sell" apparatus in every form penetrate our cultural space without resistance. [Harvest House Publishers, 1971, p. 1]

Acknowledging Canada's unique situation by virtue of geography and language (excluding Quebec), it nevertheless shares with other developed *market* economies a general vulnerability to United States communications in their many forms. Since Canadian economic activity is lubricated with the same incentives and appeals that stimulate its powerful southern neighbor, the communications instrumentalities that work so effectively in Detroit and Cleveland are not likely to be less influential when they cross over effortlessly to Toronto, Ottawa, and Vancouver.

Those advanced industrial states that rely mainly on private initiatives and rewards for their economic performance will find their efforts to partially insulate themselves from the most powerful external communications networks difficult to realize. In the market economies, and especially in the United States, communications institutions and structures have developed essentially as instruments of marketing. The mass media, advertising, public relations, market surveys, public opinion polling, and even the school system are inseparable components of a communications apparatus fashioned to accommodate a privatized production and consumption economy. Therefore, for these societies the problem is mainly *within* the domestic institutional order. External penetration acts mostly as a reinforcing agent.

It is possible to imagine the introduction of a variety of *national* communications/cultural policies in Western European and other market economies. They would signify an increasingly fragmented and very unstable nonsocialist world system. The relatively free flow of (commercial) information is but one element, however essential, in a viable international market system in which capital, technology, raw materials, and messages circulate more or less freely. When market economies begin to curtail the circulation of any of these elements essential to a dynamic world system it is an indication of crisis in the *total* system. When individual nation-states with private market structures restrict informational, capital, or resource flows, the world orbit of the system is constricted and problems for all participants are intensified. Accordingly, national communications policy that imposes limits on the importation of United States films, TV programming, magazines, books, and other cultural/informational material is disruptive without being effective *unless* it is accompanied by far-reaching structural changes in the prevailing domestic economic order. This is so because the informational prohibitions will either be circumvented by external acquisition of domestic communications facilities, or the local cultural products will be fashioned according to the same commercial characteristics of the material that is denied entry.

We can only conclude that the efforts of individual national states, which are themselves organized around the market, to restrict the inflow of commercial informational cultural products will be unavailing insofar as their own communications environments are concerned. Communications, now as in the past, originate in and reflect the basic institutional structure of the society. If that structure is commercial, the government cannot exorcize external products fashioned by similar privatistic motivations.

Ferment in the Underdeveloped World

Industrialized market economies are constrained to accept, however reluctantly, the communications flow of the principal international message makers and transmitters. Less developed societies have still fewer options. If their political evolution has not given them at least a measure of autonomy from the global market process, their immersion in the international commercial informational stream is progressively intensified.

The communications penetration of less developed states comes about easily because their institutional structures are weak and generally unable to offer resistance. Consequently, communications networks can be imposed almost intact in these societies, bringing with them both the hard and the software of relatively sophisticated and thoroughly commercialized systems. Communications satellite ground stations, for example, are constructed in the midst of still primitive economies through arrangements sometimes shrouded in mystery with no safeguard against their unrestricted use (Telecommunications Reports, 1971). Television programming designed for advanced market economy audiences appears in dozens of nonindustrial states. Local entrepreneurs learn marketing techniques from the subsidiaries of worldwide advertising agencies, although the main business of these firms is to attend to the needs of the foreign branch plants and companies situated in their areas.

Thus, the informational infrastructure of the most technically advanced market society is transported to and recreated wherever its economic influence extends. The provision of news, the character of recreation, and the very definition of reality are more or less automatically supplied by the imported informational system. In analyzing "Why Peru Seized TV," an American correspondent described a media structure typical of most poor countries:

> Concentration of broadcast ownership in the hands of a few families; only 36% of the shows of Peruvian origin; commercials accounting for almost 40% of broadcast time; stations concentrated in a few large urban centers, not serving most of the people who live in the countryside. [Maidenberg, 1971]

Prospects

Communications choices of developing states are, in fact, quite limited. Substantial economic, technical, social, and political obstacles, which interlock at many points, restrict full freedom of selection of one or another informational course. Obviously, there is no comprehensive policy that can neatly satisfy all the authentic communications needs of people facing many combinations of problems. All the same, the severity of the limitations makes it easier to consider approaches that might otherwise be regarded as too elemental.

Although total insulation from the present international message flow is, for most nations, an unrealistic and perhaps undesirable objective, some protection may be feasible in the short run and it need not have tragic consequences.

Scarcity of means may, in fact, produce alternatives that can reduce some of the unattractive features of Western mass communications. For example, the passive consumption of information that pervades the Western style of transmission raises the serious question whether such a method can ever produce a level of individual consciousness that democratic decision making requires. Direct human engagement in the communications process may be a crucial desideratum in achieving this end. Obviously, this does not require sophisticated technology and, for a time, might even be served better without it. In any case, the mix of personal involvement and media use will vary greatly, depending on a nation's developmental level.

In the most developed capitalist state, media movements are emerging that rely on individual and group involvement. The appearance of low-cost videotaping equipment has produced a number of groups that are experimenting with individualized techniques of communications exchange. Their strongest concern is with giving people an opportunity for self-expression and freeing them from the processed information that flows through the conventional channels. To be sure, it is no accident that the emergence of such activity is first to be observed in the United States. It is made possible by the already high level of industry, the relative inexpensiveness, and therefore the availability of communications equipment for *mass* participation. To a small but widening extent, individuals now can participate in their own message making and consciousness formation in a technically advanced society. The possibility of individual transcendence of the industrial order by using the most developed products of that order is truly a liberating although distant prospect. Although feasible, it is still far too early to say whether it is realizable. The institutional framework within which the current experimentation proceeds is neither hospitable to its own demise nor lacking in a proven ability to absorb or deflect technologically based threats to its own preservation.

Whatever challenges develop within the commercial communications super-state will be of enormous consequence to the rest of the world. But for most of the global population, living in impoverished countrysides and excluded from access to video recorders and monitors and TV cassettes, another communications path to self-realization must be available. It will be based of necessity on the willingness to permit large numbers of people to participate in the communications process and in all the social decision-making that flows from that participation. Accordingly it will make use of whatever communications technology is at its disposal but its ultimate reliance will be on *interpersonal* forms *and social organization*. Sometimes combining films with discussions, radio forums and mass mobilizations, the actual methods will vary with the availability of the resources. The great diversity of communications forms utilized in China during the last twenty years may offer useful examples. Included in the Chinese effort are wired radio broadcasting, 12,000 mobile film projection teams (comprised frequently of young women who exhibit and discuss films in the local dialects with their audiences across the country), and mass mobilization campaigns for popular understanding and political action (Liu, 1971).

China more than any other country demonstrates also the possibility of surviving and developing an authentic national style while denied access to the international communications flow. What was imposed as a deprivation may well have been a blessing.

To be sure, mass participation has been the necessary accompaniment of social revolution in the ex-colonial world. In its initial stages, it achieves impressive results in reducing illiteracy, establishing the national identity, and consolidating the early revolutionary changes in economic and social structure. Once these steps have been taken, precautions are in order that mass involvement does not degenerate into mass manipulation. The major structural changes that must be made at the outset are relatively well known and entirely visible to the general population. The problems and issues that follow are less familiar and usually more subtle. So there is no simple road to meaningful and permanent popular participation in the communications system. Without technology, group forms can assist the process but they, too, can be twisted into denying substantive involvement. Similarly, technology by itself is no guarantor of either participation or manipulation. Social forms are decisive. As one observer notes: "the role of communications in development is defined by the structure, not apart from it." And, "the tendency to equate communications problems with problems in disseminating technical ,information has led many extension and assistance agencies to virtually ignore social and institutional structures in promoting development," (Felstehausen, 1971, pp. 5–6).

The social structures that will best afford meaningful participation for all the people, whatever the level of their economies, are not likely to be those that have prevailed in the Western world over the last few centuries. The tasks of providing communications for authentic needs are very different in the two worlds of the advanced and the developing countries. Are there any connection points? It seems to me that there are.

The international information flow, left to itself and regulated only by the requirements of the market needs of its present powerful private transmitters and generators, will stifle whatever chance yet remains for the sovereign determination of alternative growth paths in the many weaker nations of the world. Yet the new consciousness forces in the advanced states are the natural allies of the more vulnerable societies. To link the strivings and efforts of these widely separate groupings is the task of international communicators if they were to take the mission of international well-being seriously.

Actually, there is at this time no mechanism that can incorporate this need into an institutionally functioning system. The realities are all in the other direction. The stream of global messages now and in the years ahead, to quote again the former director of the U.S. Information Services, " . . . will be private, and most of these will be commercial" (Marks, 1968, p. 141). These, it is fair to say, do not originate with the new consciousness forces in the West and will be of small benefit to the communications needs of the rest of the world that seeks new directions.

In this transitional period, while waiting for the new communications forces to successfully challenge the old in the advanced industrial societies, what can less developed states do for themselves? The path seems evident. Extricate their economies as much and as quickly as possible from the global network of market relationships and improvise means to reach their own people directly and individually.

The issue now is to create a state of mind that does not accept unquestioningly the transfer of already developed technical equipment and the accompanying processes. Communications (consciousness) policy demands, if it is to be liberating and not merely an excuse for a repetition of past historical blunders, a critical selectivity of technology, process, and information based as far as possible on self-determined national (community) needs and aspirations. Instead of the pell mell rush to accept whatever exists in the developed world as good because it is there, a searching evaluation of local need, which combines urgency for life-saving productivity with the equally vital requirement of communal conservation, will characterize this mode of thought and course of action.

For an illustration of this outlook in practice, the report of an American scientific team that recently visited North Vietnam is instructive:

Several characteristics stood out among the institutions we visited and the scientists we met. One is the attitude of independence and self-reliance present in science as in everything else Vietnamese. Even as they described to us the various kinds of aid and assistance they were receiving from foreign countries, the scientists took pains to point out that conditions in Vietnam were unique, and that their particular problems required a specifically Vietnamese solution. For example, rather than bringing in foreign medical specialists, they preferred to develop their own public health program from the beginning. [Galston and Signor, 1971, p. 384]

This is not an easy course to offer any nation and people. But if the goal is independence and self-realization, is there any other?

REFERENCES

Drucker, Peter. *The Age of Discontinuity*. New York: Harper, 1968.

Felstehausen, Herman, "Conceptual Limits of Development Communication Theory," paper prepared for the 54th annual convention of the Association for Education in Journalism, University of South Carolina, Columbia, South Carolina. August 22–25, 1971.

Furgurson, Ernest B. "Canada to U.S.: One-Way Street?" *Los Angeles Times* (November 26, 1971). Part II, p. 7.

Galston, Arthur W., and Ethan Signer. "Education and Science in North Vietnam," *Science*, *174*, 4007 (October 22, 1971).

Brief to the Ontario Royal Commission on Book Publishing, Summary and Recommendations. Harvest House Publishers of Montreal. June 15, 1971.

Illich, Ivan. "Outwitting the 'Developed' Countries," *New York Review of Books* (November 6, 1969).

Liu, Alan P. L. *Communication and National Integration in Communist China*. Berkeley: University of California Press, 1971.

Maidenberg, H. J. "Why Peru Seized TV," *New York Times* (November 14, 1971). Financial Section, p. 11.

Marks, Leonard H. "American Diplomacy and a Changing Technology," *Television Quarterly* (Spring 1968).

McNamara, Robert S. Address to the Board of Governors, World Bank Group, Washington, D.C., September 27, 1971.

Richler, Mordecai. "Maple Leaf Power Time," *New York Times Book Review*, September 26, 1971.

Sorrentino, Constance. "Comparing Employment Shifts in 10 Industrialized Countries," *Monthly Labor Review* (October 1971).

Telecommunications Reports, 37, 47 (November 22, 1971).

The Uncertain Mirror. Report of the Special Senate Committee on Mass Media (The Davey Report). Ottawa: Queen's Printer for Canada, 1970.

UNESCO. *Mass Media in Society: The Need of Research*. Reports and Papers on Mass Communication, No. 59. Paris, 1970.

United Nations Educational, Scientific and Cultural Organization, Meeting of Experts on Mass Communication. Paris, 1970.

Wilcke, Gerd. "Job Export Charge Denied," *New York Times*. (November 14, 1971). Business and Financial Section, p. 1.

PART VI

Tracking the Future

CONTENTS

INTRODUCTION

GEORGE GERBNER

If the new future of the future is to be different from the old future of the past, it should be subjected to unprecedented scrutiny. By that I mean that the use of the future as a political instrument should be explained, and even its scientific assessment demystified. Then we can move on to the consideration of the merits of social intelligence in shaping the future. That is the task of this section.

In their chapter on "The History of the Future," James W. Carey and John J. Quirk trace the work of some scientistic "oracles to the people and servants of the ruling class." The future that never comes but always is "just around the corner" or "at the end of the tunnel" serves to make the present more acceptable. Since rapid change in the "quality of life" is apparent to all, it is not difficult to claim that a change in the structure of society will come about by the same processes of extension and intensification that pervade the present. Thus the claims made for the future effects of electricity in 1852 are compared to the claims for the future effects of electronics made in 1972. Information technologists may gain a monopoly of knowledge in the profound sense of being able to define what it means to be reasonable.

Michael Gurevitch and Philip Elliott pursue the question of monopoly in their chapter on "Communication Technologies and the Future of the Broadcasting Professions." They foresee changes in the self-image of the profession (if it can be called that), and in its control over the most massive channels of communication. Ironically, they write, the public's "right to know" has been transformed into the professionals' right to define what it is that the public has a right to know. With the proliferation and diversification of channels, the status and control of the media "professional" may decline, giving way to even more overt business control exercised through technological and market "mechanisms." The disestablishment of the "New Priesthood" of media professionals may mark the end of a period of the elite standards and professional quality controls, to giving full sway to "populist commercialism" managed by technicians.

How these or other foreseeable developments will influence the political system is Forrest P. Chisman's topic in "Politics and the New Mass Commu-

483

nication." Chisman finds most current predictions questionable, if not misleading. The proliferation of channels and consequent fragmentation of audiences raises many theoretical and practical problems. The rosy myth of the future of telecommunications may lead us to ignore present warning signals and to delay needed reforms.

Even the exclusive preoccupation with technology assessment may have that effect, argues Edwin B. Parker in his chapter on "Assessment or Change in Communication Technology." "If we structure the problem as one of assessment of the technology itself, or as one of developing social indicators to better measure effects after they have happened, then the battle will have been lost before we start," Parker writes. The alternative he recommends is immediate institutional intervention and social planning, assuring maximum public utilization of the fruits of communication technology. Waiting for consensus, or even for a crisis that might precipitate consensus, is waiting too long. If the problems reside in the basic institutional structure of society, as Parker believes, crisis and concensus can be managed to support instead of change that structure. Publicity and mobilization of human energies must take advantage of the flexibilities that now exist to initiate the changes. Assessment of consequences can be more effective once institutional change is underway.

Just how such assessment might be conducted, whenever it occurs, is the topic of the last two chapters. James D. Halloran refers to the investigation of media performance as "Research in Forbidden Territory" because the organs of public consciousness have been resistant to the idea of public consciousness of their own operations. Surveying the British scene, Halloran concludes that between the clamor of governments and the self-serving proclamations of the media, a third voice is needed to track the consequences of new developments in communications.

What that third voice might be is the subject of the final chapter on "Cultural Indicators: The Third Voice," by George Gerbner. Three areas of analysis designate study of how policies are made, what message systems are produced, and what contributions they make to public conceptions of life and society. Tracking the present as it flows past might point the way to a future that is neither a mirage nor a trap.

CHAPTER 31

The History of the Future

JAMES W. CAREY AND JOHN J. QUIRK

In *The Image of the Future* (1961) F. L. Polak has traced the human preoccupations with the future to its ancient roots in Delphic oracles and astrological priesthoods. However, the moden history of the future originates with the rise of science and onset of the age of exploration. Armed with the techniques of modern science, especially the new measuring devices of precise clocks and telescopes, a secular priesthood seized hold of the idea of a perfect future, a zone of experience beyond ordinary history and geography, a new region of time blessed with a perfect landscape and a perfection of man and society. Nevertheless there exists a continuity from the ancient astrologers of the temple, tribe, and city to modern scientists, for both are elevated castes who profess special knowledge of the future, indeed establish a claim of eminent domain over the next stages of human history.

Modern oracles, like their ancient counterparts, constitute a privileged class who monopolize new forms of knowledge and alternatively panic and enrapture large audiences as they portray new versions of the future. Moreover, modern scientific elites often occupy the same double role of oracles to the people and servants of the ruling class as did the astrologers of ancient civilization. And they rely on a similar appeal to authority. Ancient astrologers used their ability to predict the behavior of planets to order social life through the calendar and to regulate agriculture. The knowledge of astronomical order in turn supported their authority as all-purpose seers capable of taming the future. Similarly, modern scientists use their capacity to predict the behavior of narrow, closed systems to claim the right to predict and order all human futures.

And yet while the future as a prophetic form has a long history, the future as a predictable region of experience never appears. For the future is always offstage and never quite makes its entrance into history; the future is a time that never arrives but is always awaited. To understand the dilemma of the future we might take a cue from the scholar reflecting on the loss of interest in history, who asked, "Does the past have a future?" and ourselves inquire,

"What sort of a past has the future had?" The future, as an idea, indeed has a definite history and has served as a powerful political and cultural weapon, particularly in the last two centuries. During this period, the idea of the future has been presented and functioned in American and British life in three quite distinct ways.

First, the future is often regarded as cause for a revitalization of optimism, an exhortation to the public to keep "faith," and is embodied in commemorative expositions of progress, world fairs, oratorical invocations, and in the declaration of national and international goals. Second, the future, in the politics of literary prophecy, is attractively portrayed as the fulfillment of a particular ideology or idealism. The past and present are rewritten to evidence a momentous changing of the times in which particular policies and technologies will yield a way out of current dilemmas, and a new age of peace, democracy, and ecological harmony will reign. Third, the future has acquired a new expression in the development of modern technologies of information processing and decision making by computer and cybernated devices. Here the future is a participation ritual of technological exorcism whereby the act of collecting data and allowing the public to participate in extrapolating trends and making choices is considered a method of cleansing confusion and relieving us from human fallibilities.

THE FUTURE AS EXHORTATION

Throughout American history, an exhortation to the future has been a standard inaugural for observing key anniversaries and renewed declarations of national purpose. At celebrations of science and industry, and orations of public officials, the invocation of a sublime technological future elevates the prosaic and pedestrian commonplaces of "the American creed" with its promises of progress and prosperity to an appeal for public confidence in established institutions and industrial practices. This exhortation to the sublime future is an attempt to ward off dissent and to embellish cosmetically the blemishes of the body politic with imagery of a greater future for all.

The strategy of the future as exhortation was exemplified by the Centennial Exhibition staged in Philadelphia in 1876. The American Centennial was observed through the preferred nineteenth century symbol of progress and optimism, the industrial exhibit. The initial purpose of the exhibit was to testify to American unity 11 years after the Civil War. However, the magnetic attraction of the exhibit was the Hall of Machinery with 13 acres of machines connected by pulleys, shafts, wheels, and belts to a giant Corliss engine in the central transcept. Symbolically, President Grant opened the Centennial by turning the levers that brought the giant engine to life, assisted by Dom Pedro, the Emperor of Brazil. The Corliss engine dominating the Centennial illustrated the giantism of nineteenth century mechanical technology, which enraptured both public and politicians. The machines were symbols of the grandeur and strength of the American people and a hopeful sign for the second century of American life. Even literary men like William Dean Howells were overcome by the Corliss engine: " ... in these things of iron and steel ... the national genius freely speaks; by and by the inspired marbles, the breathing canvases, the great literature; for the present America is voluble in the strong metals and their infinite uses" (Brown, 1966, p. 130).

While the giant hardware of the "Age of Steam" dominated the exhibit, the new electrical machines also held sway in the Centennial halls where the electric lamp and Alexander Graham Bell's telephone were on display.

In inaugurating the fair, President Grant noted that of necessity our progress had been in the practical tasks of subduing nature and building industry, yet we would soon rival the older nations in theology, science, fine arts, literature, and law. For while this was a celebration of 1876, it had an eye clearly fixed on 1976, the next centennial, progress toward which was guaranteed by native advances in mechanics and industry. However, America of the 1870s displayed numerous symptoms not altogether in harmony with the prevailing mood of the Centennial. The entire two decades following 1873 were highlighted by a worldwide depression. Earlier "improvements" in communication and transportation had led to an unprecedented degree of international integration in the economy. Failures in the economy fanned out over this international network so that the "communications revolution" of the 1830s generated, as one observer put it, three unprecedented historical phenomena: "an international agrarian market, an international agrarian depression and, as a climax, international agrarian discontent" (Benson, 1951, p. 62). Bitter discord reverberated through American society, lurking even in the shadow of the Centennial Exhibition. Labor unrest in the Pennsylvania coal fields led to strikes and union organization and to the hanging of 10 members of the Molly Maguires in 1877. During 1876, President Grant had to dispatch troops to the South to control violence in the aftermath of the disputed election of Rutherford Hayes. The Centennial itself was disrupted on the Fourth of July by Susan Anthony's presentation of the Women's Declaration of Independence. Frederick Douglass, the contemporary black leader, was an official guest at the Centennial opening, although he had difficulty getting past police to the receiving stand; however, his token presence did not retard the spread of Jim Crow legislation through the South, undoing whatever gains had accrued to blacks in the aftermath of the Civil War. Finally, nine days before the climactic Fourth of July celebration, news arrived of Custer's defeat at Little Big Horn (Brown, 1966, passim). Such realities of American life—the problems of racial and ethnic relations, of political democracy, of the industrial proletariat, and of chronic depression did not pervade the official rhetoric of the Centennial with its eyes fixed firmly on Tomorrow.

For another Centennial celebration we have dutifully created a commission on National Goals, a Bi-Centennial Committee, agencies, and commissions to foretell the year 2000. Moreover, the same problems that haunted 1876 mar the contemporary landscape. And, finally, while the favored symbols of technological progress have changed—atomic reactors, computers, and information utilities—having replaced steam engines and dynamos—the same style of exhortation to a better future through technology dominates contemporary life. This exhortation to discount the present for the future has, therefore, been a particular although not peculiar aspect of American popular culture. It is, in a trenchant phrase of Horace Kallen (1950, p. 78), "the doctrine and discipline of pioneering made art."

The reasons behind this orientation are easy enough to state, although difficult to document briefly, for the very creation of the United States was an attempt to outrun history and to escape European experience, not merely to find a new place but to found a "New World." The idea of a "new land," a virgin continent, had been part of the European utopian tradition. The discovery of America during the age of exploration removed utopia from literature and installed it in life.

This notion of our dispensation from European experience, free to realize the future without the baggage and liabilities of the past, has always been central to American belief. It first appears in a religious context, in the belief that a uniform, nonsectarian Christianity would be possible in "New England" because of the absence of European institutions and traditions. In the nineteenth century, dramatic advances in technology and industrialization were seen as an analogy to the spread of American religion so that Gardner Spring could declare in 1850 that we are living on the "border of a spiritual harvest because thought now travels by steam and magnetic wires" (Miller, 1965, p. 48). Soon the spiritual improvement wrought by Christianity was linked to those "internal improvements," particularly improvements in transportation and communication, so that canals, railways, and telegraph became the most important form of missionary activity by midcentury.

The course and domain of spiritual empire increasingly became identified with that practical enterprise, manifest destiny, the course of the American empire. America's dispensation from history gave her a missionary role in the world: to win the world to an absolute truth—at first religious, then technical; to create a radical future "of a piece with titantic entrance into the 'new world' of steam and electricity" (Miller, 1965, p. 52).

Whenever the future failed, as often it did during the nineteenth and twentieth centuries, appeal was made to a yet another new future patching up the miscarriage of previous predictions. Most importantly, preachers and politicians appealed to Americans to retain faith in the future as such; they appealed to the future as a solvent and asked the public to believe that the latest technology or social project would fully justify past sacrifices and the endurance of present turmoil.

Fifty years after the Philadelphia Centennial the foremost American historians of the period, Charles and Mary Beard, who were not unconscious of the difficulties of postwar America, were fascinated nonetheless by the vastness of the industrial inventory presented at the Sesquicentennial Exposition in contrast to what was shown in 1876. Moreover, they saw America's social destiny in "the radical departures effected in technology by electrical devices, the internal combustion engine, the wireless transmission of radio"; changes, they felt, "more momentous even than those wrought by invention in the age of Watt and Fulton." They argued that the new technology removed the gloom and depression of the age of steam and provided a new motive force to rearrange American social patterns. Electricity would emancipate mankind and integrate the city with the country as radio brought cosmopolitanism "as if on the wings of the wind." They concluded in lyrical prose that the "influence of the new motors and machines was as subtle as the electricity that turned the wheel, lighted the film and carried the song" (Beard and Beard, 1940, p. 746).

Several years later, in the midst of the Great Depression, Franklin Delano Roosevelt ritually exhorted the American people reminding them that

> We say that we are a people of the future . . . the command of the democratic faith has ever been onward and upward; never have free men been satisfied with the mere maintenance of the status quo . . . we have always held to the hope, the conviction, that there was a better life, a better world, beyond the horizon. [Nevins, 1971, pp. 400–401]

Similarly, at the 1933 Century of Progress Exposition in Chicago, where Thomas Edison was being memorialized and the electrical exhibit featured the themes of conquest of time and space, Roosevelt tried to banish doubts and fears by reference to "the inauguration of a Century of even greater progress —not only along material lines; but a world uplifting that will culminate in the greater happiness of mankind."

The function of such rhetoric was once characterized by the late C. Wright Mills (1963, p. 302): "The more the antagonisms of the present must be suffered, the more the future is drawn upon as a source of pseudo-unity and synthetic morale." The future in exhortation becomes a solvent, the very act of moving forward in time constitutes a movement away from past problems and present difficulties. The future becomes a time zone in which the human condition is somehow transcended, politics evaporated, and a blessed stage of peace and democratic harmony achieved. The historian Allan Nevins (1971, p. 398) clearly expresses this native ideology:

Unity in American life and political thought certainly does not stem from general agreement on any body of doctrines. . . . The meaning of democracy in Oregon is very different from its meaning in Alabama. We are often told that we are held together as a people not so much by our common loyalty to the past . . . as by our common faith and hopes for the future. It is not the look backward . . . but the look forward that gives us cohesion. While we share some memories, the much more important fact is that we share many expectations. . . . The great unifying sentiment of America is hope for the future. . . . For national unity it is important to maintain in the American people this sense of confidence in our common future.

These views have potent political uses. The ideology of the future can serve as a form of "false consciousness," a deflection away from the substantial problems of the present, problems grounded in conflicts over wealth and status and the appropriate control of technology, toward a future where these problems, by the very nature of the future, cannot exist. As rationalizers for the British empire in the last century urged not only recognition of but belief in the Industrial Revolution, so Nevins like other apologists asks that our "minority groups" must have their sense of deprivation relieved by partaking of "faith in sharing, on equal terms, in a happier future." Similarly, one of Richard Nixon's first acts as President was to create a National Goals Research Staff. The staff was charged with orienting Americans toward the coming bicentennial and the year 2000, so that we might "seize on the future as the key dimension of our decisions . . . " (*Futures*, 1969, p. 459).

Culturally and politically, then, the idea of the future functions in much the same way as the notion of the "invisible Hand of Providence" operating in the dreams of "heavenly cities" in the eighteenth and nineteenth centuries; it provides a basis for faith in the essential rectitude of motives and policy in the midst of the disarray of the present. The rhetoric of the future has, in the twentieth century, offered in Aldous Huxley's words (1972, p. 139), a "motivating and compensatory Future" which consoles for the miseries suffered in the present. To Huxley's critical mind, the literature of the future provided to modern generations what the Methodist sermon on hard times now and heavenly rewards later had for the first English working class at the onset of the Industrial Revolution; the rhetoric of a sublime future as an alternative to political revolution and a stimulus to acquiescence. In the new literature of the future, the salvation is not otherworldly but terrestrial revolution and its correlates in moral, social, and material betterment. As Huxley (1972, p. 140) concluded, "the thought of . . . happiness in the twenty-first century consoles the disillusioned beneficiaries of progress."

THE FUTURE AS LITERARY PROPHECY

From the enormous corpus of prophetic writing about the future, we have selected a few British and American authors who illustrate the essential features of this literature. Although the authors' motives and backgrounds differ, certain distinct common themes distinguish futurist literature. Invariably the newest technologies of communication and transportation are seen as means for the lasting solution to existing problems and a radical departure from previous historical patterns. Also, the landscape of the future is suggestively drawn as one where a sublime state of environmental balance, social harmony, and peace is achieved.

In *Futures* magazine, I.F. Clarke (1969) has identified the first major technological forecast written in the English language as the work of an anonymous author published in 1763 under the title of *The Reign of George VI, 1900-25*. This premiere utopia, which may be said to have initiated the age of extrapolation, depicted the future as a mere perfection of the ethos of the reign of George III. It projected the consolidation and expansion of the empire over the continents of Europe and North America with a Pax Brittannica of secure hegemony by means of vastly improved communication and transportation supporting commerce, foreign service, and military force. Published in the same year as the end of the French and Indian War and 13 years before the uprising of the 13 colonies, it professed to see a time when England's perennial rivals gladly accepted orders from London. Coeval with Watt's steam experiments, it suggested that the English countryside would be embellished by the waterways and routes of new industry, that cities might remain quaint, and that the society of aristocratic amenities would be perpetuated. During the predominance of the British Empire a literature of the imperial future sought to impress the reading public with such sublime reasons for continued expenditure and sacrifice on behalf of Anglo-Saxon destiny. It also became in time a ground for arguing against revolutionary ideology as Chartism, Marxism, and republicanism challenged the system.

An apotheosis of nineteenth century optimism followed in the train of the Great Exhibit of 1851 as the prevalent ethos of Victorian complacency imagined a global community of interests to be the inevitable by-product of communication and transport in the cause of trade and empire. There were some dissenters who pierced the Crystal Palace mystique and correctly read into industrialization its pernicious tendencies to dwarf man and nature under advancing machinery. The dominant note remained one of beneficent *social corollaries* to be derived from the conquests by technology of the earth and the barriers of time and space. Ironically, these included gifts for which we are still waiting, such as freedom from drudgery, a wedding of beauty and utility, and an end to warfare and cosmopolitan consciousness.

A prime document of this period is illustrative of the point that today's future is yesterday's future as well. In *The Silent Revolution: or the Future Effects of Steam and Electricity Upon the Condition of Mankind*, a projection from the perspective of 1852, Michael Angelo Garvey portrayed the world as the Great Exhibit writ large where all the problems of industrialism were finally resolved. The smoke-filled slum and the Malthusian spectre were to be eliminated as transportation redistributed population to new colonies and allowed a new and elevated working class access to "pure air and joyous landscape." Sharing the mistaken notion of most futurists that social conflict results from insufficient communication and isolation, Garvey personified the technology of travel and telegraphy. The railway was "if not the great leveller" then "the great master of ceremonies," who is "daily introducing the various classes," and "making them better acquainted in common." In a further "future period," Garvey projected a system of total communications anticipating the notions of Marshall McLuhan: "a perfect network of electric filaments" to "consolidate and harmonize the social union of mankind by furnishing a sensitive apparatus analogous to the nervous system of the living frame" (Garvey, 1852, pp. 170, 134, 103–104).

This perfect future was of a piece with other Victorian prophecies despite the proximate realities of Irish famine and labor unrest, the Crimean War, and other manifestations of discord and dispute. But the ulterior motive for the imperial era future literature was patently clear in *The Silent Revolution*. Garvey pleaded for his readers to maintain their loyalty to the regime, the proper caretaker of the future, and to avoid noisy agitation for reform or revolt. The "silent revolution" was a substitute for a social revolution, a rhetorical method to keep not only the majority but minorities silent about questions of imperial policy.

The literature of the future of the empire continued to mirror and mold prevailing opinion of the British elite well into the twentieth century. Its attitudes regularly overshadowed critical warnings about the fate awaiting overextension abroad and retention of obsolete institutions at home. Although the citations from twentieth century versions of the literature of the imperial futurists already seem arcane to us because of the depletion of English power, it is well to realize the degree to which American futurism in the present context— for instance, in Zbigniew Brezezinski's (1970) "Technetronic Society"— derives its inspiration from the British version of an imperial future: a *Pax Americana* augmented by electronic instruments of communication for the conduct of foreign policy and warfare and the pacification of the home populace.

In *The World in 2030*, a view from the year 1930, the Earl of Birkenhead tried to blend imperialism and futurism to ward off erosion of public confidence caused by the depression and the rise of dictators. To offer a relief from the over 230 years of turmoil, Birkenhead predicted a characteristic turning point identical to that delineated by current writers about the future: "Today we are witnessing the death of a society and tradition which have existed since the first French Revolution and the Industrial Revolution" (p. 116). This change, however, was not to be political or social, but technological. Electrification of the English countryside and decentralized, smokeless factories were to comprise a handsome landscape of laboratories resembling an "interminable park" and dispensing the plentitude of an "industrial Arcadia."

Public disaffection from remote government might be treated by obtaining formal participation through electronic communications, so that "it will be feasible once more to revive *that form* of democracy which flourished in the city states of Ancient Greece" (Birkenhead, 1930, pp. 8–9). Broadcasting of special debates could be followed by instant opinion polling through devices inserted in telephone exchanges. But, this meant no real transfer of power to the people, because in Birkenhead's analysis, government should probably be handed to a class of expert specialists whose electronic consultation might be a mere formality, a guise of democracy for the electronic Leviathan.

Furthermore, Birkenhead envisioned the future world as continuing management of world affairs and the evolution of international organization around the nucleus of the British Commonwealth with India, South Africa, and even Dublin again inside its orbit. The future world would be made safe for the RAF patrol over the pipelines in Iraq and for upper-class amenities of silent Rolls Royce and riding to hounds. In sum, "the world in 2030" was to be nothing more than the wishful dream of an 1830 Tory mentality given technocratic expression.

So pronounced was the tendency of the British *intelligentsia* to conceive of the future solely in terms of the empire that it affected even Liberals, Fabians, and scientific modernizers such as J.B.S. Haldane. H. G. Wells, the most inventive of the futurists during the first part of the twentieth century, was initially a member of a circle who viewed the Empire as "the pacific precursor of a practical World State" (Wells, 1929, p. 126) and the royal military equipped with the latest technology of communication and transport as the forerunners of a "world police" able to be dispatched quickly to any trouble spot to quell insurrectionary activity.

In contradistinction to the imperialist futurists, there arose an alternative view of the future genuinely dedicated to the decentralization of power and industry, a rehabilitation of the natural landscape, and a revival of regional cultures. Its major figures were the Russian anarchist and naturalist Peter Kropotkin and the Scots regionalist Patrick Geddes. Kropotkin's vision of an "industrial village" of the future foresaw the dispersion of production and population to communal and workshop levels (Kropotkin, 1913). The transmission of electricity would replace the huge steam engine, dehumanized factories, and alienation of labor. This attractive idea was further elaborated by Geddes as a theory of the reversal of the adversities of the Industrial Revolution and the arrival in the near future of a "eutopian" mode of life.

During and after World War I, Geddes and his colleague sociologist Victor Branford edited a series of books and pamphlets collectively published as

Interpretations and Forecasts, The Making of the Future, and *Papers for the Present* (Geddes, 1917). Geddes' foremost American disciple, Lewis Mumford, has credited this biologist and town planner with the introduction of "the future, as so to speak, a legally bounded terrain in social thought" (Barnes, 1966, p. 384). Geddes earned the title of "big brother of reform" through his activist field experiments from Chicago and Edinburgh to India and Palestine. His intellectual influence extended to contemporaries such as Jane Addams and John Dewey and has reemerged in updated versions in the work of recent figures such as Paul Goodman and Marshall McLuhan.

Geddes' own portrayal of the future drew a dialectical contrast between old and new forms of technology. Electricity was to be the key to a "great transition" from forms of concentration to decentralization, from pollutants to ecology, from urban congestion and false cosmopolitanism to regional and folk revival: "We may divide the age of machinery into the paleotechnic age of smoke and steam engine, and the neotechnic age of electricity and radium, the conquest of noise and the utilization of waste" (Geddes, 1917, Preface). The aim of the future for Geddes was a neotechnic "Eutopia" under a "partnership of man and nature" in a world redesigned to resemble a garden.

There is a remarkable similariy between Geddes' conception of the future and notions entertained by contemporary futurology. Geddes expected the passing away of politics, parties, and ideologies. In place of political activism, he and Branford advocated a third alternative *beyond* right and left to be carried out by "peace armies" of "university militants" going to the peoples of the world in projects of environmental reconstruction, conservation, educational reform, and civic design. Imperialism would be superceded by autonomous regional federations. This neglect of political facts and factors in Geddes' ideas has been evaluated by Lewis Mumford as the critical oversight in his view of the future world.

Geddes' future was premised on several other errors. The new technology as applied brought about increased centralization and concentration and domination over the landscape by powerhouses, and extended the range of control by imperialistic power centers over indigenous cultures and regions.

The ideas of Geddes and Kropotkin had their influence in the United States among leading conservationists, regional planners, and social critics. The transfer to the American scene of Geddes' neotechnic formulations was especially due to the works of Lewis Mumford.

In 1934, in his *Technics and Civilization,* Mumford attributed a series of "revolutionary changes" to "qualitative" effects of electricity itself, particularly hydroelectric turbines and incipient automated machinery in the factory. These he supposed to include a "tidying up of the landscape" by "Geotechnics" in the "building of reservoirs and power dams" and a lifting of the "smoke pall" as "with electricity the clear sky and clear water . . . come back again." The sublime landscape of the radiant future would be one of an inter-marriage of town and country, agriculture and industry, and even distribution of surplus population and of wealth (Mumford, 1934, pp. 255–256).

From radio and "person-to-person" electronic communication, Mumford hoped for a universal democracy by technology: "there are now the elements of almost as close a political unity as that which was once possible in the tiniest cities of Attica" (Mumford, 1934, p. 241).

Still, Mumford's Americanization of Geddes' gospel was subjected to the same irony of history as had overtaken previous projections of the future. The hydroelectric project and reservoir eventually further uprooted and eroded the environment. The air, water, and land were neither cleared nor cleansed, as we who now inhabit this future landscape well know. The megalopolis continued to grow. Total automation is still more predicted than realized, while the C.I.O.'s organizing drive began in earnest just as Mumford wrote of the end of the working class. Politically, it was an age of dictators and centralized rule.

Mumford himself was compelled to admit that there had been a "miscarriage of the machine," since civilization was still stalled in a "pseudomorphic" stage: "The new machines followed the pattern laid down by previous economic and technical standards." In subsequent reevaluation, Mumford has seen that belief in electricity as a revolutionary force was in fact mistaken even "in those plans that have been carried through, the realization has retrospectively disfigured the anticipation" (Mumford, 1959, p. 534).

Nevertheless, Mumford's themes have reappeared in future literature. It was 36 years ago that he composed a section on "shock absorbers," the essence of which reappeared in Alvin Toffler's *Future Shock* (1971). Toffler has revived themes of the sublime technological revolution of 40 years vintage as a means of peering toward the year 2000. Toffler's recent work is embellished by the same recurring symbolism of the futurist genre. There are a number of "final and qualitative" departures in store for the new millenium. According to Toffler, the new society has "broken irretrivably with the past" surpassing geography and history.

What we encounter in Toffler is a portrait of the future as a new realm of dispensation from the consequences of the industrial revolution. The era of automation is pictured as a change "more important than the industrial revolution." The new industries of electronics and space technology are characterized by "relative silence and clean surroundings" as contrasted to the imagery of "smoky steel mills or clinking machines." The end of the assembly line dispenses with classic class conflict by placing a "new organization man" in the leading role as historical protagonist. Anticipatory democracy will be instituted in "town halls of the future," where critics of technics will be outdistanced by a futurist movement. Dissident minorities and recalcitrant middle

Americans will be co-opted into programmed participation into future-planning games. Groups will be dissuaded from opposition to the space program and have their dissent funneled into support for improved technology. Cadres of specialists shall be attached to various social groups so that expertise will be married to the solicitation of consent.

Another illustrative comparison can be taken from the literature of the 1930s and 1970s. Contemporary rhetoric of a sublime national future merely places the computer and transistor where the powerhouse generators once held sway as predominant technology. A striking similarity may be seen in the parallel between the initial celebration of the Tennessee Valley Authority as a New Deal showcase and the recent projection of the electronic counterculture in *The Greening of America* (Reich, 1970).

Contrary to prevalent interpretations, the New Deal had its futurist impulses in efforts to enact projects for the construction of new communities, the decentralization of power, the reclamation of the landscape, and the electrification of the American countryside. This aspect of New Deal thought reflected the ideas of the old progressive conservationists like Gifford Pinchott, who had been influenced by Geddes and the regional planning movement.

The TVA was the subject of a vast oratorical and journalistic outpouring, centering on its image as a model of the future. For instance, it was held to be a "Revolution by Electricity" by Paul Hutchinson (1937), editor of *The Christian Century*, who lifted his idiom from Lewis Mumford. In his words, the TVA was to "fashion the future of a new America." The "real revolutionary" was the new machine "which might at last become as much of a liberating and regenerative agency as the dreamers of the early industrial revolution declared it would be." The Tennessee Valley Authority was to be marked by a complete "absence of politics" and decentralization into "factory-plus-farm villages" (pp. 83–95, passim). It would deny the iron laws of managerialism and bureaucratic revolution.

However, the TVA's own record has been a final reversal of these promises. Internally, it has developed technocratic structures and its new towns display a company town psychology. By strip-mining and other such practices it has marred the landscape. Like its technological big sister, the Atomic Energy Commission, it is aligned economically and politically with parts of the military-industrial complex. If anything, the machine again became the real counterrevolutionary.

In *The Greening of America*, Charles Reich predicts a "transformation" beyond a "mere revolution such as the French or Russian . . . ." This new form of revolution offers answers to questions of identity and community, of history and politics. In the age of the computer and counterculture, Reich's rhetoric resembles Hutchinson's of four decades ago. For instance, "The machine itself has begun to do the work of revolution." And, "Prophets and philosophers have proposed these ways of life before, but only today's technology has made them possible" (Reich, 1970, pp. 204, 383).

Reich attributes to electronics and cybernetics the social *correlates* of a higher consciousness, participation in a shared community, and renewed contacts with the land. The trouble with the Reichean formulation of a revolutionary machinery and a new cultural emergence is that its manifestations are either illusory or ephemeral.

At bottom, the counterculture is primarily an extension of the existing entertainment and leisure industries, instead of a regeneration of the humane dimension. Reich cites the devotees of Woodstock and is silent on the Altamont tragedy. The record industry lets the counterculture have the prophetic lyrics and collects the profits and the real cultural power.

There is a pronounced tendency, however, for prophets and movements to resort to incantations to reassure themselves about their cherished illusions. The enthusiasts of the Tennessee Valley Authority a generation ago, as Harold Ickes observed sardonically in 1944, began to believe that they might breathe life into a new democracy merely by intoning "TVA, TVA, TVA." Similarly the "greening of America" and like-minded counterculture scenarios impress one as nothing more than a chanting exercise for a new generation of Americans sent to their rendevous with another electric destiny. Presently, the shaping of the future remains routed along past lines. We see technological patterns and organizational forms continuing the trend toward concentration and centralization of power and control in established institutions.

THE FUTURE AS A PARTICIPATION RITUAL

The writings of Reich and Toffler are merely the outer edge of a large body of literature forecasting another technological revolution and a new future. This revolution is preeminently one in communication, for as Norbert Weiner noted some years ago, "the present time is the age of communication and control." Modern engineering is communication engineering, for its major preoccupation is not the economy of energy "but the accurate reproduction of a signal" (Weiner, 1948, p. 39).

This third communications revolution was preceded by the innovation of printing, which mechanized the production of information, extended literacy, and enlarged the domain of empire. The second revolution occurred over the last century with the marriage, through electricity, of the capacity to simultaneously produce and transmit messages—a process that extends from the telephone and telegraph to television. Now, this third communication revolution involves the linkage of machines for information storage and retrieval with the telephone, television, and computer producing new systems of "broadband" communications or "information utilities."

The revolutionary potential of these "improvements" in communication does not derive from the prosaic facts about them—more information sent faster and further with greater fidelity. Instead, their attraction resides in the supposed capacity to transform the commonplace into the extraordinary: to create novel forms of human community, new standards of efficiency and progress, newer and more democratic forms of politics, and finally to usher a "new man" into history. The printing press, by extending literacy, not only taught men to read but was expected to eradicate ignorance, prejudice, and provincialism. Similarly, the telegraph and radio were seen as magnetic forces binding people into international networks of peace and understanding. Recently the "cybernetic revolution," by increasing available information by a quantum leap, promises to make "policy options . . . clearly defined, the probable outcomes of alternative measures accurately predicted and the feedback mechanism from society . . . so effecive that man could at last bring his full intelligence to bear on resolving the central problems of society" (Westin, 1971, p. 1).

The basis of this third communications revolution, the marriage of the time-shared computer for both data analysis and information storage and retrieval with the telephone and television, is portrayed as the ultimate communications machine; it combines the speed and intimacy of dialogue, the memory of history, the variable output of sight and sound, the individuality of total information combined with totally free choice, the political awareness and control of a fully informed and participant electorate, and the analytic skill of advanced mathematics.

Despite the manifest failure of technology to resolve pressing social issues over the last century, contemporary intellectuals continue to see revolutionary potential in the latest technological gadgets that are pictured as a force *outside* history and politics. The future as it is previsioned is one in which cybernetic machines provide the dynamic of progressive change. More important, while certain groups—industrialists, technocrats, and scientists—are portrayed as the appointed guardians of the new technology, they are not ordinarily viewed as an elite usurping the power to make history and define reality. They are viewed as self-abnegating servants of power, merely accomodating themselves to the truth and the future as determined by the inexorable advance of science and technology. In modern futurism, *it is the machines that possess teleological insight.*

Moreover, the new communications technology is extended into virtually every domain of social life, invading even the family through home computer consoles for information, entertainment, education, and edification. And the public is invited to participate in a technical ritual of planning the future through World Games and electronic Delphic techniques as a rehearsal for the new stage of participatory democracy to be ushered in by communications technology.

Unfortunately, the vision of democracy by electricity has been with us since at least the telegraph and telephone and has been put forward by most writers about the future over the last century. James Russell Lowell, assessing the aftermath of the Civil War in the 1860s, felt that "the dream of Human Brotherhood seems to be coming true at last." He pinned this belief to the new form of the town meeting that technology could bring into existence:

It has been said that our system of town meetings made our revolution possible, by educating the people in self-government. But this was at most of partial efficacy, while the newspapers and telegraph gather the whole nation into a vast town-meeting where everyone hears the affairs of the country discussed and where better judgment is pretty sure to make itself valid at last. No discovery is made that some mention of it does not sooner or later reach the ears of a majority of Americans. It is this constant mental and moral stimulus which gives them the alertness and vivacity, the wide-awakeness of temperament, characteristic of dwellers in great cities. . . . [Lowell, 1871, Vol. 5, p. 239]

Despite the shortcomings of the town meetings, the newspaper, the telegraph, the wireless, and the television to create the conditions of a new Athens, Buckminister Fuller, one of the most vocal and visible of contemporary advocates, has described a form of instantaneous, daily, plebiscitory democracy through a computerized system of electronic voting and opinion polling.

Devise a mechanical means for nationwide voting daily and secretly by each adult citizen of Uncle Sam's family: then I assure you will Democracy be saved . . . This is a simple mechanical problem involving but a fractional effort of that involved in distributing the daily mails to the nation . . . Electrified voting . . . promises a household efficiency superior to any government of record, because it incorporates not only the speed of decision of the dictator . . . but additional advantages that can never be his. [Fuller, 1963, pp. 13–14]

But it is also obvious that the extraordinary demands made on the citizen by such a system would merely co-opt him into the technical apparatus with only the illusion of control.

To participate in such a system the citizen of the future will have to undergo a continuing, lifelong education in real time, the acquisition of new knowledge when it is needed in time to meet problems as they arise. Recognizing the implausibility of all this, Donald Michael has recommended a form of republicanism instead of direct democracy. He argues that specialists will have to mediate between the technology and the citizen and government. Such specialists will be retained by groups to represent them to the government. But given the engineered "complexity" of the new information systems, involvement of the public becomes a mere ritual of participation or overparticipation to legitimate rule by a new scientific elite. If either of these modes of citizen participation are seriously entertained as the way past the present crises in politics, then of only one thing may we be sure: no matter what form of government we live under in the future, *it will be called democracy.* There is in the writing on the future no consideration of the nature of the polity because, in fact, political community is today very near a total collapse by the rush upon it of the very values the new futurists represent: rationalization, centralization, and uniformity. Other writers, notably C. Wright Mills, at least recognized that the basic problem was the one of elitism. While some futurist writers recognize that we are in a situation where meritocratic elites replace the old plutocracy, they do not take the next step—the growth of technocratic elites presumes the atomization of society; the condition of their rule is the erosion of political and social community and the creation of a new monopoly of knowledge.

Many new futurists recognize that knowledge is power—they say it so often it perhaps has never occurred to them that it first of all needs to be meaningful and relevant knowledge—and that it can be monopolized like any other commodity. However, they rarely recognize that the phrase "monopoly of knowledge" has two interpretations. In the first, monopoly of knowledge simply means the control of factual information or data. Communications is crucial here because the development of more elaborate codes and storage facilities allows groups to control information and deny access to the uninitiated and disconnected. Moreover, competition for innovation in the speed of communication is spurred by the fact that if information flows at unequal rates what is still the future for one group is already the past for another. Ithiel Pool and his colleagues illustrate this meaning of monopoly of knowledge and simultaneously paint a generous portrait of the new information systems in breaking this monopoly.

The information facilities provided by the computer can . . . serve as a decentralizing instrument. They can make available to all parts of an organization the kinds of immediate and complete information that is today available only at the center. The power of top leadership today is very largely the power of their information monopoly. . . . A society with computerized information facilities can make its choice between centralization and decentralization, because it will have the mechanical capability of moving information either way. [Westin, 1971, p. 248]

There is, however, a more stringent sense of the meaning of a monopoly of knowledge. When one speaks, let us say, of the monopoly of religious knowledge, of the institutional church, one is not referring to the control of particles of information. Instead, one is referring to control of the entire system of thought, or paradigm, that determines what it is that can be religiously factual, that determines what the standards are for assessing the truth of any elucidation of these facts, and that defines what it is that can be accounted for as knowledge. Modern computer enthusiasts may be willing to share their data with anybody. What they are not willing to relinquish as readily is the entire technocratic world view that determines what it is that qualifies as an acceptable or valuable fact. What they monopolize is not the data itself but the approved, certified, sanctioned, official mode of thought, indeed the definition of what it means to be reasonable. And this is possible because of a persistent confusion between information and knowledge.

Rarely in writing about the new communications technology is the relationship between information and knowledge ever adequately worked out, because it is not recognized as a problem. Information and knowledge are generally taken as identical and synonymous. It is assumed that reality consists of data or bits of information and this information is, in principle, recordable and storable. Therefore, it is also possible, in principle, for a receiver to know everything or to at least to have access to all knowledge. But this primitive epistemology, admittedly primitively described, will not do the intellectual work or carry the argumentative freight heaped upon it. Knowledge is, after all, paradigmatic. It is not given in experience as data. There is no such thing as "information" about the world devoid of conceptual systems that create and define the world in the act of discovering it. Such paradigms are present in information systems; they are meta-informational, contained in computer programs, statistical devices, information storage, and retrieval codes, technical theories that predefine information and perhaps most importantly in systems of binary opposition, that *lingua franca* of modern science.

Moreover, as one hopes the history and sociology of science have finally established, paradigms are not independent of exterior biases and purposes; they instead express a value-laden rationale in technical language. Computer information systems are not just objective, information recording devices. They are emanations of attitudes and hopes. The subjective location of such attitudes and hopes remains vested in the servants of the institutional monopoly of foreknowledge, for instance, the Rand Corporation. The "idea of information" is another way past the real political factors of class, status, and power, but these formidable realities cannot be dissolved into a future where they are presumed not to exist because they have been absorbed and transformed by the computational machinery.

In summary, then, the "third communications revolution" has within it the same seeds of miscarriage that have historically attended innovations in communications. Instead of creating a "new future," modern technology invites the public to participate in a ritual of control where fascination with technology masks the underlying factors of politics and power. But this only brings up to date what has always been true of the literature of the future. This literature with its body of predictions, prescriptions, and prophecies is a cultural strategy for moving or mobilizing or arousing people toward predefined ends by prescribed means. It would legislate and magistrate beyond the writ of any previous parlimentary or judicial body. It presumes to arbitrarily decree what shall be done and to appeal for the enactment of the plans brought forth. In the process, parts of the past are selectively deleted and aspects of the present are ignored. If such factors contradict the desired end in view they must be proclaimed "obsolete," or examples of cultural lag.

Unlike the mere revisionist or clairvoyant, the futurist has the advantage that the future can always be rewritten for there is no record to compare it with, no systematic verification of prophecy. The futurist can keep extending the day of consummation or rely on the forgetfulness of the public when the appointed but unfulfilled day arrives.

We have been treating here what should be called the "futurean mirage"— the illusion of a future. The futurean mirage is that the future is already out there, converging with the last stage of history, the great departure from all previous stations of travail to the final "stability zone." It posits the future as more than the *next* time dimension; instead, the future is conceived as an active agent reaching back into the present and past from its own superior vantage point and revising time and ineluctably removing obstacles to the previous unachieved rendevous with destiny. However, this sublime future is definitely not an open space in time, openly arrived at; instead, it is a carefully prepared predestination determined not on the grounds of human needs but technological imperatives peculiar to the devices by which the decision making of the futurist mystique is based.

The great irony is that while we seem to be living through the anticipatory "age of the future," there is no real future left open to us as a viable site. For the past projections of the future, in their influence as an ideational power-house on the course of policy and history, have foreclosed the formerly available futures filled with variable choices and exhausted the once rich cultural and natural resources that might have provided the basis in the past for a humane future in a livable landscape.

The emphasis of the futurist cast on the instantaneous efficiencies and speed over space in communications has by its focus on vast scale and fast pace eclipsed the public vision of its own immediate and long-term community with its indigenous interests. The mythology of the powerhouse, with its promise of decentralized economies and ecological harmonies, has actually provided a glossy picture of the sublime future whose subliminal aspects really have tendencies to commercial empires and cosmetically treated landscapes engineered for exploitation.

There remain elements of cultural permanence and political vitality in the nontechnological parts of our national inheritance. To draw on these resources, is it not time for the conception of the future to be rejoined to the real past and the realities of the present?

REFERENCES

Barnes, Harry Elmer, ed. *An Introduction to the History of Sociology*. Abridged edition. Chicago: University of Chicago Press, 1966.

Beard, Charles, and Mary Beard. *The Rise of American Civilization*. New York: Macmillan, 1940. (Two volumes in one.)

Benson, Lee. "The Historical Background of Turner's Frontier Essay," *Agricultural History, 25* (April 1951), 59–82.

Birkenhead, Earl of. *The World in 2030 A.D.* New York: Brewer and Warren, 1930.

Brown, Dee. *The Year of the Century, 1876*. New York: Scribner, 1966.

Brzezinski, Zbigniew. *Between Two Ages*. New York: Viking, 1970.

Clarke, I.F. "The First Forecast of the Future,"*Futures, I,* 4 (June 1969), 325–330.

Fuller, R. Buckminster. *No More Secondhand God and other writings*. Carbondale, Ill.: Southern Illinois University Press, 1963.

"Statement by President Nixon on creating a National Goals Research Staff," *Futures. I,* 5 (September 1969), 458–459.

Garvey, Michael Angelo. *The Silent Revolution: or the Future Effects of Steam and Electricity Upon the Condition of Mankind*. London: William and Frederich G. Cash, 1852.

Geddes, Patrick. *Ideas at War*. London: Williams and Norgate, 1917.

Hutchinson, Paul. "Revolution by Electricity," *New Directions*, Warren Bower (ed.). New York: Lippincott, 1937.

Huxley, Aldous. *Tomorrow and Tomorrow and Tomorrow and Other Essays*. New York: Perennial Library, 1972.

Kallen, Horace M. *Patterns of Progress*. New York: Columbia University Press, 1950.

Kropotkin, Petr. *Fields, Factories and Workshops*. New York: Putnam, 1913.

Lowell, James Russell. *The Works of James Russell Lowell*. Standard Library Edition, Cambridge: Riverside Press, 1871.

Miller, Perry. *The Life of the Mind in America*. New York: Harcourt Brace and World, 1965.

Mills, C. Wright. *Power, Politics and People*. Irving Louis Horowitz (ed.), New York: Ballantine, 1963.

Mumford, Lewis. *The Human Prospect*. Harry T. Moore and Karl W. Deutsch (eds.), Boston: The Beacon Press, 1955.

Mumford, Lewis. "An Appraisal of Lewis Mumford's 'Technics and Civilization' (1934)," *Daedalus, 88*, 3(Summer 1959) 527–536.

Mumford, Lewis. *Technics and Civilization*. New York: Harcourt, Brace and World, Harking Book edition, 1963.

Nevins, Allan. "The Tradition of the Future," in *Now and Tomorrow*, Tom E. Kakonis and James C. Wilcox (eds.), Lexington, Mass.: D.C. Heath, 1971, pp. 396–404.

Polak, Fred L. *The Image of the Future*. New York: Oceana, 1961.

Reich, Charles. *The Greening of America*. New York: Random House, 1970.

Toffler, Alvin. *Future Shock*. New York: Bantam Books, 1971.

Weiner, Norbert. *Cybernetics*. Cambridge: M.I.T. Press, 1948.

Wells, H. G. *The Way the World is Going*. New York: Doubleday, 1929.

Westin, Alan. *Information Technology in a Democracy*. Cambridge: Harvard University Press, 1971.

CHAPTER 32

Communication Technologies and the Future of the Broadcasting Professions

MICHAEL GUREVITCH AND PHILIP ELLIOTT

New technologies often give birth to new occupations. The media professions illustrate this process, although media professionals themselves often have little contact with or understanding of the technological hardware that facilitates their professional work. The structure of these professions as a whole is determined to a considerable extent by the technological base of the media. The pace of technological change in the media may bring about new professional opportunities, or, in the extreme case may result in the attrition of a whole occupation, based on a dated technology. At the very least, technological change requires the constant adaptation of old professional routines and practices, and parallel changes in the norms, the ideology, and the self-image of the profession.

This chapter will look at the potential impact of the new technology of communication on some of the characteristic features of media professions. It sketches the relationship between present technology and the structure of these professions, with broadcasting as a case in point; it focuses briefly on elements in the new technology, and examines ways in which professional structure and ideology will be influenced by changes in the technological base of the profession.

Our basic contention is that technical changes may lead to further deprofessionalization of the media professions. To express the same point in different terms, it may lead to the disestablishment of the broadcasting organizations that most clearly epitomize a "professional" approach to broadcasting. The main factor bringing this about will be the large increase in available television outlets, promised by the new technology. Moreover, the increase in the number of communication channels will probably be accompanied by changes in the techniques of production, whose salient outcome, from our point of view, will be to simplify techniques and so make access to the means of content production easier. These new opportunities are bound to influence current professional norms, standards, and values governing the production of content, as well as the ideology of the profession, to be more in accordance with that of other semiprofessions that are primarily concerned with value maintenance and with the dissemination of symbolic-cultural products.

BROADCASTING AS A PROFESSION

One objection to our argument is that those who were never professional cannot be deprofessionalized. In this chapter, the term professional is used in a more general sense than is common in the sociological literature (Habenstein, 1963). Instead of defining profession according to the formal features of various accepted occupations and then measuring the broadcaster against this definition, our aim is to show the link between professionalism in broadcasting, a particular form of broadcasting organization, and the technical factors that have supported both. It is not so much broadcasting as a professional occupation that is under threat from technical change as broadcasting as a professional organization (Etzioni, 1964).

Two processes—specialization of knowledge and specialization of function —lie behind the increase in the number of professions during this century and the last. In the first case new occupations have emerged to take over responsibility for the development and application of a new field of knowledge. In the second, new needs and possibilities for providing particular services have emerged within the existing structure of society.

Broadcasters, defined as those producing the cultural content of broadcasting output, fit into the second pattern. The development of media of communication has involved the gradual separation of the different activities in the communication process and their association with different groups, preeminently producers, performers, and audience. Technical change has underpinned this process of specialization, so that in the present situation those who work within the mass media form a relatively distinct, bounded, introverted, and to some extent isolated group in society.

Specialization of function such as one finds in broadcasting leads to attempts to develop an occupational expertise from actual work experience (Hughes, 1958). In contrast, the problem for a profession of the first type, a new specialism in science or medicine, is how to apply new knowledge to client needs.

Once again it could be argued that this difference only underlines the point that broadcasting is more properly considered as a craft (or set of crafts) than as a profession (or set of professions). Broadcasters possess few of the characteristics commonly accepted as professional attributes—a body of theoretical knowledge, a long period of education and training, occupational organization, and control over entry, control over practice through standards, disciplinary procedures, and codes of ethics (Millerson, 1964, Hickson and Thomas, 1969). Such definitional disputation seems relatively sterile, however, compared to a consideration of the two main ways in which the concept of professionalism can be applied to broadcasters (Elliott, 1972). Professionalism is both a claim to status and autonomy and also a source of guidance for occupational performance. These two senses of professionalism are bound together in the concept of professional ideology, which is at the same time outward and inward looking—a mechanism through which the profession defines and defends its position in relation to others outside, and a mechanism through which it makes sense of ongoing tasks and problems in the work situation.

Various processes of social change have had considerable impact on the position of professions and the nature of professionalism. The range of different types of status differentiations available within society has decreased and occupation has become particularly prominent among those remaining. A related development of significance in many professions has been the widening of the clientele for professional services. Traditionally, professional clients were to be found among those of sufficient wealth and status to pay directly for the service. Widening the scope of professional clientele has given new forms to old problems among the professions—how to preserve their authority over the clients or their representatives, how to arrange the financial and organizational means for practice? All these are experienced acutely by the broadcaster, especially in controversies over quality.

The general decline in status differentiation has been paralleled in the development of the media. Each new medium has made possible a continued growth in audience size and a continued homogenization of audience taste. Note that while this process has apparently exhausted itself within the bounds of the nation-state, distribution by satellite makes the development of a multinational audience conceivable. Indeed in a sense through the distributive networks of international corporations such an audience already exists. A countervailing process can also be discerned whereby some media overtaken by newer media in the pursuit of the mass audience become instead specialized media catering to minorities. This is especially likely in those cases where the economics of the medium depend on the individual consumer, the buyer of a record or magazine, for example, making an effective choice at the point of consumption. Television especially that financed by license fees provides a polar opposite case of a medium in which the consumer's effective choice covers the whole service, not the individual items of which it is composed (although choice of these particular items may well be significant, depending on the extent of competition and other factors external to the organization). These considerations underline a point that we shall reiterate continually in this chapter: the present and future of the broadcasting professions depend on the social, organizational, and economic situation within which they work.. In considering the impact of technical change on the profession, we must remember that this impact will be mediated through these other factors.

In most countries at the present time, broadcasting is carried on by large-scale organizations that both produce and distribute their product. Commercial broadcasting systems have allowed more scope for separate production units to emerge, marketing their programs to distributors, although both are likely to be linked by patterns of common ownership. In such cases, market forces, in the shape of the beliefs and understandings of audience tastes and tolerances held by those in financial control of the industry, play the largest part in setting the situation within which the program maker works. Under government controlled broadcasting systems, the broadcaster may also be controlled and constrained, if for different purposes.

Between these two extremes, however, a third type of broadcasting system has emerged in which the broadcaster is supposedly insulated from both types of control—government and commercial. It is an open question how successful such insulation has been. Nevertheless, the aim has been to model broadcasting on the style of the traditional professions within the framework of liberal, democratic society. Responsibility for performing the particular social function is allocated to a specific group that has some appearance of autonomy and independence but that is also linked at crucial points into the structure of the society it serves. One such link is the recruitment of broadcasting staff from the educated middle-class elite and their continued contact with this group in their work and leisure activities. Another link is that in the final analysis politicians hold the ring for broadcasters. Leaders of broadcasting organizations are often appointed to represent the "public" interest, in other words, to respond to changes in the political climate. The Israeli Broadcasting Authority was set up as a public broadcasting organization modeled after the BBC, but it provides a clear case of a broadcasting organization that is closely attuned to the political structure of society. Even in Britain, where the tradition of independent public broadcasting is held up as a model of its kind, controversy over the need for censorship of the coverage from Northern Ireland has made broadcasters more sensitive to the climate of opinion within which they work.

Broadcasting professionalism can be identified in all three types of broadcasting system, but it is to be found in its most developed form in these independent, noncommercial organizations. The BBC ideal of "public service," for example, epitomizes the traditional professional tension between client service and client control. The professional broadcaster interprets public desires and public needs and decides how both should be met "in the public interest."

Professionalism is not a means to complete autonomy but another means through which the dominant elites are able to keep cultural production within an ideological framework supportive of their interests. Professionalism involves more complex, tenuous, and occasionally conflicting mechanisms than commercialism or direct government control. The important point, however, is that if technical change can be linked to deprofessionalisation, other mechanisms of social control will be necessary to ensure that cultural production continues to support dominant ideological interests.

PROFESSIONAL IDEOLOGIES AND PROFESSIONAL SITUATIONS

Two aspects of the ideology of media professionals are most directly threatened by changes in the technological base of the profession: first, the self-image of the profession in terms of its role in the cultural system of modern society and second, its function as a mechanism for the protection of the profession. The first is related to the profession's autonomy; the second is related to its exclusiveness.

Among the many professional and occupational groups engaged in the production and dissemination of culture, the role of mass communicators is ambiguous. Although their role is often described as "creative," a closer look at their activities indicates that they act predominantly as "cultural entrepreneurs." Their main tasks are to initiate cultural production, to "translate" or popularize its products in order to facilitate their distribution to a large heterogeneous audience. The communicators' right to perform these tasks depends on their professional control of the means of mass communication.

The communicators' "professional control" is not total, but is embedded in controls exerted by a variety of social and political groups outside. Nevertheless, the keenness with which media professionals insist on claiming the widest possible professional autonomy can be taken as an indication of the importance they attach to their control over the tools of their profession. It also highlights its importance for their professional self-image and for their conception of the professional status of their occupation.

Control over access to the means of mass communication also protects the exclusiveness of the profession. Many of the mechanisms used by such "classic" professions as medicine and law to protect their exclusiveness cannot be found in the media professions. Most of the media jobs require little, if any, formal training; to a large extent the skills and the qualifications required for these jobs are considered by media practitioners to be innate instead of acquired; and, ironically, the media professionals themselves are fundamentally opposed to the notion of licensing because of its potential misuse as an instrument for external, primarily political, control over the profession. Some alternative protective devices have emerged in the form of ideological legitimations for the monopoly position media professionals hold over the means of mass communication. Various examples of such ideologies might be cited, but the essence is contained in the ideology of the "freedom of the press."

Basically, the concept of the freedom of the press draws its legitimacy and its aura from the wider ideological base that defines the proper relationship between the governors and the governed in Western democratic tradition. The view of a democratic electorate as a rational, enlightened, and informed body necessarily points to the need for setting up channels of communication and information between the electorate and the elected. In a modern industrial society this is done by the media. The media's claim to freedom from external controls is thus based on (or borrowed from) the rights of the citizenry to be fully informed and to enjoy access to all relevant information. Ironically, however, this has been turned into a mechanism for protecting the professionals' exclusive access to the media. What emerged as a safeguard against undue intervention in the media by established authority operates as a mechanism for limiting access to the media by all outsiders, rulers and ruled alike.

Recruitment and Socialization

Entry into the broadcasting profession in most countries is a haphazard process depending heavily on personal particularistic factors supported by general educational and status qualifications. The "on the job" training that this involves is a particularly effective way of acquiring the ideology of the profession as well as its expertise (Breed, 1955). There have been few attempts to introduce more systematic entry procedures. In the commercial sector of the industry in Britain, unionization of the "creative" personnel has coincided with a period of stagnation in the industry, following a relatively long period of expanding opportunities and high rewards. The union is able to operate a closed shop in this section of the industry, but obtaining a union card has simply become another of the haphazard entry procedures. It does not measure qualifications or training in any professional sense. After entry to the profession, personal particularistic factors continue to be of great significance in finding work. These factors, coupled with a necessary element of public display, make for marked individualism in the pursuit of a broadcasting career.

Professionalism has traditionally been associated with independent private practice, so it may seem paradoxical to argue that the more professional the broadcaster the more he will be enmeshed in a large-scale broadcasting organization. Occupational solidarity has had little opportunity to develop among such groups as performers, whose prospects for future employment are highly insecure and who are liable to be continually moving from one employment situation to another. For some categories of staff, however, especially program makers, working within broadcasting organizations has involved a relatively unique organizational mix of professionalism, bureaucracy, and attempts to provide scope for creativity (Burns, 1970). Bureaucratic tendencies toward hierarchic control and careers within such organizations have been paralleled by the growth of a professional creed at all levels. The importance of this professional creed is that it provides common standards of judgment to facilitate work flow; but more than that, it helps the organization to articulate by providing a legitimated mechanism of control and coordination. Executives and program makers can negotiate around this creed in any dispute over control and content. Potentially disruptive conflict over program content can be neutralized into disputes over program style and quality. Furthermore, professionalism is useful externally as well as internally. The organization or its leaders can use this technique to defend their autonomy before others in society who question their rights and competence.

THE TECHNOLOGICAL INNOVATIONS

The new technology of communication will bring with it a variety of tools and technical devices that will have an effect on most aspects of broadcasting.

The most significant technical change for the structure of the media professions has to do with the change in the distributive capacity of the media. One of the most important consequences of the introduction of these new technological devices will be the proliferation in the number of communication outlets. The mass media appear to be moving from "an era of scarcity, to an era of affluence." This phrase describes the newly expected freedom of television from the limitations on the number of available television channels in the broadcasting spectrum. Both cable television and the use of audiovisual discs and cassettes basically imply a departure from the concept of broadcasting as it has been understood heretofore, that is, the transmission and reception of signals through the air.

Alongside the innovations in distribution is the wide range of technological inventions that will affect the techniques of content production. It appears that such innovations as miniaturization and further simplification of production hardware will make possible increased access to the means of content production. The easier the manipulation of photographic and recording equipment becomes, and the lower the cost of the equipment and its use, the more widespread the range of people involved in the production of transmittable material will be.

These two simultaneous developments—the opening up of access to the means of production and to the means of distribution—point in the same direction: a relaxation of the hold that media professionals currently have over the means of mass communication.

THE CONSEQUENCES OF TECHNOLOGICAL CHANGE

Contents

In discussing the effects of technological changes on the nature of the contents, a distinction should be made between straight "informational" material, that is, news material, on the one hand, and "cultural" contents on the other. A related and more useful distinction is that between "time-bound" versus "time-free" contents. These clearly should be taken as denoting opposing poles on a continuum instead of representing a dichotomy of content types. "Time-bound" contents, unlike "time-free" contents, must be disseminated immediately. The value of speed in transmitting conventional "news" material is one of the cardinal operating rules of journalism. On the other hand, the broadcasting of a concert, an opera, or a play are supposedly free from the rigid constraints of time. The more closely the material comes to the "immediate" pole of the continuum, the more costly it is, since its collection, processing, and distribution presuppose the ready availability of a large and complex man-machine system that has to be maintained at all times at a high level of readiness. These cost implications have a direct bearing on our own problem: the larger the cost, the greater the need for a large organizational framework that will be capable of bearing it, and consequently the greater is the probability that the production and dissemination of these materials will remain within the existing framework of the organizational structures that currently dominate the mass communication scene. Technological innovation will therefore have relatively small effect on the organizational structures and on the professional personnel engaged in the production and the broadcasting of "news." Indirectly, however, the possibility of a multiplicity of channels might be expected to make some of the current ideological foundations for large-scale news organizations, in particular impartiality and objectivity, less relevant and significant. On the other hand, one might conjecture that the production and distribution of "time-free" material, currently within the domain of large media organizations, might gradually break away from the hold of these organizations and be taken over by smaller, more flexible, possibly ad hoc groupings, thus facilitating the entry into these operations of a larger number of "nonprofessionals."

Political Control

Such developments will inevitably necessitate some realignment of the present system of relationships between media organizations and the political institutions of the societies in which they operate. Irrespective of other socio-political and economic factors that impinge on the problem of control, it would seem that a process of decentralization in the media might lead in either of two opposing directions. On the one hand, one might speculate that a widening of the "span of control" might lead to a reduction in the capacity to supervise or control directly each of the separate units in the network, and thus result in relatively greater freedom for the media as a whole. On the other hand, it could be argued that since one of the effects of decentralization is to leave each of the component units somewhat smaller and perhaps weaker, the opportunities for a central authority to exert effective control actually increase. Professionalism as an intermediate factor is likely to decrease. To the extent that precedents from the development of print communication are anything to go by, it seems to us that the first possibility, that of decreased direct control, is a more probable outcome. To compensate, we may expect the broadcasting media to become even more closely integrated into the commercial system.

Finance

Most professional public broadcasting organizations have been supported by license revenue, a system designed to maintain the independence of the organization from both commercialism and government. One direct consequence of technical change has been to undermine the viability of licensing revenue. Where the market for television sets has been saturated, it no longer provides a growing source of revenue. Continued technical developments, however, are among the many factors that have continued to increase costs. There may be a further period of expanding license revenues with the introduction of color receivers, but many European broadcasting organizations are already facing the need for severe economies or alternative revenue sources. Some economies already introduced, such as the increasing use of short-term contracts for program-making staff, will tend to foster the individualism and insecurity already present in broadcasting careers and alter relationships between permanent executives and transient program makers within the organizations; briefly, they begin the process of deprofessionalization envisaged in this chapter. Similar arguments could be applied to other trends, for example, toward cooperative productions and international program marketing. These so increase the range of criteria that need to be met before production starts that the role of the professional program maker is considerably reduced.

Continuing technical change, particularly the multiplication of channels, will further undermine the viability of license revenue. Complex and politically tendentious arrangements would be necessary to divide the revenue between different producing organizations. Furthermore, the more channels that are provided independent of license revenue, the more popular and commercial demand will grow for such apparently "free" sources to take over broadcasting entirely. Once engaged with commercial broadcasting organizations in a competition for mass audiences, professional broadcasting organizations find themselves in a cleft stick. If they succeed in the competition, there is no justification for their existence, as they are only providing the "same" service as the competition. If they fail, they are only addressing a minority and so there is no reason why the mass should be called on to support them. One of the key questions for later consideration is how far technical change is likely to maintain or disrupt the mass characteristics of the broadcasting media.

Recruitment and Ideology

The potential effects of the new technology on recruitment to the media professions touch on one of the core elements of the status of these professions: their ability to protect their exclusiveness. Speculation concerning the effects of the new technology on the problem of recruitment seem to go in two opposite directions. At first glance, it seems reasonable to assume that a process of decentralization in the structure of media organizations and the consequent proliferation of communication outlets would create new needs for broadcast materials in excess of the amounts produced at present and therefore will increase the requirements for man power. Thus the doors will inevitably be thrown wide open to the entry of a large number of new recruits into the profession, or at least to access in the media of many more people who are currently denied that access. Under these circumstances, the entire process of regulated recruitment to the media professions might be disrupted, and the professional essence of these occupations might be jeopardized. The anxiety that such prospects create among media professionals thus becomes obvious and understandable.

However, it is not at all clear whether the proliferation of communication outlets necessarily leads to developments along these lines. The example of American radio indicates that when the emergence of a large number of broadcasting stations is coupled with automation, the total number of personnel required for operating these stations does not necessarily go up. Clearly, the factor that leads to growth in personnel is not the increase in the number of broadcasting stations but the question of whether or not these new outlets will be willing to rely on existing broadcasting materials generated by larger centers of production, or whether they will insist on producing their own material. The need for increasing personnel stems not from the disseminating function of the media, but from its productive-creative function.

Threats to the profession are likely to stimulate responses couched in ideological terms. One example of this line of defense has been provided in a conference convened to discuss "The New Communication Technology and Its Social Implications" (Morgan, 1971). In discussing the potential impact of the new technology on diversification of output, the professionals' argument ran as follows; proliferation of channels does not necessarily lead to diversification of output, but on many occasions results merely in a multiplication of equally unsatisfactory outputs. Developments along these lines amount to intellectual and cultural pollution or, more picturesquely, to a "Lake Erie of the mind." In other words, the professionals invoked the virtues of quality and high standards in media contents to warn against potential deterioration that might come about as a result of an infringement on their exclusive competence in the media. While this argument rightly raises doubts about the relative pros and cons of competitive versus noncompetitive situations in the media, at the same time it invokes the spectre of strictly quantitative competition to protect the professionals' exclusive privileges of jurisdiction over the media.

If, however, the new technology would indeed eventually result in an opening up of access to the media, the professional ideology of the broadcasters will have to shift from legitimizing its claim to exclusive jurisdiction by invoking the values of quality and the professional standards to other bases of legitimation. One possible example that media professionals could then follow (and to which we shall return later on) is that of the clergy. In discussing the inadequacy of the ministry's "technical base" as substantiating its claim for professionalism, Wilensky (1964) argues that "the ministry has a monopoly of recognized skill not because its technical base is scientific, but because the clergy's tasks and tools belong in the realm of the sacred" (p. 139). While presumably no communicator would claim professional exclusiveness on a strictly analogous basis, the semisacred character of the concept of "the freedom of the press" has always been seen by the profession as an appropriate value on which to base a continued claim for exclusiveness.

Change in the Professional Situation

In capitalist countries, commercial applications must be discovered for technical innovations before they will be introduced. At the present time, preparation for audiovisual discs and cassettes is continuing among those corporations able to produce and market the hardware as soon as commercial applications are ready, but these preparations are not accompanied by any extensive public, professional, or political debate on how and when to introduce the new techniques to gain most benefit from them. As happened with the initial introduction of radio broadcasting, commercial application is likely to happen first, creating a situation to which other interested parties—media professionals and organizations, their audiences and political controllers—will need to respond.

If this is the likely course of the innovatory process, it follows that, initially at least, the technical changes will increase the importance of the commercial mechanisms linking audience demand and program production. The two types of distributive change, cassettes and cables, would activate rather different commercial mechanisms. Cassettes and discs make unit purchasing or hiring possible, introducing a similar economic structure into the broadcasting field as already applies in gramophone records, magazines, and, to a lesser extent, films. Cables, on the other hand, simply increase the range of outlets available to all receivers of television. If the outlets are mainly dependent on advertising revenue, specialization may be expected to develop between them only to the extent that specialization of product and market is functional for the advertiser. Otherwise specialization will have to be introduced through the application of organizational control or the introduction of effective choice at the point of consumption through a system such as pay TV.

Such considerations are crucial to any analysis of the impact of innovations on the broadcasting profession since one of the tasks of the profession up to the present time has been to generalize and popularize information and entertainment for a mass audience. The more audiences that develop that are able to support specific types of programming, the more one might expect subject expertise to replace the generalizing and popularizing abilities of the broadcaster. This will make professional mystification of the production skills a less feasible tactic.

The professional broadcaster and the professional broadcasting organizations in which he is most obviously found were both the largely fortuitous product of a unique combination of technical and social factors. As such they have few resources to fall back on in any attempt to maintain their position. It has been difficult to unionize an occupation whose members are drawn from the middle class with pretensions to professional status and whose work situation puts a premium on individualism. There has never been an adequate basis of professional expertise to foster the type of professional association and exclusiveness that in other professions has provided an alternative means of defense. We may anticipate therefore that occupational organization will be able to do little more than help people to adapt to changing situations. It cannot be expected to play much part in shaping those situations.

In the short run, at least, those working in broadcasting organizations may find themselves more directly exposed to the impact of change than the organizations for which they work. Apart from the general phenomenon of organizational inertia, there is also the more specific point that the professionalism of broadcasting organizations has been indirectly a mechanism for ensuring that the cultural output of broadcasting has been integrated into the structure of society. It seems unlikely that professionalism will be abandoned as a method of social control without some other mechanism replacing it. This is one reason why it may be overoptimistic to hope that the expansion of outlets will lead to a universal increase in freedom of expression and access. There may be less need for control given a variety of outlets than was the case with one or two broadcast channels and one or two broadcast organizations. Many voices may be expected to cancel each other out or at least to end up simply preaching to the converted. But beyond this we may point to the tendency toward commercialism already implicit in the application of new technology. This is likely to be further encouraged by political and economic elites unwilling to be seen exercising direct controls but able to rely on factors in the commercial structure to maintain control over output for them. A similar development may be traced in the press from the direct and indirect controls operated in most European countries up to the nineteenth century, to the development of a commercial press in this century and the last. One of the factors that limits access to the commercial press at the present time is the high level of capital investment necessary. Given the importance of technology in broadcasting it seems likely that this will also apply there. But those who control the technology will have little claim to the title of professional broadcaster as it has been understood in the era of broadcasting scarcity. On the contrary, they will be employees of monopoly capital, and the process of integrating the culture-producing function into the structure of complex industrial societies will have been carried a stage further.

Conclusion

The trend toward deprofessionalization in media occupations that we envisage following from the diffusion of new communication technology will affect both the character of the occupations and the structure of the organizations in which they are to be found. So far as the occupation itself is concerned, exclusiveness is the main point at risk. Mass communicators have been "generalists" instead of "specialists," but they have developed a specialized expertise through their use of the communication media. Exclusiveness has meant not so much that particular individuals or groups were kept out (although some effectively have been) but that any individual to be admitted had to conform more or less exactly to the rules of the game as recognized and operated by the professionals. These rules of the game have been enshrined in occupational understandings and professional creeds and represent the modus vivendi established between those controlling and those working within media organizations.

The element of professionalism and creative autonomy in this modus vivendi is particularly insecure in the present situation of technical change. Increased communication outlets may make it possible for transient outsiders to enter the media industries more easily. More significant, however, is the likely impact of technical change on the large-scale broadcasting organizations. As the professionals are deprofessionalized, so the broadcasting organizations, especially the European-style "professional" broadcasting organizations, will be disestablished. Whether multichannel competition leads to programming for minorities or a concerted rush for the same middle ground will depend to some extent on how the new technology is organized and financed. One thing is certain, however. There will be little room for the professional elitism that thrived in the situation when the broadcaster not only could presume he knew what was best for the public but was also encouraged to do so by those interested in keeping control over popular culture.

There is a curious parallel between deprofessionalization and disestablishment in the media and similar developments that have for a long time been present in the church. In some respects clergy and broadcasters have occupied similar roles: hence, for example, one recent description of television broadcasters as "The New Priesthood" (Bakewell and Garnham, 1970). Both have manipulated social symbols and reflected social values to perform a general system-maintenance function in society. The church lost its central position in performing this function as the development of industry and commercialism undermined appeals to the sacred with more mundane values based on worldly goals. Professionalism can be seen as a transitional attempt to interpose a fundamentally elitist method of performing social functions based on appeals to special expertise between the passing era of appeal to absolute sacred and juridical rights and the coming era of commercialism founded on economic interest. If nothing else the new communication technology may help to sharpen the contrast between elitism and populist commercialism. If it opened up some third alternative, then we would really be justified in talking about a revolution in communication technology.

REFERENCES

Bakewell, J. and N. Garnham. *The New Priesthood*. London: Allen Lane, 1970.

Bennis, W. G., and P. E. Slater. *The Temporary Society*. New York: Harper and Row, 1968.

Breed, W. "Social Control in the Newsroom: A Functional Analysis," *Social Forces*, *33* (1955), 326–335.

Burns, T. "Public Service and Private World" in J. Tunstall (ed.). *Media Sociology*. London: Constable, 1970.

Elliott, P. *The Sociology of the Professions*. London: MacMillan, 1972.

Etzioni, A. *Modern Organizations*. Englewood Cliffs, N.J.: Prentice-Hall, 1964.

Habenstein, R. W. "A critique of 'Profession' as a Sociological Category," *Sociological Quarterly, 4* (1963), 291–300.

Hickson, D. J., and M. W. Thomas. "Professionalization in Britain," *Sociology, 3* (1969).

Hughes, E. C. *Men and Their Work*. Glencoe, Ill.: The Free Press, 1958.

Millerson, G. *The Qualifying Associations*. London: Routledge, 1964.

Morgan, R. P. "The New Communications Technology and Its Social Implications." International Broadcast Institute, 1971.

Wilensky, H. "The Professionalization of Everyone?" *American Journal of Sociology, 70* (1964).

CHAPTER 33

Politics and the New Mass Communications

FORREST P. CHISMAN

It has been widely assumed that the structure and level of development of mass communications can have an important influence on political systems. It has even been argued that changes in mass communications help bring about corresponding changes in political organization. Hence, we have the belief that the development of the printing press and later the newspaper were driving forces in the growth of liberal democratic states.

Working on these assumptions, many people, especially in the United States, have expressed great interest in certain new developments in telecommunications. They feel that the growth of satellite and cable technology in particular promise to strengthen existing democratic institutions. This is because it is generally accepted that the success of liberal democracies depends on an informed public, and it is equally an article of faith that our most powerful means of communication, television, offers a diet of public information that is both too meager and too uniform. Cable and satellites can provide more channels than conventional television broadcasting and they can substantially reduce the cost of purchasing air time. Many people believe that the results will be more plentiful and diverse public affairs programs which should clearly benefit our political process. This is one reason why they suggest that satellite and cable development should be encouraged and why they propose lifting certain provisions of our present broadcasting regulatory policies, such as the "fairness doctrine."

I think that much of this optimism is premature and that some of it is clearly unjustified. We have very little way of knowing what effects cable and satellites will have on the American political process and judging from what we know, these effects will probably be a combination of good, bad, and indifferent. As a result, we must monitor developments carefully and not be overly eager to lift many of the admittedly imperfect safeguards that now exist.

To begin with, it is important to realize that we have no reliable way of forecasting the structure and effects of a mass communications system in which the access cost is low and in which, as a result, the public is offered plentiful and diverse public affairs content. We do not know, in particular, what content will be produced, how it will be used by the mass public, or what effects it will have on their knowledge and attitudes. History is clearly no guide because we have never had such a communications system reaching a truly mass audience. To be sure, the entry costs of newspaper publishing in the nineteenth century were low, although not as low as often thought, and the diversity of content in some places was great. At the same time, the literacy rate in the nineteenth century was lower than it is today, and the population was much more widely dispersed. Some large cities may have had a great number of newspapers, but most people did not live in large cities. By the time the country became urbanized, radio and then television with their uniform, mass appeal programming were upon us. Moreover, from the 1930s on, there was a steady decrease in the number and diversity of both newspapers and magazines. In short, most Americans have never been served by a mass communications system that provided diverse and abundant content. They have gone from one-newspaper towns to cities served by the mass media as we know them today.

Even if lessons from the past are relevent, however, they do not give us grounds for optimism about the new mass communications. If we say that nineteenth-century newspapers are the model that cable and satellites will imitate, we can expect their news and public affairs programming often to be superficial, partisan, and inaccurate—tending toward social and political fragmentation instead of integration. Given our present circumstances, this hardly seems something to be encouraged.

A more attractive way to predict the future of the new communications seemingly would be to look at present-day radio broadcasting and magazine publishing. Here again, entry cost is by no means as low as often suggested, but it is certainly lower than in commercial television, and a great and often admirable diversity has been the result in recent years. It is tempting to think that cable and satellites will develop in the same way.

This analogy may have some validity, particularly in the entertainment field, but it is hard to say just how far we should push it. It must be remembered that diversity in magazines and radio has developed in the shadow of the television colossus. Television is incomparably the most-used medium, especially for news and public affairs information. The psychological effects of diverse magazine and radio industries may, therefore, be considerably offset by the more powerful effect of a uniform and restricted television industry. Moreover, it is possible to speculate that the present structures of the publishing and radio broadcasting industries are to a significant extent a reaction against the uniformity of television. If television itself became more diverse, it would lack the antithesis that has helped shape radio and magazines, and it might develop very differently than they have. Finally, it has often been pointed out that the imperatives of television production and transmission differ significantly from those of the written or spoken word. For example, the effectiveness of television presentations depends to a significant extent on their visual interest as well as on their intellectual content. This means that many important events and ideas are difficult to convey by television.

I do not mean to imply that there are no lessons to be learned from magazine publishing and radio about the form that television will or should take in the future. I only wish to say that the new mass communications will be a very different animal from those we know today and that any longrange predictions about its form or effects based on present experience are bound to be perilous.

Even if there are no exact models for the new mass communications in our past or present experience, however, we might expect that social scientists would be able to coax some rough projections from existing data. Unfortunately, social scientists are notoriously bad forecasters, and as yet they have come up with very few promising results. Members of the cable television industry have been clandestinely offering a pot of gold to any market researcher who can tell them what lies at the end of their particular rainbow. Generally speaking, they want to know fairly simple things such as what the market demand is likely to be for various new services. As of yet, they have had virtually no takers who could convincingly claim to make solid projections of this sort. Because the effectiveness of market research depends heavily on our knowledge of the present media context, it seems unlikely that its techniques in their present form will be useful for forecasting the new mass communications.

Other social scientists have contributed as little to our knowledge about the probable effects of the new mass communications as the market researchers. There have been practically no serious studies of the subject, and I personally have seen few plans for them. In many ways this is not surprising. With so many unknown variables and with so little to guide them from the past, social scientists may justifiably wish to ponder at length about the new mass communications before making a commitment to specific research. Their ingenuity may eventually provide us with a reliable picture of the future, but it has not yet done so.

We do not know, therefore, what the structure or effects of the new mass communications are likely to be. As a result, general optimism about their influence on politics seems at least premature. We can, however, go farther than this in evaluating some of the predictions that are being made about cable and satellites. In light of our knowledge about present-day television and politics, there are grounds for believing that much of the optimism about the new mass communications is probably unfounded as well as premature. Cable and satellites can potentially have bad as well as good effects on democratic systems, both by raising new problems and by reinforcing old ones. I shall illustrate this by considering three examples of the benefits of the new mass communications often cited by optimists: the potential of these technologies to provide broader and cheaper candidate exposure, to offer more news and documentary coverage, and to allow for various forms of public opinion polling.

One of the most common complaints about present-day television is that it provides far too little opportunity for political candidates to discuss and debate their views seriously. Ultimately, this is because television spectrum space is limited and political programming must compete with other fare in order to present a balanced diet to the viewer and an adequate income to the broadcaster. This limits public service time and increases the cost of purchasing commercial time for political messages. As a result, wealthier candidates have an unfair advantage. In addition, campaign costs are ridiculously high and candidates for all but the most important local elections are excluded from the use of television. It is often said that by making available more channel space, the new communications technology will help to eradicate many of these problems. With a large number of channels, the cost of almost unlimited access will be within the reach of almost any candidate, and, in fact, it has been suggested that time be provided free on special channels. Moreover, with better distribution systems, candidates for local office will be able to reach only their target audiences. Members of the public will, as a result, have the opportunity to inform themselves fully about public affairs at every level.

It is possible that some of these optimistic projections are justified, but there are grounds for believing that some of them may not be. To begin with, even if we assume that channel space is free, the cost of an effective appearance on television is likely to be quite substantial. With a multiplicity of channels, political candidates will be competing against prime-time entertainment programming for an audience. Past experience indicates that in order to compete effectively, they will have to invest considerable sums in prepared newspaper and television advertisements to draw the attention of even their most loyal followers to their appearance and that they must place these ads at normal commercial rates. Moreover, sophisticated candidates will undoubtedly realize that a simple face-on presentation is not necessarily the most effective way to win votes and that, in fact, it may be counterproductive. Increasingly, candidates are reluctant to use television in this way, and the experience of the Kennedy-Nixon debates has led most of them to shy away from debate formats as well. They realize that the slick, prepackaged techniques of Madison Avenue are simply more effective television than more humble presentations, and for this reason it is not unrealistic to expect that with more spectrum space, we might get more Madison Avenue instead of more serious discussions of the issues.

I do not think that politicians are by and large immoral, but they are realistic, and their first objective at election time is to win by the most effective means possible. On television those means are expensive, regardless of how many channels we have. As a result, there are reasons to believe that while the new mass communications may reduce campaign spending somewhat, its costs will still remain very high. Admittedly, the humblest candidate will be able to have some say on television, and there is value in that, but everything we know indicates that he will be as overpowered by the wealthy in the television of the future as he is today and that competitive pressures will continue to send campaign costs skyrocketing.

All of this has special applicability to the local candidate. It may be that we are not doing him a favor by allowing him to make use of television. Local political organizations are often valued as the "farm teams" of our political system. This is in part because techniques for campaigning are comparatively primitive and their cost is low. While wealthy candidates may have some advantage, the differential is not great because pamphlets, bumper stickers, local advertisements, and doorbell ringers are just not that expensive. Introducing television into this environment may well have the effect of raising the stakes substantially. The candidate who can afford expensive, prepackaged presentations may gain a substantial advantage over the one who can only sit in front of a single camera in the local studio. In addition, the cost of even a simply produced appearance on television is not negligible. Our present experience indicates that, even given free channel time, candidates will still have to spend at least several hundred and possibly several thousand dollars to cover the cost of operating a studio. While these costs will not be prohibitive for wealthy candidates, they may unfairly limit the number of appearances that others can make.

As a result, introducing television into local politics may cause the candidate of modest means to be even more excluded from a real chance at office than he is today. True, the new mass communications may help to stimulate interest in local politics, but they may also bring to them some forms of venality that until now have been peculiar to state and national campaigning.

Aside from the economic considerations that I have mentioned, there are less tangible problems that may lead us to doubt the benefits of more mass communications to political campaigning. It seems likely that as the volume of political television increases, the overall influence of this form of political communications will increase also. If this is true, we may wonder whether we really want even more emphasis than at present placed on those superficial attributes of candidates, such as good looks, quick wits, and pretty wives, which take on special importance in television campaigning. Will there truly be no small corner of America in which an ugly, inarticulate bachelor who happens to be a political genius can hope to attain political office? Will the print media decline in importance as channels of political communication, and will this create problems? Finally, to take a different tack, perhaps all these considerations are unimportant; it may be that the fairyland of diverse entertainment programs offered by the new mass communications will be so attractive that political programming will not be able to compete even as well as it does today. As a result, we may see an overall decrease of both public information and concern about politics.

From what little we know, therefore, it seems that the new mass communications are unlikely to solve the serious problems posed by present-day television campaigning, and that they may well raise new problems. A stronger case for the benefits of improved telecommunications technology can be made by looking at its potential to deliver information about public affairs that is not strictly partisan. Here optimists generally think of the potential for more continuous news programming analogous to all-news radio and for more in-depth reporting or documentaries. To me it seems feasible that all-news television channels could be established. They would clearly be more expensive than current news operations, however, and it remains to be seen whether people would distribute their viewing time over the day sufficiently for this incremental cost to be absorbed by advertisers. I have some doubts as to whether an all-news television channel would be a significant public benefit because I doubt that it would lead to a greater volume of news any more than all-news radio has, and I think that the advantages of greater availability are strictly marginal. Nevertheless, I can see no serious problems with this idea if it finds an audience.

More documentaries and other informational programming do, however, pose difficulties. Compared with fast news, documentaries have extremely high unit costs and they appeal to a fairly small audience. Even if the advertisers of the future are more willing to sell to specialized television audiences than they are today, it seems unlikely that they will be willing to pay higher rates for that privilege, and at present rates many documentaries lose money. Those people who talk about the possibility of more documentaries on television generally realize this, and they commonly suggest that pay television may be an answer to the problem. The people who like documentaries can see them at a price. This is a reasonable suggestion if we consider documentaries to be simply another entertainment form. In that case there is no serious problem if some people can afford them more easily than others. If, on the other hand, we think that there is something substantial to be gained by having broader exposure to serious presentations of public issues, then we may be concerned about charging for it. Levying only a moderate price may mean that the less interested and the less affluent will not receive this benefit of the new mass communications. Moreover, it is conceivable that viewer-supported documentaries may drive out those supported by advertisers, and, as a result, the mass public may be less well served than they are today.

In addition to all-news or all-documentary channels providing more or less the same kind of material that we receive today, there have been some suggestions that we can expect new forms of public information. One common idea is that news may be presented from more different points of view. On the assumption that present-day news programming reflects an Eastern, white, liberal bias, we may have news and opinion shows that are avowedly biased in other geographical, racial, or political directions. Insofar as this means that we should put our biases on the table, it is certainly a good idea. Hopefully, it would lead to a form of comparison shopping in which people would learn different sides of issues by listening to a variety of news programs. This, in turn, could lead the average man to have a deeper understanding of people who look at things differently than he does.

There are reasons to be skeptical about biased news, however. We may wonder how many people have sufficient interest to watch more than one news show or the intellectual courage to systematically expose themselves to views very different from their own. Present-day reporters try to reflect different points of view in one brief presentation. This may be a difficult ideal to attain, but unless information is "packaged" in this way, the average man with a limited interest in politics may seldom be exposed to views other than his own. If the ideal of objective reporting is dropped in favor of unashamed ideological bias, each individual may expose himself only to news that is slanted to suit his prejudices, and he may learn even less than he does today about how other people view the world. This might well contribute to social fragmentation. We do not know that this would be the case because, as I have pointed out, we do not know very much about the effects that a diverse communications system might have. The possibility that the new mass communications might have socially disruptive effects, along with possibly beneficial effects, should, however, be considered seriously.

In addition to providing additional channel space for communications from political figures and commentators to members of the public, the new mass communications also promise to make it possible for members of the public to reply. In the foreseeable future, these two-way communications systems will probably allow a standard television picture to come into the home and a digital response to be sent out. Even this limited capability for interactive television has a variety of interesting uses, however, and in discussions of politics the most frequently mentioned use is some variety of public opinion polling. This interest in polling is based on the common democratic assumption that public decision makers should be responsive to the wishes of the people and that a lack of an adequate means for ascertaining those wishes on a continuing basis is a serious defect in large democratic states. I have seen at least a dozen serious proposals to experiment with the use of advanced telecommunications systems for public opinion polling. Generally, these consist of staging a television debate between proponents of different sides of an important issue and then asking viewers to push a button indicating which position they favor. It is hoped that decision makers will act in accordance with public views so recorded or at least seriously take them into account.

This use of the new mass communications raises a host of practical problems. Everyone must be hooked up to the response system, and steps must be taken to be sure that everyone has a chance to see the presentation and to register his or her vote. Because there are so many public issues and because most people have other things to do, clearly not very many problems can be resolved in this way, and some rational means must be devised for selecting the issues and the presentations about them that will be made. It is doubtful whether these practical difficulties will be resolved in the near future, but they should make us wary that the intrinsic appeal of television polling will lead to premature implementation of it. The result might be the creation of a small affluent television constituency whose views are given undue weight.

Even if these practical difficulties could be overcome, the prospect of television polling raises the spectors of biased presentations and unrealistic alternatives being offered to the public and of hasty, unconsidered opinions being sent back in return. In addition, most serious democratic thinkers have never felt that public policy issues should be decided by popular votes. Their reasons have ranged from a distrust of the sophistication and good intentions of the mass public to the belief that good decision making requires more time to study an issue than most people have. In general, the liberal ideal of democracy is that government should be a body of experts open to popular ideas that are intelligent and strongly felt and that its performance in this and other respects should be reviewed at elections. This is very different from saying that governments should act on majority votes concerning particular issues. Before deciding that television voting would be beneficial, therefore, we should look carefully at the reasons that in the past have caused our best minds to question the basic principles underlying it.

It is possible, therefore, that polling, like the other applications of the new mass communications to politics that I have mentioned, may have both bad and good effects on liberal democratic political systems. In raising these difficulties, I do not mean to suggest that they will definitely arise in the future. I do wish to assert, however, that we do not know what effects the new mass communications will have on politics and that, given our present state of knowledge, it is reasonable to think that some of those effects may not be beneficial. We do not know because we do not have the answers to a number of specific questions. How will candidates and parties make use of increased channel space? How will their decisions, together with an increase in the amount of entertainment programming, affect the viewing habits of the mass public? Will people be willing to pay more for high-quality public affairs programming? Will specialized interpretations of the news lead to comparison shopping or social isolation? Will television polling be considered a toy or taken seriously as a political tool? These are difficult questions because many of them are interrelated and depend on the economic and value judgments of many different kinds of people. Given our lack of guidance from the past and analogies in the present, it may be that these questions cannot be answered with much certainty. They are, however, important because to a significant extent the form of the new mass communications will depend on whether we make informed decisions about public policy. If we understand the likely consequences of different courses of action, we can overcome many of the potential difficulties I have discussed. If, on the other hand, we simply assume that more mass communications must be beneficial to our political system, there is a good chance that some of those difficulties will arise.

We need, therefore, continuing research efforts that will identify problems and carefully investigate significant indicators of the future as they emerge from the growth of the new telecommunications industry. There are the beginnings of such efforts at a number of centers, but their primary concerns to date have been with technical and economic issues. We need more emphasis on the social, the psychological, and the more narrowly construed political side of things.

Fortunately, the new mass communications will probably not have a significant impact on politics or other segments of American life for some years to come. As a result, we probably will have sufficient lead time to find out whether they will pose new problems and to formulate public policy for dealing with those problems. My comments about the new mass communications do, however, have some immediate implications for public policy. To begin with, I think there is some danger that overoptimism about the future of telecommunications may unconsciously lead us to delay needed reforms in present-day political broadcasting. If what I have said is true, this would be a great mistake. Despite hopeful predictions, there are grounds for believing that the new mass communications will not substantially contribute to the solution of many of the more acute problems that we now face. Television campaigning still will be enormously expensive. Ideas still will take second place to images. As a result of cost and indifference, there still will be too little sophisticated public affairs information available to most people, and the average uninvolved citizen still will have far too little influence on government. We cannot hope that the new technology by itself will solve these problems, and it may make them worse. As a result, if we do not regulate campaign spending, reform campaign practices, and encourage public affairs programming now, these same problems will be with us for many years to come.

Just as expectations about the new mass communications should not lead us to neglect new efforts to reform political television, the same expectations should not lead us to give up the safeguards that we already have. There is a great deal of talk today about whether the Federal Communications Commission's "fairness doctrine" should be applicable to the television industry of the future. This doctrine requires program originators to present all substantial viewpoints on important public issues. It is often suggested that if the new mass communications make it possible for anyone to state his own point of view on television, it is not necessary to impose this kind of responsibility on programmers. We would have, in essence, a free forum for public debate.

Attractive as this prospect may be, it is unlikely that the new mass communications will give equal exposure to different points of view any more than they will to different candidates. The expensive, well-produced presentation on a public issue placed on a frequently watched channel certainly will not be equal in impact to the soap box arguments on a public access channel. Moreover, a greater volume of programming may render the arguments of the impecunious even less powerful than they are today. Finally, doing away with the "fairness doctrine" opens the way for programming such as biased news with all the attendant difficulties I have already pointed out. Certainly the "fairness doctrine" will have to be modified to take account of the special characteristics of the new mass communications, and it may even have to be strengthened. There is no reason to suppose, however, that it should be abolished in the foreseeable future. Objectivity and balance may be difficult to achieve, but the pursuit of them is an essential part of our culture.

The new mass communications may, therefore, raise many new problems for political television and leave many existing problems unresolved. In stressing the problems, I do not mean to deny that more abundant channel space will have many benefits. It may reinforce political and social groups, widen our horizons, and stimulate our economy. But we would be wrong to be too broadly optimistic or to base our presentday decisions on promises about the distant future. For the time being, we must try to learn more and proceed with caution.

REFERENCES

Alexander, Herbert, and Ithiel de Sola Poole. "Politics in a Wired Nation," paper prepared for the Sloan Commission on Cable Television.

Bagdikian, Ben H. *The Information Machines.* NewYork: Harper & Row, 1971.

Barrett, Marvin. "The Future of News and Public Affairs Broadcasting and CATV," paper prepared for the Sloan Commission on Cable Television.

Blumler, Jay G., and Denis McQuail. *Television in Politics, Its Uses and Influence*. Chicago: University of Chicago Press, 1969.

Dordick, Herbert S., and Jack Lyle. "Access by Local Political Candidates to Cable Television: A Report of an Experiment," report prepared for the Rand Corporation, November, 1971.

Feldman, N.E. "Cable Television: Opportunities and Problems in Local Program Origination," report prepared for the Rand Corporation, September, 1970.

Lang, Kurt, and Gladys Engel Lang. *Politics & Television*. Chicago: Quandrangle Books, 1968.

Mill, John Stuart. *Utilitarianism, Liberty, and Representative Government*. New York: E.P. Dutton, Everyman's Library Edition, 1951.

Price, Monroe, and John Wicklein. *Cable Television: A Guide for Citizen Action*. Philadelphia: United Church Press, 1972.

Sloan Commission on Cable Communications. *On the Cable: The Television of Abundance*. New York: McGraw-Hill, 1971.

Toffler, Alvin. *Future Shock*. New York: Random House, 1970.

CHAPTER 34

Technology Assessment or Institutional Change?*

EDWIN B. PARKER

The argument for technology assessment is heard these days in connection with many different technologies and in connection with technology in general. There are serious dangers in the effects of highly technological societies on the ecological balance of the planet.

In the list of technologies to be assessed, communication technology should be given a special place. Changes in the capability and costs of information access, storage, and transmission will influence the distribution of who has what information at what time. These changes can alter the distribution of political power. Changes in political power or in the structure and influence of different social institutions will determine how and for what purposes other technologies will be used. Information and communication technology provides the "command and control systems" for all the other technologies. Here is where we should focus our attention if we wish to influence decision-making power.

If assessment of communication technology is to have a useful effect on how the technology is used, the task must be defined broadly. A narrow definition of the task of technology assessment could subvert the very purposes of assessment. The problem is as follows. The resources available to the developers of the technology are likely to be much greater than those available to the assessors. In order to avoid domination by the technology developers, the technology assessors need an organizational structure that is independent. But the relative strengths of the two independent social organizations will tend to be very uneven. Given the greater economic rewards, vested interests, and consequent power that will accrue to the developers of the technology relative to the assessors, it would be surprising if the weaker organization prevailed over the

* Preparation of this chapter was supported in part by National Science Foundation Grant GR–86 to Stanford University. Thanks are due to Susan Krieger for her assistance.

stronger. This is particularly true in a society such as the United States that values corporate enterprise and is still somewhat fearful of social planning.

If there is to be any hope for the technology assessors in this situation of unequal power, then the assessment task must be broadened in two dimensions. One is to define the task to include assessment of the institutions that control the technology being assessed. ("Institution" is used here to refer to any established organization within the society, including business organizations such as the Columbia Broadcasting System, trade associations such as the National Cable Television Association, governmental organizations such as the Federal Communications Commission, universities such as Stanford, etc.) The second is to expand the task assignment to include active attempts at institutional change as well as passive evaluation.

This chapter underlines the urgency of assessment and control of new communication technology and describes a preliminary exploration of how that might be accomplished. The chapter should not be judged by its specific recommendations, but by whether it succeeds in shifting the focus of attention to an alternate way of defining the problem.

NEW COMMUNICATION TECHNOLOGY

The pace of change in communication technology is quickening. After only two decades of dominance, broadcast television is threatened by cable and video cassette technologies. Computer technology and new techniques of data communication are beginning to cast a shadow of the future on the production and distribution techniques of newspapers, although major change may still be more than a decade away (Bagdikian, 1971; Parker, in press). Communication satellite technology has already been demonstrated to be technically and economically superior to much land and undersea cable communication. The fact that the satellite technology is not yet used for United States domestic communications indicates that the proponents of that technology have not yet demonstrated the political power to translate that technical capability into institutional change. New specialized common carriers with an eye on the expanding digital data communications market resulting from the continuing expansion of computer technology are beginning to pose the threat of at least token competition for the telephone companies with their voice communication technology.

The social effects of this much heralded communications revolution may be as far-reaching as the social effects of printing technology were on medieval society. Elizabeth Eisenstein (1969), writing in the British history journal, *Past and Present*, reinterprets what she calls the "problem of the Renaissance" by looking at the social effects of printing technology. She makes a very plausible case for printing technology as a key factor leading to the Renaissance and the Industrial Revolution. Now the new communication technology born of that Industrial Revolution may be the key factor shaping the emergence of post-industrial society. Like the earlier communications revolution, it is dramatically reducing the costs of storage, distribution, and access to information. As in the case of printing technology, this will lead to changes in who has access to what information at what time. It is also likely to lead to further acceleration of the irreversible process of knowledge accumulation. As with printing technology it will lead to major institutional battles over who controls the information technology and its content. Information access and control is such a significant component of institutional power that struggles over control of the new technology are as inevitable as they were over printing technology.

Both church and secular governments attempted to restrict the access to printing technology in earlier centuries because wider direct access by people to print media, whether as senders or receivers of messages, would reduce the power of the then existing institutions in their roles as filters and interpreters of information. State licensing and religious persecution (for example, by the Spanish Inquisition) were used in attempts to maintain control. But the technology of printing was such that decentralization was relatively easy. It was possible to establish clandestine presses or to cross national boundaries in search of a more favorable political climate for publishing. The decentralized nature of the technology plus the significantly lower unit costs made information more widely accessible. The previously existing "establishment," perceiving a serious threat to the established order, attempted unsuccessfully to prevent their authority from being bypassed, although they slowed down the rate of change.

Similarly, in our day we hear people with otherwise "liberal" views discussing with concern the dangers of permitting unrestricted public access to cable television. What about the danger of incitement to riot or other threats to the way the society is now organized? Some argue that access should be restricted to established corporate entities that are sure to be "responsible" because they have a financial and institutional interest in continued existence and are thus responsive to court judgments or other legal sanctions. Their concern is that if any angry young person with limited financial resources could gain access to the cameras and microphones to vent his anger against the society, no fine or court judgment for damages would provide an effective deterrent.

The fear is justified in exactly the same way that church and government authorities were threatened in the early days of print. If access to the means of effective communication is given to those outside the existing institutional structure of the society, those structures may be subjected to pressures for change too great to resist. If there is a sufficiently large segment of the society so dissatisfied with the existing social institutions that a speech on cable television will incite them to violent acts, should we blame the speaker or the social structure that led to such widespread discontent? Are we willing to maintain the freedom of speech and freedom of expression in print media that were so hard won in an earlier day? Are we going to block the next logical extension of those principles in the next generation of media? Is it worth subverting those basic principles of freedom of expression in order to weaken the assault on our existing institutions by those left out or disadvantaged by those institutions? How can we, who got our chance as a result of the previous victories for freedom of expression, justify closing the doors against those whose hopes lie in the next extension of that freedom?

The single issue of open access to cable television is symbolic of the current battle for institutional control over the new communication technology. The heart of the matter is centralized versus decentralized control. Present broadcast media come under attack from both the political right and the political left because they have centralized control over what is transmitted. Sometimes the critics are only saying that someone else (themselves) should exercise that central control. But the important question is, should any small group have that kind of centralized control? As we move into an era of computerized information retrieval systems, large data banks and a visual media system with enough capacity to permit as wide a variety of content as the print media, will we have centralized or decentralized control over access for both sending and receiving messages? It is possible to establish centralized control over almost any communication technology, as the Soviet control of print media in their society demonstrates. The new electronic communication technology will be much easier to maintain under centralized control because of the high system costs and the necessary visibility of the system. We can have clandestine presses and even clandestine radio stations, but can we seriously imagine a clandestine communication satellite or a clandestine cable television system? We are as unlikely to have competing cable television systems in the same neighborhood as we are to have competing telephone companies serving the same neighborhood, and for the same economic reasons.

Technology is not the culprit. Remote access time-sharing computer systems, shared data banks, and cable television technology can make possible decentralized access to the communication system of the society in a way not possible with previous technologies. The same technology that can bring us 1984 (Orwell, 1949) can bring a decentralized information system in which every citizen can say what he wants to say and can select what he wants to receive. The technology of national data banks accessible only to the Federal Bureau of Investigation and other "authorized" users can also make possible a mass media system in which the control over what is transmitted shifts from the sender of the messages to the receivers. The dangers of 1984 will be no less real if the central control is exercised by a consortium of large corporations (as is more plausible in the United States) or by government (as is more plausible in the Soviet Union). In either case the opportunity will be present for detailed monitoring of the behavior of all citizens as a by-product of computerizing the records of most economic and social transactions. The battle already underway is an institutional one over who will control and who will benefit from the new communication technology. The battle is a crucial one for society because the outcome will determine how decision-making power is distributed and hence how all major social decisions are made.

If we structure the problem as one of assessment of the technology itself, or as one of developing social indicators to better measure social effects after they have happened, then the battle will have been lost before we start. By the time we have definitive measures of social effects, the political, economic, and institutional structure surrounding the new technology will be well entrenched and highly resistant to change.

The comparison with the institutional structure of television broadcasting is instructive. In the earliest days of television broadcasting the potential for educational and public service uses of television were widely recognized. But the technology was developed and implemented by private corporations for private profit. The principal broadcasting institutions were not established to meet the educational or other public needs of the society. Instead, educational institutions had, in principle, the opportunity to utilize some channels that were reserved for public use. But empty channels do not an educational institution make. To utilize those channels effectively it is necessary to develop institutions with a reasonably stable financial base and the technical capability to use the channels effectively. The present financial problems of the Corporation for Public Broadcasting provide clear evidence that we have not been able to create educational broadcasting institutions that are anything more than very weak little sisters of the commercial broadcasting institutions. Whatever one thinks of the Corporation for Public Broadcasting or whether it deserves to survive, that model of building institutions to utilize new technology is not a good one. That model consists of taking technology developed for other purposes, using "reserved channels" that were usually those least desired by the commercial interests, and building financially weak institutions around it to perform the public service functions.

Much of the research that has been conducted on the effects of television on society, including some that I have done myself (Schramm et al., 1961) could be classed as technology assessment even though we were not calling it that then. The research was motivated in part by a concern for potentially harmful side effects and for missed opportunities. Although knowledge for its own sake was a strong part of the motivation, an interest in questions of public policy with respect to broadcasting technology was a significant factor in choosing that particular problem to investigate. However you judge the quality of the research, the amount and timing of the research was clearly "too little, too late" to have any effect on the structure of broadcasting in the United States. The legal and regulatory structure was firmly established, the institutions were profitable, and there was no effective political opposition. A more promising time to attempt to influence the institutional structure of broadcasting in this society would have been much earlier or now, at a later time when the coming of cable television poses an economic threat to present broadcasting interests and hence provides a point of vulnerability. The new technology and its possible economic consequences are likely to provide a more powerful tool for institutional change than weak appeals to social responsibility.

There is an alternative to letting the development and exploitation of technology be determined primarily by market factors in the private sector of the economy with the concerns of the public sector, such as education, being tacked on in a weak way at the end. Instead of having all the development proceed on the basis of profit expectations with other social effects left unplanned and largely unanticipated, we can choose to develop and utilize the technology to meet a variety of social needs, including education.

The combination of cable television and computer technology is likely to make possible (and profitable) community information utilities in which a variety of services are made available to homes. The forces of the marketplace will cause the information utility to take shape in some form without governmental initiative. Elsewhere (Parker and Dunn, 1972), I have argued the case for a federal telecommunications technology initiative in which a research, development, and demonstration program is established to develop the potential of the new technology to deliver education and information services to every home. The techniques required to deliver education and public sector information services are likely to be different from those required for many commercial and entertainment services.

The institutional structures needed to make such services widely accessible are quite different from those of the traditional marketplace. The recent Mitre Corporation report recommending cable television plans for Washington, D. C. provides an excellent example (Mason, 1972). The Mitre analysis shows that in a purely commercial cable venture for Washington, maximum profits would be obtained at about 50% saturation, while maximum public service would be obtained at closer to 100% saturation. Even though the latter goal looked economically viable, simple profit maximization would not lead to anywhere near the latter level of saturation. Similarly, the RAND Corporation report on metropolitan Dayton predicts a 40% saturation with a $6/month subscriber fee, leading to a 14% or greater rate of return on investment. The authors predict that lowering the subscriber fee would increase penetration but reduce the rate of return (Johnson, 1972). In the short run, the gap between the information-rich and the information-poor is likely to be widened instead of narrowed by the introduction of new communication technology unless pressure can be brought to bear with sufficient strength to guarantee near universal distribution of opportunity for access.

INEVITABILITY OF TECHNOLOGY DEVELOPMENT?

Apologists and pessimists share the view that technological development is inevitable, differing only in the values they place on the social consequences. Apologists argue that there is no way to stop the "progress" of technological development and that on balance the results are beneficial to mankind. If any negative side effects occur, a "technological fix" can be found to repair the damage. Whether or not you are optimistic about the consequences of technology development, the inevitability argument is plausible. Both Jacques Ellul (1964) and Herbert Marcuse (1964) seem convinced of the inevitability of technological development although they both value it negatively.

What is inevitable (or rather, irreversible) is not the particular form the physical technology takes, but the knowledge that makes it possible. Just as the biblical story of Adam and Eve has it, there is no turning back once having bitten into the figurative apple of knowledge. The only way the knowledge process can be reversed is by destroying the society and the people in whose heads and artifacts the knowledge is carried. How the society uses that knowledge is another matter. Knowing how to build "doomsday weapons" does not imply that they will be used, any more than knowing how to produce enough food to feed all humanity implies that it will be produced and distributed in ways that prevent all starvation. The defeat of the appropriations for the supersonic transport is one tiny hopeful sign that we do not have to treat the development of technology as inevitable, given the knowledge of how to develop it.

For several generations now scientists who were generating new knowledge have used the irreversability argument to defend whichever branch of knowledge they chose to pursue. All knowledge is either neutral or good, depending on which version of the argument you hear. If it is used in dangerous or harmful ways, that is the responsibility of those using the technology, not those developing the knowledge. The irreversibility provided a convenient rationalization: if they did not pursue that knowledge others would. By this argument some of the best brains in the society were able to maintain clear consciences by disclaiming responsibility for the social effects of the knowledge they produced. One is reminded of the Tom Lehrer record in which he humorously mimics Werner von Braun justifying his rocket research for the Germans and later the Americans, "I put the rockets up, where they come down is not my department." It was not until late in World War II and in the immediate postwar period that a significant group of scientists began to realize that, they as developers of the means to nuclear destruction, had a social responsibility to attempt to influence the uses of the technology they made possible. At the very least they were making a moral choice when they chose which knowledge to pursue. That social conscience has been spreading through science, albeit slowly, as the implications of the responsibility argument for other areas of knowledge, whether chemical, biological, psychological, or social, are realized. Scientists and technologists creating the knowledge that permits new communication and information technology ought to be even more concerned about the uses and consequences of their knowledge, because the information technology affects the distribution of decision-making power in the society, and hence how and for what purposes all other technologies are used.

Concurrently with this hopeful trend since World War II, a second and possibly countervailing trend has been taking place. A problem deeper than how knowledge is to be used is that of which knowledge to pursue. We have maintained an ethic of academic freedom to protect the right of scientists and scholars to make their own choices concerning which knowledge to pursue. Nevertheless, that freedom has always been easier to exercise if the resulting knowledge served the dominant institutions of the society better than it served the dissident groups. If the pursuit of knowledge led in directions that seemed threatening to the dominant institutions (e.g., Marxist economics), there was a tendency to label it "political" and not protect it with academic freedom.

In more subtle ways, the pursuit of knowledge has tended to serve one segment of society more than others. For example, much of science and technology development since World War II has been supported by the military. Academic freedom has been protected, but the scientists interested in exploring new directions that the military wanted explored were more likely to receive financial support than those the military did not want explored or did not care about. Even without a single scientist compromising his choice of problem to fit the expediency of obtaining funds, the net social effect of that pattern of funding is to further the interests of those providing the funds. Even when funds were given for pure basic research in areas of no conceivable military consequence, the net effect may have been to buy the political support of scientists and technologists as an influential segment of society, thereby increasing the overall political power of the military relative to other institutions.

Communication scientists and scholars who are willing to accept responsibility for the social consequences of the knowledge they produce should thus focus their attention more broadly on the social and institutional context in which they conduct their studies. They should concern themselves with the causes as well as the effects of the technologies they assess. They should be concerned with the assessment of the institutions managing the technology as well as assessing the effects of their uses of the technology. The alternative to succumbing to the inevitability of technology argument and hence serving the interests of the institutions now controlling the technology is to shift the focus of attention to the assessment of those institutions and to the development of strategies for institutional change.

INSTITUTIONAL CHANGE

If we are correct in our analysis indicating the core of the problem lies in the social institutions that control the development and deployment of technology, then the problem at first glance appears hopeless. The institutions that have captured or grown up around the significant technologies of our time constitute the dominant order of the society. Changing our technology of ground transportation will involve changing the automobile industry and its suppliers (including the steel industry), the oil industry, and the self-perpetuating dedicated gasoline tax system that supports the continuing cycle of highway construction at the expense of other forms of transportation.

The institutions that have grown up around particular technologies or techniques are often highly resistant to change. The motivation of self-preservation is powerful for organizations as well as individuals. Institutions utilizing or exploiting one technology may constitute the most effective force resisting introduction of new technologies that might have adverse consequences for the existing institutions. One of the major political factors accounting for the lack of a United States domestic satellite system may be opposition from existing communication carriers that do not want the new technology unless it is under their control. The broadcasting establishment, including the National Association of Educational Broadcasters, is opposing the potentially competitive technology of cable television. A case close to home for many of us is the education institutions that have grown up around the technology and techniques of books, blackboards, and face-to-face lectures. These institutions tend to resist innovation made possible by changes in communication technology.

The unwanted side effects from the exploitation of technology are thus seen to be similar in origin to the failures to utilize technology that could produce additional external benefits. Both problems stem from institutions calculating their own self-interest without adequately incorporating external costs and benefits into their decisions. The solution to the problems cannot be found either in the "technological fix" or in technology assessment. The answer to both the exploitation and resistance versions of the problem lies in change in social institutions. The goal should be to bring about the changes needed to control the technology in the interests of the society as a whole instead of one particular institution or group of institutions.

Strategies for Institutional Change

The concept of unfettered competition between individuals and institutions in a pluralistic society, with the spoils going to the stronger, has long since been discarded as an appropriate model for society. The answer seems to lie in some kind of social planning that attempts to incorporate the interests of all the people instead of accepting the social consequences of unplanned laissez-faire competition. How to accomplish that is easier said than done, of course. Those attempting such planning may themselves constitute one or more of the many institutions in the society, attempting to increase their influence relative to other institutions. Or those doing the planning may be components of larger political or economic institutions and hence largely serving those interests instead of the society as a whole. If the social planners do not have the power to implement their plans, then the planning is likely to be ineffective because the decisions will be made elsewhere. If they do have power to implement their plans, then who controls the planners to guarantee that they are, in fact, responsive to the needs and wishes of all of the society? These are questions that go to the very roots of how our society is organized, questions for which we have, at best, imperfect answers. The answers are a function of how we ask the questions. Adequate answers are unlikely to be found before we improve the phrasing of the questions. The present phrasing dates from 1880 to 1930, a period during which the Interstate Commerce Commission, the Federal Communications Commission, and other regulatory agencies were established as the appropriate planning and regulating agencies, with Congress as the agent of all the people, overseeing the planners. The history of those institutions should remind us that if there is a simple resolution of these complex problems, we have not found it yet. It is well to be humble when exploring such ground.

The problems we are grappling with will not wait until we have perfectly satisfactory proposals for resolution of the planning dilemma. If we are immobilized by the lack of a perfect resolution we are, in effect, serving the interests of those institutions already acting to influence policy in their interests. Just as scientists have had to learn that there is no ultimate *truth*, so we must accept that there is no universal *good* that we can use to resolve all the value questions that must be faced. We must set goals for the kind of change we wish in the social institutions with which we are concerned, just as we must develop hypotheses to be tested in the course of creating the instrumental knowledge needed to help us reach them. Both the goals and the hypotheses may need to be revised as we learn more about what it is we seek as well as about the consequences of various actions.

My argument is a moral one, directed at communication scholars and scientists. It assumes that people should be responsible for the effects of their actions to the extent that the consequences can be anticipated. The way we define our research problems and the kinds of knowledge we pursue can influence social change, in part through supporting or creating some institutions and weakening others. Therefore, we should take as much care in our analysis of the value context and probable social consequences of our actions (including choice of research problem) as we do in our search for knowledge itself. At the same time, we should not forget that social goals have to be arrived at collectively (just as the validation of scientific knowledge is a collective process). Our responsibility is to nudge the consensus in the direction we think is best, not to find some ultimate solution to be imposed on society in an authoritarian manner.

The timing of attempts to change social institutions is critical. Almost the only time it is possible to change institutions is when they are undergoing a period of crisis or instability. If the institutions in question are in a position that they must make some changes because of political, economic, or technical factors, then there is more chance to nudge them in a desired direction than when they are in a stable state. The political factor may be public outcry concerning pollution; the economic factor may be financial deficits; the technical factor may be new technology that threatens the old way of doing business. Occasionally, it may be possible to create a crisis; most of the time the only option is to wait for one to happen or to focus attention on some other institution that is undergoing a crisis of some kind.

The dilemma is that the time at which action is required comes before we have reached a consensus on what new or modified institutions we prefer. It also comes before we have adequate knowledge of the empirical questions involved in evaluating the probable effectiveness of alternate institutional change strategies or the probable social effects of different institutional forms. Despite these difficulties, we need to act on the basis of the best information we have and the best judgment we can make at the time the decisions are being made. It should help in the search for resolution if we can be successful in focusing attention at the place where the problems reside—in the basic institutional structure of the society.

Both apologists for the existing institutional structure and pessimists about the possibility of change are likely to bring our attention back to the central dilemma of technology assessment, however we redefine it. Even if technology assessors follow the advice of this chapter and broaden their focus to examine the institutions controlling the technology and to become active in attempting to bring about change, how can they be effective? They are still in the position of a weaker group attempting to change a stronger group. The honest answer is that there is no guarantee of success in such circumstances, even though conscience may compel us to try.

Nevertheless, there are some general strategies that should greatly increase the chances of the technology assessors influencing the direction of technology development especially if the institutions controlling the technology are themselves unstable and hence forced to change in some direction for other reasons. The general strategies that seem most effective involve sound intelligence, widespread publicity, and mobilized manpower. Sound intelligence is the part of the technology assessment problem that is best understood—conducting research to find out as much as possible about the technology and institutions being assessed. Publicity has been utilized with some success by Ralph Nader and his associates. In the long run, mobilized manpower may be the most effective way to counter superior economic forces. The goal of this or any other mix of strategies should be to change the climate of ideas so that institutional change becomes politically possible.

Publicity intended to change the climate of ideas can be a powerful weapon for social change. Those of us who hope to prevent centralized control over the new communication technology will need to find our modern day John Miltons who can as eloquently advocate freedom of expression in the new media as Milton defended freedom of expression in the print media of his time.

Freedom of access to the new information technology will make the task of assessment and control of all other technologies easier. The key strategies of intelligence, publicity, and manpower all involve access to information technology. The technology will make it possible to lower the cost of access to whatever information is available; the key policy question concerns whether the legal and institutional barriers to free access will be removed. But for such "intelligence" to be trusted, there also must be open access to the sending or the storing end of the information systems so that people can select their information from sources they consider trustworthy. Open access is also required to be able to use publicity to influence the climate of ideas. One potential feature of the new information technology that may prove to be very significant will be the capacity to use the information system to locate other people with similar interests, so that it can be used as a tool to aid in mobilizing manpower, on short notice, if necessary, to counter attempts by existing institutions to use their technologies for their own benefit when that benefit is at the expense of the rest of the people in the society. Organized social action is needed to counter organized social action. If those who are disadvantaged by the present institutions have the information resources necessary to organize themselves to defend their interest, then there is a stronger chance that technology assessment can be effective.

My conclusion, then, is to reemphasize the urgency of the need for assessment of communication technology and institutions. Effective assessment of communication technology leading to institutional changes permitting open access may be just the bootstrap we need to make possible effective assessment of other technologies. Both the problem and the resolution of social effects of technology lie in the way our social institutions are structured instead of in technology per se. A time of technological change provides opportunity for institutional change. Let us not miss the opportunity in communication technology by focusing too narrowly on technology assessment instead of institutional change.

REFERENCES

Bagdikian, Ben H. *The Information Machines*. New York: Harper & Row, 1971.

Eisenstein, Elizabeth. "The Advent of Printing and the Problem of the Renaissance," *Past and Present, 45* (November 1969), 19–89.

Ellul, Jacques. *The Technological Society*. New York: Knopf, 1964.

Johnson, Leland L., et al. "Cable Communications in the Dayton-Miami Valley: Basic Report," RAND Corporation Report R–943 KF/FF, January 1972.

Johnson, Leland L. "Cable Communications in the Dayton-Miami Valley: Summary Report," RAND Corporation Report R–942 KF/FF, January 1972.

Marcuse, Herbert. *One-dimensional Man*. Boston: Beacon Press, 1964.

Mason, William F. "Urban Cable Systems," Mitre Corporation Report, M72-57, May 1972

Orwell, George. *Nineteen Eighty-Four*. New York: Harcourt, 1949.

Parker, Edwin B. "Technological Change and the Mass Media," in *Handbook of Communication*, Schramm, Wilbur, et al. (eds.), Chicago: Rand McNally, in press.

Parker, Edwin, and Donald Dunn. "Information Technology: Its Social Potential," *176*, (June 30, 1972), 1392-1399.

Schramm, Wilbur, et al. *Television in the Lives of Our Children*. Stanford: Stanford University Press, 1961.

CHAPTER 35

Research in Forbidden Territory

JAMES D. HALLORAN

From time to time we may need to put the media under the microscope, but when we do, we should be careful to maintain the wider perspective. We are always studying the mass communication process, the operation of the media *in society*. All aspects of the mass communication process should be seen in the wider economic, technological, political, and social settings.

A related point is that technological innovation is usually accompanied by some form of institutional change aimed at facilitating the adoption and development of the innovation. But this change may lead to other changes in the institution's internal structure and external relationships. As a result, new objectives are set and pursued, but some well-established services cease to be available. Someone benefits and someone suffers. Often the decision makers and the organizations benefit, while the unprotected public has to meet the social cost (Mesthene, 1971, pp. 51–62).

In thinking about the future, we also need to bear in mind the social costs that stem from the restricted or inadequate exploitation of an innovation. Existing interests and structures act as inhibiting factors. Consequently, in addition to studying technological change we must also look at related institutional changes. In particular, we should see whether it is possible to bring about specific forms of institutional change to cope with technological change, with a view to minimizing social costs and maximizing social benefits.

Problems of media development have wide social repercussions. This is a public matter, something that falls within the sphere of public interests as distinct from private interests, and the relevant institutions and the decision-making process should be geared accordingly. In fact, in some countries it has been accepted for some time that decisions in this vital area should be taken away from the free-for-all of the marketplace.

But what about mass media *research*? Here it seems that laissez-faire is more the order of the day. Can we afford this? What has been the payoff in the past? There is an interesting tension here. The need for collective, focused, problem-oriented action in research is felt by some not to be in the best interests of social science; they feel that it clashes with the traditional freedom of the social scientist "to follow truth wherever it may lead," even though some of the outcomes of this pursuit may be detrimental to the freedom of others. It is interesting to note that there are those who will support the call for "action and decision in the public interest" as far as media policy is concerned but who do not quite see it that way when it comes to research policy. We shall return to this later.

A document issued by Unesco (1971, p. 23) sets out proposals for an international program of communication research and, in doing so, touches on several relevant questions. The general strategy follows from an earlier document (Unesco, 1970, p. 33) on the need for research, and may be summarized as follows:

In brief, the research approach should be motivated by two overriding considerations. We need research that covers all aspects of communication as a total process. We need research that studies the media and the communication process in general within the wider social, political and economic setting. Only in this way is it possible to avoid the fragmentation and imbalances of the past. Control, ownership, support, resources, production presentation, content, availability, exposure, consumption, use, influence and overall consequences can all be incorporated within such a general research framework.

It is also emphasized that the research should be *problem and policy oriented.* It is unnecessary to reawaken sterile debates based on false dichotomies such as "pure" and "applied." The need for theoretical sophistication is recognized as also is the need for improved methodology but the main emphasis must be that the research is directed towards the solution of social problems. There is no reason why theoretical, practical and normative considerations should be regarded as incompatible. [Unesco, 1971, p. 6, author's italics]

This may be thought to be laboring the point a little in 1972, but in view of the fragmented, atheoretical, conceptually crude, socially irrelevant nature of so much mass communication research, I doubt if this argument can be repeated too often.

Unfortunately, this approach would still be regarded by some researchers as controversial, but perhaps not quite as controversial as some of the other passages from the report that are quoted below:

The main problems lie, however, in the fact that governments usually do not know enough about the nature of the communication process in society and have too little knowledge about the capacities and orientation of the public and private communication systems. Thus, too often, *they have not been able to set up systems that correspond to their actual developmental needs.* This report, inter alia, attempts to *lay at the disposal of policy-makers a programme of research which incorporates those projects for communication research that would appear to be most useful in the service of national development activities.* [Unesco, 1971, p. 5, author's italics]

Having outlined *what* needs to be done, the report goes on to indicate *how* these objectives might be accomplished. The main instrument would be:

A national communication policy council, composed of decision makers and adminis-trators advised by those active in various aspects of communication as well as by researchers, which could be set up to act as a coordinating machinery providing plan-ning advice to the ministry responsible for the economic and social development of the nation.

The research task will have to be shared by a number of institutional teams within the media, the universities, the government ministries and by individual researchers assuming responsibility for specific projects and inquiries. *The co-ordination of this research effort at the national level and the allocation of scarce financial resources to research institutes would then be one of the prime functions of such a national commu-nication policy council. It should here be stressed that research scientists rarely are responsible for the formulation of policy or the carrying out of plans: their contribution to the policy council should be the systematic collection and processing of data on all the aspects of communication in order to provide the base knowledge by which policy-makers and administrators can make their decisions.* [Unesco, 1971, p. 8, author's italics]

These are clear enough statements that mass communication research should be a public concern, policy oriented, firmly tied to the national interests and not subject to either the idiosyncratic fancies of the individual researcher or the vested interests of the research institution. The Unesco proposal is obviously written with the needs of developing countries very much in mind, but there is nothing in the document to suggest that the passages quoted above are not meant to apply to the more advanced societies as well.

It would appear from the document that national interests must predomi-nate over social scientific interests. Does this mean that the ideology of science must be harnessed to the ideology of the nation-state, or at least that national policy must dictate both the character and the programs of social science in this area? Has the mass communication researcher any alternative but to cooperate and plug into the national system and hope for the best? Does sur-vival depend on the amount of satisfaction given to those who determine national policy?

There must be many mass communication researchers and other social sci-entists, particularly those who insist that in the final analysis social science can never accept an exclusively therapeutic or problem-solving role, who would like to have the answers to these and other related questions. If both the aims and instruments of research are controlled, how can there be the autonomy and independence of inquiry that, so it is argued, is the sine qua non of any scientific endeavor?

We are, of course, discussing policy research, the pros and cons of which have been argued many times before. With the nation, as with the single institution or organization, we still have to face the apparently inevitable clash between national interests and sovereignty on the one hand, and the requirements of social scientific inquiry on the other. Irving Horowitz (1968) has argued that where policy needs rule, the critical aspects of social science will be minimized.

Relationships between social science and policy vary from country to country. In some countries the research effort is geared entirely to national policy. In others, the two spheres might be formally regarded as completely independent of each other. In practice, different parts of the research sphere will probably have different relationships with the policy sphere. The pattern can vary from complete servitude to genuine critical independence, but there is more than a suspicion that independence and purity are usually inversely related to power, status, and influence in decision making. In this sort of situation there is usually considerable confusion and uncertainty about the role of social science with regard to policy.

The history of mass communication research in Great Britain, to which I shall confine myself, shows that if we hold that the provision of relevant information on vital matters of public concern is an acceptable criterion of success, then our free, plural-based system has failed. It could be that the policy makers at both government and institutional levels do not want to be troubled by the sort of information that independent, autonomous, social scientific inquiries are likely to produce. We have seen cases where nonconfirmatory research has been ignored (this happens even when the research has been commissioned by the institution), and we know that ignorance has its uses and that knowledge can be embarrassing. Even national inquiries on the future of broadcasting (e.g., Committee on Broadcasting [Pilkington] 1960) apparently have little use for social science. With one or two notable exceptions, which will be examined later, what little social scientific information we have in mass communications in Great Britain has come from the efforts of individual scholars, and has been produced in the course of their normal academic work. Funding agencies have given some support, as in recent years has the Social Science Research Council, but there is nothing to suggest that mass communication research has ever been more than a passing or marginal interest.

On the other hand, a great amount of money and effort have been expended on service, administrative, market, and commercial research by the various media institutions. Whether this work can be properly called social scientific is a debatable point that need not be followed here. It is not autonomous, nor does it usually claim to be. It tends to be atheoretical and, at its best, it is more concerned with sampling requirements than with refinements in conceptualization. Not all of the work is published, but the work that is published suggests that it rarely goes beyond a direct servicing of the institution concerned and when it does venture into "social research," perhaps under pressure from outside the institution, the research will be addressed to problems as defined by the institution.

It may be argued that an institution should not be criticized for failing to do something that it never sets out to do. This is true up to a point, but there is more to it than this. First there is the wasteful expenditure of what in the national context must be regarded as scarce resources and, second, there is the danger that research will be sold short and its potential underestimated, with damaging consequences all around. Media administrators and producers, not surprisingly, come to identify mass communication research with what goes on in their audience research departments. When there is a call for more research, the reply is that a great deal is already being done. The public (including politicians) are not capable of making distinctions between different types of research, that is, between service and critical research. We have an excellent example here of "value free" research in the faithful and unquestioning service of the values of the establishment.

In 1969, the Television Research Committee (TVRC, 1969, pp. 46–47)—further details of whose work will be given later—recommended amongst other things that:

The BBC, ITA, the independent television companies, newspaper, publishing, cinema, advertising and other media interests, should provide financial support in the shape of research grants, fellowships, scholarships etc., to enable independent research to be carried out in universities and other institutes of higher education.

Three years after the publication of the TVRC report, it is fair to state that very little has been done by the media industries about any of its recommendations. Perhaps the blame is not entirely the media's, for the government could have given more backing to the recommendations. Is it unthinkable that all media institutions should be required to make annual financial contributions to autonomous mass communication research to be carried out in three or four independent research institutions?

For a sum less than half of the BBC's annual expenditure in audience research, three university-based research institutions could carry out a very full and challenging independent research program. The Television Research Committee recommended in 1969 that the government, through the Social Science Research Council and the University Grants Committee, should provide such long-term support. It also proposed that what resources were made available should not be dissipated on numerous unrelated projects, but should be used to strengthen the existing institutional developments in a few "centers of excellence."

The Social Science Research Council responded favorably to the work of the Committee and apparently accepted the general tone and direction of the report, although it did not approve of the concentration of the research effort in a few centers. The center established by the Television Research Committee and Leicester University remains the only center of its kind in the country, although there is a television research unit at another university. For the most part, what work is done is still piecemeal and uncoordinated.

The main reason for this is that the traditional approach to research generally accepted by most of the bodies concerned is individualistic. Grants are usually awarded to individual scholars, mostly teachers in a university, to enable them to carry out research in their chosen field.

In the social sciences it is rarely possible to pose questions and provide answers in the manner of some of the natural sciences, and it is a refusal to recognize this that has often led us up the wrong path. It is the nature of most of our work that it tends to produce useful ideas with an increasingly firm factual base instead of clear-cut answers to major policy questions. We must try to tease out the relationships that have a crucial effect on policy, and in doing so provide not so much widely applicable generalizations as a sound, informed basis for decision making and, at the same time, cut down the area of reliance on guesswork and prejudice (Shonfield, 1970–1971, pp. 3–7).

It is unlikely that we have used our freedom in the best possible way. But we would say that in examining social problems and current issues we need not accept the prevailing or conventional definitions at face value. In fact, it often seems that one of our main functions is to get people in the media and others responsible for media policy to search for new definitions and understandings, to question their basic assumptions, and to reexamine the current "common-sense" explanations of the nature and effects of their work.

But, assuming that our claim is a valid one and that on the whole we have avoided the major pitfalls usually associated with problem orientation, how long are we going to be allowed to continue with this sort of work? Will the paymasters and others who grant the research facilities continue to do so if they consider that our research results threaten their cherished values? Are we not naive to think that we shall be allowed—still less positively encouraged— to probe and question as we have done?

The major contributions to the British mass communication research scene over the past decade have largely been a matter of chance. Can we afford to leave research developments to chance in the future? There are signs that the media are becoming increasingly cautious and reluctant to cooperate in any research that attempts to go beyond their own narrow and self-interested definitions of what is important. On the other hand, there is no sign that research will be established on a basis that will enable the necessary long-term comprehensive programs to be developed.

In the present circumstances, my preference is to work the existing system to the utmost. Three or four well-established, independent research institutions within the university structure would fulfill the requirement. The problem orientation of these centers need not conflict with their social scientific autonomy. It would not be a panacea but it would go a long way toward providing the information on which intelligent policy formulation depends.

The problems are not insuperable, and money need not be the main stumbling block. Indeed, if we take into account all the money at present spent by the media on all types of commercial, service, and administrative research, then additional funds would not be required. The answer lies in a redistribution that would give us a more intelligent and socially relevant use of existing resources.

REFERENCES

Horowitz, I. L. *Professing Sociology*. Chicago: Aldine, 1968.

Mesthene, E. J. "Prolegomena to the Study of the Social Implications of Technology," in *The New Communications Technology and its Social Implications*, Roger P. Morgan (ed.), London: International Broadcast Institute, 1971.

Shonfield, Andrew. Introduction to the Annual Report, in *Social Science Research Council Newsletter. Special*. London: S.S.R.C., 1971.

Television Research Committee. *Second Progress Report and Recommendations*. Leicester: Leicester University Press, 1969.

Unesco. *Proposals for an International Programme of Communication Research* Paris: Unesco, September 1971.

Unesco. "Mass Media in Society. The Need of Research," *Unesco Reports and Papers on Mass Communication*, No. 59. Paris: Unesco, 1970.

CHAPTER 36

Cultural Indicators:
The Third Voice

GEORGE GERBNER

Private and governmental commissions, congressional committees, and founda-tion-supported studies since the early 1930s have called for some surveillance of media performance and effects. But none of these proposals spelled out how that might be done or limited the scope to manageable proportions relevant to scientific purpose and public policy. Consequently, there is probably no area of significant social policy in which far-reaching decisions are made with as little reliable, systematic, cumulative, and comparative information about the actual trends and state of affairs as in the sphere of the mass production and distribution of the most broadly shared messages of our culture.

We are only vaguely aware of the fact that decisive policy making *is* going on, and that cultural politics is as much a part of the fabric of modern life as economic, welfare, or military politics. Abstract conceptions of "censorship" obscure the realities of direction, constraints, and controls in any mass produc-tion. Formal aesthetic categories derived from other times and places ignore social functions, relationships, and power, which lie at the heart of the cultural policy process.

We know very little about trends in the composition and structure of the mass-produced systems of messages that define life in urbanized societies. We know no more about the institutional processes that compose and structure those systems. Much of our research on how people respond and behave in specific situations lacks insight into the dynamics of the common cultural con-text in which and to which they respond.

Economists, anthropologists, and other social scientists have long been searching for measures of cultural differentiation and diversity. Citizens con-cerned with public issues such as health, education, delinquency, aging, gener-ational conflict, group relations, drugs, and violence point to cultural "trends" to support their case. But there is no convincing evidence to support any case.

Educators increasingly wonder about the consequences inherent in the compulsion to present life in salable packages. We harness the process of acculturation to consumer markets of instant gratification. Is "Enjoy now, pay later" the prescription for healthy impatience with empty promises or for selfishness and irresponsibility?

Claims come from every vested interest in society. They generally fall into two categories. One is the voice of the political agent or agency staking a claim to "issues" dear to the heart of a political clientele. The other consists of media voices speaking for industrial and business clients. Lacking is a "third voice" of independent research building a continuing and cumulative factual basis for judgment and policy. That is what a scheme of cultural indicators is designed to do.

Cultural indicators will not resolve the issues. Policy making is the task of citizen judgment and responsible authority. However, cultural indicators can illuminate the aspects that relate more to institutional policy than to personal choice or taste, more to general trends and configurations than to specific items, works, and qualities. In so doing, cultural indicators can assist those responsible for making and implementing policy, as well as the general public, in arriving at sounder judgments concerning the role of mass communications in the cultivation of public policy alternatives.

Cultural indicators will help close the "intelligence gaps" created by historic changes in institutionalized public acculturation. These are changes in the technologically based and collectively managed production of messages. To go McLuhan's half-truth one better, society is the message. Corporate, technological, and other collective processes of message production short-circuit former networks of social communication and superimpose their own forms of collective consciousness—their own publics—on other social relationships. The purpose of a scheme of cultural indicators is to monitor the aspects of our system of generating and using bodies of broadly shared messages that are most relevant to social issues and public policy decisions.

STOCK-TAKING IN A CHANGING WORLD

"Our progress as a nation depends today, as it has in the past, on meeting our national challenges with knowledge and reason," wrote the Secretary of Commerce in sending Americans their first census form by mail in 1970. "To do so, we must constantly take stock of ourselves."

But the ways in which "we must constantly take stock of ourselves" change as society changes. Article 1, Section 2 of the U.S. Constitution directed that "Representatives and direct Taxes shall be apportioned among the several States . . . according to their respective Numbers, which shall be determined by adding to the whole Number of free persons . . . three-fifths of all other Persons." It took the Constitution 100 years to recognize each and every person as a "whole Number," and the Census another 100 years to publish, in collaboration with the Social Science Research Council, the first *Historical Statistics of the U.S.*. In the meantime, the census went far beyond head counting. It became a chief source of regular, periodic, and cumulative demographic information essential to government, schools, business, and industry.

Economic accounting has also become a national responsibility. The President's Council of Economic Advisers prepares an annual report on the nation's economic health. The social and cultural transformations of our society have made economic and labor statistics and census information less than adequate to "meeting our national challenges with knowledge and reason." "Indeed," comments *Toward a Social Report* (1969), "economic indicators have become so much a part of our thinking that we have tended to equate a rising National Income with national well-being. Many are surprised to find unrest and discontent growing at a time when National Income is rising so rapidly . . . .Why have income and disaffection increased at the same time?"

Research that might shed light on such problems has been piecemeal, sporadic, uncoordinated, and rarely comparable over time and across cultures. Much of it has been conducted in response to "crises" and forgotten when interest (and funds) declined. Rarely have such studies made contributions to policy. Seldom did they continue long enough or in a broad enough framework to add much to the orderly accumulation of social intelligence.

The recognition of such deficiencies and waste has led social scientists and government officials to propose various remedies. Proposals call for a Council of Social Advisers, an annual Social Report, a National Institute for the Social Sciences, and other forms of "social accounting" or "social indicators." A review by Land (1971) argued for going beyond the accumulation of output data to gather intelligence about some "conceptualization of a social process."

President Johnson directed the Secretary of Health, Education, and Welfare "to search for ways to improve the Nation's ability to chart its social progress." The Department's response in 1969 illustrated the "intelligence gap" in health, mobility, public order and safety, learning, science, art, and in "information not only on objective conditions, but also on how different groups of Americans perceive the conditions in which they find themselves" (*Toward a Social Report*, 1969).

President Nixon set up a National Goals Research Staff charged with developing "indicators that can reflect the present and future quality of American life, and the direction and rate of its change." Although the group was disbanded without fulfilling its charge, another commission was appointed to review the information needs of government, and the National Science Foundation launched its problem-oriented program of Research Applied to National Needs (RANN). In 1972, the National Institute of Mental Health awarded the first research grant to Dr. Larry P. Gross and myself to conduct a pilot study leading to a full-fledged Cultural Indicators project. This chapter is largely a discussion of assumptions and concepts underlying that project. It draws from and develops previous statements on that subject (Gerbner, 1966a, 1969a, 1969b, 1970).

AREAS AND TERMS OF ANALYSIS

The reliable observation of regularities in the production, composition, structure, and image-cultivation characteristics of large message systems is a specialized research enterprise. Selective habits of personal participation limit even the sophisticated practitioner to risky extrapolation about the cultural experience of different or heterogeneous communities.

The areas and terms of analysis leading to cultural indicators stem from a conception of communication and its institutionalized role in society. They were developed through prior studies (most of which are listed under my name in the bibliography). The studies demonstrated that the mass cultural presentations of many aspects of life and types of action teach lessons that serve institutional purposes. People do not have to accept these lessons but cannot escape having to deal with the social norms, the agenda of issues, and the calculus of life's chances implicit in them.

I have defined communication as interaction through messages bearing man's notion of existence, priorities, values, and relationships. Codes of symbolic significance conveyed through modes of expression form the currency of social relations. Institutions package, media compose, and technologies release message systems into the mainstream of common consciousness.

How is this massive flow managed? How does it fit into or alter the existing cultural context? What perspectives on life and the world does it express and cultivate? How does it vary across time, societies, and cultures? Finally, how does the cultivation of collective assumptions relate to the conduct of public affairs, and vice versa?

The questions designate three areas of analysis: how mass media relate to other institutions, make decisions, compose message systems, and perform their functions in society are questions for *institutional process analysis*; how large bodies of messages can be observed as dynamic systems with symbolic functions that have social consequences is the question of *message system analysis*; and what common assumptions, points of view, images, and associations do the message systems tend to cultivate in large and heterogeneous communities, and with what public policy implications, are problems for *cultivation analysis*.

INSTITUTIONAL PROCESS ANALYSIS

How do media managers determine and perform the functions their institutions, clients, and the social order require? What is the overall effect of corporate controls on the basic terms of symbolic output? What policy changes do, in fact, alter those terms and how?

Mass media policies reflect not only a stage in industrial development and the general structure of social relations but also particular types of institutional and industrial powers and pressures. Mass communicators everywhere occupy sensitive and central positions in the social network. They have suppliers, distributors, and critics. Other organizations claim their attention or protection. They have associations of their own. They have laws, codes, and policies that channel and constrain them. And they have patrons who, as in any industrial production, supply the capital, the facilities, and the authority (or at least opportunity) to address mass publics.

Decision Making

Any enterprise may appear free to those who run it. But in a more objective sense, all mass production, including that of messages, is managed. Only a small portion of all potential messages can be formulated and even fewer can be selected for mass distribution. Therefore, research cannot realistically focus on whether or not there is "suppression"; selective suppression is simply the other side of the mass communication coin. The analysis must consider all major powers, roles, and relationships that have a systematic and generalized influence on how messages will be selected, formulated, and transmitted.

Some studies (e.g., Gerbner, 1961b, 1964) suggest that systems of messages produced by any institutional source, commercial as well as overtly partisan, have some ideological orientation implicit in selection, emphasis, treatment. Other research, such as that by Warren Breed (1960), Pool and Shulman (1964), David Manning White (1964), and Walter Geiber (1960, 1964), show that most newsmen respond more to the pressures and expectations of the newsroom than to any generalized concept of audience or public interest. A study of newsroom decisions (Bowers, 1967) found that three out of four publishers are active in directing news decisions, with their influence greatest in news of the immediate market area, and in subjects that affect the revenue of the paper.

The systematic exercise of powers resides in institutional roles and in relationships to centers of power. A scheme designed to analyze this process needs to identify the power roles, suggest some sources of their powers, and specify those functions that affect what the media communicate. Power and its application become relevant to this scheme as they affect what is being communicated to mass media publics. Figure 1 outlines nine types of power roles or

Power Roles (Groups)	Types of Leverage	Typical Functions
1. AUTHORITIES. Make and enforce legally binding decisions	Political and military	Arbitrate, regulate, legitimize power relations; demand service
2. PATRONS. Invest, subsidize	Control over resources	Set conditions for the supply of capital and operating funds
3. MANAGEMENT	Control over personnel	Set and supervise policies; public relations
4. AUXILIARIES. Supplement and support management	Access to specialized services	Provide supplies, services
5. COLLEAGUES	Solidarity	Set standards; protection
6. COMPETITORS	Scarcity	Set standards; vigilance
7. EXPERTS. Talent, technicians, critics, subject specialists	Skill, knowledge, popularity, prestige	Provide personal creative, performing, technical services, advice
8. ORGANIZATIONS	Pressure through representation, boycott, appeal to authorities	Demand favorable attention, portrayal, policy support
9. PUBLICS. Groups created or cultivated (or both) by media	Individual patronage	Attend to messages; buy products

Figure 1 **Major power roles, types of leverage, and typical functions directing the formation of mass-produced message systems.**

groups, and briefly notes the types of leverage and typical functions attached to each role.

1. *Authorities* possess legal powers to enact and enforce demands or impose sanctions on communicators. Legislative, executive, judicial bodies, regulatory commissions, public administrators, the police, and the military may have such authority. Authorities may assume rights patrons ordinarily have, and may impose sanctions (such as for seditious or criminal acts) that patrons cannot. Authorities may also depend on the support of communicators for much of their authority; the "regulated" have been known to regulate the regulators.

2. *Patrons* are those who directly invest in or subsidize media operations in exchange for economic, political, or cultural benefits. Their clients are the media that provide such benefits in exchange for discretionary patronage. Media patrons may be banks, advertisers, other corporate or civic organizations, religious or military bodies, or governments. The principal types of patrons and the major client relationships determine the role of media management in the power scheme of every society. The client relationship also affects the institution's approach to most issues and problems, and permeates the climate of communicator decision making.

3. *Management* consists of executives and administrators who make up the chain of command in the organization. They formulate and supervise the implementation of policies intended to fulfill the terms of patron and other power group support. They engage and control all personnel. Management's chief functions are *(a)* to cultivate client relations and *(b)* to conduct public relations. From the management's point of view, the messages that the institution produces must serve these two functions.

4. *Auxiliaries* provide supplies and services necessary to management's ability to perform its tasks. They are distributing organizations, networks, agencies, and syndicates; suppliers of raw materials, talent, artistic properties, and logistical services; wholesale and retail outlets; related manufacturing and trade concerns, associations, unions; and the holders of patents, copyrights, or other property rights.

5. *Colleagues* are communicators whose status, sense of direction or standards, and solidarity with one another can exert leverage on the formation and selection of messages.

6. *Competitors* are other professionals or media whose claims on scarce resources or ability to innovate can force the institution to exercise vigilance and either innovate or emulate in order to maintain its relationships with patrons and publics.

7. *Experts* possess needed personal skills, knowledge, critical abilities, or other gifts. They are writers, editors, creative talent, technicians, critics, researchers, subject matter specialists, consultants, and others who can give (or withhold) personal services necessary for communication.

8. *Organizations* are other formally structured or corporate groups that may claim attention, protection, or services. They may be business, political, religious, civic, fraternal, or professional associations. Inasmuch as some sort of public visibility has become a virtual requirement for organizational viability and support, the competition for attention is intense. Large organizational investments in public relations through the media exert pressure on media content and make media dependent on freely available (and self-serving) organizational resources.

9. *Publics*, finally, are the products of media output—groups created and cultivated through the messages. They are loose aggregations of people who may have little in common. But the symbols they share cultivate a community of meaning and perspective despite other differences. Management's task of "public relations" is to develop this sense of community into material value for the institution and its patrons.

The Exercise of Power

Institutional power is exerted through the leverage built into power roles. Authorities can apply political or policy pressure; patrons can provide or withdraw subsidy; managements can hire and fire; auxiliaries can work over-time or quit servicing; colleagues can strike; competitors can raid, scoop, or corner the market; experts can refuse to serve; organizations can support, protest, or boycott; and publics can patronize or stop reading, viewing, buy-ing, or voting.

These are forms of leverage rooted in the structure of institutional roles and relations. Power applied to communications usually involves the demands that such and such be (or not be) communicated or altered in certain ways.

The demand may be ad hoc, that is, pertaining to a particular message, subject, or policy. When a system of "do's" and "don'ts" is to be regularly applied, it is usually *codified* (as in codes, regulations, and laws). The force or weight behind the leverage is a measure of institutional power. The test is what happens if the demand is not obeyed or the code not observed. That test is applied in what I call a critical incident.

If nothing happens, there has been no power, or at least no display of force to indicate power. (The ability and willingness to display force by applying sanctions is taken as an indication of institutional power.) Sanctions are *substantive* if they pertain to the substance or content of the communication itself, as in the order to revise or omit (or print) a story, add or delete a scene, with-hold necessary information, or jam a broadcast. The force behind a substantive demand or sanction is procedural. Sanctions are *procedural* if they pertain to the procedure by which communication is created. Revoking a license, firing or blacklisting, denying equipment or raw materials, discriminatory taxation or rates, strikes, boycotts, and imprisonment are procedural sanctions.

While analytically distinct, neither power roles nor types of leverage are in reality separate and isolated. On the contrary, they often combine, overlap, and telescope in different configurations. The accumulation of power roles and possibilities of leverage give certain institutions dominant positions in the mass communication of their societies.

Institutional process analysis seeks, through interviews, participant observa-tion, and the study of records, to amplify this scheme of power roles, functions, and leverage, and to apply it to the investigation of decision making in com-munications. (Gerbner, 1958a, 1958b, 1959, 1969a, 1972b.) Critical incidents (a clash of powers, when things "go wrong") may set the lines of powers and influence for more routine control of media content. The investigation contrib-utes to cultural indicators in account of the interplay of roles and powers that direct the formation of mass-produced message systems. These directions can then be related to the analysis of the message systems and to their cultivation functions. For example, how does management respond to the pressure of authorities or organizations concerning some sensitive aspects of content, and how does that response affect the frequency and symbolic functions of that particular content configuration across media and over time?

MESSAGE SYSTEM ANALYSIS

The material for analysis is taken from the massive flow of symbols produced by mass media for large and heterogeneous (usually national) audiences. Unlike most social science data, these are not symbols used to make inferences about largely hidden processes. They are visible and manifest sources of public acculturation. They provide direct access to the specific imagery, context, and content of a relatively centralized and institutionally managed release of symbolic materials into the common cultural environment.

The Systems

The most popular products of mass-produced culture provide special opportunities for the study of socially potent message systems. In these systems— popular fiction, drama, and news—aspects of life are recreated in significant associations with total human situations. An area of knowledge or the operation of a social enterprise would appear only when dramatic or news values— that is, social symbolic functions—demand it.

Dramatic and fictional entertainment especially exhibit ritualistically repetitive social symbolic mechanisms that reveal conventionally cultivated approaches toward people and life. Unlike life, the bulk of popular fiction and drama is an "open book." Facts do not get in the way of its reality, which is the reality of values. Characterizations are usually apt, motivations are transparent, problems and conflicts are explicit, and the interplay of forces that determines the outcome, and outcome itself, are usually clear. Of all the products of mass-produced culture, these appeal to the widest and most heterogeneous publics. Most people, especially the young and the less educated, encounter most subjects and ideas in the form of "incidental" treatment in the course of their relatively nonselective leisure time entertainment. In that way, "entertainment" can force attention to what most people would never seek out as "information."

The symbolic composition and structure of the message system of a mass medium defines its own synthetic "world." Only what is represented exists. All that exists in that "world" is represented in it. "Facts" reflect not opaque reality but palpable design. Focus directs attention, emphasis signifies importance, "typecasting" and fate accent value and power and the thread of action or other association ties things together into dynamic wholes. The "world" has its own time, space, geography, demography, and ethnography, bent to institutional purpose and rules of social morality. What policies populate, actions animate, fates govern, and themes dominate this "world"? How do things work in it, and why do they change from time to time?

The "system" in message systems is that of institutional design and purpose. The systematic functions of and trends in the composite "message" can be made visible through the "decomposition" of the presentations into units and categories relevant to investigative purpose, and their "recomposition" in the form of social symbolic functions.

The Analysis

The analysis is designed to investigate aggregate and collective premises presented in samples of material. It deals with the "facts of life" and dynamic qualities represented in the systems. Its purpose is to describe the symbolic "world," sense its climate, trace its currents, and identify its functions.

The results make no reference to single communications. They do not interpret selected units of symbolic material or draw conclusions about artistic style or merit. The findings represent what large and heterogeneous communities absorb; they do not necessarily resemble what specific individuals or groups select.

The premises defining life in the symbolic world provide common imagery and a basis for interaction among separate and disparate groups. That common basis forms the agenda of public discourse and a starting point for individual conclusions and interpretations. The analysis of message systems pivots on the determination of those common terms and is limited to clearly perceived and reliably coded items.

The reliability of the analysis is achieved by multiple codings and the measured agreement of trained analysts. If one were to substitute the perceptions and impressions of casual observers, no matter how sophisticated, the value of the investigation would be reduced, and its purpose confounded. Only an analysis of unambiguous message elements and their separation from personal impressions left by unidentified clues can provide a baseline for comparison with the intentions of policy makers and the perceptions or conceptions of audiences. No such relationships can be established as long as the actual common terms and their symbolic functions are unknown, are derived from unexamined assumptions, or are inferred from subjective verbalizations of uncertain and ambiguous origin.

Dimensions and Measures

The study of a system *as system* notes processes and relationships expressed in the whole, not in its parts. Unlike literary or dramatic criticism, or, in fact, most personal cultural participation and judgment, message system analysis observes the record of institutional behavior in message mass-production for large and heterogeneous communities. The reliable observation of that record of institutional behavior reveals collective and common instead of individual and unique features of public image formation and cultivation. The scheme and methods of analysis are designed to inquire into selected dimensions of the process composing and structuring message systems.

These dimensions stem from aspects of communication that we have previously identified as the cultivation of assumptions about *existence, priorities, values,* and *relationships.* Figure 2 summarizes the questions, terms, and measures of analysis relevant to each dimension.

The dimension of assumptions about existence deals with the question "What is?", that is, what is available (referred to) in public message systems, how frequently, and in what proportions. The availability of shared messages defines the scope of public attention. The measure of attention, therefore, indicates the presence, frequency, rate, complexity, and varying distributions of items, topics, themes, and so on represented in message systems.

The dimension of priorities raises the question, "What is important?" We use measures of *emphasis* to study the context of relative prominence and the order or degrees of intensity, centrality, or importance. Measures of attention and emphasis may be combined to indicate not only the allocation but also the focusing of attention in a system.

The dimension of values inquires into the point of view from which things are presented. It rates certain evaluative and other qualitative characteristics, traits, or connotations attached to different items, actions, persons, groups,

Dimensions:	EXISTENCE	PRIORITIES	VALUES	RELATION-SHIPS
Assumptions about:	WHAT IS?	WHAT IS IMPORTANT?	WHAT IS RIGHT OR WRONG, GOOD OR BAD, ETC.?	WHAT IS RELATED TO WHAT, AND HOW?
Questions:	What is available for public attention? How much and how frequently?	In what context or order of importance?	In what light, from what point of view, with what associated judgments?	In what overall proximal, logical or casual structure?
Terms and measures of analysis:	ATTENTION Prevalence, rate, complexity, variations	EMPHASIS Ordering, ranking, scaling for prominence, centrality, or intensity	TENDENCY Measures of critical and differential tendency; qualities, traits	STRUCTURE Correlations, clustering; structure of action

Figure 2   Dimensions, questions, terms, and measures of message system analysis.

and so on. Measures of *tendency* are used to assess the direction of value judgments observed in messages.

The dimension of relationships focuses on the more complex associations within and among all measures. When we deal with patterns instead of only simple distributions, or when we relate the clustering of measures to one another, we illuminate the underlying *structure* of assumptions about existence, priorities, and values represented in message systems.

The four dimensions, then, yield measures of attention, emphasis, tendency, and structure. One or more of these measures can be applied to any unit of analysis. We have studied trends in the distribution of *attention* devoted to the subject of mental illness, of education, and of violence (Gerbner, 1961a, 1966b, 1972a). *Emphasis* was measured in the investigation of comparative press perspectives in world communication (Gerbner, 1961a). Research on political tendencies in news reporting and on the characterizations of violents and victims in television drama focused on measures of differential *tendency*. The study of the "film hero" utilized all dimensions of analysis (Gerbner, 1969c).

The analysis may record topics, themes, persons, and types of action represented in the material. It may touch on the history, geography, demography, and ethnography of the symbolic "world." The symbolic population and its interpersonal and group relationships may be observed. Themes of nature, science, politics, law, crime, business, education, art, illness and health, peace and war, sex, love, friendship, and violence may be coded. The roles, values, and goals of the characters who populate the symbolic "world" may be related to the issues with which they grapple and to the fates to which they are destined.

The scheme provides a conceptual framework and practical instrumentation for the systematic gathering and periodic reporting of comprehensive, cumulative, and comparative information about mass-mediated message systems. Content indicators can include measures specific to given issues, policies, or symbolic functions, such as the television "violence index" (Gerbner, 1972a). Or they can deal with general features of the symbolic world—census figures ranging over time, space, personality types, and social roles. Indicators can also trace the presentation of heroes and villains, victors and victims, fair means or foul, or the configuration of certain themes, actions, and values over time and across cultures.

Content indicators tell us not so much what individuals think or do as what most people think or do something *about* in *common*. They will tell us about the shared representations of life, the issues, and the prevailing points of view that capture public attention, occupy people's time, and animate their imagination. They will help us understand the impact of communication media development and social change on the symbolic climate that affects *all* we think and do. We can then inquire into the institutional aspects and the cultural consequences in sharper awareness of the currents that tug and pull us all.

CULTIVATION ANALYSIS

The most distinctive characteristics of large groups of people are acquired in the process of growing up, learning, and living in one culture rather than in another. Individuals make their own selection of materials through which to cultivate personal images, tastes, views, and preferences, and they seek to influence those available to and chosen by their children. But they cannot cultivate that which is not available. They will rarely select what is scarcely available, seldom emphasized, or infrequently presented. A culture cultivates not only patterns of conformity but also patterns of alienation or rebellion after its own image. Its affirmations pose the issues most likely to be the targets of symbolic provocation or protest.

The message systems of a culture not only inform but form common images. They not only entertain but create publics. They not only satisfy but shape a range of attitudes, tastes, and preferences. They provide the boundary conditions and overall patterns within which the processes of personal and group-mediated selection, interpretation, and image-formation go on.

Cultivation analysis begins with the insights of the study of institutions and the message systems they produce, and goes on to investigate the contributions that these systems and their symbolic functions make to the cultivation of assumptions about life and the world. Style of expression, quality of representation, artistic excellence, or the quality of individual experience associated with selective exposure to and participation in mass-cultural activity are not considered critical variables for this purpose. What is informative, entertaining (or both), good, bad, or indifferent by any standard of quality are selective judgments applied to messages quite independently from the social functions they actually perform in the context of large message systems touching the collective life of a whole community. Conventional and formal judgments applied to specific communications may be irrelevant to general questions about the cultivation of assumptions about what is, what is important, what is right, and what is related to what.

Message systems cultivate the terms on which they present subjects and aspects of life. There is no reason for assuming that the cultivation of these terms depends in any significant way on agreement or disagreement with or belief or disbelief in the presentations, or on whether these presentations are presumably factual or imaginary. This does not mean, of course, that we do not normally attach greater credibility to a news story, a presumably factual report, a trusted source, a familiar account, than to a fairy tale or to what we regard as false or inimical. It does mean that in the general process of image formation and cultivation both "fact" and "fable" play significant and interrelated roles.

The Problem of Effects

The bulk of experimental and survey research on communications "effects" has contributed little to our understanding of the mass cultural process. The reason is that most of it stemmed from disciplinary and theoretical perspectives that did not consider that process the principal criterion of relevance.

Mass communications research should be concerned with mass communications and not with assorted tactics of manipulating behavior. The preoccupation with such tactics is itself a reflection of manipulative pressures in a culture in which the "behavior" that ultimately counts (and pays most of the research costs) is that at the cash register, box office, or ballot box. But the concern with tactics at the expense of strategy has been self-defeating. It has neglected the steady cultivation of issues, conceptions, and perspectives that gives meaning to all ideas and actions. All animals "behave" but only humans *act* in a symbolic context. The tactical preoccupation has generally ignored that context and obscured the basic functions of communications—to cultivate, conserve, support, and maintain. The "effects" of communications are not primarily what they make us "do" but what they contribute to the meaning of all that is done—a more fundamental and ultimately more decisive process. The consequences of mass communications should be sought in the relationships between mass-produced and technologically mediated message systems and the broad common terms of image cultivation in a culture. If the citizens of a self-governing community do not like those terms, they cannot satisfy themselves by injecting a few messages of a different sort. They must attend to the structures and policies that produce most messages in ways functional to their institutional purposes.

The Question of Change

The principal "effect" of mass communications is to be found in coming to terms with the fundamental assumptions and premises they contain, and not necessarily in agreeing or disagreeing with their conclusions or in acting on their specific propositions at any one time. Communication is the nutrient culture and not just the occasional medicine (or poison) of mental life. The most critical public consequences of mass communications are in defining and ordering issues, and not just in influencing who will buy what in the short run.

Change can be evaluated and even noticed best in light of the massive continuities that systems of communications typically cultivate. "No change" may be a startlingly effective result of communications sustaining a belief against the cultural current. One cannot really compare a person swimming upstream with another drifting downstream and yet others straining in other directions. To compare and measure their "progress," all speeds and directions must be related to the current itself. If that were to change, all directions and even the meaning of "progress" would change without any change in "behavior" on the part of the swimmers. Similarly, the meaning and measure of communication "effects" are relative to the general flow, composition, and direction of the message-production and image-cultivation processes. It means little to know that "John believes in Santa Claus" until we also know in what culture, at what time, and in the context of what message systems cultivating or inhibiting such beliefs.

A culture cultivates the images of a society. The dominant communication agencies produce the message systems that cultivate the dominant image patterns. They structure the public agenda of existence, priorities, values, and relationships. People use this agenda—some more selectively than others—to support their ideas, actions, or both, in ways that, on the whole, tend to match the general composition and structure of message systems (provided, of course, that there is also other environmental support for these choices and interpretations). There is significant change in the nature and functions of that process when there is change in the technology, ownership, clientele, and other institutional characteristics of dominant communication agencies. Decisive cultural change does not occur in the symbolic sphere alone. When it occurs, it stems from a change in social relations that makes the old patterns dysfunctional to the new order. Such a change changes the relative meanings and functions of the existing images and practices even before these are actually altered. When altered, the new cultural patterns restore to public communications their basic functions: the support and maintenance of the new order.

The strategic approach to mass communications research considers an understanding of the mass cultural process, instead of other aspects of human behavior, the principal criterion of relevance. Institutional process and message system analyses generate the framework of terms and functions for cultivation analysis. Short-terms or campaign-type "effects" studies, responses to messages elicited in unknown or uncertain symbolic contexts, or research concerned with "success" or "failure" of preconceived communication objectives are not adequate to the task. The dynamics of continuities, rather than only of change, need to be considered in the examination of mass-produced message systems and their symbolic functions. Such examination is necessarily longitudinal and comparative in its analysis of the processes and consequences of institutionalized public acculturation.

In a general sense there are no communication failures, only failures of intentions and campaigns. All systems of communications may cultivate terms and assumptions implicit in them whether or not those were intended or consciously recognized. We usually communicate more than we intend or know about, and often not what we wish. Many "communication failures" can be interpreted as the success of the receivers to understand the messages better than those who designed them, but in ways they did not intend. Message systems perform symbolic functions that may be apparent to *none* of the parties engaged in the communication. Communications research attempts to reconstruct these functions. Cultivation analysis seeks to discover their contributions to knowledge and meaning.

Symbolic Functions

Symbolic functions are intimately involved in and govern most human activity. The human meaning of an act stems from the symbolic context in which it is embedded. The significance of a person's life or death rests in some conception of role, personality, goals, and fate. When the symbolic context changes, the significance of acts changes. A structure may shift to accommodate the change and to preserve—or even enhance—the symbolic functions of the act. For example, in the TV violence study we found that as the proportion of violent characterizations was cut, the imbalance in the risks of victimization between groups of unequal social power increased, thereby strengthening the symbolic function of violence as a demonstration of relative social powers (Gerbner, 1972a). Such observations enable us to ask questions about what *that* message might cultivate in public conceptions and behavior. Thus we would relate the viewing of television violence to the cultivation of certain conceptions of goals, values, people, and power, instead of only to notions about "violent behavior."

In another study (Nunnaly, 1960), the opinions of experts on 10 information questions concerning the mentally ill were compared with mass media (mostly fictional and dramatic) representations of mentally ill characters. The mass media image was found to diverge widely from the expert image. The "public image," as determined by an attitude survey along the same dimensions, fell between the expert and the media profiles. Thus, instead of "mediating" expert views, the media tended to cultivate conceptions far different from and in many ways opposed to those of the experts. What may be seen in isolation as "ineffective" communication was, on the contrary, powerful media cultivation "pulling" popular notions away from expert views. The reason is not necessarily ignorance or intentional obscurantism, but instead the difference between semantic labels for a certain type of behavior (such as "mental illness") and the symbolic functions of the dramatic representations of that behavior. The symbolic functions of mental illness in popular drama may be primarily those of indicating a dramatically convenient resolution of certain problems or of designating a morally appropriate "punishment" for certain sins. The dramatic associations with the personality traits that define mental illness *in the plays* should provide a basis for the further investigation of what that implicit message might cultivate in viewer conceptions.

The study of specific message structures and symbolic functions reveals how these communications help define, characterize, and decide the course of life, the fate of people, and the nature of society in a symbolic world. The symbolic world is often very different from the "real" world. Symbolic behavior usually bears little resemblance to everyday actions. The power and significance of symbolic functions rests in the *differences*. Fiction, drama, and news depict situations and present action in those realistic, fantastic, tragic, or comic ways that provide the most appropriate symbolic context for the emergence of some human, moral, and social significance that could not be presented or would not be accepted (let alone enjoyed) in other ways. (For a further discussion of symbolic functions, see Chapter 17.

The Cultivation Process and Its Analysis

Symbolic structures may cultivate certain premises about the world and its people and about the rules of the game of life. These premises are not necessarily embodied in overt prescriptions (which may, in fact, be quite different), but are implicit in the way things are presented, and in the way they function in the symbolic context. The same premises may lend themselves to a range of conclusions, depending on who draws them, why, when, and how. But the range of conclusions is held together by the definitions implicit in the premises. Should the premises change, the range and complexion of conclusions might also shift.

The cultivating effects of common communications patterns are typically those of selective maintenance on a certain level. General cultural patterns do not "cause" but support or weight or skew tendencies also functional to other (but not necessarily all) aspects of the social and institutional order.

Cultivation analysis starts with the patterns found in the "world" of public message systems. The common structures composing that world present images of life and society. How are these reflected in the expectations, definitions, interpretations, and values held by their "consumers"? How are the "lessons" of symbolic behavior derived from other times and places, and presented in synthetic contexts, applied to assumptions about life? In order to investigate the relationship between message systems and the views and expectations of audiences it is necessary to evolve and adapt a set of measures and investigative techniques.

The principal approaches employed in the cultivation analysis are projective techniques, depth interviews, and periodic questions on sample surveys. Adult and child panels provide subjects for projective and interview work. Projective techniques can structure situations in which respondents tend to reveal views, expectations and values of which they may not be consciously aware, or which they would not verbalize if asked directly. Techniques of depth interviewing can isolate and highlight views, expectations, and values, and relate these to media exposure patterns and to demographic and other characteristics. Questions selected from the projective tests and interviews and others designed especially for survey use are to be submitted periodically to a national adult probability sample of respondents.

The impact of television and of its further development by cable and other technologies is of special concern, as is the cultivation of social concepts among children. For most people television *is* popular culture. Social symbolic patterns established in childhood are the most easily cultivated throughout life. Longitudinal and cross-cultural research is needed to follow the lead of message analysis into the living laboratory of popular cultures.

We need to know general trends in the cultivation of assumptions about problems of existence, priorities, values, and relationships before we can validly interpret specific relevant facts of individual and social response. The interpretation of public opinion (i.e., published responses to questions elicited in specific cultural contexts), and of many media and other cultural policy matters, require cultural indicators similar to the accounts compiled to guide economic decisions and to other indicators proposed to inform social policy making.

Technological developments in communications hold out the possibility of greatly enhancing culture-power on behalf of existing social patterns—or of their transformation. A modern Socrates might say, "know thy communications to know thyself." He would probably add that under conditions of symbolic mass production, the unexamined culture may not be fit to live in.

REFERENCES

Bowers, David R. "A Report on Activity by Publishers in Directing Newsroom Decisions," *Journalism Quarterly* (Spring 1967).

Breed, W. "Social Control in the News Room," in *Mass Communications*, Wilbur Schramm (ed.), Urbana, Ill.: The University of Illinois Press, 1960.

Gerbner, George. "The Social Role of the Confession Magazine," *Social Problems* (Summer, 1958a).

———. "The Social Anatomy of the Romance-Confession Cover Girl," *Journalism Quarterly* (Summer 1958b).

———. "Mental Illness on Television: A Study of Censorship," *Journal of Broadcasting* (Fall 1959).

————. "Psychology, Psychiatry and Mental Illness in the Mass Media: A Study of Trends, 1900–1959," *Mental Hygiene* (January 1961a).

————. "Press Perspectives in World Communications: A Pilot Study," *Journalism Quarterly* (Summer 1961b).

————. "Regulation of Mental Illness Content in Motion Pictures and Television," (with Percy H. Tannenbaum). *Gazette, 6* (1961c).

————. "Ideological Perspectives and Political Tendencies in News Reporting," *Journalism Quarterly* (Autumn 1964).

————. "An Institutional Approach to Mass Communications Research," in *Communication: Theory and Research*, Lee Thayer (ed.), Springfield, Ill.: Charles C. Thomas, 1966a.

————. "Education About Education by Mass Media," *The Educational Forum* (November 1966b).

————. "Institutional Pressures Upon Mass Communicators," in *The Sociology of Mass Media Communicators*, Paul Halmos (ed.), *The Sociological Review Monograph* No. 13, pp. 205–248. University of Keele, England, 1969a.

————. "Toward 'Cultural Indicators'; The Analysis of Mass Mediated Systems," *AV Communication Review* (Summer 1969b).

————. "The Film Hero: A Cross-Cultural Study," *Journalism Monograph* No. 13, 1969c.

————. "Cultural Indicators: The Case of Violence in Television Drama." *The Annals* of the American Academy of Political and Social Science, March, 1970.

————. "Violence in Television Drama: Trends and Symbolic Functions," in *Television and Social Behavior*, G. S. Comstock and E. A. Rubinstein (eds.), Vol. 1. *Content and Control*. Washington: Government Printing Office, 1972a.

————. "The Structure and Process of Television Program Content Regulation in the U.S." in *Television and Social Behavior*, G. S. Comstock and E. A. Rubinstein (eds.), Vol. 1. *Content and Control*. Washington: Government Printing Office, 1972b.

Gieber, Walter. "Two Communicators of the News: A Study of the Roles of Sources and Reporters," *Social Forces* (October 1960).

————. "News is What Newspaper Men Make It," in *People, Society, and Mass Communications*, Lewis A. Dexter and David M. White (eds.), New York: The Free Press, 1964.

Land, Denneth C. "On the Definition of Social Indicators," *The American Sociologist* (November 1971).

Nunally, Jum C., Jr. *Popular Conceptions of Mental Health; Their Development and Change.* New York: Holt, Rinehart and Winston, 1960.

Pool, Ithiel De Sola, and Irwin Shulman. "Newsmen's Fantasies, Audiences, and Newswriting," in *People, Society and Mass Communications*, Lewis A. Dexter and David M. White. (eds.), New York: The Free Press, 1964.

Toward a Social Report, U.S. Department of Health, Education and Welfare, Washington: Government Printing Office, 1969.

White, David M., "The Gatekeeper: A Case Study in the Selection of News," in *People, Society and Mass Communications*, L. A. Dexter and D. M. White (eds.), New York: The Free Press, 1964.